Probability and Mathematical Statistics (Continued)

WILLIAMS • Diffusions, Markov Processes, and Martingales, Volume I: Foundations

ZACKS • Theory of Statistical Inference

AF682783

Applied Probability and Statistics

ANDERSON, AUQUIER, HAUCK, OAKES, VANDAELE, and WEISBERG • Statistical Methods for Comparative Studies

ARTHANARI and DODGE • Mathematical Programming in Statistics

BAILEY • The Elements of Stochastic Processes with Applications to the Natural Sciences

BAILEY • Mathematics, Statistics and Systems for Health

BARNETT and LEWIS • Outliers in Statistical Data

BARTHOLOMEW • Stochastic Models for Social Processes, *Second Edition*

BARTHOLOMEW and FORBES • Statistical Techniques for Manpower Planning

BECK and ARNOLD • Parameter Estimation in Engineering and Science

BELSLEY, KUH, and WELSCH • Regression Diagnostics: Identifying Influential Data and Sources of Collinearity

BENNETT and FRANKLIN • Statistical Analysis in Chemistry and the Chemical Industry

BHAT • Elements of Applied Stochastic Processes

BLOOMFIELD • Fourier Analysis of Time Series: An Introduction

BOX • R. A. Fisher, The Life of a Scientist

BOX and DRAPER • Evolutionary Operation: A Statistical Method for Process Improvement

BOX, HUNTER, and HUNTER • Statistics for Experimenters: An Introduction to Design, Data Analysis, and Model Building

BROWN and HOLLANDER • Statistics: A Biomedical Introduction

BROWNLEE • Statistical Theory and Methodology in Science and Engineering, *Second Edition*

BURY • Statistical Models in Applied Science

CHAMBERS • Computational Methods for Data Analysis

CHATTERJEE and PRICE • Regression Analysis by Example

CHERNOFF and MOSES • Elementary Decision Theory

CHOW • Analysis and Control of Dynamic Economic Systems

CLELLAND, BROWN, and deCANI • Basic Statistics with Business Applications, *Second Edition*

COCHRAN • Sampling Techniques, *Third Edition*

COCHRAN and COX • Experimental Designs, *Second Edition*

CONOVER • Practical Nonparametric Statistics, *Second Edition*

CORNELL • Experiments with Mixtures: Designs, Models and The Analysis of Mixture Data

COX • Planning of Experiments

DANIEL • Biostatistics: A Foundation for Analysis in the Health Sciences, *Second Edition*

DANIEL • Applications of Statistics to Industrial Experimentation

DANIEL and WOOD • Fitting Equations to Data: Computer Analysis of Multifactor Data, *Second Edition*

DAVID • Order Statistics, *Second Edition*

DEMING • Sample Design in Business Research

DODGE and ROMIG • Sampling Inspection Tables, *Second Edition*

DRAPER and SMITH • Applied Regression Analysis, *Second Edition*

DUNN • Basic Statistics: A Primer for the Biomedical Sciences, *Second Edition*

DUNN and CLARK • Applied Statistics: Analysis of Variance and Regression

ELANDT-JOHNSON • Probability Models and Statistical Methods in Genetics

ELANDT-JOHNSON and JOHNSON • Survival Models and Data Analysis

continued on back

Abstract Inference

Abstract Inference

ULF GRENANDER

Brown University

John Wiley & Sons

New York · Chichester · Brisbane · Toronto

Library of Congress Cataloging in Publication Data:

Grenander, Ulf.
Abstract inference.

"A Wiley-Interscience publication."
Bibliography: p.
Includes index.
1. Mathematical statistics. I. Title.

QA276.G729 519.5 80-22016
ISBN 0-471-08267-8

Printed in the United States of America

10 9 8 7 6 5 4 3 2 1

Preface

During the preparation of the manuscript for this book I have benefited from innumerable discussions with colleagues in the Division of Applied Mathematics at Brown University. I would like to mention especially Walter Freiberger, Stuart Geman, Chii-Ruey Hwang, Donald McClure, and Philip Thrift.

I received advice, criticism, and references to the literature from, among others, R. Bahadur, A. Holevo, I. Ibragimov, L. LeCam, E. Parzen, and D. Slepian.

The work has been supported by the Division of Mathematical and Computer Sciences of the National Science Foundation, the Directorate of Mathematical and Information Sciences of the Air Force Office of Scientific Research, and the Mathematical and Information Sciences Division of the Office of Naval Research. I should like to thank Emmon Barrett, I. N. Shimi, and Robert Grafton for their understanding and positive approach.

I am indebted to E. Fonseca for typing the several versions of the manuscript with skill and patience, and to E. Addison for her careful job in preparing the diagrams.

ULF GRENANDER

Djursholm, Sweden
August 1980

Contents

Abstract Inference

Introduction

1 THE TWO OBJECTS OF STUDY

The title of this book is ambiguous, and intentionally so. On the one hand, *Abstract Inference* can refer to problems of statistical inference when *the sample space is an abstract space*. On the other, it can refer to the case when *the parameter space is an abstract space*.

Abstract sample spaces have been dealt with in considerable depth in the literature, and especially in the important case when they form function spaces over some given set T of arguments: *statistical inference in stochastic processes*. Beginning with our 1950 monograph (see the Bibliography at the end of this book), many mathematicians have studied this topic. Today several of the basic problems have been solved, resulting in a vast literature, in part motivated by applications among which communication engineering should be mentioned first. The theory is still in active development, as can be seen in the leading journals of mathematical statistics and probability theory. This research effort has resulted in an elegant theory, nearing completion in at least in some respects, and with a multitude of applications.

Chapters 3–6 will present an outline of this theory as it exists today. To keep the volume to a reasonable size, we have not been able to include all the results that deserve to be reported. The criterion for selection has usually been potential for further extensions and nontrivial generalization. The applied aspects had to be left out, unfortunately, as did the related computational problems.

For the second topic, when *the parameter takes values in an abstract space*, the situation is just the opposite. The published literature is quite fragmentary, with some scattered results, but notably lacking completeness and coherence. Some of the existing results have both substance and depth, but a comprehensive theory is definitely missing. We offer the beginning of such a theory in Part III. Up till now the study of abstract parameter spaces has not been spurred by applied problems to the same extent as for abstract sample spaces. There is a recent exception to this, however, namely our theory of patterns leading to statistical inference with parameters taking values in image algebras [see Grenander (1978)]. In pattern analysis the pure images play the role of parameters, and since the images need not be finite dimensional, classical inference is not applicable. Therefore it is clear that pattern analysis will require the development of mathematical tools capable of handling many types of infinite-dimensional parameter spaces. This was actually our motivation for introducing the *sieves and general regularization methods* discussed in Part III.

The reader will become aware of the fact that no complete theory is offered for abstract parameter spaces. Chapters 7 and 8 present, however, from a unifying perspective, many new results for special cases. Again, we have selected these cases because they seem to be capable of genuine extensions and point toward the beginning of a general theory. We hope that some readers will respond to this challenge and try to extend and strengthen the partial results reported in these chapters. More general results in this direction are given in Chapter 9.

2 OBSERVING GENERAL OBJECTS

As in other areas of mathematics, progress in statistical inference has been made by the complementary processes of *deepening* and *widening*. During the first process special cases are studied in depth, giving body to the theory to be built. In the second process the knowledge gained about special cases is then unified into a generalizing doctrine, whereupon the first process takes over again, on a higher level of abstraction, and so on. The classical theory of statistical inference with a fixed number of observations, say before 1930, dealt with samples from $\mathbb{R}^n$, or from a subset of $\mathbb{R}^n$, and had its parameter space the real line, sometimes $\mathbb{R}^d$, or subsets of them. Its logical support in probability theory dealt entirely with finite-dimensional probability distributions.

Thus the basic concept was the probability distribution in $\mathbb{R}^n$, in particular the independent and identically distributed (i.id.) distributions. At that stage the idea of a stochastic variable was not precisely defined, although the term was used. There seems to have existed some doubt whether arguments were valid when operating directly on stochastic variables rather than on their distributions. With the appearance of Kolmogorov's seminal *Grundbegriffe* in 1933 it became clear that stochastic variables could be viewed as measurable functions in measure theory, and hence could be handled rigorously, and not just a finite number of them at a time. This opened the field of stochastic processes for serious study.

Measure theory itself did not yet play an important role for statistical inference. Indeed, if the sample size was fixed, measurability questions were seldom crucial; the major difficulties lay elsewhere.

Asymptotic problems of testing and estimation, when the sample size was potentially infinite, needed measure theoretic justification, however, since their rigor. us treatment involved probabilistic limit theorems. This tendency culminated in Cramér's *Mathematical Methods of Statistics* in 1945, representing a water shed in statistical theory, from which time it has been generally accepted that statistical theory can and should be based on as rigorous foundations and derivations as any other branch of mathematics.

Statistical inference typically starts from a family $\{P_\alpha\}$ of probability measures given on a fixed σ-algebra $\mathcal{Q}$ of a given sample space X. The P_α are indexed by a parameter α, belonging to some parameter space A.

In *point estimation* we study maps (estimates) $\alpha^*: X \to A$ that possess optimality properties, often expressed through an extremum principle. Alternatively, we start from a family of estimators $\{\alpha^*\}$ and study their analytical properties, whether optimal or not.

In *hypothesis testing* we study maps (tests) $t: X \to \{A_0, A_1\}$, where A_0 and A_1 are subsets of A, the null hypothesis and the alternative hypotheses, respectively. Again we search for optimal tests or analyze given ones.

When dealing with *confidence regions* we examine maps $c: X \to 2^A$ and their properties. Other types of inference could be added to the above three, but these are the ones that we will be concerned with here (see Note I.A).

Much of *standard inference* methodology can be said to be *likelihood based*. A first step in solving a problem is the calculation of the likelihood function. This is obviously so for maximum likelihood estimation, for the Neyman-Pearson approach to testing hypotheses, and for the notion of confidence regions. Actually the same holds for an important *nonstandard* methodology, namely, Bayesian inference. Here the likelihood function is also needed, although it is treated differently and is combined with the *a priori* distribution.

An exception is linear inference, when we only allow methods based on linear functions, and the likelihood function need not be calculated. Such functions have variances that can be calculated in terms of the covariance matrix of the sample, which is why we need not introduce the likelihood function. Many other exceptions could be cited.

Leaving aside such situations, which are not likelihood based, it is clear that an attempt to extend classical methods of inference to abstract spaces should start with the determination of likelihood functions. For example, if the sample space is a Hilbert space, how should the likelihood function be defined, and by what method can it be calculated?

In the late 1940s, when the first attempts were made to study such questions, some doubt existed whether the notion of a likelihood function was the right one to start from. After all, in the spaces of interest there exists no natural invariant, "Lebesgue measure," which played such an important role in classical inference. Today the answer is clear, however. We should look for the *Radon-Nikodym derivative* of one hypothetical measure with respect to another and let this derivative play the role of the likelihood function.

When we do this it will become apparent that *singularity* of the hypothetical measures with respect to one another can occur in case when it would not have been expected intuitively. For classical inference singularity represents a trivial exception easily dealt with. For abstract sample spaces this need not be so. The problem requires more careful study, and we shall pay a good deal of attention to this question.

The main task is, however, to *calculate the Radon-Nikodym derivative* (when it exists) and employ it for the construction of tests and estimates. This should be done for the important classes of probability measures, such as, Gaussian, stationary, or Markovian ones.

It was shown in our 1950 monograph how singularity/nonsingularity could be established in many situations and how to calculate the corresponding Radon-Nikodym derivatives. Toward the end of that decade a number of mathematicians extended and deepened the results. Much of this early theoretical work was done by Soviet mathematicians, as can be seen in the Bibliography, while another active group at this early stage, mainly in the United States, was partly motivated by problems in communication engineering and had a more practical flavor.

As a result of this work, a well-established theory of inference in abstract sample spaces can be said to have existed from the mid 1960s on. The sample space that had received most attention was of course the Hilbert space of L_2-form, but more recent work has extended the theory to other abstract spaces. This process is still going on. Indeed, it seems to accelerate, and we can expect continued advances in this direction. As a mathematical theory matures, it tends toward generalization and refinement. One tries to prove earlier results by methods that are more generally applicable and also to simplify and unify the methodology. This tendency in statistical inference has led to some remarkable achievements, which we study in detail in Part II.

As mentioned above, the study of the likelihood function, that is, Radon-Nikodym derivatives, should come first. It is therefore studied in Chapter 3 under general conditions and with the emphasis on explicit calculations.

Linear methods, not requiring the explicit determination of the likelihood function, are presented in Chapter 4. Some of the elegant solutions that we shall encounter also have a good deal of applied interest, but this is not elaborated in the text. The case when the time parameter is discrete, but with a sample size tending to infinity, also belongs here: *a potentially infinite argument space*. The methodology of the potentially infinite case has much in common with the infinite one, and is therefore studied together with it.

Gaussian measures deserve much of our attention, just as for classical inference, and they are analyzed in Chapter 5. There exist many fundamental results for testing Gaussian hypotheses in various function spaces. Among them one should mention the mean value function alternative, the Hájek-Feldman alternative, as well as singularity/perpendicularity results for stationary processes.

Chapter 6 is less homogeneous. It incorporates scattered results for another important class of stochastic processes, those of the Markov type. It is not quite clear why the study of inference of Markov processes has not led to as complete a theory as for Gaussian or stationary processes. One reason may be that there are "so many" Markov processes, with widely varying structure, compared to, for example, Gaussian processes. The latter ones are, after all, completely determined just by their mean values and covariance functions. For Markov processes the infinitesimal generator of the corresponding semigroup of probability operators, which is then the basic concept, can take on widely varying forms.

3 PARAMETRIC STUDY OF NONPARAMETRIC PROBLEMS

In Part III we turn to the second object of study; an abstract parameter space. As mentioned previously, the theory is not as finished and elegant as for abstract sample spaces. The literature on the subject is also much less voluminous, the results are less definitive, the tools involved in the analysis are more diverse. As a consequence the presentation in Part III is more rhapsodic than in the earlier parts of the book.

The starting point for this study was again to extend the classical likelihood based method to general parameter spaces. What happens, for example, with maximum likelihood estimates of an abstract parameter? Do they still have desirable qualities similar to those possessed by maximum likelihood (ML) estimates in finite-dimensional Euclidean spaces?

The usual techniques employed by many mathematicians such as Fisher, Fréchet, Cramér, and Rao require regularity conditions for the frequency functions. Smoothness, say differentiality with respect to the parameter, could be translated into Fréchet differentiality, and so on. This will no doubt be of use for further work, say the abstract Cramér-Rao inequality, but it is not used in Chapters 7 and 8.

It will be more convenient to base some of our analyses on an ingenious method due to Wald. This is built on an extremal property of the mutual entropy and requires no strong smoothness assumptions.

In Chapter 8 such problems are studied for several abstract parameter spaces, and some remarkable, although limited, results are obtained. The special cases studied are intended as nuclei around which a general theory is expected to crystallize. Some unification of these cases is obtained later in Chapter 9, and we expect a further advance in this direction in the near future.

The basic theme in Part III is to treat the *nonparametric cases in a parametric manner*. Therefore the results are expressed in the topology of the parameter space. For example, when the parameter is a function in L_2, then the statistical properties are expressed in terms of the L_2-norm of the error. Pointwise properties are studied only occasionally.

One could approach problems of inference in an abstract parameter space without likelihood based arguments. This is the case for the important concept of *regularization of parameter estimates*, which is discussed in Chapter 7. It connects some seemingly unrelated problems and deserves further study in a more general setting.

The main topic in Part III is, however, that of *parameter sieves*. By this we mean the explicit restriction of the parameter space to subsets intended to improve the properties of maximum likelihood estimates when the maximization is restricted to the subset. We show how sieves can be constructed in many situations and also consider the analytic properties of the resulting estimates obtained from this powerful method. The general study of sieves in Chapter 9 is based on an idea of Wald refined by Bahadur.

Part III can be seen as an attempt to generalize classical inference. Such a generalization is certainly a legitimate mathematical activity on its own merit, but this has *not* been our only purpose.

Instead we have been motivated especially by current progress in *pattern analysis*. Typically this leads to probability measures over sample spaces consisting of deformed images, and where the measure depends upon an unknown (pure) image [see Grenander (1976) for details].

The pure image therefore plays the role of a statistical parameter. Having observed the deformed image (with the role of the sample), we want to infer the pure image (see Note I.B).

The pure images form algebraic structures, the image algebras, whose properties are now beginning to be well understood. It ought to be possible to exploit these properties to obtain good inferences.

The image algebras vary a lot in their structure. In simple cases they are, or can be embedded in, Hilbert or Banach spaces. On the other hand the pure images can be mathematical objects such as sets, grammars, group elements, or other mathematical objects which have so far seldom appeared as statistical parameters.

To learn about pattern analytic inference we have studied a number of special cases, and this gave rise to much of the work reported in Part III.

The first conscious attempts to allow more than finite-dimensional parameter spaces occurred in *nonparametric inference*, when the unknown is, for example, a distribution function specified only by qualitative constraints such as continuity. The term "nonparametric" was not in common use until the 1950s, but the idea goes further back in the history of statistics.

Even such an elementary and fundamental notion as a histogram is based on the assumption that the frequency function is not restricted to any finite-dimensional parametric family. In the same vein procedures such as the sign test, the Smirnov test, and numerous others were introduced in order to free us from assumptions that the probability measures have to be given in any particular analytic form.

This does not mean that such ideas were spelled out in terms of functional analysis: the parameter space was seldom specified as some topological space. This was not done until later with the emergence of a rigorous mathematical theory of statistics and, in particular, in recent work on robust methods.

One can see a beginning tendency to regard general parameters as elements in topological spaces in the Cramér-von Mises ω^2-criterion [see Cramér (1927) and von Mises (1931)] and in the Kolmogorov-Smirnov's tests [see Kolmogorov (1933) and Smirnov (1939)]. Note that these criteria are global and lead to topologies distinct from that of pointwise convergence. Another global measure for estimates was used in Grenander (1951a) for empirical spectral analysis and in Rosenblatt (1956b) and Parzen (1962) for the study of the estimation of frequency functions. Many more recent instances could be mentioned, some of which are referred to in the Bibliography.

So far we have barely mentioned Bayesian methods with parameters in various function spaces, and we admit that we do not give these attempts the space and attention that they merit. Only occasionally do we point to some particularly appealing Bayes solution.

The *a priori* measure in a Bayesian analysis can be chosen so that it gives special weight to the "most regular" elements in the sample space. This idea is certainly related to the notion of a sieve. Another related notion is that of penalty function, which attaches a penalty to the "less regular" elements in A. The interested reader is referred to the valuable monograph by Tapia-Thompson (1978). A third idea related to sieves is to assume that our observations do not come directly from $P_\alpha(dx)$ but also contain errors of observation.

The main object of study in Part III is the sieve, however, and less attention is given to the other alternative approaches mentioned.

Most of Part III is new so that the results cannot always be expected to have reached their definitive forms.

4 GENERAL APPROACH

Although operating in abstract spaces we attempt to carry out the analysis in as concrete terms as possible; this involves no contradiction. Generalizations for their own sake are avoided. They are undertaken only when they offer something essentially new, in addition to what happens in the classical cases.

This implies that we are mainly concerned with particular spaces like l_p, L_p, C, C_p, and BV. We do not hesitate to impose conditions like separability, or real scalars, or σ-finite measures, and so on. In this way we do not arrive at a high degree of generality, but we gain in concreteness and perhaps in lucidity.

Our general approach differs somewhat from what is customary. Much of contemporary mathematics deal with structural and existential questions at the expense of analytical calculations. We pay some attention to, for example, the existence of consistent estimators and BLUEs, but more to the determination of these estimators and to their analytical properties. In the same spirit we would sometimes attribute greater significance to the many special cases examined in the text than to some of the seemingly more general theorems in Parts II and III.

The reader is supposed to have a good background in general probability theory as presented, for example, in Feller's two volumes, and in statistical theory, as in Cramér's well-known text. Real and complex analyses are the main analytical tools, and we assume that the reader has the background of, for example, Riesz-Nagy (1952) and Titchmarsh (1939). No previous knowledge is required concerning inference in abstract spaces.

To make the exposition as self-contained as possible, we have included in the first two chapters of Part I some of the probabilistic background material needed. Chapter 1 presents a brief outline of probability measures in abstract spaces. This is needed for Part II, but it will be useful also when reading Part III, for example, when A is an abstract space carrying an *a priori* measure. No proofs are given in this chapter.

A reader familiar with these topics can skip Chapter 1 and start with Chapter 2 directly.

Chapter 2 is more special and is devoted to the Hilbert space study of stochastic processes. The results quoted are used later on, but the methods of proof are also exploited for inference problems. For this reason Chapter 2 has full proofs and many examples.

As far as is possible, we adhere to a systematic notation, see Notation, and exceptions are pointed out in the text when they occur.

Equation (i) in the jth section of the kth chapter is referred to as (i) in the same section, as $(j.i)$ in another section of the same chapter, and as $(k.j.i)$ in another chapter.

5 NOTATION

A	parameter space
$A(f)$	arithmetic mean of A
a.c.	almost certainly, with probability 1
a.e.	almost everywhere
α	parameter value
α^*	estimator
$B(x, r)$	ball of all x with $\|x\| \leqslant r$ or $< r$; closed or open
$BV(E)$	set of functions of bounded variation over E
C	superscript, denotes complement
$\mathbb{C}$	complex field
$C(E)$	space of continuous functions over set E with sup norm
$C_p(E)$	space of functions over E with continuous pth derivatives
dt	fixed measure on T, often Lebesgue measure
dx	fixed measure on X
$E(\lambda)$	resolution of the identity forming spectral family
$F(d\lambda)$	measure $\|Z(d\lambda)\|^2$
$f(x, \alpha)$	Radon-Nikodym derivative of P_α with respect to given measure dx
$G(f)$	geometric mean of f
$G(s, t)$	Green's function for given boundary value problem of some L
$H(f)$	harmonic mean of f
H_α	hypothesis corresponding to measure P_α

i.o.	infinitely often
L	linear operator
$L_2(x_t; t \in E)$	subspace spanned by x_t in Hilbert space for all t in set E
$L_p(E, \mu)$	function space of functions over E with respect to μ-measure
L_α	likelihood function for fixed sample
l_p	space of sequences with summable pth power
λ	eigenvalue of R
ML	maximum likelihood
m	mean value vector
$m(t)$	mean value function
μ	mesh size of sieve
$\mathbb{N}$	set of natural numbers
$PC(E)$	space of piecewise continuous functions on E
P_S	projection operator associated with subspace S
P_α	probability measure on X for fixed parameter value α
ϕ	eigenfunction (normalized) of R
$\phi(z)$	characteristic function of auxiliary argument z
$\psi_\nu(t)$	regression functions
Q	infinitesimal generator of given one-parameter semigroup
R	covariance operator, covariance matrix
RKHS	reproducing kernel Hilbert space
$\mathbb{R}^d$	d-dimensional Euclidean space
$r(s, t)$	covariance function
$\rho(s, t)$	correlation function
S_d	d-dimensional sphere
$\mathbb{S}(\mu)$	sieve in A
T	argument space
T	superscript denoting transpose
$\mathbb{T}^d$	d-dimensional torus
t	value of argument, "time"
tr	trace
$\mathbb{U}^d$	d-dimensional unitary space
V	value space
$V(t)$	variance function
$W(t)$	Wiener process

X	sample space
x	observed sample
$Z(d\lambda)$	stochastic set function over set Λ with orthogonal increments
$\mathbb{Z}$	set of integers
$\mathbb{Z}_n$	set of integers modulo n
Ω	reference space
ω	point in Ω
*	superscript denoting dual space; sometimes denotes set of all finite strings from given set
$\#(S)$	number of elements in set S

PART 1

Some Probability Theory on Abstract Spaces

CHAPTER 1

Probabilistic Background

1 PROBABILITIES IN FUNCTION SPACES

Most of the results in Chapters 2–6 take place in the setting of function spaces, sometimes treated as abstract mathematical structures, sometimes as concretely defined spaces. It is natural to start by introducing probability measures in the space of *all* functions, taking values in some *value set* V, say $V=\mathbb{R}$, and with some *argument set* T so that x: $T \rightarrow V$, or equivalently $X=V^T$

Definition 1 A stochastic process is an indexed family of stochastic variables $\{x(t); t \in T\}$ so that we can write $x(t)=x(t, \omega)$ in terms of a probability space $(\Omega, \mathcal{A}, P)$, $\omega \in \Omega$.

Here Ω is the reference space that introduces randomness via elements $\omega \in \Omega$, $\mathcal{A}$ is a σ-algebra of subsets of Ω, and P is a probability measure defined for all sets belonging to $\mathcal{A}$.

The fundamental construction of such probability measures, which is due to Kolmogorov (1933), ensures the existence of such a measure when all the finite-dimensional marginal distributions are given. This means the following.

For any n and $t_1, t_2, \ldots, t_n \in T$ specify a probability distribution in $\mathbb{R}^n$, an n-dimensional marginal distribution, by the joint distribution function $F(x_1, x_2, \ldots, x_n; t_1, t_2, \ldots, t_n)$, where F should of course have the properties that belong to distribution functions. Then F defines a probability measure on the σ-algebra of Borel sets in $\mathbb{R}^n$. These sets are now viewed as the bases of (finite-dimensional) cylinders in the function space V^T consisting of all real-valued functions defined on the argument set T. The general form on such a cylinder is therefore

$$C=\{x|(x(t_1),x(t_2),\ldots,x(t_n))\in B\} \tag{1}$$

where B is a Borel set in $\mathbb{R}^n$.

Kolmogorov's consistency theorem guarantees that a probability measure exists on the σ-algebra generated by all cylinders as in (1), such that its restriction to cylinder sets has the probabilities specified by the finite-dimensional distribution functions.

THEOREM 1 KOLMOGOROV'S CONSISTENCY THEOREM

The family of finite-dimensional distribution functions $F(x_1, x_2,\ldots,x_n; t_1,t_2,\ldots,t_n)$; $n=1,2,\ldots$; $x_k\in\mathbb{R}$, $t_k\in T$, *can be extended to the* σ-*algebra generated from all cylinders whose bases are finite-dimensional Borel sets if and only if*:

(a) *Any* F *is symmetric with respect to any permutation of the vector* $(x_1,x_2,\ldots,x_n)$ *and the same permutation of the vector* $(t_1,t_2,\ldots,t_n)$.

(b) $\lim_{x\to+\infty}F(x_1,x_2,\ldots,x;t_1,t_2,\ldots,t_n)=F(x_1,x_2,\ldots,x_{n-1};t_1,t_2,\ldots,t_{n-1})$.

When we apply Kolmogorov's theorem, the argument set T will often be $\mathbb{R},\mathbb{Z},\mathbb{R}^d,\mathbb{Z}^d$, or subsets of them. Occasionally T will be a torus or a finite cyclic group. It should be noted that the theorem holds generally for any set T. In the following T will always be a separable metric space.

The value space V was chosen as $\mathbb{R}$ in the formulation above, but it is obvious that it can be directly extended to $\mathbb{C}$, $\mathbb{R}^d$, and $\mathbb{C}^d$. In the quite general case, however, some caution is needed {see Note 1.1.A).

It will be instructive to consider some examples. Let us start with two trivial ones and then go ahead to more substantial ones.

Example 1 Given a distribution function F on $\mathbb{R}$, define

$$F(x_1,x_2,\ldots,x_n;t_1,t_2,\ldots,t_n)=\prod_{k=1}^{n}F(x_k)$$

for any n and distinct $t_1,t_2,\ldots,t_n$. In other words, let all the $x(t_k)$ be (i.i.d.). This represents *complete chaos* with no stochastic dependence between the ordinates of the random function $x(\cdot)$. The i.i.d. condition is a standard assumption in classical inference, but its extension to function spaces is of limited interest.

Example 2 Given again a distribution function F on $\mathbb{R}$, define

$$F(x_1, x_2, \dots, x_n; t_1, t_2, \dots, t_n) = F(x_1) \prod_{k=2}^{n} H(x_k; x_1)$$

where $H(y; x) = 1$ if $y \geqslant x_1$, and 0 otherwise. This represents *complete dependence* so that $x(t_k) = x(t_1)$ with probability 1. Again, this is too extreme to be of real interest. The next case will be encountered many times in later chapters.

Example 3 For $T = V = \mathbb{R}$ start from two real-valued functions $m: \mathbb{R} \to \mathbb{R}$ and $r: \mathbb{R}^2 \to \mathbb{R}$. Here m is completely arbitrary, while r should satisfy the following:

r should be symmetric: $r(s, t) = r(t, s)$,
r should be nonnegative definite,

where the second condition means that for any vector $(t_1, t_2, \dots, t_n)$ the matrix

$$R(t_1, t_2, \dots, t_n) = \{r(t_i, t_j); i, j = 1, 2, \dots, n\} \tag{2}$$

should be nonnegative definite. Then define $F(x_1, x_2, \dots, x_n; t_1, t_2, \dots, t_n)$ as the distribution function of an n-dimensional Gaussian random vector with mean value vector $(m(t_1), m(t_2), \dots, m(t_n))$ and covariance matrix $R(t_1, t_2, \dots, t_n)$. The consistency conditions (a) and (b) obviously hold so that Kolmogorov's theorem defines a probability measure as that of the important *Gaussian stochastic process* with *mean value function* m and *covariance function* r.

In addition to the mean value function and covariance function we occasionally use the *variance function* $V(t) = r(t, t)$, the *standard deviation function* $\sigma(t) = +\sqrt{V(t)}$, and the *correlation function* $\rho(s, t) = r(s, t)\sigma^{-1}(s)\sigma^{-1}(t)$.

Example 4 The last example can be extended to $V = \mathbb{R}^d$ by letting $m: \mathbb{R} \to \mathbb{R}^d$, and by letting the values of r be $d \times d$ matrices such that R in (2) takes on block form and remains a symmetric and nonnegative definite matrix, but now of dimension $(n \times d) \times (n \times d)$. This leads to the Gaussian vector valued stochastic process.

Example 5 To deal with $V = \mathbb{C}$, which often occurs in the following, we can just refer to Example 4 with $d = 2$ by separating x into real and imaginary parts. We meet, however, a minor difficulty concerning terminology for complex-valued stochastic variables.

If x_1 and x_2 are complex-valued stochastic variables, both with mean value 0, it is customary to define their covariance as

$$\operatorname{Cov}(x_1, x_2) = E(x_1\bar{x}_2) = E(u_1u_2) + E(v_1v_2) + iE(v_1u_2 - u_1v_2)$$

Here we have written $x_1 = u_1 + iv_1$, $x_2 = u_2 + iv_2$ in terms of their real and imaginary parts. In particular

$$\operatorname{Var}(x_1) = E(u_1^2 + v_1^2) = E(|x_1|^2)$$

The joint Gaussian distribution of x_1 and x_2 is specified if we are given the 4×4 covariance matrix of (u_1, u_2, v_1, v_2), which requires the specification of ten real parameters. Knowing the values of $\operatorname{Var}(x_1), \operatorname{Var}(x_2), \operatorname{Cov}(x_1, x_2)$ means fixing only four real parameters and is therefore not enough.

To arrive at a unique, but somewhat arbitrary definition, we can assume, for example, that for $s, t = 1, 2$,

$$E(u_su_t) = E(v_sv_t) = \frac{1}{2}\operatorname{Re}\{\operatorname{Cov}(x_s, x_t)\}$$

$$E(u_sv_t) = -\frac{1}{2}\operatorname{Im}\{\operatorname{Cov}(x_s, x_t)\} \tag{3}$$

With this convention, which can be replaced by others, the whole probability measure is defined for the *complex-valued Gaussian process* one the mean value and covariance functions are given [see Doob (1953), p. 73].

Example 6 Given nonnegative functions $f(x; t)$ and $g(y, x; s, t)$ such that $f(\cdot\,; t)$ and $g(y, \cdot\,; s, t)$ for all y are frequency functions in $\mathbb{R}^d$ with respect to a measure μ, except possibly for a set of y-values of μ-measure zero. We also demand that for $s < t < u$,

$$\int_{\mathbb{R}^d} f(y; s)g(y, x; s, t)\mu(dy) = f(x; t)$$

$$\int_{\mathbb{R}^d} g(y, x; s, t)g(x, z; t, u)\mu(dx) = g(y, z; s, u) \tag{4}$$

hold for almost all x, y, z. For $t_1 < t_2 < \dots < t_n$ define $F(x_1, x_2, \dots, x_n; t_1, t_2, \dots, t_n)$ as the joint distribution function corresponding to the joint frequency function

$$f(x_1;t_1)\prod_{k=2}^{n} g(x_{k-1},x_k;t_{k-1},t_k)$$

Using (4) one can verify that the consistency conditions hold so that Kolmogorov's construction applies. The result is the *Markov process* with values in $\mathbb{R}^d$ and with marginal frequency functions $f(\cdot\,;t)$ and transition probability kernel $g(\cdot,\cdot\,;s,t)$.

Example *7* Let $\lambda(u)$, $u\in\mathbb{R}^d$, be nonnegative and integrable over any compact subset of $\mathbb{R}^d$. We let T consist of all compact subsets of $\mathbb{R}^d$. To any finite family $(t_1,t_2,\ldots,t_n)$ of compact sets in $\mathbb{R}^d$ we associate stochastic variables $x(t_1), x(t_2),\ldots, x(t_n)$ in the following way. We can introduce new compact sets $\{s_\nu\}$ such that they are all disjoint and

$$t_1=\bigcup_{\nu\in A_1} s_\nu$$

$$t_2=\bigcup_{\nu\in A_2} s_\nu$$

$$\vdots$$

where $A_1, A_2,\ldots$ are sets of natural numbers. Associate with s_ν a Poisson stochastic variable y_ν with mean

$$m_\nu=\int_{s_\nu}\lambda(u)\,du$$

where du is Lebesgue measure, and let all y_ν be independent of each other. Put

$$x(t_1)=\sum_{\nu\in A_1} y_\nu$$

$$x(t_2)=\sum_{\nu\in A_2} y_\nu$$

$$\vdots$$

and define $F(x_1,x_2,\ldots,x_n;t_1,t_2,\ldots,t_n)$ as the joint distribution function of $x(t_1), x(t_2),\ldots, x(t_n)$. It can be verified that this construction does not depend upon how the subsets s_ν are selected. Again it can be verified directly that the consistency conditions hold so that the probability measure can be extended to the σ-algebra generated from the cylinder sets. This is the *Poisson process with intensity* λ.

In this example the interpretation of t is different from the one used earlier. Here t stands not for a point in $\mathbb{R}$, or $\mathbb{R}^d$, but for a subset.

Returning to the case $T=V=\mathbb{R}$ we briefly discuss some measure theoretic difficulties concerning the behavior of sample functions. How can we deal with *regularity properties* such as continuity, differentiality, or Lipschitz *for the sample functions* of the stochastic process $x(t)$?

Denote by ω one of the functions belonging to $V^T=\Omega$. The value of $x(t)$ at t if ω is the selected sample function can then be written as $x(t,\omega)$ to bring out clearly the dependence upon both the argument t and the random choice in Ω.

Now we can view the stochastic process in three ways (see Figure 1). The first is when we fix t and look at $x(t,\cdot)$ as a function of ω. The way we have constructed a probability measure say P, on Ω makes $x(t,\cdot)$ a measurable function so that we can view $x(t)=x(t,\cdot)$ as a stochastic variable: the vertical dashed line in the figure. When t varies, we get a one-parameter family of stochastic variables. If these stochastic variables are treated as elements of some abstract space, *our stochastic process takes the form of a curve in an abstract space* whose elements are stochastic variables. This perspective will be applied often in what follows.

If, on the other hand, we want to study the behavior of sample functions, for example, the regularity properties mentioned above, we must fix ω and consider the result as a function of t: the horizontal dashed line in the figure. Then when ω is also allowed to vary, we get *a family of random sample functions*.

Finally, we can view the stochastic process as a function of two variables, as indicated by the notation $x(t,\omega)$, treated on equal bases. This will

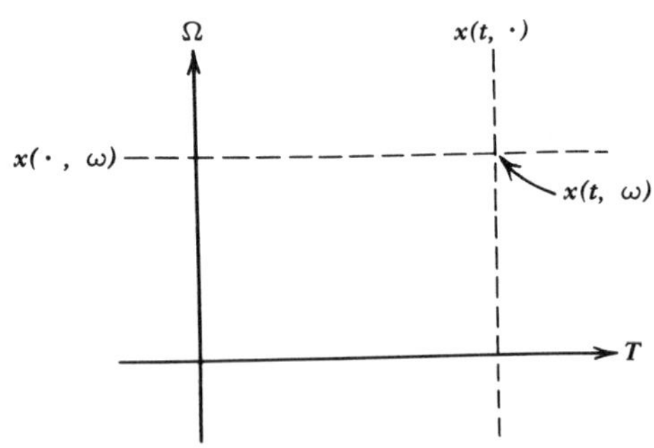

Figure 1

be useful when we want to integrate the process and apply Fubini's theorem, as will be done below.

Of course these perspectives are logically equivalent, and convenience alone will dictate which one to choose.

The measure theoretic difficulty arises from the fact that the property that sample functions be continuous on T, or other similar requirements, cannot be expressed in terms of denumerably many t-points. If T has a cardinality greater than that of $\mathbb{N}$, we cannot immediately assert that the event is measurable.

A simple, but not always sufficient, way of dealing with the continuity of stochastic processes is by the concept of continuity of probability.

Definition 2 A stochastic process $x(t)$ is continuous in probability at t if $x(t+h)$ tends to $x(t)$ in probability as $h \to 0$, so that for every $\varepsilon > 0$,

$$\lim_{h \to 0} P\{|x(t+h) - x(t)| > \varepsilon\} = 0$$

If $x(t)$ is continuous in probability for every t, we call $x(t)$ just continuous in probability.

Similarly one could say that $x(t)$ is *differentiable in probability* with derivative $x'(t)$ if there is a stochastic variable denoted by $x'(t)$ such that

$$\frac{x(t+h) - x(t)}{h} \to x'(t)$$

in probability as $h \to 0$.

Continuity in probability is an important concept, but it is not to be confused with continuity of sample functions. The following example, which will also be used in later chapters, brings out this point clearly.

With $T = [0, 1]$ and with a Poisson process with constant intensity λ over T, denote by $s_1, s_2, \ldots, s_n$ the random points generated by the process, ordered $0 < s_1 < s_2 < \ldots < s_n < 1$. Strict inequality can be assumed with probability 1. Note that both n and the s_k are stochastic variables. Define

$$\begin{aligned} x(t) &= z_0; \qquad \text{for } t \in [0, s_1) \\ x(t) &= z_1; \qquad \text{for } t \in [s_1, s_2) \\ &\vdots \\ x(t) &= z_n; \qquad \text{for } t \in [s_n, 1] \end{aligned} \tag{5}$$

where $z_0, z_1, \ldots, z_n$ is an i.i.d. sample from $N(m, \sigma^2)$. This stochastic process can be viewed as a special case of Example 6, slightly modified to allow for conditional probability distributions with a discrete component.

It is continuous in probability over T since for any t

$$P[x(t+h)=x(t)]=P(\text{no } s_k\in[t,t+h])=e^{-\lambda h}\to 1$$

as $h\to 0$. On the other hand the definition in (5) makes it appear, at least heuristically, that a sample function is discontinuous somewhere in T as soon as $n>0$, which has probability $1-e^{-\lambda}>0$.

In this example we have given a direct representation of the stochastic process in terms of the random quantities n, s_k, z_k. With Ω we now introduce the reference space

$$\Omega=R\cup(\mathbb{R}^2\times j)\cup(\mathbb{R}^2\times j^2)\cup\ldots$$

$j=[0,1]$ and

$$\omega=(z_0,z_1,\ldots,z_n,s_1,s_2,\ldots,s_n)\in\Omega$$

Then (5) expresses the way $x(t)$ depends upon the random element ω. The probability measure on Ω induces one for $x(t)$. In this special case all sample functions are simple functions and easy to deal with as far as regularity conditions are concerned.

From now on we shall not insist on any particular reference space Ω; it will be chosen as conveniently as possible. As usual we then assume given a probability space $(\Omega, \mathcal{Q}, P)$.

To be able to resolve regularity questions, we follow Doob (1953), chap. 2, §2, and for a slightly different treatment Neveu (1964), chap. III, and we use the concept of *separable versions of stochastic processes*.

Definition 3 A stochastic process $x(t,\omega)$ is called separable if there is a sequence of $t_k\in T$, $k=1,2,\ldots,$ and a set $N\in\mathcal{Q}$ with $P(N)=0$ such that for any open interval I and closed set A on $\mathbb{R}$

$$\{\omega|x(t,\omega)\in A, t_k\in I\}-\{\omega|x(t,\omega)\in A, t\in I\}\subseteq N$$

If a stochastic process is separable, then regularity questions for sample functions simplify as far as measurability is concerned. Indeed, consider properties such as:

$x(\cdot,\omega)$ is continuous everywhere in T,

$x(\cdot,\omega)$ is differentiable everywhere in T,

$x(\cdot,\omega)$ is measurable in T.

These events in the reference space Ω can now be made to belong to the σ-algebra $\mathcal{Q}$ under mild conditions.

First, it is known that separable versions can be obtained [Doob (1953)].

THEOREM 2 EXISTENCE OF SEPARABLE VERSIONS

Given any stochastic process $x(t)$ one can construct a separable process $\tilde{x}(t)$ such that

$$P[x(t)=\tilde{x}(t)]=1; \qquad \forall t \in T$$

This implies that $x(t)$ has the same finite-dimensional probability distributions as $\tilde{x}(t)$, the separable version of $x(t)$ [see Doob (1953), p. 57, and Note 1.1.B].

Separable versions will be used, explicitly or not, in Part II when T is the real line or an interval on it. In Part III the set is the parameter space. The theorem holds for a general set T under mild conditions, for example, if it is a separable metric space.

Second, one can deal with measurability of sample functions.

THEOREM 3

If $x(t)$ is continuous in probability a.e. on T, there exists a separable version of it which is measurable in the sense that $x(t, \omega)$ is measurable on the product space $T \times \Omega$ [see Doob (1953), *p.* 61].

This makes it possible to talk to the integral of sample functions as stochastic variables. Say that

$$\int_T E|x(t)|\, dt < \infty$$

and apply Fubini's theorem [see, e.g., Halmos (1950), chap. VII, §36]. Then the integral

$$I(\omega) = \int_T x(t, \omega)\, dt$$

exists a.e. and the function $I(\omega)$ is measurable. It is indeed a stochastic variable.

Also, if $x(t)$ is a real-valued process with

$$\int_T E[x^2(t)]\, dt < \infty \tag{6}$$

then the sample functions belong a.c. to $L_2(T)$, the Hilbert space consisting of functions quadratically integrable over T with respect to Lebesgue measure dt. This fact will be useful in Chapter 3 and later.

Third, one can make informative statements concerning the continuity of sample functions.

THEOREM 4

If T is a finite interval on $\mathbb{R}$ and

$$E[|x(t+h)-x(t)|^p]=0(|h||\log|h||^{-1-r}) \tag{7}$$

with positive p and r, $p<r$, then there is a separable version $\tilde{x}(t)$ for which sample functions are continuous on T, [*see Cramér-Leadbetter* (1967), *pp.* 65–66, *and Neveu* (1964), *pp.* 90–93].

Hence if (7) holds, we can consider the sample functions as elements of the Banach space $C(T)$: the set of continuous functions on T with the topology induced by the sup norm.

Somewhat stronger, if (7) is replaced by

$$E[|x(t+h)-2x(t)+x(t-h)|^p]=0(|h|^{1+p}|\log|h||^{-1-r})$$

then there exists a separable version whose sample functions a.c. have continuous derivatives [see Cramér-Leadbetter (1967), pp. 69,70]. Hence we can treat the sample functions as elements of the Banach space $C_1(T)$.

Many other results of this type exist, but these will suffice for the present purpose.

2 PROBABILITIES IN TOPOLOGICAL VECTOR SPACES

We have seen that we can often consider the sample functions as elements in some abstract space, which will always be separable in what follows. When we approach problems of inference, it will often be helpful to express the information inherent in a sample function in terms of a denumerable coordinate system with coordinates $x_1, x_2, x_3, \ldots$. These *observables*, the x_k, must be chosen so that they determine the sample function uniquely.

If the instrumentation is of limited accuracy, the observer may have access only to a finite number n of observables $x_1, x_2, \ldots, x_n$, and one would hope that his inferential power would be almost as high as if he knew all the observables. To make such statements precise, we need convergence results which are presented in Chapter 3. Considerable care is needed to do this, and the results are sometimes surprising.

To make the reader familiar with the way observables may be chosen, let us consider three special cases of considerable significance.

Example 1 Consider a Gaussian stochastic process with $V=\mathbb{R}$ and T=closed finite interval on $\mathbb{R}$, and with mean value function identically zero and a continuous covariance function $r(s,t)$ satisfying, uniformly in t,

$$r(t,t+h)-r(t,t)=0(|h|); \qquad t,t+h\in T$$

Then we can satisfy the bound (1.7) in the previous section by putting $p=3$. Indeed, we have

$$\begin{aligned} E[x(t+h)-x(t)]^2 &= r(t+h,t+h)-2r(t,t+h)+r(t,t) \\ &= r(t+h,t+h)-r(t,t+h)-[r(t,t+h)-r(t,t)]=0(|h|) \end{aligned} \tag{1}$$

A Gaussian stochastic variable $N(0,\sigma^2)$ has absolute third moment $=\sqrt{(2/\pi)}\ \sigma^3$ so that

$$E[|x(t+h)-x(t)|^2]=0(|h|^{3/2})$$

and (1.7) holds. Hence we can select a separable version of the process whose sample functions are continuous in T with probability 1.

Let us introduce therefore a sequence $t_1, t_2, t_3, \ldots$ everywhere dense in T, and define the observables $x_k = x(t_k)$, $k=1,2,\ldots$. It is clear that these observables can be used since they determine the sample function with probability 1.

Note that our sample space $X=C(T)$ and that each $x(t_k)$ is a linear bounded functional in $C(T)$.

Example 2 Let $x(t)$ be a stochastic process with the same V and T as in the previous example, but assume now only that its mean value function $m(t)$ is continuous on T as well as its covariance function $r(s,t)$ on $T\times T$. Then $x(t)$ is continuous in probability since the left-hand side of (1) tends to zero with h (although not necessarily with the bound indicated), and applying Chebyshev's inequality we see that $x(t)$ is continuous in probability. According to Theorem 3 of Section 1 we can select a separable version $x(t,\omega)$ measurable on $T\times\Omega$. But (1.6) obviously holds so that the sample functions are a.c. in $L_2(T)$.

This is a separable Hilbert space and hence possesses a denumerable basis. Select a complete orthonormal (CON) system $(z_1, z_2, \ldots)$. For the moment we avoid the crucial question of which CON system is most suitable.

Introduce the observables by the inner product in $L_2(T)$:

$$x_k = (x, z_k)$$

where x stands for the sample function viewed as an element in $L_2(T)$. Since $\{z_k\}$ was assumed to be complete, these observables suffice to determine the sample functions a.e. with probability 1.

Again our observables turn out to be bounded linear functionals, but this time in $L_2(T)$.

Example 3 Consider a Poisson process defined on a compact subset $C \subset \mathbb{R}^d$ and a sequence of Borel subsets $C_1, C_2, C_3, \ldots$ such that for any $x \in C$ and $\varepsilon > 0$ there exists an n with $x \in C_n$ and diameter $(C_n) \leqslant \varepsilon$. Introduce the observables

$$x_k = \#\{z_\nu \in C_k\} = x(C_k)$$

using $x(\cdot)$ as a random-valued set function as in Example 7 of Section 1. They determine the sample element uniquely with probability 1. The observables are again bounded linear functionals, this time in $BV(C)$.

Here we let the sample functions be measures. Since these measures are extremely simple and have finite point sets as supports with probability 1, no essential measure theoretic difficulties arise. For more general random measures a more careful analysis is needed; [see Kallenberg (1975)].

An alternative way of looking at Example 3 would be to use as sample elements the point set mentioned. Such *random sets* will be encountered occasionally.

These examples illustrate how observables may be introduced in function spaces for use in inference. We could, however, also approach the problem in abstract terms and assume only that the sample space X is, for example, a Hilbert or Banach space. We would then introduce the probability distribution directly into the space.

We assume throughout the following that X is a separable Banach space with a separable dual space X^* whose elements are denoted by x^*. Our probability measure P should be defined on the σ-algebra $\mathscr{Q}$ of Borel sets in X.

Such a P can be determined from the cylinder sets, say with real scalars,

$$\{x | (x_1^*(x), x_2^*(x), \ldots, x_n^*(x)) \in B\}$$

whose finite-dimensional base B is a Borel set in $\mathbb{R}^n$. Indeed, the σ-algebra generated by such cylinder sets coincides with the Borel σ-algebra $\mathcal{Q}$ as shown by Mourier (1953) (see Note 1.2.A). It is therefore enough to specify all the finite-dimensional distributions of $(x_1^*(x), x^*(x), \ldots, x_n^*) \in \mathbb{R}^n$, which amounts to the same thing as to specify the probability distribution of any single $x^*(x)$ on $\mathbb{R}$.

This can be done by the characteristic function for a real-valued stochastic variable. This leads naturally to the definition of the *characteristic functional*

$$\phi(x^*) = E\left[e^{ix^*(x)}\right] \tag{2}$$

which has several of the basic properties of the characteristic functions in $\mathbb{R}$ (and $\mathbb{R}^n$). For example:

(a) $\phi(0) = 1$.

(b) $\phi(x^*)$ is uniformly strongly continuous in x^*.

(c) $\phi(x^*)$ is a nonnegative definite function of x^*.

(d) If $P = P_1 * P_2 =$ convolution of P_1 and P_2, then $\phi(x^*) = \phi_1(x^*)\phi_2(x^*)$, where ϕ, ϕ_1, ϕ_2 are the characteristic functionals corresponding to P, P_1, and P_2, respectively.

These statements are easy to prove with the same methods as in the real line. We have already mentioned that $\phi(x^*)$ for all $x^* \in X^*$ determines P uniquely. The uniqueness theorem for characteristic functions generalizes directly from the case $\mathbb{R}^n$ [for the latter see Cramér (1945), p. 93].

How about the classical continuity theorem [see Cramér (1945), p. 96]? One part of this question is easy to answer. If a sequence P_n converges weakly to a probability distribution P so that

$$\lim_{n\to\infty} \int_X f(x) P_n(dx) = \int_X f(x) P(dx)$$

for any continuous and bounded real-valued function f on X, then it follows immediately from the bounded convergence theorem that convergence holds for the corresponding characteristic functionals:

$$\phi_n(x^*) \to \phi(x^*); \qquad \forall x^* \in X^*$$

The converse is more sensitive and requires the notion of a *tight* family of measures.

Definition 1 A sequence $\{P_n\}$ of probability measures is said to be tight if for any $\varepsilon > 0$ there exists a compact set $C_\varepsilon \subseteq X$ such that for all n

$$P_n(C_\varepsilon) \geqslant 1 - \varepsilon$$

One can then give a satisfactory answer to the question raised.

THEOREM 1

In order that $P_n \to P$ weakly, it is necessary and sufficient that $\{P_n\}$ is tight and that $\phi_n(x^*) \to \phi(x^*), \forall x^*$.

If P_n and P are probability measures in $\mathbb{R}$, it is necessary and sufficient for $P_n \to P$ weakly that $\hat{P}_n \to \hat{P}$. That this is not always so in infinite-dimensional sample spaces can be seen by a simple example. Let x_n be a sequence in a Hilbert space such that $x_n \to x$ weakly but not strongly. Let P_n and P be the probability measures concentrated at x_n and x, respectively. Then $\hat{P}_n(x^*) = x^*(x_n)$, $P(x^*) = x^*(x)$, so that $\hat{P}_n \to \hat{P}$. However, P_n does not tend weakly to P. Here tightness does not apply.

It is also nontrivial to extend Bochner's theorem that continuous nonnegative definite functions on $\mathbb{R}$ are Fourier transforms of probability measures [see, e.g., Feller (1966), p. 585 for the classical case]. For a Hilbert space the answer is given by the following result [see Sazonov (1958)].

THEOREM 2

Consider a complex-valued function of ϕ of x^* on X^* with $\phi(0) = 1$. Then ϕ is the Fourier transform of a probability measure on the Hilbert space X if and only if ϕ is nonnegative definite and continuous in the S-topology.

The *S-topology* is defined through all neighborhoods in X^* of the form in (3) and their translates:

$$(Sx^*, x^*) < 1 \tag{3}$$

where S is an arbitrary nonnegative definite and self-adjoint operator belonging to the finite-trace type. Some authors use the term *nuclear* for such operators (see Note 1.2.B for further references).

By far the most studied case of probability measures in abstract spaces is that of a Gaussian measure in a real separable Hilbert space. The classical definition of a Gaussian measure in $\mathbb{R}^n$ can be formulated like this without making use of densities which would neglect the singular case: a probability measure in $\mathbb{R}^n$ is said to be Gaussian if any real linear combination

(c, x), $c_1x_1+c_2x_2+\ldots+c_nx_n$ of the stochastic variable $x=(x_1, x_2,\ldots, x_n)$ has a Gaussian distribution on the real line. Equivalently, x should have a characteristic function of the form

$$\phi(z)=E[\exp i(z, x)]=\exp\left[i(m, z)-\frac{1}{2}z^TRz\right] \tag{4}$$

where m is a fixed vector, the mean value vector, and R is a symmetric nonnegative definite matrix, the covariance matrix.

In a Hilbert space, however, we cannot just ask that (4) hold with obvious modifications:

$$\phi(x^*)=E[\exp ix^*(x)]=\exp\left[ix^*(m)-\frac{1}{2}(x^*, Rx^*)\right]; \qquad x^*\in X^* \tag{5}$$

Instead we have to go back to Theorem 2 and *add the condition that R in* (5) *be of trace class* (or nuclear). The fixed m is just an arbitrary element in X. According to the theorem then ϕ defines a legitimate probability measure in X, and this defines the Gaussian probability measure (see Note 1.2.B). To show that ϕ in (5) is S-continuous, it is enough to choose $S=P_m+R$, where P_m is the projection operator to the one-dimensional subspace spanned by m.

A simple but illustrative example, $m=0$, $R=$identity operator, does not define a legitimate Gaussian measure. Indeed, this would mean (in a weak sense that we shall not pursue here) that all the components $x_1, x_2,\ldots$ of the stochastic x-vector would be i.i.d. with $N(0, 1)$, and this does not correspond to a completely additive set function in X.

The expected value of a random element $x=x(\omega)$ in the Banach space X with respect to P is defined, assuming that $\|x(\omega)\|$ and that all $x^*[x(\omega)]$, $x^*\in X^*$, are P integrable, as the unique element m satisfying

$$E\{x^*[x(\omega)]\}=x^*(m) \tag{6}$$

This is the *Pettis integral* [see, e.g., Hille-Phillips (1957), pp. 76–84], which has many of the properties of ordinary Lebesgue integrals. We then write

$$m=E(x)=\int_\Omega x(\omega)P(d\omega)$$

One useful bond is

$$\left\|\int_\Omega x(\omega)P(d\omega)\right\|\leqslant\int_\Omega\|x(\omega)\|\,P(d\omega)$$

We also have for a bounded operator $L: X \to X$

$$E(Lx) = LE(x) \tag{7}$$

and

$$\mathrm{Cov}(Lx) = L^T \mathrm{Cov}(x) L \tag{8}$$

where the covariance operator $\mathrm{Cov}(x)$ is defined as in (9) below.

In the case of a Gaussian measure the element m in (5) is just the expected value of the random element x, while the operator R satisfies

$$(x_1^*, Rx_2^*) = \mathrm{Cov}[x_1^*(x), x_2^*(x)] \tag{9}$$

Other properties of measures in vector spaces are studied as they are needed in later chapters.

3 PROBABILITIES IN LOCALLY COMPACT SPACES

Occasionally we study inference problems for probability measures in topological spaces that lack linear properties. The methods of the proceeding section are then not applicable, and we have to look for alternative techniques.

In this case X will be assumed to be locally compact, and any probability space $(X, \mathcal{Q}, P)$ will still be based on the σ-algebra $\mathcal{Q}$ of Borel sets.

Weak convergence of probability measures is still defined as above, but we no longer have access to Fourier transforms (characteristic functions) via linear functionals, as defined in (2.2). However, if X forms a group with respect to a binary operation, Fourier analysis will still be available in terms of group representations.

First, assume that X is an *Abelian topological group*, and introduce the *group characters* as continuous complex-valued solutions of the equation

$$\gamma(x_1 + x_2) = \gamma(x_1)\gamma(x_2); \qquad x_1, x_2 \in X \tag{1}$$

where $\gamma(\cdot)$ takes as values complex numbers with modulus 1. Define a binary operation $\gamma_1(x)\gamma_2(x)$ and the topology by the neighborhoods $\{\gamma \| 1 - \gamma(x)| < \varepsilon\}$ for x in some compact set and $\varepsilon > 0$. These characters also form a locally compact Abelian group Γ, the dual group of X. For a given probability measure in X its Fourier transform is now defined as

$$\hat{P}(\gamma) = \int_X \gamma(x) P(dx)$$

as a function $\hat{P}: \Gamma \to \mathbb{C}$. It satisfies conditions quite similar to those of characteristic functions in finite-dimensional Euclidean spaces.

THEOREM 1

(a) *On a locally compact Abelian topological group P is uniquely determined by* $\hat{P}$.

(b) $|\hat{P}(\gamma)| \leqslant 1$, $\hat{P}(\gamma_e) = 1$ *for the identity character* γ_e.

(c) $\hat{P}(\gamma)$ *is continuous on* Γ.

(d) *The convolution* $P_1 * P_2$ *of two probability distributions on X has as its Fourier transform the product* $\hat{P}_1 \hat{P}_2$ *of the Fourier transforms of* P_1 *and* P_2*, respectively.*

(e) *If* $P_n \to P$ *weakly, then* $\hat{P}_n \to \hat{P}$*. Also if* $\hat{P}_n \to P$*, continuous on* Γ*, then* $\hat{P}$ *is the Fourier transform of a probability measure on X, and* $P_n \to P$ *weakly.*

There is also a characterization of Fourier transforms as nonnegative definite functions.

THEOREM 2

If $p(\gamma)$ *is a complex, continuous, nonnegative definite function on* Γ*, it is the Fourier transform of some probability measure on X, and vice versa.*

For a proof of this theorem, see, e.g., Grenander (1963), pp. 73–75.

For the cases that occur most often in the following, the characters are given in Table 1.

Table 1

Group G	Characters $\gamma(g)$; $g \in G$; $\gamma \in \Gamma$	Character Group Γ
$\mathbb{R}$	$\exp ig\gamma$	$\mathbb{R}$
$\mathbb{R}^n$; $g=(g_1, g_2, \ldots, g_n)$	$\exp[i \sum_1^n g_k \gamma_k]$	$\mathbb{R}^n$; $\gamma=(\gamma_1, \gamma_2, \ldots, \gamma_n)$
$\mathbb{T}$ = one-dimensional torus (circle)	$\exp ig\gamma$	$\mathbb{Z}$
$\mathbb{T}^n$; $g(g_1, g_2, \ldots, g_n)$	$\exp[i \sum_1^n g_k \gamma_k]$	$\mathbb{Z}^n$
$\mathbb{Z}$	$\exp ig\gamma$	T^n
$\mathbb{Z}^n$	$\exp[i \sum_1^n g_k \gamma_k]$	T
$\mathbb{Z}_k$ = cyclic group over k object; $g = 1, 2, \ldots, k$	$\exp ig\gamma$	$\mathbb{Z}_k$; $\gamma = \frac{2\pi}{n}, \frac{4\pi}{n}, \ldots, 2\pi$

This was for X locally compact and Abelian. If X is not necessarily Abelian, but *local compactness is strengthened to compactness*, Fourier transforms take the following form.

Consider all solutions of the equation (with the group operation now written in multiplicative form):

$$M(x_1x_2)=M(x_1)M(x_2); \qquad x_1, x_2\in X$$

where the *group representation* $M(\cdot)$ is a continuous function taking as values unitary $n\times n$ matrices, $n=0,1,2,3,\ldots$. In other words, we consider a homomorphism from G into a topological group of $n\times n$ matrices. If for two representations M_1 and M_2 there exists a matrix A such that $M_1(x)=AM_2(x)A^{-1}$ for all $x\in X$, then the two representations are said to be equivalent. From each equivalence class select one element. A representation is called irreducible if there is no proper subspace of n-dimensional unitary space left invariant by $M(x)$, $\forall x\in X$.

Consider a set $\{M^{(r)}(\cdot);\ r=0,1,2,\ldots\}$ of irreducible and mutually nonequivalent group representations of X, reserving the symbol $M^{(0)}(\cdot)$ for the identity. Given a probability measure on X, its Fourier transform is now defined as the sequence of matrices

$$P_r=\int_X M^{(r)}(x)P(dx); \qquad r=0,1,2,\ldots$$

Such a Fourier transform has properties as follows.

THEOREM 3

On a compact topological group X the Fourier transform $\{\hat{P}_r\}$ satisfies:

(a) *P is uniquely determined by $\{\hat{P}_r\}$.*

(b) $\|\hat{P}_r\|\leqslant 1$.

(c) *The convolution of two probability measures $P^{(1)}$ and $P^{(2)}$ has as its Fourier transform the product $\{\hat{P}_r^{(1)}\times\hat{P}_r^{(2)}\}$ of the Fourier transforms of $P^{(1)}$ and $P^{(2)}$, respectively.*

(d) *Consider a sequence P, $P^{(1)}$, $P^{(2)},\ldots$ of probability measures on X and their Fourier transforms $\hat{P}_r$, $\hat{P}_r^{(1)}$, $\hat{P}_r^{(2)},\ldots$. Then $P^{(n)}\to P$ weakly is equivalent to $\hat{P}_r^{(n)}\to\hat{P}_r$ as $n\to\infty$, for all r.*

Example 1 A simple case, both compact and Abelian, is when $X=T^1$. Then the group characters are of the form [solve (1) continuously]:

$$\gamma(x)=e^{inx}; \qquad n\in\mathbb{Z}$$

where we have identified the torus with the half-open interval $[0,2\pi)$ so that the Fourier transform is just the sequence of complex numbers

$$\hat{P}_n = \int_0^{2\pi-0} e^{inx} P(dx)$$

A reasonable definition of a Gaussian distribution would be obtained by asking that

$$\hat{P}_n = \exp\left(imn - \frac{\sigma^2}{2} n^2\right); \qquad m\mathbb{R}^1, \quad \sigma \in \mathbb{R}_+ \tag{2}$$

If $\sigma = 0$, this is obviously the Fourier transform of the measure with all its mass at $x = m_0$, $m_0 \in [0, 2\pi)$, $m \equiv m_0 \pmod{2\pi}$.

If $\sigma > 0$, we see that P is absolutely continuous with the frequency functions

$$f(x) = \frac{P(dx)}{dx} = \frac{1}{\sqrt{2\pi}\sigma} \sum_{k=-\infty}^{\infty} \left\{ \exp - \frac{1}{2\sigma^2} (x - m - 2\pi k)^2 \right\}$$

Indeed, direct integration gives, after exchanging the order of integration and summation, which is obviously legitimate here,

$$\int_0^{2\pi} e^{inx} f(x)\, dx = \sum_{k=-\infty}^{\infty} \frac{1}{\sqrt{2\pi}\sigma} \int_0^{2\pi} \exp\left[inx - \frac{1}{2\sigma^2}(x - m - 2\pi k)^2 \right] dx$$

and substituting $x - 2\pi k = y$,

$$= \sum_{k=-\infty}^{\infty} \frac{1}{\sqrt{2\pi}\sigma} \int_{-2\pi k}^{-2\pi(k-1)} \exp\left[iny - \frac{1}{2\sigma^2}(y - m)^2 \right] dy$$

$$= \frac{1}{\sqrt{2\pi}\sigma} \int_{-\infty}^{\infty} \exp\left[iny - \frac{1}{2\sigma^2}(y - m)^2 \right] dy = \exp\left(inm - \frac{\sigma^2}{2} n^2 \right)$$

as given in (2).

On the multi-dimensional torus T^d the above generalizes directly, and the result need not be given here.

If X is locally compact but neither compact nor Abelian, Fourier theory is still available, although in a much more complicated way. The interested reader should consult Grenander (1963), chap. 5.

Sometimes we can embed our X- or A-spaces in some well understood topological vector space. A very special but typical instance is offered by the following example.

Example 2 Consider the family $\mathcal{K}$ of all convex compact sets K in R^2. In $\mathcal{K}$ we define multiplication by a positive number c as

$$cK = \{cz \mid z \in K\}$$

and Minkowski addition of two sets

$$K_1 + K_2 = \{z \mid z = z_1 + z_2,\ z_1 \in K_1,\ z_2 \in K_2\}$$

$\mathcal{K}$ is topologized by the *Hausdorff distance*

$$d(K_1, K_2) = \inf\{c \mid K_1 \subseteq K_2 + cD,\ K_2 \subseteq K_1 + cD\}$$

where D is the unit disk centered at the origin

$$D = \{z \mid \|z\| \leqslant 1\}$$

Convex sets are conveniently described by their support functions

$$s(\phi) = \max_{(x, y) \in K} \{x\cos\phi + y\sin\phi\}$$

whose geometric interpretation is shown in Figure 1.

Then it is easily seen that, with obvious notation:

(a) $s(\cdot)$ is continuous and periodic 2π.

(b) cK has support function cs.

(c) $K_1 + K_2$ has support function $s_1 + s_2$.

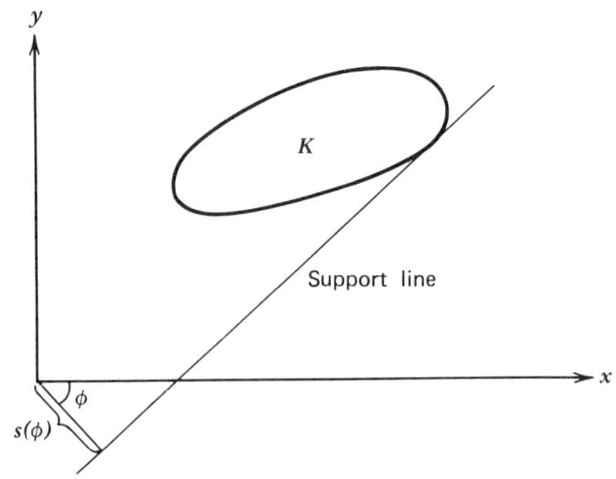

Figure 1

(d) $K_1 \subseteq K_2$ if and only if $s_1 \leqslant s_2$.

(e) $d(K_1, K_2) = \sup_\phi |s_1(\phi) - s_2(\phi)|$.

When $\mathcal{K}$ appears as parameter or sample space in a problem of inference, properties (a)–(e) make it natural to embed in $C(T^1)$. However, we could also employ the following representation result due to Vitale (1974) (see Note 1.3.A).

THEOREM 4

In order that $s(\cdot)$ be the support function of a set in $\mathbb{R}^2$ it is necessary and sufficient that

$$s(\phi) = a\cos\phi + b\sin\phi + \int_0^\phi \sin(\phi - \psi) R(d\psi)$$

where a and b are real constants and R is a bounded measure on $[0, 2\pi)$, *satisfying the orthogonality relations*

$$\int_0^{2\pi} \cos\phi R(d\phi) = \int_0^{2\pi} \sin\psi R(d\psi) = 0 \tag{3}$$

This makes it possible to deal with $\mathcal{K}$ as represented by bounded measures on T^1 restricted by (3).

In other situations we may exploit the special structure of our sample/parameter space, with or without group structure. An interesting case, which we return to in Chapter 7, is the following example.

Example 3 Consider a syntax-controlled probability model for a context free language L generated by the unambiguous grammer G over a terminal vocabulary V_T and a nonterminal vocabulary $V_N = \{1, 2, \ldots, R\}$. For any nonterminal $i \in V_N$ consider the possible rewriting rules into finite strings over V_N and V_T:

$$r: i \rightarrow s_{ij} \in (V_N \cup V_T)^*; \qquad j = 1, 2, \ldots, n_i \tag{4}$$

Associate probabilities $p^r = p_{ij}$ to the rewriting rule r in (4) with

$$\sum_{j=1}^{n_i} p_{ij} = 1, \qquad p_{ij} \geqslant 0$$

Consider a sentence $s \in L$ and a derivation of it employing the rewriting rules $r_1, r_2, \ldots,$ possibly with repetition. Associate the probability

$$P(s) = \prod_k p^{r_k}$$

to the sentence s, where r_k is the number of applications of outer k. It has been shown in Grenander (1976), pp. 82–87, that under certain conditions this defines a nondefective probability measure on the language L [see Grenander (1976) for full details and proof of statement].

The set of context free grammars over a given terminal vocabulary is denumerable. Let us enumerate them as $G_1, G_2, \ldots$. The lth grammer G_l has a number, say n_l, of rewriting rules associated with n_l probabilities. A complete specification can then be denoted by $G_l(p_1, p_2, \ldots, p_{n_l})$.

Form the parameter space as a disjoint union

$$A = \dot{\bigcup} S_l$$

where S_l is a copy of the simplex in R^{n_l} containing the admissible probability vectors $(p_1, p_2, \ldots, p_{d_l})$. Note that if one simplex is repeated several times, the copies should be considered disjoint from each other. The dimension d_l of S_l is always less than n_l.

In A we introduce the natural topology making the different sets S_l isolated from each other, and with Euclidean d_l -space topology in S_l.

Then A is locally compact, but of course without any natural group structure. To introduce a probability measure in A, we proceed as follows. First define the marginal distribution of l by a probability distribution $\{g_l;\ l = 1, 2, \ldots\}$. For each l define a conditional probability measure Q_l over S_l. The P-measure on A is defined by

$$P(E) = \sum_{l=1}^{\infty} q_l Q_l[E \cap S_l]$$

for any Borel set E.

Other examples are encountered later on in the text.

CHAPTER 2

Geometry of Stochastic Processes

1 CURVES IN HILBERT SPACE

When stochastic processes are studied in terms of their second moments—variances and covariances—they can be approached by purely geometric methods. In this chapter we pursue this idea to arrive at conclusions and techniques that are indispensable for the later chapters.

Given a probability space $(\Omega, \mathcal{A}, P)$ consider a stochastic process $x(t, \omega) = x(t)$, viewed as a set of stochastic variables, (Chapter 1, Section 1). Here $t \in T$, so far only assumed to be a *separable metric space*. The values of $x(t, \omega)$ will for the most part be scalar, either real or complex, but occasionally we allow them to be d-dimensional in $\mathbb{R}^n$ or $\mathbb{U}^n$.

The choice between real and complex-valued $x(t)$ is a matter of convenience, and we work with both. In this chapter it is convenient to allow complex values, but everything goes over into the real case with simple modifications.

It is assumed that our $x(t)$ have *finite second moments*

$$E\left[|x(t)|^2\right] < \infty; \qquad \forall t \in T$$

According to Schwarz' inequality, the mean value function $m(t)$ is then also defined. We assume that $m(\cdot) \equiv 0$; if not we should apply the reasoning to the reduced process $x_0(t) = x(t) - m(t)$. For probabilistic considerations this suffices, but later, in Chapter 4 especially, $m(t)$ will not always be known, so that $x_0(t)$ cannot be formed, which leads to completely different procedures.

Given a set of stochastic variables $\{x_\lambda;\ \lambda \in \Lambda\}$ with finite second moments and a set Λ of arbitrary cardinality, form the set of all linear combinations, the linear closure:

$$\left\{ \sum_{k=1}^{n} c_k x_{\lambda_k} \right\} \tag{1}$$

for any $n \in N, c_k \in \mathbb{C}$. Close the set in (1) topologically according to convergence in the mean using the norm

$$\| y - z \| = \sqrt{E[|y-z|^2]} \tag{2}$$

and denote the result by $L_2(x_\lambda, \lambda \in \Lambda)$.

It is well known from the theory of quadratically integrable functions that $L_2(x_\lambda, \Lambda)$ forms a Hilbert space if the inner product is the one corresponding to the norm in (2), namely,

$$(y, z) = E(y\bar{z})$$

In particular this construction applied to our stochastic process $x(t)$ for $\Lambda = T$ gives us the Hilbert space $L_2(x_t, t \in T) = L_2(x)$.

We assume, unless the opposite is stated, *that $x(t)$ is continuous in the mean*.

Definition 1 The stochastic process $x(t)$ is said to be continuous in the mean at t_0 if

$$\lim_{h \to 0} \| x(t_0 + h) - x(t) \| = 0 \tag{3}$$

It is called continuous in the mean on T if it is continuous in the mean for all $t \in T$.

If (3) holds, then Chebyshev's inequality implies that $x(t)$ is continuous in probability at t_0.

Continuity in the mean is essentially the same as continuity of the covariance function.

THEOREM 1

The stochastic process is continuous in the mean on T if and only if the covariance function $r(s, t)$ is continuous jointly in s and t on the diagonal set of the product set $T \times T$; then $r(s, t)$ is also continuous on the entire set $T \times T$.

Proof Assume r to be continuous at the point $s=t=t_0$ on the diagonal.

$$\|x(t_0+h)-x(t_0)\|^2$$
$$=r(t_0+h,t_0+h)-r(t_0+h,t_0)-r(t_0,t_0+h)+r(t_0,t_0) \qquad (4)$$

As $h\to 0$, all four terms tend to the same value, namely, $r(t_0, t_0)$, so that the limit of (4) is zero, and hence $x(t)$ is continuous in the mean at $t=t_0$.

On the other hand, assume that $x(t)$ is continuous in the mean on T. We then have

$$\begin{aligned}|r(s,u)-r(t,v)| &= |E[x(s)\overline{x(u)}]-E[x(t)\overline{x(v)}]|\\ &\leqslant |E[x(s)\overline{x(u)}]-E[x(s)\overline{x(v)}]|+|E[x(s)\overline{x(v)}]-E[x(t)\overline{x(v)}]|\\ &= |E[x(s)\overline{(x(u)-x(v))}]|+|E[(x(s)-x(t))\overline{x(v)}]|\end{aligned}$$

and with Schwarz's inequality we see that this is at most equal to

$$\|x(s)\|\cdot\|x(u)-x(v)\|+\|x(s)-x(t)\|\cdot\|x(v)\|$$

Now keep s, u fixed as an arbitrary point in $T\times T$ and let $(t,v)\to(s,u)$. Then $\|x(u)-x(v)\|\to 0$, $\|x(s)-x(t)\|\to 0$. Also $\|x(u)\|$ is a fixed constant and, for $\varepsilon>0$ and using the triangle inequality, we have

$$\|x(v)\|\leqslant\|x(v)-x(u)\|+\|x(u)\|\leqslant\varepsilon+\|x(u)\|$$

if v is close enough to u. Hence the covariance function is continuous at (s,u). Q.E.D.

What is the "dimension" of the Hilbert space $L_2(X)$? If $x(t)$ is continuous in the mean, the answer is given by the following theorem.

THEOREM 2

Assume that $x(t)$ is continuous in the mean on T. Then $L_2(X)$ is separable.

Proof The argument set has been assumed to be a separable metric space, so that we can select a sequence $s_1, s_2,\ldots$ everywhere dense in T. Consider the stochastic variables $x(s_1)=x_1, x(s_2)=x_2,\ldots,$ and an arbitrary linear combination

$$L=\sum_{k=1}^{n} c_k x(t_k)$$

Each $x(t_k)$ can be approximated by some x_ν by choosing s_ν sufficiently close to t_k so that $\|x(t_k)-x_\nu\| \leqslant \varepsilon$. Hence L can be approximated by a linear combination of x_ν's with an approximation error at most $\varepsilon\Sigma_k|c_k|$. Hence the sequence $\{x_\nu\}$ is everywhere dense in $L_2(X)$ so that this Hilbert space is separable. Q.E.D.

Note that $L_2(X)$ is included in the Hilbert space of all quadratically integrable complex-valued functions on $(\Omega, \mathcal{Q}, P)$,

$$L_2(X) \subseteq L_2(\Omega, \mathcal{Q}, P)$$

It can be a *proper* subset, as can be shown by simple examples.

Let $x(t)$ be real-valued for a moment, and consider the *geometric significance of the standard deviation* of some stochastic variable x in $L_2(X)$. It is clear that it coincides with the norm

$$\sigma(x) = \text{standard deviation of } x = +\sqrt{E[x^2]} = \|x\|$$

If we study two stochastic variables x_1 and x_2 simultaneously, all linear-quadratic statements can be interpreted in the subspace S spanned by x_1 and x_2:

$$S = \text{all } c_1x_1 + c_2x_2; \qquad c_1, c_2 \in \mathbb{R}$$

It then makes sense to speak of the angle ϕ between x_1 and x_2 (see Figure 1).

We get from plane trigonometry

$$\sigma^2(x_2 - x_1) = \sigma^2(x_1) + \sigma^2(x_2) - 2\sigma(x_1)\sigma(x_2)\cos\phi \tag{5}$$

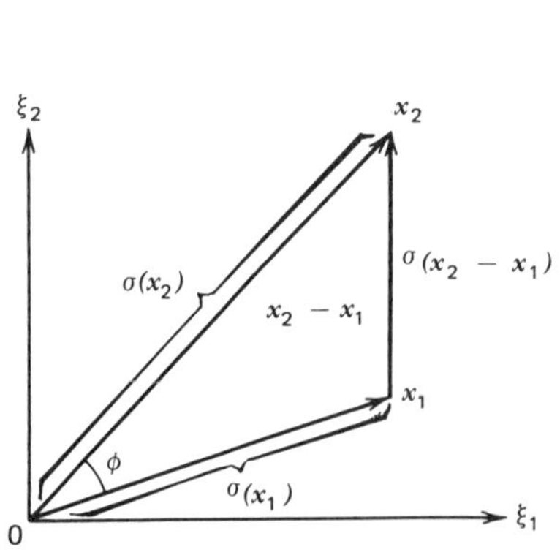

Figure 1

which is the same as

$$\|x_2-x_1\|^2=\|x_2\|^2+\|x_1\|^2-2(x_2,x_1)$$

which can be written, using (5), as

$$\|x_1\|^2+\|x_2\|^2-2\|x_1\|\cdot\|x_2\|\cdot\cos\phi$$

so that

$$\cos\phi=\frac{(x_1,x_2)}{\|x_1\|\cdot\|x_2\|}=\rho(x_1,x_2)$$

with ρ being *the correlation coefficient, geometrically meaning the cosine of the angle between the two "vectors."* In particular, if this angle is 90°, *orthogonality*, the stochastic variables are uncorrelated. If it is 0°, they are a.c. proportional with a positive proportionality constant; if it is 180°, they are a.c. proportional in the opposite sense.

Now consider a stochastic variable x and a family $\{x_\lambda, \lambda\in\Lambda\}$, all in $L_2(X)$. It is well known that there is a unique element x^* in the closed subspace $L=L_2(x_\lambda, \lambda\in\Lambda)$ which solves the minimum problem

$$\min_{x^*\in L}\|x-x^*\|$$

Here x^* is the projection of x down to L, $x^*=P_L x$ written in terms of the projection operator P_L. Since the approximation error $x-x^*\perp x_\lambda$, all λ, we have the *normal equations*

$$E(x^*x_\lambda)=E(xx_\lambda); \qquad \text{all } \lambda\in\Lambda \tag{6}$$

THEOREM 3

The element x^* is uniquely determined by the system of equations (6).

Proof For an indirect proof assume that x^{**} is another solution of (6) so that with $d=x^{**}-x^*\in L$,

$$E(dx)=0; \qquad \text{all } \lambda\in\Lambda \tag{7}$$

But d can be approximated in the norm by linear combinations d_ε of $x_\lambda' s$, $\|d_\varepsilon-d\|\to 0$. Using (7) for the inner product, we get

$$\|d\|^2=\lim_{\varepsilon\downarrow 0}(d,d_\varepsilon)=0$$

so that $d=0\Rightarrow x^{**}=x^*$. Q.E.D.

In particular if Λ is finite, $\Lambda=\{1,2,\ldots,n\}$, and if we write

$$x^*=\sum_1^n c_k x_k = c^T\xi$$

$c=\text{col}(c_1,\ldots,c_n)$, $\xi=\text{col}(x_1,\ldots,x_n)$, then (6) reduces to

$$Rc=\rho$$

where R is the $n\times n$ covariance matrix of the random x-vector and $\rho=\text{col}[\text{cov}(x,x_1),\ldots,\text{cov}(x,x_n)]$. If R is nonsingular, the projection can be expressed as

$$x^*=\rho^T R^{-1}\xi$$

with the approximation error given by

$$\begin{aligned}\|x^*-x\|^2&=(x-x^*,x)=(x-\rho^T R^{-1}\xi,x)\\&=\|x\|^2-\rho^T R^{-1}\rho\end{aligned} \tag{8}$$

The statistical reader recognizes *this way of writing the projection operator as the regression of* x *upon* $\{x_\lambda, \lambda\in\Lambda\}$ *with* the residual variance given by (8) [see, e.g., Cramér (1945), pp. 302–308].

Returning to $x(t)$, when $L_2(X)$ is separable we can introduce a CON (complete orthonormal) system $\{\xi_1,\xi_2,\ldots\}$:

$$\begin{aligned}&(\xi_\nu,\xi_\mu)=0; \qquad \text{if } \nu\neq\mu\\&\|\xi_\nu\|=1\\&\{\xi_\nu\} \text{ complete in } L_2(X)\end{aligned}$$

This can be done in many ways, and we shall return to the crucial question of the most "economical" way of doing it. One important procedure is the celebrated *Gram-Schmidt orthogonalization* which can be applied when $T=\{1,2,3,\ldots\}$. Define

$$\xi_1=\frac{x(t)}{\|x(t)\|}$$

with the smallest t for which $x(t)\neq 0$. Iteratively, when $\xi_1,\xi_2,\ldots,\xi_\nu$ have been constructed, define

$$y(\nu+t)=x(\nu+t)-P_{L_\nu}x(\nu+t) \tag{9}$$

and

$$\xi_{\nu+1} = \frac{y(\nu+t)}{\|y(\nu+t)\|}$$

where t is the smallest natural number for which $y(\nu+t) \neq 0$. In (9) P_{L_ν} stands for the projection operator associated with the subspace L_ν spanned by the ν first ξ_k.

It is obvious that $\{\xi_\nu\}$ is orthonormal, and completeness follows from the fact that each $x(t)$ can be calculated as a linear combination of ξ_ν's.

If Λ is a continuum, the Gram-Schmidt orthogonalization is not directly applicable, and other constructions are described in Section 4.

Once we have constructed a CON system, it can be used for representing the stochastic process. Consider the formal series

$$\sum_{\nu=1}^{\infty} (x(t), \xi_\nu)\xi_\nu = \sum_{\nu=1}^{\infty} f_\nu(t)\xi_\nu \tag{10}$$

THEOREM 4

The series in (10) *converges in the mean to* $x(t)$.

Proof With the partial sum

$$s_n(t) = \sum_{\nu=1}^{n} f_\nu(t)\xi_\nu$$

we have by simple calculations, familiar from expansion in orthogonal series,

$$\|x(t) - s_n(t)\|^2 = \sum_{\nu=n+1}^{\infty} |f_\nu(t)|^2 \to 0; \qquad n \to \infty$$

since the Fourier coefficients f_ν are quadratically summable. Q.E.D.

This is just Parseval's relation adapted to the present case.

The projection operator solving (6) can then be expressed in series form via the CON system. Indeed, essentially repeating the proof of Theorem 4 we get

$$x^* = \sum_{\nu=1}^{\infty} (x, \xi_\nu)\xi_\nu$$

with the approximation error

$$E\left[\|x-x^*\|^2\right]=\|x\|^2-\sum_{\nu=1}^{\infty}|(x,\xi_\nu)|^2$$

In this perspective *the second-order properties of stochastic processes are those of curves in a Hilbert space*. For each value of the argument t, the stochastic variable $x(t)$ is seen as a point in the space $L_2(X)$ generated by the process $\{x(t), t\in T\}$. (see Note 2.1.A). Statements about any finite number of stochastic variables are reduced to problems in n-dimensional Euclidean or unitary geometry.

To make the above more concrete, let us discuss a few simple cases.

Example 1 For $V=\mathbb{C}$, $T=\mathbb{R}$, introduce the *random phase model*

$$x(t)=\sum_{k=1}^{\infty}A_k e^{it(\lambda_k+\phi_k)};\qquad \sum_{k=1}^{\infty}A_k<\infty \tag{11}$$

where the amplitudes A_k are positive constants, the frequencies λ_k are real constants, and $(\phi_1,\phi_2,\ldots,\phi_n)$ is an i.i.d. sample from the rectangular distribution $R(0,2\pi)$.

The mean value function is by the theorem on bounded convergence

$$E[x(t)]=m(t)=\sum_{k}^{\infty}A_k\frac{1}{2\pi}\int_0^{2\pi}e^{i(t\lambda_k+\phi_k)}\,d\phi_k=0$$

and the covariance function

$$\begin{aligned}(x(s),x(t))&=E\left[x(s)\overline{x(t)}\right]\\&=\sum_{k\neq l}A_k\bar{A}_l\frac{1}{2\pi}\int_0^{2\pi}\int_0^{2\pi}e^{i(s\lambda_k+\phi_k-t\lambda_l-\phi_l)}\,d\phi_k\,d\phi_l\\&\quad+\sum_k|A_k|^2\frac{1}{2\pi}\int_0^{2\pi}e^{i\lambda_k(s-t)}\,d\phi_k\\&=\sum_k|A_k|^2e^{i\lambda_k(s-t)}\end{aligned}$$

Note that these inner products of vectors on the curve depend only upon the difference of the arguments $s-t$. Hence the geometry is invariant with respect to translations of the t-axis.

In this case Ω is the denumerable Cartesian product of copies of $[0,2\pi]$ with P given as proportional to Lebesgue measure to any coordinate of Ω.

The real version of this model is the same, except as that the exponential function in (11) should be replaced by $\cos(\lambda_k t+\phi_k)$.

Example 2 With $V=\mathbb{R}$, $T=\mathbb{R}_+$, define an *orthogonal process* indirectly through the properties

$$\begin{gathered} m(t)\equiv 0 \\ r(t,t)=V(t); \qquad V(0)=0 \\ x(t_4)-x(t_3)\perp x(t_2)-x(t_1) \text{ if } (t_1,t_2)\cap(t_3,t_4)=\phi;\ t_1\leqslant t_2\leqslant t_3\leqslant t_4 \end{gathered} \tag{12}$$

where $V(t)$ is a given nondecreasing and nonnegative function (see Note 2.1.B). Of course we must show that there exists (at least one) stochastic process satisfying (12).

To do this, define for any finite set $t_1, t_2, \ldots, t_n$, ordered increasingly, the n-dimensional distribution function $F(x_1, x_2, \ldots, x_n; t_1, t_2, \ldots, t_n)$ corresponding to the stochastic variables

$$\begin{aligned} x_1&=z_1 \\ x_2&=z_1+z_2 \\ &\vdots \\ x_n&=z_1+z_2+\cdots+z_n \end{aligned} \tag{13}$$

where the random vector $(z_1, z_2, \ldots, z_n)$ is Gaussian with mean value vector zero, uncorrelated components, and variances

$$E(z_k^2)=V(t_k)-V(t_{k-1}); \qquad t_0=0$$

It is not difficult to show that Kolmogorov's consistency conditions hold for this family of distribution functions. Hence they define a stochastic process $\{x(t), t\geqslant 0\}$, satisfying (12) (see Note 2.1.C).

The covariance function is then for $s\leqslant t$

$$r(s,t)=E[x(s)x(t)]=(x(s),x(t))=(x(s),x(t)-x(s))+(x(s),x(s))$$

and using the orthogonality relation in (12),

$$r(s,t)=V(s)$$

so that, for general s and t,

$$r(s,t)=V[\min(s,t)] \tag{14}$$

This process is then continuous in the mean at $t=t_0$ if and only if V is continuous at t_0; use Theorem 1. Then, Example 1 of Chapter 1, Section 2, tells us that there is a separable version whose sample functions are continuous with probability 1.

The curve $x(t)$ in $L_2(X)$ has the peculiar geometry that any chord $x(t_2)-x(t_1)$, $t_1<t_2$, is orthogonal to all $x(t)$ for $t\leqslant t_1$. This chaotic behavior of the curve will be of importance later on.

Example 3 For $V=\mathbb{C}$, $T=Z$, start from an orthonormal system $\{\xi_\nu;\, \nu\in \mathbb{Z}\}$ of stochastic variables with mean zero and define for a sequence $\{c_\nu;\, \nu=\ldots,-1,0,1,\ldots\}\in l_2$ the *moving average process*

$$x(t)=\sum_{\nu=-\infty}^{\infty} c_{t-\nu}\xi_\nu \tag{15}$$

The series converges in the mean (see the proof of Theorem 4). Its covariance function is easily calculated as

$$r(s,t)=(x(s),x(t))=\sum_{\nu=-\infty}^{\infty} c_{s-\nu}\bar{c}_{t-\nu}$$

which can be seen to depend only upon the difference $s-t$, just as in Example 1.

If $c_p=0$ for $p<0$, we have a *one-sided moving average* and can give (15) a *causal explanation*: the value of $x(t)$ is "caused" by the values $\xi_t, \xi_{t-1}, \xi_{t-2},\ldots$. It is then obvious that $L_2(x(t);\, t\leqslant a)\subseteq L_2(\xi_t;\, t\leqslant a)$, but equality does not hold automatically. This fact is of greater significance than one may think at first glance and leads to some difficulties later on.

2 CALCULUS OF SECOND-ORDER PROCESSES

Using convergence in the mean, we can now define derivatives, Riemann integrals, Stieltjes integrals for $x(t)$, and it is comparatively easy to develop the calculus needed for second-order processes.

With some loss of generality let T be an interval on the real line.

Definition 1 A second-order process $x(t)$ is said to be differentiable in the mean at t with the derivative $x'(t) \in L_2(X)$ if

$$\lim_{h\to 0}\left\| \frac{x(t+h)-x(t)}{h} - x'(t)\right\| = 0$$

THEOREM 1

The process is differentiable in the mean if and only if the covariance function $r(s,t)$ has a generalized second derivative so that

$$\frac{\partial^2 r}{\partial s \partial t} = \lim_{h,k\to 0} \frac{r(t+h,t+k)-r(t+h,t)-r(t,t+k)+r(t,t)}{hk} \tag{1}$$

exists at the point (t,t).

Proof If $x(t)$ is differentiable in the mean with the derivative $x'(t)$, we have

$$y(h) = \frac{x(t+h)-x(t)}{h} \to x'(t)$$

in the mean as $h \to 0$, so that the inner products

$$(y(h), y(k)) = \frac{r(t+h,t+k)-r(t+h,t)-r(t,t+k)+r(t,t)}{hk}$$

$$\to \|x'(t)\|^2 \tag{2}$$

But (2) implies that the generalized second derivative exists at the point (t,t).

On the other hand, if the generalized derivative exists, it follows that $(y(h), y(k)) \to$ limit c, so that

$$\|y(h)-y(k)\|^2 = \|y(h)\|^2 - (y(h), y(k)) - (y(k), y(h)) + \|y(k)\|^2 \to 0.$$

Since this Cauchy condition holds, the limit of $y(h)$ exists as an element in $L_2(X)$, and the proposition is true. Q.E.D.

If $x(t)$ is differentiable in the mean for all t, it follows by the same sort of proof as in (2) that $x'(t)$ has the covariance function

$$(x'(s), x'(t)) = \frac{\partial^2 r}{\partial s \partial t} \tag{3}$$

Of course a process can be differentiable in the mean without having its sample functions differentiable a.c. For the latter, stronger assumptions are needed (see Chapter 1, Section 1).

Higher order derivatives are dealt with in the same way so that the existence of the pth derivative is related to the existence of a generalized mixed derivative of order $2p$ of r.

If $x(t)$ is continuous in the mean on $[a, b]$ its Riemann integral is easily introduced.

THEOREM 2

For a stochastic process $x(t)$, continuous in the mean, consider the Riemann sums

$$S = \sum_{k=0}^{n} x(\theta_k)(t_{k+1} - t_k)$$

with $a = t_0 < t_1 < t_2 < \cdots < t_{n+1} = b$ and $\theta_k \in (t_k, t_{k+1})$. As the partition becomes finer, S converges in the mean to a unique limit in $L_2(X)$ that will be denoted as

$$\int_a^b x(t)\, dt$$

Proof Consider S belonging to the stated partition and S' defined analogously by a second partition. Then

$$\|S - S'\|^2 = \|S\|^2 - (S, S') - (S', S) + \|S'\|^2 \tag{4}$$

But

$$\|S\|^2 = \sum_{k,l=0}^{n} r(\theta_k, \theta_l)(t_{k+1} - t_k)(t_{l+1} - t_l) \to \int_a^b \int_a^b r(s,t)\, ds\, dt \tag{5}$$

as the partition becomes finer, since r is continuous and hence Riemann integrable on $T \times T$. The other three terms on the right-hand side of (4) tend to the same limit, so that (4) tends to zero, the Cauchy condition holds, and the limit exists.

Also, for any limit I obtained through some partitions, we get by the same sort of argument

$$(I, x(t)) = \int_a^b r(s,t)\, dt; \qquad \forall t \in T$$

This determines I uniquely (see proof of Theorem 3 of Section 1). Q.E.D.

Note that (5) means that the variance of the integral is given as the double integral of the covariance function

$$\operatorname{Var}\left[\int_a^b x(t)\,dt\right]=\int_a^b\int_a^b r(s,t)\,ds\,dt$$

The usual properties for ordinary integrals hold for Riemann integrals of second-order processes continuous in the mean, and we mention a few, leaving the proofs to the reader.

$$(a)\qquad \int_a^b [c_1x_1(t)+c_2x_2(t)]\,dt=c_1\int_a^b x_1(t)\,dt+c_2\int_a^b x_2(t)\,dt$$

(b) For $a<b<c$ we have the additive property

$$\int_a^c x(t)\,dt=\int_a^b x(t)\,dt+\int_b^c x(t)\,dt$$

(c) The definite integral has a derivative in the mean with respect to the upper limit equal to the usual value

$$\frac{d}{dt}\left[\int_a^t x(s)\,ds\right]=x(t)$$

(d) The standard deviation of the integral can be bounded by the integral of the standard deviation of the process

$$\sigma(I)=\left\|\int_u^b x(t)\,dt\right\|\leqslant\int_a^b \sigma[x(t)]$$

(apply the triangle inequality to the Riemann sums.)

If $x(t)$ is not only continuous in the mean but also a measurable process (see Chapter 1, Section 1), then the sample functions are a.c. integrable, and these integrals J must then coincide a.c. with the Riemann integral since

$$E[Jx(t)]=\int_a^b r(s,t)\,ds;\qquad \forall t\in[a,b]$$

determines J.

Stieltjes integrals are defined first for orthogonal processes and are later given a more general definition.

THEOREM 3

Let $x(t), t \in [a,b]$ *be an orthogonal process with variance function* V *and* $f \in c([a,b])$. *Introduce the Riemann-Stieltjes sums for a partition as in Theorem* 2:

$$S = \sum_{k=0}^{n} f(\theta_k)[x(t_{k+1}) - x(t_k)] \tag{6}$$

The unique limit of S *exists as the partition becomes finer and will be denoted by*

$$\int_a^b f(t)x(dt)$$

Proof Consider two partitions and the corresponding sums S and S' as in (6) and calculate

$$\begin{aligned}(S, S') = \sum_{k,l=0}^{n} f(\theta_k)\overline{f(\theta_l')}\,[&r(t_{k+1}, t'_{l+1}) \\ &- r(t_{k+1}, t'_l) - r(t_k, t'_{l+1}) + r(t_k, t'_l)]\end{aligned} \tag{7}$$

where r is given in (1.14). Since V is nondecreasing in $[a,b]$, it follows that $r(s,t)$ is of bounded variation on $[a,b]^2$. Actually all its variation is located on the diagonal $s=t$. Hence (7) converges to the limit

$$\int_a^b \int_a^b f(s)\overline{f(t)}\,r(ds, dt) = \int_a^b f(t)V(dt)$$

and we get the same limit $\|S\|^2, \|S'\|^2$ so that the limit of S exists. Q.E.D.

As a by-product we get

$$\left(\int_a^b f(t)x(dt), \int_a^b g(t)x(dt)\right) = \int_a^b f(t)\overline{g(t)}\,V(dt)$$

and in particular the important relation

$$\left\|\int_a^b f(t)x(dt)\right\|^2 = \int_a^b |f(t)|^2 V(dt)$$

Stochastic Stieltjes integrals obey laws quite similar to ordinary ones, corresponding to properties a and b above. It should be mentioned, however, that stochastic Stieltjes integrals cannot be interpreted as ordinary Stieltjes integrals of sample functions in general. The sample function need not be of bounded variation a.c.

It is easy to extend the above definition to the case when $x(t)$ is not necessarily orthogonal but has a covariance function $r(s, t)$ of bounded variation over $[a, b]$. The discussion can easily be modified to handle this case too with obvious modifications, which will not be given here.

We now have all the building blocks needed for the calculus of second-order processes. Many of the statements valid in standard calculus remain valid, and we mention a few.

Example 1 The fundamental theorem of calculus holds. In other words, if $x(t)$ is differentiable in the mean in $[a, b]$ with $x'(t) \equiv 0$, then $x(t) \equiv x(a)$. Indeed, the process $y(t) = x(t) - x(a)$ has also derivative zero so that for any $s \in [a, b]$

$$\left(\frac{y(t+h)-y(t)}{h}, x(s)\right) \to 0 \qquad \text{as } h \to 0$$

which implies, writing r_y for the covariance function of $y(t)$,

$$\frac{r_y(t+h, s) - r_y(t, s)}{h} \to 0 \qquad \text{as } h \to 0$$

or

$$\frac{\partial r_y(t, s)}{\partial t} \equiv 0$$

so that $r_y(t, s) = \text{constant}$. But $r_y(t, a) = 0$ so that $\|y\| = 0$ and $x(t) \equiv x(a)$.

Hence we can immediately assert that if

$$\frac{dx(t)}{dt} = z(t); \qquad t \in [a, b] \tag{8}$$

as derivatives in the mean, if $z(t)$ is continuous in the mean, (8) has only solutions of the form

$$x(t) = x(a) + \int_a^t z(s)\, ds$$

Example 2 This extends immediately to linear ordinary differential equations. As a simple, but typical case consider the equation

$$\frac{dx(t)}{dt}+\beta x(t)=y(t) \tag{9}$$

where again $y(t)$ is continuous in the mean. Then, with a stochastic Riemann integral, processes of the form

$$x_1(t)=x(0)e^{-\beta t}+\int_0^t e^{-\beta(t-s)}y(s)\,ds \tag{10}$$

are the only solutions to (9) since, with derivatives in the mean,

$$\begin{aligned}\frac{dx_1(t)}{dt}&=-\beta x(0)e^{-\beta t}+y(t)-\beta\int_0^t e^{-\beta(t-s)}y(s)\,ds\\&=-\beta x_1(t)+y(t)\end{aligned}$$

so that it is the unique solution of (9) with the initial value $x(0)$.

Example 3 The differential equation (9) presents some problems in the important case when, formally, $y(t)=\dfrac{dw(t)}{dt}$, where $w(t)$ is an orthogonal process with $w(0)=0$ and variance function $V(t)=\sigma^2 t$. If the process $w(t)$ is also Gaussian, it is known as the *Wiener process*.

The covariance function $\sigma^2\min(s,t)$ (see Example 1 in Section 1), does not satisfy (1), so that $y(t)$ is not defined as a derivative in the mean. Nor are the sample functions of the Wiener process differentiable a.c. Hence the equation is not really of the type in (9).

One way out is to *interpret* (9) *weakly*. We say that a second-order process $x(t)$, continuous in the mean, satisfies (9) on $[a,b]$ if for any continuous function c the equation multiplied by c can be integrated formally in terms of w:

$$\int_a^b c(t)x(dt)+\beta\int_a^b c(t)x(t)\,dt=\int_a^b c(t)w(dt)$$

Here $r(s,t)$ must be of bounded variation so that the first integral is defined as mentioned above.

If we replace (10) by

$$x_1(t)=x(a)e^{-\beta t}+\int_a^t e^{-\beta(t-s)}w(ds) \tag{11}$$

where the integral is a random Stieltjes integral, we can calculate

$$\int_a^b c(t)x_1(dt) \tag{12}$$

as

$$-\beta x(a)\int_a^b c(t)e^{-\beta t}\,dt+\int_a^b c(t)w(dt)-\beta\int_a^b c(t)\int_a^t e^{-\beta(t-s)}w(ds) \tag{13}$$

To verify (13), write down the Riemann-Stieltjes sums corresponding to (11) and (12); the limit will be (13). We also have

$$\int_a^b c(t)x_1(t)\,dt=x(a)\int_a^b c(t)e^{-\beta t}+\int_a^b c(t)\int_a^t e^{-\beta(t-s)}w(ds) \tag{14}$$

Adding (12) and (14) multiplied by β, we see that

$$\int_a^b c(t)x_1(dt)+\beta\int_a^b c(t)x_1(t)\,dt=\int_a^b c(t)w(dt)$$

so that $x_1(t)$ is a *weak solution* of (9).

If $x_2(t)$ is another weak solution, continuous in the mean and with covariance function r_y of bounded variation, with the same initial value $x(a)$ the difference $y(t)=x_1(t)-x_2(t)$ satisfies

$$\int_a^b c(t)y(dt)+\beta\int_a^b c(t)y(t)\,dt=0 \tag{15}$$

Choose a continuous function with

$$c(t)=\begin{cases}1; & a\leqslant t\leqslant s\\ 0; & s+\varepsilon<t\leqslant b\\ \text{linear}; & s\leqslant t\leqslant s+\varepsilon\end{cases} \tag{16}$$

so that the stochastic variable

$$y(s)+\beta\int_a^s y(\tau)\,d\tau \tag{17}$$

has a norm bounded by [use (15) and (16)]

$$\left(\int_s^{s+\varepsilon}\int_s^{s+\varepsilon}|r_y(du,dv)|\right)^{1/2}+\beta\left(\int_s^{s+\varepsilon}\int_s^{s+\varepsilon}|r_y(u,v)|\,du\,dv\right)^{1/2}$$

Hence with the exception of at most denumerably many s-values the norm of (17) is zero since r_y is of bounded variation. Since it depends continuously in the mean upon s, it is identically zero. But then $y(s)$ is differentiable in the mean with derivative $-\beta y(s)$, and Example 2 shows that $y \equiv 0$ so that the weak solution (11) is unique when $x(a)$ is fixed (see Note 2.2.A).

Linear equations, also of higher order, driven by the "derivative" $y(t)$ of the Wiener process, can be handled in a similar way.

By the same technique we can study the process given as an infinite Stieltjes integral:

$$y(t) = \int_{-\infty}^{t} e^{-\beta(t-s)} w(ds) \tag{18}$$

which also is a weak solution of the same equation. Its covariance function is for $h > 0$

$$r(t, t+h) = \sigma^2 \int_{-\infty}^{t} e^{-\beta(2t+h-2s)}\, ds = \frac{\sigma^2}{2\beta} e^{-\beta h}$$

Hence $r(s, t)$ can be written, since it is an even function of $s-t$,

$$r(s, t) = \frac{\sigma^2}{2\beta} e^{-\beta|s-t|} \tag{19}$$

If this process is also assumed to have Gaussian distributions, one can show that it has the Markov property [see, e.g. Feller (1966), pp. 93–98]. Whether Gaussian or not, the covariance function (19) can be viewed as "linearly Markov" since the optimal predictor of $x(t+h)$, $h > 0$, given all $x(s)$ for $s \leqslant t$, depends only upon $x(t)$ and is a linear function of this variable.

3 REPRESENTATIONS OF PROCESSES

The fundamental problem in this section is how to represent a stochastic process, perhaps with a complicated dependence structure, as linear combinations of "simple" elements. "Simple" here means orthogonal (uncorrelated).

We base our analysis on a *general theme*: *linear representations of processes induce bilinear representations of the covariance functions, and vice versa*. The first part is easy; the second one requires a deeper study (see Note 2.3.A).

For didactic reasons let us first look at the simplest case, when T is finite, $T=\{1,2,3,\ldots,n\}$, so that our "process" is just a stochastic n-vector $x=(x_1,x_2,\ldots,x_n)$. Consider a linear representation

$$x=Az; \qquad z=\mathrm{col}(z_1,z_2,\ldots,z_m),$$
$$A=\{a_{t\lambda};t=1,2,\ldots,n;\lambda=1,2,\ldots,m\}, \quad m\leqslant n$$

where $\{z_k;\ k=1,2,\ldots,m\}$ is a set of uncorrelated stochastic variables with mean 0 and variance 1. In order to use an *economical representation* with a small number m it is clear that we need never use more than n z-variables to span the n x-variables. The covariance matrix of the x-vector is

$$R=E\left[xx^T\right]=AE\left[zz^T\right]A^T=AA^T \tag{1}$$

Written in components it gives us the bilinear form

$$r_{ts}=\sum_{\lambda=1}^{m}a_{t\lambda}\bar{a}_{s\lambda}; \qquad m\leqslant n \tag{2}$$

On the other hand assume that (12) is given, and *we seek a corresponding linear representation for the x vector*. Separate cases!

First, assume that R is nonsingular. Then $m<n$ is not possible, since A would have rank less than n and R would be singular. Hence $m=n$, and since

$$0<\det(R)=|\det(A)|^2$$

the A matrix must also be nonsingular. Note that since R is nonnegative definite, we can always find (many) matrices A satisfying (1) (see below). Define the stochastic n-vector z by

$$z=A^{-1}x \tag{3}$$

which has the covariance matrix

$$E\left[zz^T\right]=A^{-1}E\left[xx^T\right](A^{-1})^T=A^{-1}R(A^{-1})^T=A^{-1}AA^T(A^{-1})^T=I$$

Hence the z-components are orthonormal, and inverting (3) gives us the desired linear representation for x.

Second, let R be singular, say of rank $r<n$. Then we can find a matrix B such that the new matrix C, made up of A as its first rows and of B as the remaining rows, is nonsingular. We build the column vector of B from the

orthogonal complement of the range space of A. Then

$$R_+ = CC^T = \begin{Bmatrix} AA^T & AB^T \\ BA^T & BB^T \end{Bmatrix} \tag{4}$$

is a covariance matrix of full rank. Let ξ be a stochastic n-vector with covariance matrix BB^T so that the concatenated vector

$$x_+ = x, \xi$$

has covariance matrix R_+. Then we are back in the nonsingular case for R_+, and we can find an orthonormal sequence y such that

$$x_+ = Cy$$

Projecting (4) down to the range space of A gives us the desired representation, if we use range $(A) \perp$ range (B), $x = Ay$.

Note, however, that in this case the y-vector has been formed not just from the x-vector, but also from the new vector ξ which has been introduced in a way that may seem artifical. This cannot be avoided, however.

Note that the bilinear representation $R = AA^T$ amounts to *factoring the covariance*. Since R is Hermitian, it can be diagonalized $R = UDU^T$, $D = \text{diag}[\lambda_1, \lambda_2, \ldots, \lambda_n]$, with the nonnegative eigenvalues λ_k. Hence we can choose a factorization with $A = UD^{1/2}$, with $D^{1/2} = \text{diag}\left[+\sqrt{\lambda_1}, +\sqrt{\lambda_2}, \ldots, +\sqrt{\lambda_n}\right]$, which corresponds to the linear representation

$$x_t = \sum_{\mu=1}^{n} \sqrt{\lambda_\mu}\, u_{t\mu} y_\mu; \qquad U = \{u_{t\mu};\, t, \mu = 1, 2, \ldots, n\} \tag{5}$$

an expansion that will appear in other forms in the next section.

This was the finite case. In order to deal with linear representations of general form we first extend the definition of stochastic Stieltjes integrals and, at the same time, obtain a fundamental isometry theorem.

Let the argument space, now called Λ, be a separable metric space with a measure μ defined on the σ-ring of bounded Borel sets. Only such sets will be used below unless the opposite is stated. The measure μ need not be finite.

Definition 1 Z is an orthogonal process on Λ if $Z(S)$ is a stochastic variable in $L_2(\Omega, \mathcal{C}, P)$ for any bounded Borel set and $Z(S)$ satisfies:

(a) $Z(S_1 \cup S_2) = Z(S_1) + Z(S_2)$ if $S_1 \cap S_2 = \phi$

(b) $(Z(S_1), Z(S_2)) = \mu(S_1 \cap S_2)$

(As usual the stochastic variables should have mean zero.)

In particular it follows that $Z(S_1)\perp Z(S_2)$ if S_1 and S_2 are disjoint. The variance of $Z(S)$ is $\mu(S)$.

We now extend the notion of random Stieltjes integral given in Theorem 3 to the present case and *give a sense to the formalism*

$$Zf=\int_\Lambda f(\lambda)Z(d\lambda)\in L_2(Z)$$

for any complex valued function $f\in L_2(\Lambda,\mu)$.

To do this we begin by simple functions in $L_2(\Lambda,\mu)$:

$$f(\lambda)=f_k; \qquad \lambda\in S_k, \qquad k=1,2,\ldots,n \tag{6}$$

where $\{S_k;\ k=1,2,\ldots,n\}$ is a partition of Λ. If μ is infinite, then f_k must of course be zero if $\mu(S_k)=+\infty$, in order that $f\in L_2(\Lambda,\mu)$. Define, for such a simple function,

$$Zf=\sum_{k=1}^{n} f_k Z(S_k) \tag{7}$$

For two simple functions, f as in (6) and g defined through

$$g(\lambda)=g_l; \qquad \lambda\in T_l, \qquad l=1,2,\ldots,m$$

where $\{T_l;\ l=1,2,\ldots,m\}$ is also a partition of Λ, we can easily calculate the inner product:

$$(Zf,Zg)=\left(\sum_{k=1}^{n} f_k Z(S_k),\ \sum_{l=1}^{m} g_l Z(T_l)\right)$$

$$=\sum_{k=1}^{n}\sum_{l=1}^{m} f_k\bar{g}_l(Z(S_k),Z(S_l)) \tag{8}$$

Using (b) in Definition 2, the expression in (8) can be written as

$$\sum_{k=1}^{n}\sum_{l=1}^{m} f_k\bar{g}_l\mu(S_k\cap T_l)=\int_\Lambda\sum_{k=1}^{m}\sum_{l=1}^{n} f_k\bar{g}_l 1_{S_k\cap T_l}\mu(d\lambda)$$

with the indicator function of a set A denoted by 1_A. Hence since $1_{S_k\cap T_l}=1_{S_k}\cdot 1_{T_l}$,

$$(Zf,Zg)=\int_\Lambda f(\lambda)\overline{g(\lambda)}\mu(d\lambda)=(f,g) \tag{9}$$

and then we have the important *isometry relation*

$$\| Zf - Zg \| = \| f - g \| \tag{10}$$

with the convention that $\| \cdot \|$ is the symbol for the norm in both $L_2(Z)$ and $L_2(\Lambda, \mu)$. It will be clear from context which one is meant.

So far (9) is restricted to the class F of finite functions, and to the class Z of stochastic variables Z which can be obtained as Zf, $f \in F$. It is clear that the map m: $F \rightarrow Z$ is bijective. Indeed if the stochastic variable z can be obtained in two ways, $z = Zf = Zg$, $f, g \in F$, then using (10) we get

$$0 \| Zf - Zg \| = \| f - g \| \Rightarrow f = c \text{ a.e.}$$

so that f and g are identified.

Since m is bijective, and of course linear and isometric, we can now extend it by completing F and Z in their respective topologies with convergence in the mean. It is well known that this leads to $L_2(\Lambda, \mu)$ for F. As a matter of fact this is a classical way of constructing the space of quadratically integrable functions. For Z we get similarly a space $L_2(Z)$. Using the same symbol m for the extended mapping, we have proven the following theory.

THEOREM 1 THE ISOMETRY THEOREM, KARHUNEN (1947)

The mapping m is linear, bijective, and isometric between the two Hilbert spaces $L_2(\Lambda, \mu)$ and $L_2(Z)$.

The map m will often be written symbolically as

$$m(f) = \int_\Lambda f(\lambda) Z(d\lambda) \tag{11}$$

and it is clear, since m is linear, that if $f_1, f_2 \in L_2(\Lambda, \mu)$,

$$\int_\Lambda [c_1 f_1(\lambda) + c_2 f_2(\lambda)] Z(d\lambda) = m(c_1 f_1 + c_2 f_2)$$

$$= c_1 m(f_1) + c_2 m(f_2) = c_1 \int_\Lambda f_1(\lambda) Z(d\lambda) + c_2 \int_\Lambda f_2(\lambda) Z(d\lambda)$$

Also, because of the isometry,

$$\left(\int_\Lambda f(\lambda) Z(d\lambda), \int_\Lambda g(\lambda) Z(d\lambda) \right) = \int_\Lambda f(\lambda) \overline{g(\lambda)} \mu(d\lambda)$$

in particular for $f=g$,

$$\left\| \int_\Lambda f(\lambda) Z(d\lambda) \right\|^2 = \int_\Lambda |f(\lambda)|^2 \mu(d\lambda)$$

But remember that for an indicator function 1_S we have

$$m(f) = \int_S Z(d\lambda) = Z(S)$$

using (7).

We are now ready to state and prove the fundamental result on the linear representation of second-order processes, which will be applied numerous times later on for special representations.

THEOREM 2 GENERAL PROCESS REPRESENTATION, KARHUNEN (1947)

If the covariance function can be expressed as

$$r(s,t) = \int_\Lambda f(s,\lambda)\overline{f(t,\lambda)}\mu(d\lambda); \qquad f(t,\cdot) \in L_2(\Lambda,\mu), \qquad s,t \in T \tag{12}$$

then there exists an orthogonal process Z on Λ with the associated measure μ such that

$$x(t) = \int_\Lambda f(t,\lambda) Z(d\lambda); \qquad t \in T \tag{13}$$

Necessary and sufficient for $L_2(X) = L_2(Z)$ is that there exist no nontrivial function $\phi \in L_2(\Lambda,\mu)$ with

$$\int_\Lambda f(t,\lambda)\phi(\lambda)\mu(d\lambda) \equiv 0 \tag{14}$$

Remark If (13) holds, then certainly the bilinear representation (12) is valid. The proof is an immediate consequence of the isometry theorem since

$$r(s,t) = (x(s), x(t)) = (f(s,\cdot), f(t,\cdot)) = (12)$$

Proof of Theorem 2 To start with let us assume that there is no ϕ satisfying (14). The assumed bilinear relation (12) defines an isometry $m: x(t) \leftrightarrow f(t, \lambda)$ which can be extended immediately to the set of linear combinations of stochastic variables and f-functions, respectively:

$$\sum_k c_k x(t_k) \leftrightarrow \sum_k c_k f(t_k) \tag{15}$$

This linear map can be shown to be bijective as in the proof of the isometry theorem. Now we extend the map to the completion of the two sets and get in this way a isometric linear map m between $L_2(X)$ and $L_2(f)$, where the latter stands for the Hilbert space spanned by the linear combinations on the right of (15) with convergence in the mean as given in $L_2(\Lambda, \mu)$. Obviously $L_2(f) \subseteq L_2(\Lambda, \mu)$. We can claim more, however, using the assumption that no ϕ exist such that (14) holds. If $L_2(f)$ is strictly included in $L_2(\Lambda, \mu)$, there exists a $\phi \not\equiv 0$ a.c. in the latter space, but orthogonal to the first. This contradicts the assumption, so that we conclude that $L_2(f) = L_2(\Lambda, \mu)$.

Consider a bounded Borel set S in Λ and its indicator function 1_S. Since $1_S \in L_2(\Lambda, \mu)$, because $\mu(S)$ is finite for bounded S, it follows that the map m defines a stochastic variable, to be denoted by $Z(S)$:

$$Z(S) = m 1_S$$

The isometry of m gives us immediately

$$\|Z(S)\|^2 = \|1_S\|^2 = \int_\Lambda |1_S(\lambda)|^2 \mu(d\lambda) = \mu(S)$$

and for two S-sets their inner product

$$(Z(S_1), Z(S_2)) = (1_{S_1}, 1_{S_2}) = \int_\Lambda 1_{S_1}(\lambda) 1_{S_2}(\lambda) \mu(d\lambda)$$
$$= \mu(S_1 \cap S_2)$$

satisfying (b) in Definition 1. Also, if $S_1 \cap S_2 = \phi$, it is clear that for all λ

$$1_{S_1 \cup S_2}(\lambda) = 1_{S_1}(\lambda) + 1_{S_2}(\lambda)$$

Since m is linear, it follows that

$$Z(S_1 \cup S_2) = Z(S_1) + Z(S_2)$$

satisfying (a) in Definition 1: Z is an orthogonal process. But $x(t)$ is mapped to $f(t,\lambda)$ in our construction so that we can write, using the symbolism in (11),

$$x(t)=\int_{\Lambda} f(t,\lambda)Z(d\lambda); \qquad t\in T \tag{16}$$

and we have proved the first part of the general process representation.

Now assume that there exists a ϕ as described in the statement of the theorem. Then $L_2(f)\subset L_2(\Lambda,\mu)$ strictly, so that the orthogonal complement of $L_2(f)$ in $L_2(\Lambda,\mu)$ is not empty. Span it with functions $g(t,\lambda)$, where t ranges over some new set T', and consider the new function

$$\rho(s,t)=\int_{\Lambda} g(s,\lambda)\overline{g(t,\lambda)}\mu(d\lambda); \qquad s,t\in T' \tag{17}$$

It is obviously Hermitian and also nonnegative definite since

$$\sum_{k,l} c_k\bar{c}_l\rho(t_k,t_l)=\int_{\Lambda}\Big|\sum_k c_k g(t_k,\lambda)\Big|^2\mu(d\lambda)\geqslant 0$$

Hence it is the covariance function of some process $y(t)$, $t\in T'$, for example, a Gaussian one (see Chapter 1, Section 1), and we can make $y(t)$, $t\in T'$, independent and hence uncorrelated of $x(t)$, $t\in T$. Introducing the new parameter set $T''=T\cup T'$, we define a second-order process:

$$\xi(t)=\begin{cases} x(t); & t\in T\\ y(t); & t\in T'\end{cases}$$

and can write its covariance function defining $f(t,\lambda)$ on T'' as $f(t,\lambda)$, $t\in T$, and as $g(t,\lambda)$, $t\in T'$,

$$r_\xi(s,t)=\int_{\Lambda} f(s,\lambda)\overline{f(t,\lambda)}\mu(d\lambda); \qquad s,t\in T'' \tag{18}$$

Indeed if $(s,t)\in T^2$, (18) reduces to (12), and if $(s,t)\in(T')^2$, it reduces to (17). If $s\in T$ and $t\in T'$, we have $r_\xi(s,t)=0$, because of $x(t)$, $y(t)$ being uncorrelated. But $f(s,\lambda)\perp g(t,\lambda)$, so that (18) holds in this case, too; similarly for $s\in T'$, $t\in T$.

But now the introduction of the extended set of f-functions, $t\in T''$, leads to $L_2(f)=L_2(\Lambda)$ so that no ϕ-function as described can exist. Hence we can represent

$$\xi(t)=\int_{\Lambda} f(t,\lambda)Z(d\lambda); \qquad t\in T'' \tag{19}$$

and restricting (19) to $t\in T$, we have the desired representation of $x(t)$.

Q.E.D.

In the latter case $L_2(X)$ is a proper subspace of $L_2(Z)$. The reader may find it illuminating to compare the result with the special, finite case in the beginning of this section.

4 SPECIAL REPRESENTATIONS

We already know how to expand a process in terms of orthogonal stochastic variables ξ_ν [see (1.10)]; but the time-dependent coefficients $f_\nu(t)$ have no particularly attractive properties. *It is possible to choose the orthogonal stochastic variables ξ_ν such that the time functions $f_\nu(t)$ also form an orthogonal system?*

The answer to this crucial problem is affirmative and given by the famous Karhunen-Loéve expansion. To establish it, we need a classical result on integral kernels.

THEOREM 1 MERCER'S THEOREM (1909)

Consider a symmetric nonnegative definite and continuous function $r(s,t)$ on $[a,b]^2$ and the integral equation

$$\lambda\phi(s)=\int_a^b r(s,t)\phi(t)\,dt \tag{1}$$

The eigenvalues $\lambda_1,\lambda_2,\ldots$ and associated normalized eigenfunctions $\phi_1,\phi_2,\ldots$ satisfy

$$r(s,t)=\sum_{\nu=1}^{\infty}\lambda_\nu\phi_\nu(s)\,\overline{\phi_\nu(t)} \tag{2}$$

with absolute and uniform convergence.

Remark Since $r(\cdot,t)\in L_2([a,b])$, it is easily seen that

$$r(\cdot,t)=\sum_{\nu=1}^{\infty}\overline{\rho_\nu\phi_\nu(\cdot)} \tag{3}$$

with ρ_ν = Fourier coefficient of $r(\cdot, t)$ and with convergence in the mean. But

$$\rho_\nu = \int_a^b r(s,t)\phi_\nu(s)\,ds = \lambda_\nu \phi_\nu(t)$$

so that

$$r(\cdot, t) = \sum_{\nu=1}^{\infty} \lambda_\nu \phi_\nu(\cdot)\overline{\phi_\nu(t)} \tag{4}$$

This is with convergence in the mean Mercer's theorem strengthens this to absolute and uniform convergence in $[a,b]^2$.

Proof Since $r \in L_2([a,b]^2)$, we are dealing with a Hilbert-Schmidt kernel so that the spectrum is denumerable (or finite) with 0 as the only possible clustering point. Since r is nonnegative definite, we have $\lambda_\nu \geqslant 0$. We treat the case when the spectrum is denumerable; the finite case needs a simple modification only.

For any $t \in [a,b]$ the value $r(t,t)$ must be nonnegative. If not, $r(t,t) < 0$, and since r is continuous, we can find a neighborhood $|s-t_0| < \varepsilon$, $|t-t_0| < \varepsilon$ so that $r(s,t) < 0$ in it. Define

$$\phi(t) = \begin{cases} 1, & \text{if } |t-t_0| < \varepsilon \\ 0, & \text{otherwise} \end{cases}$$

We get

$$\int_a^b \int_a^b \phi(s)\overline{\phi(t)}\,r(s,t)\,ds\,dt < 0 \tag{5}$$

against the assumption that r was nonnegative definite [approximately (5) by Riemann sums]. Hence $r(t,t) \geqslant 0$.

Consider the new kernel

$$r_n(s,t) = r(s,t) - \sum_{\nu=1}^{n} \lambda_\nu \phi_\nu(s)\overline{\phi_\nu(t)}$$

This kernel is also nonnegative definite. Indeed for any $f \in L_2([a,b])$ and using (4),

$$\begin{aligned}\int_a^b \int_a^b f(s)\overline{f(t)}\,r_n(s,t)\,ds\,dt &= \sum_{\nu=1}^{\infty} \lambda_\nu |f_\nu|^2 - \sum_{\nu=1}^{n} \lambda_\nu |f_\nu|^2 \\ &= \sum_{\nu=n+1}^{\infty} \lambda_\nu |f_\nu|^2 \geqslant 0\end{aligned}$$

where

$$f_\nu = \int_a^b f(t)\phi_\nu(t)\,dt$$

Since r_n is also a nonnegative definite kernel, $r_n(t,t) \geqslant 0$, the previous step implies

$$r(t,t) \geqslant \sum_{\nu=1}^{n} \lambda_\nu |\phi_\nu(t)|^2$$

Hence

$$\sum_{\nu=1}^{\infty} \lambda_\nu |\phi_\nu|^2(t) < \infty \tag{6}$$

so that (3) converges absolutely (use Schwarz's inequality).

If a series of nonnegative continuous functions converges to a continuous function, the convergence must be uniform so that (6) converges uniformly in t. This is Dini's theorem [see, e.g., Pólya-Szegő (1972), Vol. I, p. 81]. Hence, again using Schwarz's inequality, (2) converges uniformly in (s,t) jointly, which completes the proof of Mercer's theorem. Q.E.D.

Now let $x(t)$ be a second-order process continuous in the mean on $T=[a,b]$. Then its covariance function satisfies the assumptions in Mercer's theorem. Choose $\Lambda=\{1,2,\ldots\}$ with the μ-measure $\mu(\{\nu\})=\lambda_\nu$, $\nu=1,2,\ldots,$ and apply the general process representation from Section 3.

THEOREM 2 KARHUNEN-LOÈVE EXPANSION, (KARHUNEN (1946), LOÈVE (1948))

Introduce the stochastic variables

$$z_\nu = \sqrt{\lambda_\nu} \int_a^b x(t)\phi_\nu(t)\,dt$$

which are orthonormal and form the expansion

$$x(t) = \sum_{\nu=1}^{\infty} \sqrt{\lambda_\nu}\, \phi_\nu(t) z_\nu \tag{7}$$

It converges in the mean in $L_2(X)$ for any $t \in T$.

Proof The general process representation theorem applied to the bilinear representation (2) tells us that there exists an orthogonal process $Z(\lambda)$ on Λ such that

$$x(t)=\int_{\Lambda} f(t,\lambda)Z(d\lambda) \tag{8}$$

with $f(t,\lambda)=\phi_\nu(t)$, $\nu\leftrightarrow\lambda$, where Λ is at most denumerable. Introduce the normalized stochastic variable

$$z_\nu=\frac{1}{\sqrt{\lambda_\nu}}Z(\{\nu\})$$

and recall the stochastic Stieltjes integrals were defined in Section 3 as limits in the mean of integrals of simple functions. That means that (8) can be interpreted as a limit in the mean of the expansion (7), as stated in the theorem.

Q.E.D.

The expression in (3.5) can be viewed as a Karhunen-Loève expansion for finite T.

Example 1 Consider a process with a continuous and stationary covariance so that $r(s,t)$ can be written as $r(s-t)$ with a function of a single argument, also denoted by the same symbol r. This important case will be treated in depth later on. In the more special case when $x(t)$ is also periodic with some period P, $x(t+P)=x(t)$ a.c., considerable simplification occurs since the eigenfunctions of the covariance function $r(s,t)$ take a simple form.

Indeed, consider *Egerváry's integral equation*

$$\lambda\phi(s)=\int_0^{2\pi} r(s-t)\phi(t)\,dt \tag{9}$$

where we have chosen the period $P=2\pi$, which is no essential restriction. Multiplying (9) by the Fourier characters $\{e^{iks};\, k\in\mathbb{Z}\}$ of the circle group (see Chapter 1, Section 3), and integrating, we get, after a change of the order of integration,

$$\lambda\hat{\phi}_k=\int_0^{2\pi}\int_0^{2\pi} e^{ik(s-t)}r(s-t)e^{ikt}\phi(t)\,ds\,dt=\hat{r}_k\hat{\phi}_k; \qquad k\in Z$$

with the Fourier coefficients

$$\phi_k = \int_0^{2\pi} e^{ikt}\phi(t)\,dt$$

$$\hat{r}_k = \int_0^{2\pi} e^{ikt} r(t)\,dt \tag{10}$$

Hence since $\hat{\phi}_k \equiv 0$ can be ruled out, we have

$$\lambda = \text{one of the } \hat{r}_k$$

Put $\lambda_\nu = \hat{r}_\nu$ with the associated normalized eigenfunctions

$$\phi_\nu(t) = \frac{1}{\sqrt{2\pi}} e^{i\nu t}$$

This means that we get the Karhunen-Loève expansion

$$x(t) = \frac{1}{\sqrt{2\pi}} \sum_{\nu=-\infty}^{\infty} \sqrt{\hat{r}_\nu}\, e^{i\nu t}; \qquad t \in \mathbb{R} \tag{11}$$

Of course all $\hat{r}_\nu \geqslant 0$.

Just as (3.5) is the finite analog of (7), we can get a finite analog of (11). Consider $x = (x_1, x_2, \ldots, x_n)$ a stochastic vector with a covariance matrix of the form

$$E[xx^T] = \begin{Bmatrix} r_0, r_1, r_2, \ldots, r_n \\ r_{-1}, r_0, r_1, \ldots, r_{n-1} \\ \cdots\cdots\cdots\cdots \\ r_{-n}, \ldots, r_{-1}, r_0 \end{Bmatrix}$$

with a sequence r_t which is Hermitian, $r_{-t} = \bar{r}_t$, and *periodic* $r_{t+n} = r_t$. Then the covariance matrix is a *circulant*, and we get the representation

$$x_t = \frac{1}{\sqrt{n}} \sum_{\nu=1}^{n} \sqrt{\hat{r}_\nu}\, e^{2\pi i\nu t/n} z_\nu$$

with

$$\hat{r}_\nu = \sum_{t=1}^{n} e^{2\pi i\nu t/n} r_t$$

Example 2 For the Wiener process on $T=[0,2\pi]$ (see Example 3 in Section 2), $r(s,t)=\sigma^2\min(s,t)$, say with $\sigma=1$. We can write

$$1_{[0,s]}(t)=\frac{1}{\sqrt{2\pi}}\sum_{\nu=-\infty}^{\infty} f_\nu e^{i\nu t}$$

with convergence in the mean in $L_2([0,2\pi])$ and

$$f_\nu=\frac{1}{\sqrt{2\pi}}\int_0^{2\pi} 1_{[0,s]}(t)e^{-i\nu t}\,dt=\frac{1}{\sqrt{2\pi}}\frac{1-e^{-i\nu s}}{i\nu}$$

with f_0 interpreted as $s/\sqrt{2\pi}$. Parseval's relation gives us then

$$\min(s,t)=(1_s,1_t)=\frac{1}{2\pi}\sum_\nu \frac{(1-e^{-i\nu s})(1-e^{i\nu t})}{\nu^2}$$

We now use the general process representation with $\Lambda=Z$; $\mu(\{\nu\})=1$, $\nu\in Z$, and

$$f(t,\nu)=\frac{1}{\sqrt{2\pi}}\frac{1-e^{-i\nu t}}{i\nu}$$

Hence there exist orthonormal stochastic variables such that the Wiener process

$$W(t)=\frac{1}{\sqrt{2\pi}}\sum_{\nu=-\infty}^{\infty}\frac{1-e^{i\nu t}}{i\nu}z_\nu \tag{12}$$

with convergence in the mean in $L_2(X)$. As a matter of fact (12) converges a.c., but we do not make use of this property.

The series in (12) is related to but not the same as the Karhunen-Loève expansion.

It is, of course, not always possible to solve the integral equation (1) in closed form. Let us mention one interesting case where this can be done, and refer the reader to Grenander-Pollak-Slepian (1959) for many others.

Example 3 Consider the stationary process at the end of Example 3, Section 2. Let the covariance function be $r(s,t)=e^{-\beta|t-s|}$, $\beta>0$. The multiplicative constant has been made equal to 1 with no real loss of generality. Consider the integral equation

$$\lambda\phi(s)=\int_0^s e^{\beta(t-s)}\phi(t)\,dt+\int_s^1 e^{\beta(s-t)}\phi(t)\,dt \tag{13}$$

Since $\phi \in L_2$, we also have $\phi \in L_1$. But then ϕ must be continuous for any $\lambda > 0$. Hence it is differentiable, and so on for derivatives of all orders. Differentiating we get

$$\lambda \phi'(s) = -\beta \int_0^s e^{\beta(t-s)} \phi(t)\, dt + \beta \int_s^1 e^{\beta(s-t)} \phi(t)\, dt$$

and once more

$$\lambda \phi''(s) = -2\beta\phi(s) + \beta^2 \int_0^s e^{\beta(t-s)} \phi(t)\, dt + \beta^2 \int_s^1 e^{\beta(s-t)} \phi(t)\, dt$$

Substituting (13) in the right-hand side, this gives us

$$\phi'' + \left(\frac{2\beta}{\lambda} - \beta^2 \right) \phi = 0$$

so that the eigenfunctions have the form

$$\phi(t) = c_1 e^{i\omega t} + c_2 e^{-i\omega t}; \qquad \omega = +\sqrt{\frac{2\beta}{\lambda} - \beta^2}\,, \quad \lambda = \frac{2\beta}{\beta^2 + \omega^2} \tag{14}$$

Substituting (14) into (13) we get, after some straightforward manipulations (see Note 2.4.A),

$$c_1\left(1 - \frac{\omega i}{\beta}\right) + c_2\left(1 + \frac{\omega i}{\beta}\right) = 0$$

$$c_1 e^{i\omega}\left(1 + \frac{\omega i}{\beta}\right) + c_2 e^{-i\omega}\left(1 - \frac{\omega i}{\beta}\right) = 0 \tag{15}$$

so that the determinant must vanish and

$$e^{2i\omega} = \left(\frac{\beta - \omega i}{\beta + \omega i} \right)^2$$

or, equivalently,

$$\cos 2\omega = \frac{\beta^4 - 6\beta^2\omega^2 + \omega^4}{(\beta^2 + \omega^2)^2}$$

Denoting the positive zeros of this transcendental equation by $\omega_1, \omega_2, \ldots$

we get the eigenvalues as

$$\lambda_\nu = \frac{2\beta}{\omega_\nu^2 + \beta^2} \tag{16}$$

and the associated eigenfunctions in (14) and (15). It can be shown by a variation of the above argument that $\lambda=0$ is impossible, so that (14) and (16) give all the eigenelements.

We have repeatedly met the notion "stationary stochastic process." The formal definition consists of two parts.

Definition 1

(a) A second-order process is called (weakly) stationary if for all h

$$m(t+h)=m(t)$$
$$r(s,t)=r(s+h,t+h)$$

(b) A stochastic process is called strictly stationary if for all $n, t_1, \ldots, t_n, h$

$$F(x_1, x_2, \ldots, x_n; t_1+h, t_2+h, \ldots, t_n+h) = F(x_1, x_2, \ldots, x_n; t_1, t_2, \ldots, t_n)$$

The first condition says that all moments of order 1 and 2 are invariant with respect to translations of the time axis. The second condition requires that all finite-dimensional probability distributions be invariant.

In the important special case when the process is real valued and Gaussian it is clear that weak stationarity implies strict stationarity. For a complex-valued Gaussian process one has to exercise some caution. The implication does not hold without further restrictions (see Note 2.4.B).

If $T = \mathbb{R}^d, \mathbb{Z}^d, T^d$, the definitions extend naturally to the usual translation groups acting on these argument spaces.

Theorems 3 and 4 give *bilinear Fourier representations for the stationary covariances, and Theorems 5 and 6 linear representations for the stationary processes.*

THEOREM 3 TRIGONOMETRIC MOMENT PROBLEM, HERGLOTZ (1911)

Consider a Hermitian sequence $\{r_t, t\in Z\}$, $r_{-t}=\bar{r}_t$. In order that it be non-negative definite it is necessary and sufficient that it can be expressed as a Fourier integral:

$$r_t = \int_{-\pi}^{\pi} e^{it\lambda} F(d\lambda); \qquad t\in Z \tag{17}$$

where F is a finite measure on $[-\pi, \pi)$.

Proof Sufficiency is easy, since if (17) holds, then, for any n and complex numbers $z_1, z_2, \ldots, z_n$,

$$\sum_{s,t=1}^{n} r_{s-t} z_s \bar{z}_t = \int_{-\pi}^{\pi} \left| \sum_{t=1}^{n} z_t e^{it\lambda} \right|^2 F(d\lambda) \geqslant 0$$

To prove necessity we "regularize" the r-sequence, making it converge fast to zero. Put $z_t = \rho^t e^{-it\lambda}$ for $t \geqslant 0$ and 0 otherwise, with $0 < \rho < 1$. Then, with a simple rearrangement,

$$0 \leqslant \sum_{s,t=0}^{\infty} r_{s-t} z_s \bar{z}_t = \sum_{s,t=0}^{\infty} r_{s-t} \rho^{s+t} e^{i(-s+t)\lambda}$$

$$= \sum_{n=-\infty}^{\infty} r_n e^{-in\lambda} \sum_{h=0}^{\infty} \rho^{|n|+2h} = \sum_{n=-\infty}^{\infty} r_n e^{-in\lambda} \frac{\rho|n|}{1-\rho^2}$$

Hence

$$r_t \rho^{|t|} = \int_{-\pi}^{\pi} e^{it\lambda} F_\rho(d\lambda) \tag{18}$$

where the measure F_ρ is absolutely continuous and its density given by the uniformly convergent sum

$$\frac{1}{2\pi} \sum_{n=-\infty}^{\infty} r_n |\rho|^n e^{-in\lambda} \geqslant 0$$

The total variation of F_ρ is r_0, uniformly bounded, actually constant, when we make $\rho \uparrow 1$. Since the set $\{F_\rho\}$ is compact, we can select a weakly convergent subsequence tending to some measure F. Then (18) leads to (17), as stated. Q.E.D.

It is clear that F is uniquely determined since $\{e^{it\lambda}; t \in Z\}$ is complete in sup norm.

This result, sometimes known as *Herglotz' theorem*, leads to a direct consequence of great importance.

THEOREM 4

A (weakly) stationary sequence x_t, $t \in Z$, can be expressed as a stochastic Fourier transform:

$$x(t) = \int_{-\pi}^{\pi} e^{it\lambda} Z(d\lambda); \qquad t \in Z \tag{19}$$

where Z is an orthogonal process associated with the measure F in (17).

Proof The equation (17) can be rewritten as

$$r(s,t)=\int_{-\pi}^{\pi} e^{is\lambda}(\overline{e^{it\lambda}})F(d\lambda)$$

so that the general process representation can be applied by choosing $\Lambda=[-\pi,\pi)$, $f(t,\lambda)=e^{it\lambda}$, and $\mu=F$. Q.E.D.

Since $\{e^{it\lambda};\ t\in\mathbb{Z}\}$ is complete in the current $L_2(\Lambda,\mu)$, it also follows that $L_2(Z)=L_2(X)$.

The spectrum F determines completely the geometry of a stationary process. If F is absolutely continuous, $f=F'\in L_1$, the *spectral density* f contains all the geometric information. For further reference we state the following lemma.

Lemma 1 If the stationary process $x(t)$, $t\in\mathbb{Z}$, has a continuous spectrum its covariance matrix $R=\{r(s-t);\ s,t=1,2,\ldots,n\}$ is non singular.

Proof If R is singular, there exists a nontrivial n-vector z such that $Rz=0$. Therefore

$$0=z^TRz=\sum_{s,t=1}^{n} r(s-t)z_s\bar{z}_t$$

or, using (17),

$$0=\int_{-\pi}^{\pi}\sum_{s,t=1}^{n} e^{i(s-t)\lambda}z_s\bar{z}_tF(d\lambda)=\int_{-\pi}^{\pi}\left|\sum_{t=1}^{n} e^{it\lambda}z_t\right|^2 f(\lambda)\,d\lambda \qquad (20)$$

But the trigonometric polynomial in the integrand has at most a finite number of zeros in $(-\pi,\pi)$. Hence the integral cannot vanish except in the trivial case $f\equiv 0$. Q.E.D.

When we go from $T=\mathbb{Z}$ to $\mathbb{R}$, similar representations can be obtained.

THEOREM 5 BOCHNER'S THEOREM, BOCHNER (1932)

A continuous and Hermitian complex-valued function $r(t)$ is nonnegative definite if and only if it can be written as a Fourier integral

$$r(t)=\int_{-\infty}^{\infty} e^{it\lambda}F(d\lambda);\qquad t\in R \qquad (21)$$

in terms of a bounded measure F.

Proof Sufficiency as before. Introduce now for $\varepsilon>0$ the Fourier transform of the "regularized" function $(1-\varepsilon|t|)r(t), |t|<1/\varepsilon$ and 0, otherwise,

$$f_\varepsilon(t)=\frac{1}{2\pi}\int_{-1/\varepsilon}^{1/\varepsilon}(1-\varepsilon|t|)e^{-it\lambda}r(t)\,dt$$

For any $A>0$ we have the familiar formula for the Féjer kernel

$$\int_{-2A}^{2A}\left(1-\frac{|\lambda|}{2A}\right)e^{-it\lambda}d\lambda=2A\left(\frac{\sin At}{At}\right)^2$$

If we multiply this by $(1/2\pi)(1-\varepsilon|t|)r(t)$ and integrate over $[-1/\varepsilon, 1/\varepsilon]$, we get

$$\int_{-2A}^{2A}\left(1-\frac{|\lambda|}{A}\right)f_\varepsilon(\lambda)\,d\lambda=\frac{1}{\pi}\int_{-A/\varepsilon}^{A/\varepsilon}\left(\frac{\sin t}{t}\right)^2\left(1-\frac{\varepsilon}{A}|t|\right)r\left(\frac{t}{A}\right)dt$$

and, using the bound $|r(t)|\leqslant r(0)$,

$$\left|\int_{-2A}^{2A}\left(1-\frac{|\lambda|}{A}\right)f_\varepsilon(\lambda)\,d\lambda\right|\leqslant r(0)\frac{1}{\pi}\int_{-A/\varepsilon}^{A/\varepsilon}\left(\frac{\sin t}{t}\right)^2dt<r(0) \qquad (22)$$

We can write $f_\varepsilon(\lambda)$ as

$$\frac{1}{2\pi\varepsilon}\int_{-1}^{1}(1-|s|)e^{-is\lambda/\varepsilon}r\left(\frac{s}{\varepsilon}\right)ds$$

$$=\frac{1}{2\pi\varepsilon}\int_0^1\int_0^1 e^{-iu\lambda/\varepsilon+iv\lambda/\varepsilon}r\left(\frac{u-v}{\varepsilon}\right)du\,dv\geqslant 0$$

The last inequality can be verified by approximating the double Riemann integral by Riemann sums and appealing to the nonnegative definiteness of $r(\cdot)$. Let $A\to+\infty$ in (22),

$$\int_{-\infty}^{\infty}|f_\varepsilon(\lambda)|d\lambda\leqslant r(0) \qquad (23)$$

so that the Fourier inversion formula is valid [use (20)] and yields, for $|t|\leqslant 1/\varepsilon$,

$$(1-\varepsilon|t|)r(t)=\int_{-\infty}^{\infty}e^{it\lambda}f_\varepsilon(\lambda)\,d\lambda$$

Now we let $\varepsilon\downarrow 0$. The left-hand side tends to a continuous limit, and the continuity theorem for characteristic functions tells us that the (uniformly bounded) measure with density f_ε converges weakly to some limiting measure F. An immediate consequence is that (21) holds. Again the measure F can be seen to be uniquely determined by the knowledge of the function $r(\cdot)$. Q.E.D.

The linear representation associated with the bilinear one in (21) is one of the most useful results on weakly stationary processes.

THEOREM 6 CRAMÉR'S SPECTRAL REPRESENTATION, CRAMÉR (1942)

A stochastic process $x(t)$, weakly stationary and continuous in the mean, can be expressed as a stochastic Fourier integral

$$x(t)=\int_{-\infty}^{\infty} e^{it\lambda}Z(d\lambda); \qquad t\in R \tag{24}$$

where Z is an orthogonal process with the measure F.

Proof Since $x(t)$ is continuous in the mean, its covariance function is continuous (see Theorem 1 of Section 1). We can use Bochner's theorem in the general process representation with $\Lambda=\mathbb{R}$, $f(t,\lambda)=e^{it\lambda}$, $\mu=F$, and

$$r(s,t)=\int_{-\infty}^{\infty} e^{is\lambda}(\overline{e^{it\lambda}})F(d\lambda)$$

This gives us directly (23). Again $L_2(X)=L_2(Z)$. Q.E.D.

Because of the great importance of Cramér's spectral representation we also look at it from another perspective. In $L_2(X)$ let us consider the element

$$z=\sum_{k=1}^{n} c_k x(t_k) \tag{25}$$

and define its translate

$$T_h z=\sum_{k=1}^{n} c_k x(t_k+h)$$

We have

$$\|z\|^2=\sum c_k\bar{c}_l r(t_k,t_l)=\sum c_k\bar{c}_l r(t_k-t_l)$$
$$=\sum c_k\bar{c}_l r(t_k+h,t_l+h)=\|T_h z\|$$

Reasoning as in Section 3, we can then extend T_h to all of $L_2(X)$ and get a *linear isometric mapping* that will be denoted by the same symbol T_h.

It is also clear that, with z as in (25),

$$T_{h_1}(T_{h_2}z) = \sum_{k=1}^{n} c_k x(t_k + h_1 + h_2) = T_{h_1+h_2}z$$

so that $\{t_h; h \in R\}$ forms a *group of unitary transformations* with T_0 as the identity element. We can then apply a fundamental theorem of Stone [see, e.g., Riesz-Nagy (1952), p. 380]. It asserts that if $\{T_t; t \in R\}$ is a group of unitary operators in a Hilbert space, such that $(T_t z_1, z_2)$ is continuous for every pair z_1, z_2, there exists a resolution of the identity $E(\lambda)$ such that

$$T_t = \int_{-\infty}^{\infty} e^{it\lambda} E(d\lambda) \tag{26}$$

In the present case we can of course write

$$x(t) = T_t x(0); \qquad t \in R$$

so that, using (26), we get

$$x(t) = \int_{-\infty}^{\infty} e^{it\lambda} Z(d\lambda)$$

where $Z(\lambda) = E(\lambda)x(0)$. Using the orthogonality property of the resolution of the identity, we have for any bounded Borel set $S_1, S_2 \in R$, and putting

$$z_1 = \int_{-\infty}^{\infty} 1_{S_1}(\lambda) Z(d\lambda)$$

$$z_2 = \int_{-\infty}^{\infty} 1_{S_2}(\lambda) Z(d\lambda)$$

the inner product

$$\begin{aligned}(z_1, z_2) &= \int_{-\infty}^{\infty} 1_{S_1}(\lambda) 1_{S_2}(\lambda) \|E(d\lambda)x(0)\|^2 \\ &= \int_{-\infty}^{\infty} 1_{S_1}(\lambda) 1_{S_2}(\lambda) F(d\lambda)\end{aligned} \tag{27}$$

with $F(\lambda) = \|E(\lambda)x(0)\|^2$. Hence (27) gives us the relation needed for the Definition in Section 3, and $Z(\lambda)$ is indeed an orthogonal process and (24) is established again.

Remark 1 In Theorems 3–6 the value space $V=\mathbb{C}$, but little is changed if we choose $V=\mathbb{R}$. For example, (17) should be rewritten as

$$r_t=\int_0^{\pi}\cos t\lambda\, H(d\lambda)$$

and (19) becomes

$$x(t)=\int_0^{\pi}\cos t\lambda\, Z_1(d\lambda)+\int_0^{\pi}\sin t\lambda\, Z_2(d\lambda) \tag{28}$$

where the two orthogonal processes Z_1 and Z_2 are also mutually orthogonal, $L_2(Z_1)\perp L_2(Z_2)$, both with the associated measure H. This is similar for $T=\mathbb{R}$ in (21) and (24).

The spectral representations are extremely useful for the analytical study of (weakly) stationary processes, and we illustrate this by two applications.

Example 4 Consider the linear difference operator L defined by

$$(Ly)_t=Ly_t=a_0y_{t+p}+a_1y_{t+p-1}+\ldots+a_py_t \tag{29}$$

for any vector $y=(\ldots,y_{-1},y_0,y_1,\ldots)$. Let us solve the *linear stochastic difference equation*

$$Lx=e$$

where the stochastic vector $e=\{e_t;\,t\in\mathbb{Z}\}$ has uncorrelated components with variance $E[|e_t|^2]=\sigma^2$. Then $\{e_t;\,t\in\mathbb{Z}\}$ is weakly stationary and has, according to Theorem 4, a spectral representation, with a spectral measure F_e such that the covariance function of e_t is

$$r_e(t)=\int_{-\pi}^{\pi}e^{it\lambda}F_e(d\lambda)=\begin{cases}\sigma^2; & \text{if } t=0\\ 0; & \text{otherwise}\end{cases}$$

Hence F_e must be absolutely continuous with the spectral density $\sigma^2/2\pi$. If we require that the solution $x=\{x_t;\,t\in Z\}$ be (weakly) stationary, it can be expressed as in (19) so that

$$\int_{-\pi}^{\pi}e^{it\lambda}P(e^{i\lambda})Z(d\lambda)=\int_{-\pi}^{\pi}e^{it\lambda}Z_e(d\lambda)$$

where Z_e is the orthogonal process associated with F_e, and we have introduced the *characteristic polynomial* for the operator L:

$$P(z)=a_0z^p+a_1z^{p-1}+\cdots+a_p$$

Since Z-processes are uniquely determined, we then get, using an easily understood formalism,

$$Z(d\lambda)=\frac{1}{P(e^{i\lambda})}Z_e(d\lambda) \tag{30}$$

The solution process x_t then has the spectral density $f(\lambda)$ given by

$$\|Z(d\lambda)\|^2=\frac{\sigma^2}{2\pi|P(e^{i\lambda})|^2}d\lambda=f(\lambda)\,d\lambda \tag{31}$$

Of course $P(z)$ must have no zero on $|z|=1$ in order that a solution exist and $f\in L_1([-\pi,\pi])$.

Conversely if a weakly stationary process x_t has a spectral density as in (31), does it satisfy a linear stochastic difference equation? Say that P has the zeros $z_1, z_2,\ldots, z_n$ in the complex plane, all of modulus different from 1. In (31) it is only the values of $P(z)$ at the boundary of the unit circle $|z|=1$ that matter. We can write

$$|P(e^{i\lambda})|^2=a_0\prod_{k=1}^{n}|e^{i\lambda}-z_k|^2$$

If some zero $|z_k|>1$, we can change it, with $z=e^{i\lambda}$, without affecting $|P(e^{i\lambda})|$,

$$|z-z_k|=|z_k|\,|z_k^{-1}-\bar{z}|=|z_k|\,|\bar{z}_k^{-1}-z|$$

into the zero $(\bar{z}_k)^{-1}$, reflected into the unit circle, now with modulus less than 1. After we have done this for all such zeros, consider the stochastic process

$$y_t=Lx_t=\int_{-\pi}^{\pi}e^{it\lambda}P(e^{i\lambda})Z(d\lambda) \tag{32}$$

which is easily seen to have covariance function

$$r(s,t)=\int_{-\pi}^{\pi}e^{i(s-t)\lambda}|P(e^{i\lambda})|^2F_e(d\lambda)=\frac{\sigma^2}{2\pi}\int_{-\pi}^{\pi}e^{i(s-t)\lambda}\,d\lambda \tag{33}$$

so that it is (weakly) stationary with spectral density $\sigma^2/2\pi$ [use (31)]. Hence it has exactly the covariance structure of the e_t process.

So far we have not really used the fact that we can choose $|z_k|<1$, but this is needed to get a convenient representation for the solution expressed in the e_t-process. Indeed, we can expand $1/P$ into a Laurent series:

$$\frac{1}{P(z)}=\sum_{\nu=0}^{\infty} b_\nu z^{-\nu}; \qquad |z|\geqslant 1 \tag{34}$$

uniformly convergent outside the unit disk, since $1/P(z)$ is bounded and analytic in this region. Multiply (34) by $z^t Z_e(d\lambda)$ and integrate along the boundary of the unit circle [use (30)], and we get

$$x(t)=\sum_{\nu=0}^{\infty} b_\nu \int_{-\pi}^{\pi} e^{i(t-\nu)\lambda} Z_e(d\lambda)=\sum_{\nu=0}^{\infty} b_\nu e_{t-\nu}$$

where we have changed the order of summation and integration which is allowed due to the uniform convergence of (32). The series (33) converges in the mean in $L_2(X)$. Hence we have expressed the solution as a one-sided moving average (see Example 3 in Section 1).

Example 5 For $T=\mathbb{R}$ the above goes through with some slight modifications that we will sketch only. Instead of (29) we should now use a linear differential operator:

$$Ly_t=a_0 y_t^{(p)}+a_1 y_t^{(p-1)}+\ldots+a_p y_t$$

and we ask for a stationary solution to the *stochastic linear differential equation*

$$Lx_t=\dot{w}_t$$

where the right-hand side is the formal derivative of an orthogonal process with

$$E\left[|w(dt)|^2\right]=\sigma^2\,dt$$

We now meet the same difficulty as in Example 3 in Section 2 and have to interpret x_t as a weak solution. It can then be shown by a similar argument that we can write

$$x_t=\int_{-\infty}^{\infty} e^{it\lambda}\frac{1}{P(\lambda)} Z_e(d\lambda) \tag{35}$$

where

$$P(\lambda)=a_0(i\lambda)^p+a_1(i\lambda)^{p-1}+\ldots+a_p$$

and Z_e is an orthogonal process with

$$\|Z_e(d\lambda)\|^2=\sigma^2\,d\lambda$$

Further x_t will have the covariance function

$$r(t)=\sigma^2\int_{-\infty}^{\infty}e^{it\lambda}\frac{1}{|P(\lambda)|^2}\,d\lambda$$

and $P(\lambda)$ must have no real zero. Again we can modify the zeros to be situated in the upper half-plane and can then expand (35) in the form

$$x_t=\int_{-\infty}^{t}g(t-s)\xi(ds) \tag{36}$$

where the function g is quadratically integrable and is a linear combination of exponential function (with the usual degeneracies possible) and $\xi(s)$ is an orthogonal process with

$$\|\xi(ds)\|^2=ds$$

The expression (36) is a one-sided moving average in continuous time.

The moving average representations in Example 4 and 5 are attractive but limited to special cases. Is there in general a moving average representation? A partial answer is given by the following theorem.

THEOREM 7 TWO-SIDED MOVING AVERAGE REPRESENTATION, KARHUNEN (1947)

In order that the weakly stationary process $\{x(t);t\in Z\}$ *have a moving average representation*

$$x(t)=\sum_{\nu=-\infty}^{\infty}c_{t-\nu}\xi_\nu;\qquad c=\{c_t;t\in\mathbb{Z}\}\in l_2 \tag{37}$$

with $\{\xi_\nu\}$ *orthonormal, it is necessary and sufficient that* $x(t)$ *have an absolutely continuous spectrum.*

Proof If (37) holds, we have

$$r(s,t)=\sum_{\nu=-\infty}^{\infty}c_{s-\nu}\bar{c}_{t-\nu} \tag{38}$$

Since the sequence $c=\{c_k\}$ is quadratically summable, the Fourier series

$$\frac{1}{\sqrt{2\pi}}\sum_{\nu=-\infty}^{\infty} c_{-\nu}e^{i\nu\lambda}$$

converges in $L_2([-\pi,\pi])$ in the mean and defines a function ϕ in the same Hilbert space. Using Parseval's relation, (38) can be rewritten:

$$\begin{aligned} r(s,t) &= \int_{-\pi}^{\pi} e^{is\lambda}\phi(\lambda)\overline{e^{-It\lambda}\phi(\lambda)}\,d\lambda \\ &= \int_{-\pi}^{\pi} e^{i(s-t)\lambda}f(\lambda)\,d\lambda \end{aligned}$$

with $f=|\phi|^2\in L_1([-\pi,\pi])$. Hence the spectrum of the process is absolutely continuous, actually with the density f.

Conversely assume that the spectrum is absolutely continuous and call its spectral density f. Introduce the function ϕ in $L_2([-\pi,\pi])$ as the square root $+\sqrt{f}$ and expand it in a Fourier series:

$$\phi(\lambda)=\frac{1}{\sqrt{2\pi}}\sum_{-\infty}^{\infty} c_{-\nu}e^{i\nu\lambda}$$

converging in the mean. Then, again with Parseval's relation,

$$\begin{aligned} r(s,t) &= \int_{-\pi}^{\pi} e^{i(s-t)\lambda}|\phi(\lambda)|^2 d\lambda = \int_{-\pi}^{\pi} e^{is\lambda}\phi(\lambda)\overline{e^{it\lambda}\phi(\lambda)}\,d\lambda \\ &= \sum_{\nu=-\infty}^{\infty} c_{s-\nu}c_{t-\nu} \end{aligned}$$

In the general process representation choose $\Lambda=\mathbb{Z}$, $\mu=$counting measure, and $f(t,\lambda)=c_{t-\lambda}$. The representation theorem then tells us that there exists an orthogonal process Z on Λ so that if we put $\Xi_\nu=Z(\{\nu\})$, we get (36) as stated.

Q.E.D.

Let us mention briefly that Theorem 7 has a complete analogue for continuous time. Indeed if

$$x(t)=\int_{-\infty}^{\infty} g(t-u)\xi(du) \tag{39}$$

where ξ is an orthogonal process with $\|\xi(du)\|^2=du$ and g is in

$L_2[(-\infty,\infty)]$, then

$$r(s,t)=\int_{-\infty}^{\infty} g(s-u)\overline{g(t-u)}\,du \tag{40}$$

Let γ be the Fourier transform of g:

$$\gamma(\lambda)=\frac{1}{\sqrt{2\pi}}\int_{-\infty}^{\infty} e^{-i\lambda v}g(v)\,dv$$

interpreted in the L_2-sense. Due to the isometry of the L_2-Fourier transform, (40) can be written as

$$r(s,t)=\int_{-\infty}^{\infty} e^{i(s-t)\lambda}|\lambda(\lambda)|^2\,d\lambda$$

so that our process has an absolutely continuous spectrum with density $|\gamma(\lambda)|^2$. On the other hand if we start with a density $f\in L_1$ and introduce $\gamma=+\sqrt{f}$, then the above derivation also works backwards, appealing to the general representation theorem. Hence the representation in (39) is necessary and sufficient for a stationary process to have an absolutely continuous spectrum.

We now return to discrete time and ask whether we can also claim that $x(t)$ can be expanded in a one-sided moving average? This question, which has considerable and perhaps unexpected analytical ramifications, is treated in Section 5 from another point of view. Here we present a partial answer of geometric nature.

We need the notion of a *deterministic process*. Consider a weakly stationary process z_t and the subspaces $L_2(z_s;\, s\leqslant t)$. If these subspaces are the same for all t, the process is said to be *deterministic*. In general, when $t\downarrow-\infty$, this monotonic sequence of subspaces has a limit. *If this limiting subspace, called the distant past, is the null space, the process is said to be completely nondeterministic.*

Intuitively this notion of determinism means that z_t can be approximated with arbitrarily small error by linear combinations of x_s, where the s-values can be chosen as far back in time as we choose. We can predict z_t accurately using only the distant past. This represents a degenerate case of limited interest, but it should not be neglected.

THEOREM 8 WOLD'S PREDICTIVE DECOMPOSITION, WOLD (1938)

A (weakly) stationary process $\{x_t;\, t\in\mathbb{Z}\}$ can be decomposed into two (weakly) stationary processes orthogonal to each other, $y_t\perp z_s$:

$$x_t=y_t+z_t$$

such that

$$y_t = \sum_{\nu=0}^{\infty} c_{t-\nu}\xi_\nu \tag{41}$$

where the innovations $\{\xi_\nu\}$ *are orthonormal,* $\xi_t \in L_w(x_s; s \leqslant t)$, *and* z_t *is deterministic, while* y_t *is completely nondeterministic.*

Proof Form the subspace $L_2(x_s; s \leqslant t) = H_t$ spanned by all x_s with $s \leqslant t$. Of course $H_t \subseteq H_{t+1}$ so that the limit of H_t exists when t tends to $-\infty$; call it $H_{-\infty}$. When $t \to +\infty$, the limit is just $L_2(X)$.

If $H_t = H_{t+1}$, then because of the stationary geometry it follows that all H_t spaces are the same and $H_{-\infty} = H_t = L_2(X)$. In this case define $y_t = 0$, $z_t = x_t$. It is clear that $z_t \in \lim_{t\to-\infty} L_2(z_s; s \leqslant t)$ so that it is deterministic.

On the other hand if H_t is strictly included in H_{t+1}, then we can write

$$x_{t+1} = \eta_t + \sigma_1 \xi_{t+1}; \qquad \sigma_1 > 0, \eta_t \in H_t \tag{42}$$

where ξ_{t+1} has variance 1 and is orthogonal to H_t.

Introduce the subspace K_t spanned by all ξ_s for $s \leqslant t$. Then $K_t \subset K_{t+1}$ with strict inclusion, and it is easily seen that the limit of K_t, as $t \to -\infty$, is the null space. We can write

$$H_t = K_t \oplus H_{-\infty}$$

so that if we define

$$\begin{aligned} y_t &= P_{K_t} x_t \\ z_t &= P_{H_{-\infty}} x_t \end{aligned} \tag{43}$$

we have $x_t = H_t x_t = y_t + z_t$, and the y-process is orthogonal to the z-process.

The first relation in (43) implies

$$\begin{aligned} y_t &= P_{K_{t-n-1}} x_t + P_{\{\xi_{t-n}, \xi_{t-n+1}, \ldots, \xi_t\}} x_t \\ &= u_{t,n} + \sum_{\nu=0}^{n} (x_t, \xi_{t-\nu})\xi_{t-\nu} \end{aligned}$$

Denote the inner product $(x_t, \xi_{t-\nu})$ by c_ν (note that it does not depend upon t due to the stationary geometry) so that

$$y_t = u_{t,n} + \sum_{\nu=0}^{n} c_\nu \xi_{t-\nu}$$

Since $\{c\}\in l_2$, and since $\{\xi_t\}$ is orthonormal, the sum converges in the mean as $n\to\infty$. Also, since $K_{-\infty}$ is the null space, it follows that $u_{t,n}\to 0$ in the mean as $n\to\infty$. Hence we have shown the one-sided moving average representation for the y-process, as stated in (41).

On the other hand we have

$$H_t = L_2(y_s;\, s\leqslant t)\oplus L_2(z_s;\, s\leqslant t)$$

and when $t\downarrow -\infty$, the first subspace on the right-hand side of this equation tends to the null space. Hence

$$z_t \in H_{-\infty} = \lim_{t\downarrow -\infty} L_2(z_s;\, s\leqslant t)$$

so that the z-process is deterministic and the proof is complete. Q.E.D.

Remark 2 The representation of (41) as a one-sided moving average also helps us for *predicting a process by linear combinations*. Indeed, say that $z_t\equiv 0$ and that we have access to all x_s, $s\leqslant t$, and want to find an element x^* in $L_2(x_s;\, s\leqslant t)$ such that the *prediction error*

$$\|x_{t+1}-x^*\| = \min$$

This has of course the unique solution as a projection, which can now be calculated from (40) for $t+1$:

$$x^* = P_{K_t} x_{t+1} = \sum_{\nu=1}^{\infty} c_\nu \xi_{t+1-\nu}$$

with the minimum prediction error [see (42)], given by

$$\|x_{t+1}-x^*\|^2 = |c_0|^2 = \sigma_1^2$$

The prediction h steps ahead can be handled analogously.

Prediction theory is an important tool for linear estimation in stochastic processes. We use it repeatedly in Chapter 4, although often the prediction is based only on a finite past, for example, $T=\{1,2,\ldots,n\}$. In that connection we need a modified concept, unbiased prediction, (see Notes 2.4.C).

For continuous time there exists a similar result.

THEOREM 9

Consider a (weakly) stationary stochastic process $x(t)$, continuous in the mean. Then there exists an orthogonal process

$$\xi(t), \xi(t) \in L_2(x(s); s \leqslant t)$$

and $\|\xi(dt)\|^2 = dt$, *such that* $x(t) = y(t) + z(t)$ *with*

$$y(t) = \int_{-\infty}^{t} g(t-s)\xi(ds)$$

and $z(t)$ is deterministic, while $y(t)$ is completely nondeterministic.

We do not give a proof. The reader is referred to Hanner (1950) for the first geometric proof, and to Karhunen (1950) for the first analytic proof. An up-to-date and very general treatment of these and related questions can be found in Rozanov (1977).

Instead we return to the discrete case and point out an important fact that could be easily missed. Let $\{\xi_t\}$ be real valued and form an orthonormal system and consider a constant $a \in (-1, 1)$. Define the stationary process

$$x_t = \xi_t - a\xi_{t-1}$$

Since $|a| < 1$, we get the convergent series

$$\xi_t = x_t + ax_{t-1} + a^2 x_{t-2} + \cdots \in L_2(x_s; s \leqslant t)$$

Let

$$\eta_t = x_{t+1} + ax_{t+2} + a^2 x_{t+3} + \cdots$$

which has exactly the same geometric properties as $\{\xi_t\}$. Then

$$\eta_{t-1} - a\eta_t = x_t + ax_{t+1} + a^2 x_{t+2} + \cdots - ax_{t+1} - a^2 x_{t+2} + \cdots \\ - x_t$$

so that we have another one-sided moving average representation for our x-process. Note, however, that

$$\eta_t \notin L_2(x_s; s \leqslant t)$$

so that the condition in (41) is not satisfied. We return to this difficulty later.

We have assumed that the stationary process $x(t)$ is scalar valued. If $x(t)$ takes as vectors d-dimensional complex vectors $V=\mathbb{C}^d$, Cramér's spectral representation is still valid in slightly modified form.

To see this, note first that the natural extension of the notion of (weak) stationarity to the vector valued case is to ask that all components $x_k(t)$, $k=1,2,\ldots,d$, be *stationarily correlated*. By this we mean that for all k, l the covariance

$$\operatorname{cov}[x_k(s), x_l(t)]$$

depends only upon $s-t$ so that it can be denoted by $r_{kl}(s-t)$. In particular for $k=l$ the components of $x(t)$ should be stationary, so that we can apply the one-dimensional spectral representation theorem

$$x_k(t)=\int_{-\infty}^{\infty} e^{it\lambda} Z_k(d\lambda) \tag{44}$$

In (44) the Z_k are orthogonal processes with some spectral measure F_k. For an arbitrary $v\in\mathbb{C}^d$ introduce the scalar stationary process

$$y(t)=v^T x(t)=\int_{-\infty}^{\infty} e^{it\lambda} Y(d\lambda)$$

Then the orthogonal process $Y=v^T Z=v^T \operatorname{col}[Z_1, Z_2,\ldots, Z_d]$, since it is uniquely determined and since (44) holds.

Let i_1 and i_2 be two disjoint intervals and consider the increments of the Y-process over them:

$$Y(i_1)\perp Y(i_2) \Rightarrow \sum_{k,l} v_k \bar{v}_l (Z_k(i_1), Z_l(i_2))=0$$

Since this holds for any vector v we must have $Z_k(i_1)\perp Z_l(i_2)$ for all k, l.

On the other hand for $i_1=i_2=i$ we get in the same way

$$\|Y(i)\|^2=\sum_{k,l} v_k \bar{v}_l F_{kl}(i) \tag{45}$$

where $F_k(i)$ is the covariance matrix of the stochastic d-vector $Z(i)$. With a natural notation we therefore see that

$$x(t)=\int_{-\infty}^{\infty} e^{it\lambda} Z(d\lambda)$$

where Z is a *vector valued orthogonal process whose spectral measure is matrix valued*:

$$E\left[Z(d\lambda)Z^T(d\lambda)\right]=\{F_{kl}(d\lambda);\, k,l=1,2,\ldots,d\}=F(d\lambda)$$

Note that (45) implies that the values of F are nonnegative definite.

5 TOEPLITZ THEORY OF STATIONARY PROCESS

Theorem 8 of Section 4 is fundamental for the understanding of the linear structure of (weakly) stationary processes. It is a proof of existence, however, rather than a construction, and it leaves open the problem when a stationary process with absolutely continuous spectrum can be represented as a one-sided moving average.

To answer these questions we appeal to the theory of Toeplitz forms, initiated by Szegö in his seminal papers of 1920–1921. A detailed account, including its application to stochastic processes, can be found in Grenander-Szegö (1957).

Consider a (weakly) stationary process x_t, $t\in\mathbb{Z}$, with absolutely continuous spectrum with the density f. To obtain more attractive formulas as the end result, we express the results in terms of the renormalized density $w=2\pi f$ so that the covariances are expressed as

$$r(h)=\frac{1}{2\pi}\int_{-\pi}^{\pi}e^{it\lambda}w(\lambda)\,d\lambda$$

Our prediction problem from Section 4 then takes the form

$$\sigma_1^2=\lim_{n\to\infty}\mu_n;\qquad \mu_n=\min\left\|x_t-\sum_{\nu=1}^{n}c_\nu x_{t-\nu}\right\|^2$$

over arbitrary $c_1,c_2,\ldots,c_n$. Since $e^{it\lambda}\leftrightarrow x_t$ defines an isometric map between $L_2[(-\pi,\pi)]$ with the inner product

$$(f,g)=\frac{1}{2\pi}\int_{-\pi}^{\pi}f(\lambda)\overline{g(\lambda)}w(\lambda)\,d\lambda$$

and $L_2(X)$ with its usual geometry, wc can write, with $u_0=1$ and $u_\nu=-\overline{c_\nu}$,

$$\mu_n=\min\frac{1}{2\pi}\int_{-\pi}^{\pi}|u_0+u_1z+\ldots+u_nz^n|^2w(\lambda)\,d\lambda;\qquad u_0=1,\quad e^{i\lambda}=z$$

For which $w \in L_1 = L_1([-\pi, \pi])$ does $\mu_n \to 0$? It is clear that $\{\mu_n\}$ is nonincreasing so that its limit μ exists. But when is it zero? A beautiful answer is given by the following theorem.

THEOREM 1 SZEGÖ'S ALTERNATIVE, SZEGÖ (1920, 1921)

Either $\ln w \in L_1$, *and we get the minimum prediction error as the geometric mean*

$$\mu = G(w) = \exp\left(\frac{1}{2\pi}\int_{-\pi}^{\pi} \ln w(\lambda)\, d\lambda\right)$$

or $\ln w \notin L_1$, *and* $\mu = 0$.

Remark Given a nonnegative function f on $(-\pi, \pi)$, we use the standard definition of the arithmetic mean:

$$A(f) = \frac{1}{2\pi}\int_{-\pi}^{\pi} f(\lambda)\, d\lambda$$

the geometric mean:

$$G(f) = \exp[A(\ln f)]$$

and the harmonic mean:

$$H(f) = \frac{1}{A(1/f)}$$

We also allow the values 0 and $+\infty$ for these means.

Proof If we impose the restriction in our minimum problem that all the zeros of the polynomial

$$P(z) = \sum_{\nu=0}^{n} u_\nu z^\nu, \qquad u_0 = 1$$

be outside $|z| = 1$, it will not affect the value of the resulting minimum [see the procedure after (4.29)]. But then $\ln P^2(z)$ is analytic in $|z| \leqslant 1$ so that its real part $\log|P(z)|^2$ is harmonic in $|z| \leqslant 1$ and the mean value theorem for harmonic functions implies for the geometric mean $G(P)$

$$\frac{1}{2\pi}\int_{-\pi}^{\pi} \log|P(z)|^2\, d\lambda = \log|P(0)|^2 = 0; \qquad G(P) = 1$$

Any nonnegative trigonometric polynomial in λ can be written as the square of the modulus of a polynomial in z of the same degree and with the zeros outside $|z|=1$ [see, e.g., Grenander-Szegö (1957), p. 20]. Hence μ_n is the minimum of

$$\frac{1}{2\pi}\int_{-\pi}^{\pi} k(\lambda)w(\lambda)\,d\lambda$$

for any nonnegative trigonometric polynomial k of degree n with $G(k)=1$. Equivalently it is the minimum of

$$\frac{1}{2\pi G(k)}\int_{-\pi}^{\pi} k(\lambda)w(\lambda)\,d\lambda \tag{1}$$

for any nonnegative trigonometric polynomial k of degree n. Using Weierstrass' approximation theorem, we see that the minimum problem (with inf rather than min) will have the same value μ if we only ask that k be a nonnegative continuous function.

Using the well-known relation between the arithmetic and the geometric mean, we get therefore the inequality

$$\frac{1}{2\pi}\int_{-\pi}^{\pi} k(\lambda)w(\lambda)\,d\lambda \geqslant G(kw)=G(k)G(w)$$

which implies that $\mu_n \geqslant G(w)$ and hence the limit $\mu \geqslant G(w)$. But μ is *at most* equal to (1), where k is an arbitrary nonnegative continuous function.

Let us assume momentarily that $w(\lambda)\geqslant w_0>0$ so that $G(w)\geqslant w_0$. Approximate w in L_1-norm by a positive trigonometric polynomial $\pi(\lambda)$ such that

$$\frac{1}{2\pi}\int_{-\pi}^{\pi} |w(\lambda)-\pi(\lambda)|\,d\lambda<\varepsilon; \qquad \pi(\lambda)\geqslant w_0$$

Now in (1) let us choose $k(\lambda)=1/\pi(x)$ so that

$$\mu \leqslant \frac{1}{2\pi G(\pi^{-1})}\int_{-\pi}^{\pi}\frac{w(\lambda)}{\pi(\lambda)}\,d\lambda$$

$$=\frac{G(\pi)}{2\pi}\int_{-\pi}^{\pi}\frac{w(\lambda)}{\pi(\lambda)}\,d\lambda < G(w)G\left(\frac{\pi}{w}\right)$$

$$\left\{1+\frac{1}{2\pi w_0}\int_{-\pi}^{\pi}|w(\lambda)-\pi(\lambda)|\,d\lambda\right\} \leqslant G(w)G\left(\frac{\pi}{w}\right)\left(1+\frac{\varepsilon}{w_0}\right)$$

On the other hand

$$\log G\left(\frac{\pi}{w}\right)=\frac{1}{2\pi}\int_{-\pi}^{\pi}[\log \pi(\lambda)-\log w(\lambda)]\,d\lambda$$
$$\leqslant\frac{1}{2\pi w_0}\int_{-\pi}^{\pi}|\pi(\lambda)-w(\lambda)|d\lambda<\frac{\varepsilon}{w_0}$$

using the elementary inequality for $0<u<v$

$$\ln v-\ln u=\ln\left(1+\frac{v-u}{u}\right)<\frac{v-u}{u}$$

Hence, making ε small, we have $\mu\leqslant G(w)$, which combined with the opposite inequality derived above shows that $\mu=G(w)$.

To get rid of the condition that $w(x)$ be bounded away from zero is comparatively straightforward. Let $w\in L_1^+$. Then, for any $\varepsilon>0$, we have

$$\mu=\mu(w)\leqslant\mu(w+\varepsilon)=G(w+\varepsilon)$$

so that for arbitrary $\eta>0$, with the same reasoning as above,

$$\int_{w(\lambda)>\eta}\log[w(\lambda)+\varepsilon]\,d\lambda-\int_{w(\lambda)>\eta}\log w(\lambda)\,d\lambda<\frac{2\pi\varepsilon}{\eta}$$

This shows that if $\eta+\varepsilon<1$,

$$\int_{-\pi}^{\pi}\log[w(\lambda)+\varepsilon]\,d\lambda\leqslant\int_{w(\lambda)>\eta}\log[w(\lambda)+\varepsilon]\,d\lambda$$
$$<\int_{w(\lambda)>\eta}\log w(\lambda)\,d\lambda+\frac{2\pi\varepsilon}{\eta}\tag{2}$$

Hence

$$\overline{\lim_{\varepsilon\downarrow 0}}\int_{-\pi}^{\pi}\log[w(\lambda)+\varepsilon]\,d\lambda\leqslant\int_{w(\lambda)>\eta}\log w(\lambda)\,d\lambda$$

so that, if we let $\eta\downarrow 0$,

$$\overline{\lim_{\varepsilon\downarrow 0}}\int_{-\pi}^{\pi}\log[w(\lambda)+\varepsilon]\,d\lambda\leqslant\int_{-\pi}^{\pi}\log w(\lambda)\,d\lambda$$

if the last integral exists; otherwise the left-hand side of (2) is $-\infty$. In both cases

$$\lim_{\varepsilon\downarrow 0} G(w+\varepsilon)=G(w)$$

and the proof is complete. Q.E.D.

Remark 2 In this proof we have utilized the quadratic form

$$\frac{1}{2\pi}\int_{-\pi}^{\pi}\left|\sum_{\nu=0}^{n} u_\nu z^\nu\right|^2 w(\lambda)\,d\lambda=\sum_{\nu,\mu=1}^{n} u_\nu \bar{u}_\mu r_{\nu-\mu} \tag{3}$$

A quadratic form is called a *Toeplitz form* if the entries in the associated matrix depend only on the difference $\nu-\mu$ as in (3).

The theorem implies that if the logarithm of the spectral density is integrable, the process can be expressed as a one-sided moving average since it is completely nondeterministic, but how do we calculate the coefficients c_n? Going back to the proof of Theorem 7 in Section 4, it is clear that the problem amounts to factoring the density as $|\phi(\lambda)|^2$ in such a way that ϕ has a one-sided Fourier transform in the L_2-sense. To answer the question, introduce the function defined via the *Poisson kernel* [see Titchmarsh (1939), p. 124]:

$$\frac{1}{2\pi}\int_{-\pi}^{\pi}\log w(\theta)\frac{1-r^2}{1-2r\cos(\lambda-\theta)+r^2}\,d\theta \tag{4}$$

which is harmonic in the polar coordinates r,λ for $r<1$. Let $h(z)$ be the analytic function whose real part is given by (4) and has $h(0)$ real. Then, with

$$g(z)=e^{(1/2)h(z)}=\sum_{n=0}^{\infty} d_n z^n;\qquad |z|<1 \tag{5}$$

we have

$$\begin{aligned}|g(re^{i\lambda})|^2&=\exp\{\operatorname{Re}[h(re^{i\lambda})]\}\\&=\exp\left\{\frac{1}{2\pi}\int_{-\pi}^{\pi}\ln w(\theta)\frac{1-r^2}{1-r\cos(\lambda-\theta)+r^2}\,d\theta\right\}\end{aligned} \tag{6}$$

Using Jensen's inequality, the equality in (6), treating the Poisson kernel as the weight function, gives us the bound

$$|g(re^{i\lambda})|^2 \leqslant \frac{1}{2\pi}\int_{-\pi}^{\pi} w(\theta)\frac{1-r^2}{1-2r\cos(\lambda-\theta)+r^2}\,d\theta \tag{7}$$

Integrate (7) over $\lambda\in(-\pi,\pi)$ and use the fact that the integral of the Poisson kernel is 1. We get

$$\frac{1}{2\pi}\int_{-\pi}^{\pi}|g(re^{i\lambda})|^2\,d\lambda \leqslant \frac{1}{2\pi}\int_{-\pi}^{\pi} w(\theta)\,d\theta = \int_{-\pi}^{\pi} f(\theta)\,d\theta = r_0 \tag{8}$$

The uniform bound (8) says that g is a *Hardy class* function, $g\in H_2$. Note that the left-hand side can also be written in terms of the Taylor coefficients of g and implies, $r\uparrow 1$, that

$$\sum_{n=0}^{\infty}|d_n|^2 \leqslant r_0 \tag{9}$$

Using (7) it follows, using a well-known property of the Poisson singular kernel, that w has been factored as (see Note 2.5A)

$$\lim_{r\uparrow 1}|g(re^{i\lambda})|^2 = \exp\{\ln w(\lambda)\} = w(\lambda) \qquad \text{a.e.} \tag{10}$$

which implies, using (8) and Fatou's lemma,

$$\frac{1}{2\pi}\int_{-\pi}^{\pi} w(\lambda)\,d\lambda \leqslant \lim_{r\uparrow 1}\frac{1}{2\pi}\int_{-\pi}^{\pi}|g(re^{i\lambda})|^2\,d\lambda = \sum_{n=0}^{\infty}|d_n|^2$$

Together with (9) this shows that

$$r_0 = \frac{1}{2\pi}\int_{-\pi}^{\pi} w(\lambda)\,d\lambda = \sum_{n=0}^{\infty}|d_n|^2$$

which we now apply to the function $w(\lambda)|1+\eta e^{ik\lambda}|^2$, where η is a complex number and $k\in\mathbb{N}_+$. It gives us

$$\frac{1}{2\pi}\int_{-\pi}^{\pi} w(\lambda)|1+\eta e^{ik\lambda}|^2\,d\lambda = \lim_{r\uparrow 1}\frac{1}{2\pi}\int_{-\pi}^{\pi}|g(re^{i\lambda})|1+\eta r^k e^{ik\lambda}|^2\,d\lambda$$

from which follows (look at small values of η)

$$\begin{aligned} r_{-h} &= \int_{-\pi}^{\pi} f(\lambda)e^{-ih\lambda}\,d\lambda = \frac{1}{2\pi}\int_{-\pi}^{\pi} w(\lambda)e^{-ih\lambda}\,d\lambda \\ &= \lim_{r\uparrow 1}\frac{1}{2\pi}\int_{-\pi}^{\pi}|g(re^{i\lambda})|^2 e^{-ih\lambda}\,d\lambda \end{aligned}$$

or

$$r_{-h}=\sum_{\nu=0}^{\infty}\bar{d}_\nu d_{\nu+h}$$

If we define $c_\nu=\bar{d}_\nu$ for $\nu\geqslant 0$, and zero otherwise, we have, for $s\leqslant t$ (actually for all s, t by symmetry),

$$r(s,t)=\sum_{\nu=0}^{\infty}\bar{d}_\nu d_{\nu+t-s}=\sum_{\mu=-\infty}^{\infty}c_{s-\mu}\bar{c}_{t-\mu}$$

The general process representation can now be applied with $\Lambda=\mathbb{Z}$, μ as counting measure, and $f(t,\lambda)=c_{t-\lambda}$. Hence there exists an orthonormal sequence $\{\xi_\nu\}$ such that

$$x_t=\sum_{-\infty}^{t}c_{t-\mu}\xi_\mu \tag{11}$$

with convergence in the mean and we have obtained the following theorem.

THEOREM 2

If the logarithm of the spectral density is integrable, the construction in (4) *and* (5) *gives us the coefficients* $c_\nu=d_\nu$ *of the coefficients in a one-sided moving average* (11).

Is the c-sequence identical to the one in the Wold decomposition so that we can claim $L_2(\xi_s;\, s\leqslant t)=L_2(x_s;\, s\leqslant t)$? The answer is positive, as can be shown in the following.

Introduce the function

$$c(z)=\sum_{n=0}^{\infty}c_n z^n; \qquad |z|<1$$

which is analytic in $|z|<1$ since $\{c_n\}\in l_2$. Assume that $c(z)$ has a zero z_0 in $|z|<1$. Then we can write

$$\sum_{n=0}^{\infty}\bar{c}_n z^n=(z-\bar{z}_0)\sum_{n=0}^{\infty}\bar{c}'_n z^n; \qquad c'_0=-\frac{c_0}{z_0} \tag{12}$$

or, with convergence in the mean,

$$\sum_{n=0}^{\infty}\bar{c}_n e^{in\lambda}=(e^{i\lambda}-\bar{z}_0)\sum_{n=0}^{\infty}\bar{c}'_n e^{in\lambda}$$

If z_0 happens to be zero, a simple modification is needed in the last relation of (12). But

$$\left|\sum_0^\infty c_n e^{-in\lambda}\right| = \left|\sum_0^\infty \bar{c}_n e^{in\lambda}\right| = \left|\frac{1-e^{i\lambda}z_0}{e^{i\lambda}-\bar{z}_0}\right|\left|\sum_{n=0}^\infty \bar{c}_n e^{in\lambda}\right|$$

$$= |1-e^{i\lambda}z_0|\left|\sum_{n=0}^\infty \bar{c}'_n e^{in\lambda}\right| = \left|\sum_{n=0}^\infty \bar{c}''_n e^{in\lambda}\right| = \left|\sum_{n=0}^\infty c''_n e^{-in\lambda}\right|$$

with convergence in the mean and $c''_0 = -c_0/z_0$. The process

$$x'_t = \sum_{n=0}^\infty c''_n \xi_{t-n}$$

has the same spectral density, hence the same geometry, as the x_t-process. Therefore they have the same prediction error. But

$$E\left[\left|x_t - \sum_{n=1}^N a_n x_{t-n}\right|^2\right] = \int_{-\pi}^{\pi}\left|e^{it\lambda} - \sum_{n=1}^N a_n e^{i(t-n)\lambda}\right|^2 f(\lambda)\,d\lambda$$

$$= \frac{1}{2\pi}\int_{-\pi}^{\pi}\left[\left|1-\sum_{n=1}^N a_n e^{in\lambda}\right|\cdot\left|\sum_{n=0}^\infty c''_n e^{-in\lambda}\right|\right]^2 d\lambda$$

$$= \frac{1}{2\pi}\int_{-\pi}^{\pi}|c''_0 + b_1 e^{-i\lambda} + b_2 e^{-2i\lambda} + \cdots|^2\, d\lambda \geqslant |c''_0|^2.$$

Hence the minimum prediction error is at least equal to

$$|c''_0| = \left|\frac{c_0}{z_0}\right|^2 > |c_0|^2 \tag{13}$$

which shows that no zero exists in $|z|<1$, since the prediction error in the right-hand side of (13) is realized for the Wold decomposition.

Hence $\ln|c(z)|^2$ is harmonic in $|z|<1$, and the construction in (5) gives us essentially the coefficients corresponding to the Wold decomposition.

Using the spectral representation of the x- and ξ-processes, we have, with the usual formalism,

$$Z_x(d\lambda) = \phi(\lambda) Z_\xi(d\lambda) \tag{14}$$

with

$$\phi(\lambda) = \sum_{n=0}^\infty \bar{d}_n e^{in\lambda}\lambda \tag{15}$$

interpreted as the boundary values of $\overline{g(z)}$ at $|z|=1$ from inside the unit circle. Since $\phi \neq 0$ a.e., relation (14) determines the orthogonal Z_ξ-process. This shows that the ξ_t-process is also uniquely determined so that we have obtained analytically the innovation process $\{\xi_t\}$.

A reader who is not already familiar with this theory will find it instructive to compare the general construction in Theorems 1 and 2 with the special case in Example 4 of Section 4. Also see Notes.

For continuous time a similar result holds.

THEOREM 3

If $x(t)$, $t \in \mathbb{R}$, is a weakly stationary process with mean zero continuous in the mean, and with an absolutely continuous spectrum density f, it can be written as

$$x(t)=\int_{-\infty}^{t} g(t-s)\xi(ds)$$

with $g \in L_2[0,\infty]$, and an orthogonal process ξ with $\|\xi(ds)\|^2=ds$, and $L_2(\xi_s;\, s \leqslant t)=L_2(x_s;\, s \leqslant t)$, if and only if

$$\int_{-\infty}^{\infty} \frac{\ln f(\lambda)}{1+\lambda^2} d\lambda$$

converges.

This result is due to Karhunen (1950), where an explicit construction of g can also be found. It is closely related to Theorem 9 of Section 4.

6 REPRODUCING COVARIANCE KERNELS

Linear representation theory of stochastic processes aims at expressing the process as a linear combination, in a general sense, of a set of uncorrelated stochastic variables. The cardinality of the set varies from case to case. The $x(t)$ are looked at as elements of the Hilbert space $L_2(X)$, and the $f(t,\cdot)$ in the general process representation are elements of the Hilbert space $L_2(\Lambda, \mu)$.

Can the latter Hilbert space be chosen in a natural way, simplifying the ensuing analysis? We have already seen in Sections 4 and 5 that the answer to this question is sometimes yes, and we now look deeper into it.

The discussion is based on the notion of a *kernel function*, whose main role has been played in partial differential equations and harmonic functions, but which has also been successfully applied to stochastic processes.

The main advocate for this elegant approach has been Parzen, and the reader is referred to Parzen (1961, 1963) for further information.

The general definition of a reproducing kernel will be needed [see Aronszajn (1950) an Meschkowski (1962)].

Definition 1 A function defined on $T\times T$ is said to be a reproducing kernel of the Hilbert space H of functions defined on T if $K(\cdot,t)\in H$, $\forall t\in T$, and $(g, K(\cdot,t)_H=g(t), \forall g\in H$.

The existence and uniqueness of such a Hilbert space H is guaranteed under wide conditions by the following fundamental theorem.

THEOREM 1 CHARACTERIZATION OF REPRODUCING KERNELS, ARONSZAJN (1950)

In order that K be a reproducing kernel it is necessary and sufficient that K be nonnegative definite.

Proof If K is a reproducing kernel in some Hilbert space H consisting of certain functions defined on T, we have for any nontrivial complex vector $(z_1, z_2,\ldots, z_n)$ and $t_1, t_2,\ldots, t_n\in T$

$$\sum_{\nu,\mu=1}^{n} \bar{z}_\nu z_\mu K(t_\nu, t_\mu)=\sum_{\nu,\mu=1}^{n} \bar{z}_\nu z_\mu\big(K(\cdot,t_\mu), K(\cdot,t_\nu)\big)$$

$$=\left\|\sum_{\mu=1}^{n} z_\mu K(\cdot,t_\mu)\right\|^2 \geqslant 0$$

On the other hand if K is known to be nonnegative definite, form the linear closure $L(K)$ of the functions $K(\cdot,t), \forall t\in T$, and define the inner product of two functions $f, g\in L(K)$:

$$f(t)=\sum_{\nu=1}^{n} a_\nu K(t,t_\nu)$$

$$g(t)=\sum_{\nu=1}^{m} b_\nu K(t,t_\nu) \tag{1}$$

by the expression

$$(f,g)_H=\sum_{\nu,\mu=1}^{n} a_\nu \bar{b}_\mu K(t_\mu,t_\nu) \tag{2}$$

We can include any t-value we wish among the $\{t_\nu\}$, since we allow the a- or b-coefficient to be zero. If f can also be expressed as in (1) but with coefficients a'_ν, it follows that

$$(f, g)_H = \sum_{\nu,\mu=1}^{n} a_\nu \bar{b}_\mu K(t_\mu, t_\nu)$$

$$= \sum_{\mu=1}^{n} \bar{b}_\mu \sum_{\nu=1}^{n} a_\nu K(t_\mu, t_\nu) = \sum_{\mu=1}^{n} \bar{b}_\mu f(t_\mu)$$

But this can be rewritten as

$$\sum_{\mu=1}^{n} \bar{b}_\mu \sum_{\nu=1}^{n} a'_\nu K(t_\mu, t_\nu) = \sum_{\nu=1}^{n} a'_\nu \overline{\sum_{\mu=1}^{n} b_\mu K(t_\mu, t_\nu)}$$

$$= \sum_{\nu=1}^{n} a'_\nu \overline{g(t_\nu)} = \sum_{\nu=1}^{n} a'_\nu \overline{\sum_{\mu=1}^{n} b'_\mu K(t_\mu, t_\nu)}$$

which rearranged is just

$$\sum_{\nu,\mu=1}^{n} a'_\nu \overline{b'_\mu} K(t_\mu, t_\nu)$$

so that the definition of inner product in (2) is unique.

Since K is nonnegative definite, (2) will have the usual properties of an inner product in a Hilbert space. Moreover, with $b_r = \delta_{r\mu}$, (2) reduces to

$$(f, K(\cdot, t_\mu))_H = \sum_{\nu=1}^{n} a_\nu K(t_\mu, t_\nu) = f(t_\mu) \tag{3}$$

and we should recall that any t can be included among the t_μ (see above).

Now we complete $L(K)$ and call the resulting Hilbert space H. If f_n is a Cauchy sequence, using (3) and Schwarz's inequality,

$$|f_m(t) - f_n(t)| = |((f_m - f_n), K(\cdot, t))_H| \leqslant \| f_m - f_n \|_H \cdot \| K(\cdot, t) \tag{4}$$

Note that $K(\cdot, t) \in L(K)$ with

$$\| K(\cdot, t) \|_H^2 = (K(\cdot, t), K(\cdot, t))_H = K(t, t)$$

But (4) shows that $\{f_n(t)\}$ converges pointwise to some limit f. Also

$$|\| f_m \|_H - \| f_n \|_H| \leqslant \| f_m - f_n \|_H$$

and since we can define the norm of f as

$$\| f \|_H = \lim_{n\to\infty} \| f_n \|_H$$

which is uniquely determined, it is seen that the completed space H is indeed a Hilbert space with the reproducing kernel K. Use (3) together with pointwise convergence of any sequence f_n converging in the topology of H. Q.E.D.

For any nonnegative definite covariance function $r(\cdot,\cdot)$ we can then construct the space H on which r is a reproducing kernel. Since

$$r(s,t) = (r(\cdot,s), r(\cdot,t))_H \tag{5}$$

we can apply the method used for the isometry theorem in Section 3. This gives the following attractive result of great generality.

THEOREM 2 REPRODUCING KERNEL REPRESENTATION, PARZEN (1961)

To each $g \in H$ there corresponds a stochastic variable $x \in L_2(x)$, where $x(t)$ is a process associated with the reproducing kernel r, denoted by (X, g), such that

$$x(t) = (X, r(\cdot,t))_H$$
$$((X,g),(X,h)) = (g,h)_H; \qquad g,h \in H \tag{6}$$

Proof Define a mapping between finite linear combinations in $L_2(X)$ and H by

$$\sum_{k=1}^{n} c_k x(t_k) \rightarrow \sum_{k=1}^{n} c_k r(\cdot, t_k)$$

The squared norms of both sides of (6)—in their respective spaces—are both equal to

$$\sum_{k=1}^{n} c_k \bar{c}_l r(t_k, t_l) \tag{7}$$

Because of this isometry the mapping (6) is bijective and can be extended, as was done several times before in the text, to $L_2(X) \leftrightarrow H$. Inner products as in (5) will be preserved. Q.E.D.

Remark 1 To realize the significance of the representation in (5) consider a real-valued process $x(t)$ with a real argument $t \in (0,1) = T$, for example. Let it generate the Hilbert space $L=L_2[x(t), t\in T]$. Then we often need an expression for an *arbitrary* element z in L_2. We could try to represent z by an "ordinary" inner product in $L_2(T)$:

$$z=\int_0^1 x(t)f(t)\,dt$$

where f is an arbitrary quadratically integrable function. But *the z we get from* (7) *do not in general exhaust L_2.*

The inner product representation in (5) *exhausts L_2*, and this is an important reason why the RKHS approach is useful in our context.

Of course one can also express the z by integrals of the form (3.11) in terms of an orthogonal process $Z(\lambda)$. This is a general representation but requires knowledge of the Z-process. Such representations and RKHS methods complement each other and are closely related.

The notation (X, g) should not be understood as an inner product between sample functions of $x(t)$ with g; it is only a suggestive symbolism.

In particular if a measure space (Λ, μ) exists such that with the same notation as in Section 3,

$$r(s,t)=\int_\Lambda f(s,\lambda)\overline{f(t,\lambda)}\mu(d\lambda)$$

then H will consist of all functions g defined on T expressible as

$$g(t)=\int_\Lambda g^*(\lambda)\overline{f(t,\lambda)}\mu(d\lambda) \tag{8}$$

for some g^* in the space $L_2(f)$ and with

$$\|g\|_H^2=\int_\Lambda |g^*(\lambda)|^2\mu(d\lambda)$$

Indeed, it is clear that this H is a Hilbert space and that Definition 1 applies since $r(\cdot, t)$ can be written in the form (8) with $g^*(\lambda)=f(t,\lambda)$ so that $r(\cdot,t)\in H$. Also for any g expressed as (8) we have

$$(g, r(\cdot,t))_H=\int_\Lambda g^*(\lambda)\overline{f(t,\lambda)}\mu(d\lambda)=g(t)$$

which completes the proof.

Further if $L_2(f)=L_2(\Lambda,\mu)$ so that $x(t)$ can be expressed as in (3.16), then

$$(X,g)_H=\int_\Lambda g^*(\lambda)Z(d\lambda)$$

with the same orthogonal process Z as in (3.16).

In Chapter 4 we use this approach on inference problems, and here we just give a couple of examples.

Example 1 Let $T=\{1,2,\ldots,n\}$ with the nonsingular covariance matrix R. Define the inner product, say with real scalars,

$$(f,g)_H=\sum_{s,t=1}^{n} f_s m_{st} g_t; \qquad f,g\in\mathbb{R}^n \tag{9}$$

with $\{m_{st};\, s,t=1,2,\ldots,n\}=M=R^{-1}$. Then R is a reproducing kernel on H, consisting of $\mathbb{R}^n$ equipped with the inner product (9). Indeed, the first condition in Definition 1 is trivial and the second is easy:

$$(f,R(\cdot,u))_H=\sum_{s,t=1}^{n} f_s m_{st} r_{tu}=\sum_{s=1}^{n} f_s\delta_{su}=f_u$$

Example 1 is of little interest and is given only for didactic reasons. The next one contains more.

Example 2 Consider an *autoregressive process* $x(t)$ with discrete time $t\in\mathbb{Z}$, that is, a weakly stationary solution of a stochastic difference equation

$$(Lx)(t)=a_0x(t)+a_1x(t-1)+\ldots+a_px(t-p)=e_t; \qquad t\in\mathbb{Z} \tag{10}$$

with $\{e_t\}$ forming an orthonormal sequence of stochastic variables, $e_t\perp L_2(x(s);\, s<t)$. In Example 4 of Section 4 we learned how to study such equations by Fourier analysis. The spectrum was seen to be absolutely continuous, which implies that the covariance matrix is nonsingular (see Lemma 1 in Section 4).

Consider a finite length sample $x(t), t=1,2,\ldots,n$. The reproducing kernel space H will now be shown to be given by the inner product

$$(f,g)_H=\sum_{t=p+1}^{n}(Lf)(t)\cdot(Lg)(t)+\sum_{s,t=1}^{p} f(s)d_{st}g(t); \qquad f,g\in\mathbb{R}^n \tag{11}$$

and $D=\{d_{st}\}$ is the inverse of $\{r(s,t);\, s,t=1,2,\ldots,p\}$, and $n\geqslant 2p$.

Following the proof in Parzen (1960), let us assume that the process $x(t)$ is Gaussian with the covariance matrix R so that its joint frequency function is

$$f(x)=\frac{1}{(2\pi)^{n/2}\sqrt{\det R}}\exp-\frac{1}{2}x^TR^{-1}x \tag{12}$$

This is no restriction. We are only concerned with second-order properties, and the distribution assumption is made only to simplify the following derivation.

But using Example 1 we know that $x^TR^{-1}x=(x,x)_R$. Now we use (10) together with the orthogonality relations following it, which tells us that $x(1),x(2),\ldots,x(p),e(p+1),\ldots,e(n)$ have a joint frequency function equal to

$$\frac{1}{(2\pi)^{n/2}\sqrt{\det R_p}}\exp\left\{-\frac{1}{2}\left[(u,u)_{R_p}+\sum_{p+1}^{n}v^2\right]\right\}$$

with $(x(1),\ldots,x(p))=u,(e(P+1),\ldots,e(n))=v$, and R_p as the $p\times p$ covariance matrix of u. But $(x_1,x_2,\ldots,x_n)$ is linearly related to (u,v) with the Jacobian a_0^{n-p} so that

$$f(x)=\frac{1}{(2\pi)^{n/2}\sqrt{\det R_p}}a_0^{n-p}\exp\left\{-\frac{1}{2}(u,u)_{R_p}+\sum_{p+1}^{n}[(Lx)(t)]^2\right\} \tag{13}$$

Using (12) this tells us that

$$(x,x)_R=(u,u)_{R_p}+\sum_{p+1}^{n}[(Lx)(t)]^2 \tag{14}$$

Define the vector δ_s with components $\delta_{s\nu},\nu=1,2,\ldots,n$, and use the convention $a_t=0$ for $t<0$ and $t>p$. We then get for $1\leqslant s,t\leqslant p$, using (12), and $n\geqslant 2p$,

$$\begin{aligned}(\delta_{n-s+1},\delta_{n-t+1})_R&=\sum_{\nu=p+1}^{n}(L\delta_{n-s+1})(\nu)(L\delta_{n-t+1})(\nu)\\&=\sum_{\nu=p+1}^{n}a_{s+\nu-n-1}a_{t+\nu-n-1}=\sum_{\nu=1}^{n-p}a_{s-\nu}a_{t-\nu}\end{aligned} \tag{15}$$

Also

$$(\delta_s,\delta_t)_R = d_{st} + \sum_{\nu=p+l}^{n} (L\delta_s)(\nu)(L\delta_t)(\nu)$$

$$= d_{st} + \sum_{\nu=p+1}^{n} a_{\nu-s}a_{\nu-t} = d_{st} + \sum_{\nu=1}^{n-p} a_{\nu+p-s}a_{\nu+p-t} \tag{16}$$

Combining (15) and (16) with (14), we see that (11) holds for $n \geqslant 2p$ with

$$d_{st} = \sum_{\nu=1}^{\min(s,t)} (a_{s-\nu}a_{t-\nu} - a_{\nu+p-s}a_{\nu+p-t})$$

The full power of the approach via reproducing kernels becomes apparent first when the t-argument is continuous as in the next case.

Example 3 Consider the Wiener process $W(t), \sigma^2 = 1, t \in T[a,b], a > 0$ (see Example 3 in Section 3) and introduce the Hilbert space H of functions f on t expressible as

$$f(t) = f(a) + \int_a^t f'(s)\,ds; \qquad t \in T$$

with a quadratically integrable f'. Define the inner product as

$$(f,g)_H = \frac{1}{a} f(a)g(a) + \int_a^b f'(s)g'(s)\,ds$$

To see that this is the reproducing kernel space for the covariance function $r(s,t) = \min(s,t)$, note that with the indicator functions $1_t(u)$ corresponding to the set $\{u \mid u \leqslant t\}$ we have

$$r(s,t) = a + \int_a^b 1_s(u)1_t(u)\,du$$

so that $r(\cdot,t) \in H$ with $f' = 1_t$ and $f(a) = a$. Also, for any f in H,

$$(f, r(\cdot,t))_H = \frac{1}{a} f(a)a + \int_a^b f'(u)1_t(u)\,du$$

$$= f(a) + \int_a^t f'(u)\,du = f(t)$$

which completes the proof.

Example 4 The continuous time analogue to Example 2 is of importance for the following. Consider the stochastic linear differential equation

$$(Lx)(t)=\sum_{k=0}^{p} a_k x^{(k)}(t)=\dot{w}; \qquad t\in T=[a,b]$$

(see Example 5 in Section 4). Then the reproducing kernel space associated with the covariance function $r(s,t)$ can be obtained explicitly similarly to (13), and we state the result without proof. It consists of functions with the inner product

$$(f,g)_H=\int_a^b (Lf)(t)(Lg)(t)\,dt+\sum_{\nu,\mu=0}^{p-1} f^{(\nu)}(a)d_{\nu\mu}f^{\mu}(a)$$

where $\{d_{\nu\mu}\}$ is the inverse of the $p\times p$ matrix with entries

$$\left(\frac{\partial^{\nu+\mu}r(s,t)}{\partial^{\nu}s\,\partial^{\mu}t}\right)_{s=t=a}$$

In particular if

$$\dot{x}(t)+\beta x(t)=\dot{w},$$

the process is expressed as a one-sided moving average

$$x(t)=\int_{-\infty}^{t} e^{-\beta(t-s)}w(ds)$$

with covariance function [see (2.19)]

$$r(s,t)=\frac{1}{2\beta}e^{-\beta|s-t|}$$

Then H consists of functions f with the inner product

$$\begin{aligned}(f,g)_H&=\int_a^b (f'+\beta f)(g'+\beta g)\,dt+2\beta f(a)g(a)\\&=\int_a^b (f'g'+\beta^2 fg)\,dt+\beta[f(a)g(a)+f(b)f(g)]\end{aligned}$$

Thus an arbitrary element in $L_2(x(t);t\in T)$ can be associated with an $f\in H$ and expressed as

$$(X,f)_H=\int_a^b f'(t)x(dt)+\beta^2\int_a^b f(t)x(t)\,dt+\beta[f(a)x(a)+f(b)x(b)]$$

where the first integral is interpreted weakly as described earlier in this chapter.

PART 2

Inference in Abstract Sample Space

CHAPTER 3

General Inference in Abstract Sample Space

1 LIKELIHOOD RATIOS IN ABSTRACT SPACE

Given a sample space X, as discussed in Chapter 1, and with the simplest possible parameter space $A=\{0,1\}$, we have two probability measures P_0 and P_1, both defined on the Borel sets of X. A fundamental problem in classical inference is how to *test the null hypothesis* H_0 that $\alpha=0$, corresponding to the measure P_0, against the *alternative hypothesis* H_1 that $\alpha=1$, corresponding to the measure P_1.

In the standard Neyman-Pearson approach to this problem we ask for a subset W, the *critical region*, with the decision rule that if the observed sample $x\in W$, we reject H_0; otherwise we accept it. To measure the performance of the test we use the

$$\textit{level of significance of the test}: \varepsilon=P_0(W)$$

which expresses the probability of an *error of the first kind*: reject H_0 when it is true. Also,

$$\textit{power of the test}: P_1(W)=1-\delta$$

where δ means the probability of an *error of the second kind*: accept H_0 when it is false.

Of course we would like both ε and δ to be small, but they are usually antagonistic to each other, so that we have to be satisfied with balancing one against the other. The famous Neyman-Pearson lemma tells us how to minimize δ for fixed ε.

How can this be carried over to our case when X is some abstract space? The Neyman-Pearson lemma, as well as many other classical procedures, is *based on the likelihood function* and, particularly, the likelihood ratio. In a general abstract space we have no direct equivalent to the Lebesgue measure in $\mathbb{R}^n$ so that we must circumvent this difficulty before we can begin to study special inference problems.

Our guiding principle will be to *let the Radon-Nikodym derivative play the role of the likelihood ratio*, and we see that this will immediately solve the *conceptual problem*: concepts and ideas from classical inference can be applied to the abstract sample spaces once we have the Radon-Nikodym derivatives. This holds both for standard inference and for methods based on Bayesian ideas. The *analytical problem*, on the other hand, may be formidable, but that is another matter.

The basic problem is therefore to calculate the Radon-Nikodym derivative and to analyze its properties. We start to do this by proving a convergence theorem for likelihood ratios.

For given P_0 and P_1 there exists a set H with $P_0(H)=0$ and a nonnegative function f integrable with respect to P_0 such that for any Borel set $E\subset X$ we have [see e.g., Halmos (1950), sec. 30]

$$P_1(E)=\int_E f(x)P_0(dx)+P_1(E\cap H) \tag{1}$$

The function f is the *Radon-Nikodym derivative* and will be denoted by

$$f(x)=\frac{P_1(dx)}{P_0(dx)}$$

Two extreme cases will be encountered repeatedly.

Definition 1 If the set H has $P_1(H)=1$, the two measures P_0 and P_1 are said to be perpendicular to each other.

If this happens, the critical region $W=H$ will allow perfect (probability 1) discrimination between H_0 and H_1. The test will yield the correct result, with errors of the first and second kinds being zero.

Definition 2 If P_1 and P_2 are both absolutely continuous with respect to each other, they are said to be equivalent.

If so, then $P_1(H)=0$ and $f\neq 0$ with P_0-probability 1. This could be said to constitute the nonpathological case, but we will see that counterintuitive situations exist when it is not true, although it may seem so at first glance.

Intermediate cases also occur when P_1 and P_2 are neither perpendicular nor equivalent, but surprisingly seldom for cases of practical interest.

Express the sample points by observables $x_1, x_2, x_3, \ldots$; (see Chapter 1, Section 2). The values of the x_ν will be assumed to be real, but the following holds equally well if they are complex or vector valued. Define the *finite dimensional section* as $x^n = (x_1, x_2, \ldots, x_n) \in \mathbb{R}^n$ and assume that x^n has joint frequency functions with respect to Lebesgue measure in $\mathbb{R}^n$, g_0^n, and g_1^n under the hypotheses P_0 and P_1, respectively. $\mathcal{Q}$ shall be the σ-algebra generated by all cylinder sets in X with finite-dimensional Borel base and $\mathcal{Q}_n$ the σ-algebra in $\mathbb{R}^n$ generated by Borel conditions on x^n. Then we have the following theorem, which will play a fundamental role for the results in the rest of this chapter and in the later chapters.

THEOREM 1 LIKELIHOOD RATIO CONVERGENCE, GRENANDER (1950)

If x^n has absolutely continuous distributions with respect to Lebesgue measure in $\mathbb{R}^n$ under P_0 and P_1, and if the frequency functions are g_n^0, g_n^1, form the likelihood ratio

$$f_n(x) = \frac{g_n^1(x^n)}{g_n^0(x^n)}$$

assumed to be defined a.e. P_0. Then

$$\begin{aligned} &f_n(x) \to f(x) && \textit{in probability } (P_0) \\ &f_n(x) \to f(x) && \textit{in probability } (P_1) \textit{ in } H^C \\ &f_n(x) \to +\infty && \textit{in probability } (P_1) \textit{ in } H \end{aligned}$$

Proof Let us first deal with the *regular case* $P_1(H) = 0$. Consider an arbitrary cylinder set C with base $B \subset \mathbb{R}^n$ so that we have

$$\begin{aligned} P_1(C) &= \int_C f(x) P_0(dx) \\ &= \int_B g_n^1(x^n)\, dx^n = \int_B f_n(x^n) g_n^0(x^n)\, dx^n \end{aligned}$$

But this implies, using Kolmogorov's definition of conditional expectation with respect to P_0,

$$f_n(x) = E_0[\, f(x) | \mathcal{Q}_n]; \qquad \text{a.c. } P_0 \tag{2}$$

Letting n tend to infinity, we can then appeal to a well-known limit theorem for conditional expectations [see Doob (1953), p. 319]

$$f(x) = \lim_{n \to \infty} f_n(x); \qquad \text{a.c. } P_0$$

a bit stronger than the convergence probability in the statement of the theorem.

Now assume instead that we have the *intermediate case* $0<P_1(H)<1$ and fix $\varepsilon>0$. Since $P_0(H)=0$, we can cover it by a denumerable and disjoint union of cylinder sets C_ν with finite-dimensional bases, and with

$$P_0\left(\bigcup_\nu C_\nu\right)<\varepsilon$$

But

$$\sum_{\nu=1}^{\infty} P_1(C_\nu)\leqslant 1$$

so that for a given $\varepsilon>0$ we can find an integer N such that

$$\sum_{\nu=N+1}^{\infty} P_1(C_\nu)<\varepsilon$$

Hence

$$\begin{aligned}P_1[H\cap(x\,|\,f_n(x)\leqslant k)]&\leqslant P_1\left[\left(\bigcup_{\nu=1}^{\infty} C_\nu\right)\cap(x)|f_n(x)\leqslant k)\right]\\&\leqslant P_1\left[\left(\bigcup_{\nu=1}^{N} C_\nu\right)\cap(x\,|\,f_n(x)\leqslant k)\right]+\varepsilon\end{aligned}$$

Now we choose n greater than the subscript of any coordinate appearing in the conditions defining $C_1, C_2,\ldots, C_N$. The set

$$C=\left(\bigcup_{\nu=1}^{N} C_\nu\right)\cap(x\,|\,f_n(x)\leqslant k)\subset X$$

is a cylinder set with a base B in $\mathbb{R}^n$ and

$$\begin{aligned}P_1[H\cap(x\,|\,f_n(x)\leqslant k]&\leqslant\varepsilon+\int_B f_n(x)g_n^0(x^n)\,dx^n\\&\leqslant\varepsilon+kP_0\left(\bigcup_{\nu=1}^{N} C_\nu\right)\leqslant\varepsilon(1+k)\end{aligned}$$

For fixed k this can be made arbitrarily small by choosing ε small enough, which implies that f_n tends to $+\infty$ in H in probability with respect to P_1.

Introduce the probability measure

$$Q(S)=\frac{P_1(S\cap H^c)}{P_1(H^c)}=\int_S \frac{f(x)}{P_1(H^c)}P_0(dx); \qquad S\in\mathcal{Q}$$

which is possible since $P_1(H^c)>0$. In other words we put $Q(S)=P_1(S|H^c)$, and x^n has a frequency function g with respect to Q which can be written as

$$g(x^n|H^c)=\frac{P_1(H^c|x^n)}{P_1(H^c)}g_n^1(x^n) \tag{3}$$

Indeed, relation (3) is a direct consequence of

$$Q(B)=\frac{P_1(B\cap H^c)}{P_1(H^c)}=\int_B \frac{P_1(H^c|x^n)}{P_1(H^c)}g_n^1(x^n)\,dx^n$$

for any Borel set B in $\mathbb{R}^n$. But the probability Q is regular with respect to P_0 so that we can use the first part of the theorem which was proved above, which tells us that a.c. with respect to Q, we have

$$\frac{f(x)}{P_1(H^c)}=\lim_{n\to\infty}\frac{g_n^1(x^n)}{g_n^0(x^n)}\frac{P_1(H^c|x^n)}{P_1(H^c)}$$

But if $x\in H^c$, we can use the same sort of reasoning as after (2) [see Doob (1953), p. 343]:

$$\lim_{n\to\infty}P_1(H^c|x^n)=1; \qquad \text{a.c. } P_1$$

so that

$$\lim_{n\to\infty}\frac{g_n^1(x^n)}{g_n^0(x^n)}=f(x); \qquad \text{a.c. } P_1 \text{ in } H^c \tag{4}$$

Can there exist a set $E\in\mathcal{Q}$ with $P_0(E)>0$ such that (4) does not hold? We already know that $P_1(E)$ must be zero. But if $P_0(E)>0$, then we can apply the previous result, but with P_0 and P_1 interchanged, and get

$$\lim_{n\to\infty}\frac{g_n^0(x^n)}{g_n^1(x^n)}=+\infty \qquad \text{in probability } P_0 \text{ in } E$$

or

$$\lim_{n\to\infty} \frac{g_n^1(x^n)}{g_n^0(x^n)} = 0 = f(x) \qquad \text{in probability } P_0 \text{ in } E$$

It only remains to deal with the most *singular case* when $P_1(H)=1$. But we already know that $f_n \to +\infty$ in probability with respect to P_1. Now f must be zero a.c. P_0 and $1/f_n \to +\infty$ in probability P_0. Hence $f_n \to 0 = f$ in probability P_0, so that the statement in the theorem holds. Q.E.D.

Corollary 1 If the measures P_0 and P_1 restricted to x^n are denoted by P_0^n, P_1^n and if P_1^n is absolutely continuous with respect to P_0^n, then Theorem 1 remains valid if f_n is replaced by the Radon-Nikodym derivative

$$\frac{P_1^n(dx^n)}{P_0^n(dx^n)}$$

The proof goes through as above.

Corollary 2 Under the conditions of Theorem 1, in the regular case

$$\lim_{n\to\infty} E_0[\,|f_n - f|\,] = 0 \tag{5}$$

The first part of the proof of Theorem 1 is essentially a martingale proof. Since f_n forms a martingale closed by f, we can see that (5) holds by appealing to a well-known theorem for martingales [see Doob (1953), p. 319, and Note 3.1.A].

Theorem 1 tells us how to determine f and H in (1) from finite-dimensional likelihood ratios using a limit procedure in general. To actually calculate the limit in special cases is a different story, sometimes requiring considerable ingenuity. We give some examples of how to do it in Section 2 and more follow in Chapters 5 and 6.

Before turning to examples, however, it should be stressed that with f playing the same role as the likelihood ratio in finite-dimensional statistics, many of the classical statistical procedures generalize immediately to abstract sample spaces. Of course it should be shown carefully that the old proofs do not exploit the finite-dimensional character of X, and we give two instances of how it is done, both of them straightforward generalizations of classical results.

THEOREM 2 NEYMAN-PEARSON LEMMA, NEYMAN-PEARSON (1933)

If the critical region W is of the form

$$W = \{x \mid f(x) \geq c\} \subset X \tag{6}$$

no other critical region at the same level of significance has greater power.

Proof In exactly the same way as in $\mathbb{R}^n$ we have for any $V \subset \mathbb{R}^n$, $P_0(V) = P_0(W)$,

$$P_0(W \cap V^c) = P_0(W) - P_0(W \cap V)$$
$$= P_0(V) - P_0(W \cap V) = P_0(V \cap W^c) \tag{7}$$

But using (7), we get

$$P_1(W \cap V^c) = \int_{W \cap V^c} f(x) P_0(dx) \geq c \int_{W \cap V^c} P_0(dx)$$
$$= cP_0(W \cap V^c) = cP_0(V \cap W^c) \tag{8}$$

Again using (6), but in the opposite direction,

$$P_1(V \cap W^c) = \int_{V \cap W^c} f(x) P_0(dx) \leq cP_0(V \cap W^c) \tag{9}$$

Combining (8) and (9), we find

$$P_1(W \cap V^c) \geq P_1(V \cap W^c)$$

and adding $P_1(W \cap V)$ to both sides,

$$P_1(W) \geq P_1(V)$$

Q.E.D.

Similarly we can extend the classical construction of the best Bayesian test.

THEOREM 3

If P_1 is absolutely continuous with respect to P_0 and if the a priori probabilities of the two hypotheses H_0 and H_1 are π_0 and π_1, respectively, the best test is given by the critical region

$$W = \{w \mid f(x) > \pi_0 / \pi_1\} \tag{10}$$

Remark 1 The term "best test" should here be understood as the one minimizing the probability of any error. We do not have to distinguish between errors of the first and second kinds when a Bayesian approach is used.

Proof The probability of the test leading to the wrong results is

$$\begin{aligned}\pi_0 P_0(\text{reject } H_0)+\pi_1 P_1(\text{reject } H_1)&=\pi_0 P_0(W)+\pi_1 P_1(W^c)\\&=\pi_0\int_W P_o(dx)+\pi_1\int_{W^c} f(x)P_0(dx)\\&=\pi_1+\int_W[\pi_0-\pi_1 f(x)]\,P_0(dx)\qquad(11)\end{aligned}$$

To minimize (11), we should choose W in such a way that the integral is as small as possible. This is done by choosing W as the set where the integrand is negative, in other words,

$$W=\{x\,|\,\pi_0-\pi_1 f(x)<0\}$$

and this is the same as (10). The best critical region need not be unique.

Q.E.D.

Given two probability measures P_0 and P_1, they are both absolutely continuous to the measure $Q=P_0+P_1$ so that the corresponding Radon-Nikodym derivatives exist. Introduce the functions in $L_2(X,Q)$:

$$\psi_0(x)=\sqrt{\frac{P_0(dx)}{Q(dx)}}$$

$$\psi_1(x)=\sqrt{\frac{P_1(dx)}{Q(dx)}}$$

Of course we have $\|\psi_0\|=\|\psi_1\|=1$ with $L_2(X,Q)$ norm.

If P_0 and P_1 are equivalent with $f=dP_1/dP_0$, we introduce, following Kakutani (1948),

$$\rho=\rho(P_0,P_1)=\int_X\sqrt{f(x)}\;P_0(dx)\qquad(12)$$

According to Schwarz's inequality we must have $0<\rho\leq 1$ and $\rho=1$ if and only if $f\equiv 1$ a.e., which is equivalent to $P_1=P_0$. The function ρ is symmetric in the arguments P_1 and P_0, since

$$\rho(P_1,P_0)=\int_X\sqrt{g(x)}\;P_1(dx)$$

with $g = dP_0/dP_1$, and this integral can be rewritten as

$$\int_X \frac{1}{\sqrt{f(x)}} f(x) P_0(dx) = \int_X \sqrt{f(x)}\; P_0(dx) = \rho(P_0, P_1)$$

If we introduce

$$\sigma = \sigma(P_0, P_1) = -\ln \rho(P_0, P_1)$$

we have:

(a) $0 \leqslant \sigma < \infty$.
(b) $\sigma(P_0, P_1) = \sigma(P_1, P_0)$.
(c) $\sigma(P_0, P_1) = 0$ if $P_0 = P_1$.

This is not a metric, however, since the triangle inequality is not valid.

Also,

$$\rho(P_0, P_1) = \int_X \sqrt{f(x)}\; P_0(dx) = \int_X \sqrt{\frac{P_0(dx)}{Q(dx)}} \sqrt{\frac{P_1(dx)}{Q(dx)}}\; Q(dx)$$

$$= (\psi_0, \psi_1) = \int_X \sqrt{P_0(dx) P_1(dx)} \tag{13}$$

in the inner product of $L_2(X, Q)$ and where the symbolism in the last integral is similar to that in a *Hellinger integral* [see, e.g., Hobson (1918)]. Note that the definition of ρ in (13) is independent of the choice of Q as long as both P_0 and P_1 are absolutely continuous with respect to it. Also this definition does not require P_0 and P_1 to be equivalent as we assumed earlier.

The relation (13) gives us

$$\|\psi_0 - \psi_1\|^2 = \|\psi_0\|^2 + \|\psi_1\|^2 - 2(\psi_0, \psi_1) = 2[1 - \rho(P_0, P_1)]$$

so that the probability measures absolutely continuous with respect to Q are isometrically embedded in $L_2(X, Q)$ with

$$d(P_0, P_1) = \|\psi_0 - \psi_1\| = +\sqrt{2[1 - \rho(P_0, P_1)]}$$

which metric does not depend upon the choice of Q.

If $\rho = 1-\varepsilon$, for small ε, the functions $-\ln\rho$ and $2[1-\rho]$ are both asymptotically linear in ε. This implies that σ and d^2 behave asymptotically in the same way and lead to the same topology.

It should be remarked that the probability measures $\{P_\alpha;\ \alpha\in A\}$ appearing in an inference problem should be defined in the *same sample space X and the same σ-algebra $\mathcal{Q}$ in this sample space*. This is to guarantee that, for example, an estimate $\alpha^*(x)$ be a legitimate stochastic variable for *any* α, or that a critical region W be measurable for *any* α.

2 CALCULATING RADON-NIKODYM DERIVATIVES

We now make ourselves familiar with the use of Theorem 1 of Section 1. To do this we study a series of examples, beginning with two innocuous ones, which will show up later on in several more sophisticated versions (see Note 3.2.A).

Example 1 With $T=\mathbb{N}$ and all $x(t)$ i.i.d. according to $\mathbb{N}(\alpha,1)$, consider P_1 and P_0 associated with the parameter values α_1 and α_0, $\alpha_1\neq\alpha_0$. We get with a well-known manipulation, starting with

$$g_i(x^n)=\frac{1}{(2\pi)^{n/2}}\exp-\frac{1}{2}\sum_{\nu=1}^{n}(x_\nu-\alpha_i)^2;\qquad i=0,1$$

the expression

$$f_n(x)=\exp\left[(\alpha_1-\alpha_0)\sum_1^n x_\nu+\frac{n}{2}(\alpha_0^2-\alpha_1^2)\right]$$

In this example it is intuitively obvious that the least regular situation is at hand: P_0 and P_1 perpendicular to each other. To show this, consider the distribution of f_n under P_1. We know that the exponent can be written in terms of the average $\bar{x}$:

$$n\left[(\alpha_1-\alpha_0)\bar{x}+\frac{1}{2}(\alpha_0^2-\alpha_1^2)\right]\tag{1}$$

But under P_1 the law of large numbers tells us that $\bar{x}\to\alpha_1$ in probability so that the expression in brackets in (1) tends to

$$(\alpha_1-\alpha_0)\alpha_1+\frac{1}{2}(\alpha_0^2-\alpha_1^2)=\frac{1}{2}(\alpha_1-\alpha_0)^2>0$$

so that

$$f_n(x) \to +\infty \qquad \text{in probability } P_1$$

According to Theorem 1 of Section 1, the likelihood ratio convergence result, the set $H = X$ so that P_1 is singular with respect to P_0. With the same proof we see that P_0 is singular relative to P_1. The two measures are perpendicular, as could be guessed.

Example 2 The same as Example 1, except that now all the $x(t)$ are $N(0, \alpha)$ with two values α_1 and α_0, $\alpha_1 \neq \alpha_0$. In the same way

$$f_n(x) = \left(\frac{\alpha_0}{\alpha_1}\right)^{n/2} \exp\left[\frac{1}{2}\left(\frac{1}{\alpha_0} - \frac{1}{\alpha_1}\right)\sum_1^n x^2\right] \tag{2}$$

so that

$$\ln f_n = \frac{n}{2}\left[\ln\frac{\alpha_0}{\alpha_1} + \frac{\alpha_1 - \alpha_0}{\alpha_0\alpha_1}\frac{1}{n}\sum_1^n x^2\right]$$

Under P_1 the bracketed quality tends in probability to

$$\ln\frac{\alpha_0}{\alpha_1} + \frac{(\alpha_1 - \alpha_0)\alpha_1}{\alpha_0\alpha_1} > 0 \tag{3}$$

note that the function $x - 1 - \ln x$, $x = \alpha_1/\alpha_0$, is convex with the minimum value zero at $x = 1$. Hence

$$f_n(x) \to +\infty \qquad \text{in probability } P_1$$

and the rest of the argument is obvious. Again P_1 and P_0 are perpendicular.

Now we turn to a slightly more difficult case where the answer is less intuitive.

Example 3 Let $x(t)$, $T = [0, 2\pi]$, be the Wiener process with $\alpha = \sigma^2$. How should we choose the observables in order to simplify the analysis? We do it in two ways.

First let us recall the representation

$$W(t) = \frac{1}{\sqrt{2\pi}} \sum_{\nu=-\infty}^{\infty} \frac{1 - e^{-i\nu t}}{i\nu} z_\nu$$

established in Chapter 2, Section 4, with all the z_ν i.i.d. according to $N(0,\alpha)$, and where the stochastic Fourier coefficients can be written as

$$z_\nu = \frac{1}{\sqrt{2\pi}} \int_0^{2\pi} e^{i\nu t} W(dt) \in L_2[W(t); t \in T]$$

But that means that with the observables z_ν we get back to Example 2, and we can claim immediately that P_1 and P_0, corresponding to parameter values α_1 and α_0, $\alpha_1 \neq \alpha_0$, are perpendicular.

Second, let us use the fact that the increments of $W(t)$ over disjoint intervals are independent. Recall that according to Example 1 in Chapter 1, Section 2, the sample functions are continuous a.c. For an arbitrary N introduce the observables

$$z_\nu^N = W\left(2\pi \frac{\nu+1}{N}\right) - W\left(2\pi \frac{\nu}{N}\right); \qquad \nu = 0, 1, \ldots, N-1$$

They are all stochastically independent, and $N(0, \alpha/N)$. The finite-dimensional likelihood ratio is then similar to (2), namely,

$$f_N = \left(\frac{\alpha_0}{\alpha_1}\right)^{N/2} \exp\left[\frac{N}{2}\left(\frac{1}{\alpha_0} - \frac{1}{\alpha_1}\right) \sum_1^N (z_\nu^N)^2\right]$$

Hence

$$\ln f_N = \frac{N}{2}\left[\ln \frac{\alpha_0}{\alpha_1} + \frac{\alpha_1 - \alpha_0}{\alpha_0 \alpha_1} S_N\right] \tag{4}$$

with the quadratic form

$$S_N = \sum_1^N (z_\nu^N)^2 \tag{5}$$

Under P_1 the term $(z_\nu^N)^2$ has mean α_1/N and variance $3(\alpha_1/N)^2$, using the fourth moment of a Gaussian distribution, so that

$$E[S_N] = \alpha_1$$

$$\operatorname{Var}[S_N] = 3\frac{\alpha_1^2}{N} \tag{6}$$

The moments in (6) imply that $S_N \to \alpha_1$ in probability P_1. Hence the bracket in (4) tends to the limit (3), and we arrive at the same conclusion as before (see Note 3.2.B).

It should be remarked that we have used a doubly indexed family of observables z_k^N which makes for no complication. It is enough to let N run through the values $1,2,4,8,\ldots$ so that $N=2^p$; use only the z_k^N for $k=0,1,2,\ldots,N-1$. The observables for the other N-values are determined by these and can be left out with no loss of information. The sample functions are determined by these observables since they are continuous a.c.

This has important statistical implication if we look at it as a *problem of estimation*. Let $A=\mathbb{R}^+$ and ask for an estimate $\alpha^*: C(T)\to\mathbb{R}^+$ after having observed $W(t), t\in T$. Since all the P_α measures on $C(T)$ are perpendicular to each other, *one should be able to estimate* α *with no error*. We can actually give a simple explicit construction of such an α^* using the quantity S_N used in the above derivatives.

Indeed, we know that for an arbitrary $\alpha\in A$ we have $S_N\to\alpha$ in probability P_α. If we consider the normalized process $\alpha^{-1/2}W(t)$, we know that $S_N\to 1$ in probability P_1, and we can extract a subsequence $S_{N_1}, S_{N_2},\ldots$ conveying to 1 a.c. P_1. But all the processes $\alpha^{-1/2}$ $W(t)$, for different $\alpha\in\mathbb{R}^+$, differ only by a multiplicative constant. Hence $S_{N_1}, S_{N_2},\ldots$ converge a.c. P_α, for all α, to the constant α, and *the estimate*

$$\alpha^* = \lim_{i\to\infty} S_{N_i}$$

is perfect: $P_\alpha(\alpha^*=\alpha)=1$ for all α. Actually the selection of a subsequence is not needed since the S_N sequence itself converges a.c. [see, e.g., Doob (1953), p. 395].

This is related to the notion of *consistent estimation*. A sequence $\{\alpha_n^*; n\in\mathbb{N}\}$ is said to be consistent if $\alpha_n^*\to\alpha$, for all α, in probability P_α. In Chapter 4 this definition is modified somewhat. Classically n means sample length, so that consistency is then an asymptotic property. In infinite-dimensional sample spaces we can, however, find perfect estimators sometimes already for fixed length of the sample function.

Similarly in Example 1 a perfect estimator is given by the law of large numbers:

$$\alpha^* = \lim_{n\to\infty} \frac{1}{n}\sum_1^n x_k$$

and in Example 2 by

$$\alpha^* = \lim_{n\to\infty} \frac{1}{n}\sum_1^n x_k^2$$

We return to consistency in more general situations in Section 3.

Example 4 Consider the stationary Markovian and Gaussian process in Example 3 of Chapter 2, Section 2, with constant mean and covariance function $r(t)=\sigma^2/(2\beta)e^{-\beta|t|}$. Then

$$|r(t,t+h)-r(t,t)|=\frac{\sigma^2}{2\beta}\left[1-e^{-\beta|h|}\right]=0(h)$$

uniformly in t, so that Example 1 in Chapter 1, Section 2, guarantees that there exists a version with continuous sample functions. We could then choose the observables as $x(t_k)$ with the t everywhere dense in $T=(0, L)$.

An elegant way of calculating the Radon-Nikodym derivative exploits the fact that the stochastic variables

$$x(0),\, y_\nu=x(t_{\nu+1})-\rho_\nu x(t_\nu); \qquad \nu=0,1,2,\dots,n-1$$

are orthogonal, where we have rearranged momentarily the t in their natural increasing order and $\rho_\nu=e^{-\beta(t_{\nu+1}-t_\nu)}$. To see the orthogonality, calculate for $h>0$

$$\operatorname{Cov}[y_{\nu+h}, y_\nu]=\frac{\sigma^2}{2\beta}\left[e^{-\beta(t_{\nu+h+1}-t_{\nu+1})}-\rho_\nu e^{-\beta(t_{\nu+h+1}-t_\nu)}-\rho_{\nu+h}e^{-\beta(t_{\nu+h}-t_{\nu+1})}\right.$$
$$\left.-\rho_{\nu+h}\rho_\nu e^{-\beta(t_{\nu+h}-t_\nu)}\right]$$

which is zero for $h>0$, and for $h=0$ we get the variance

$$\operatorname{Var}(y_\nu)=\frac{\sigma^2}{2\beta}(1-\rho_\nu^2)$$

In the same way it is seen that $x(0)$ is orthogonal to all the y_k and has, of course, the variance $\sigma^2/2\beta$. Hence the finite-dimensional frequency function is

$$g_n(x^n)=\text{constant}\cdot\exp\left[-\frac{1}{2}\left\{\frac{2\beta}{\sigma^2}[x(0)-\alpha]^2+\frac{2\beta}{\sigma^2}\sum_{\nu=0}^{n-1}\frac{[y_\nu-(1-\rho_\nu)\alpha]^2}{1-\rho_\nu^2}\right\}\right]$$

where the constant does not depend upon α. Then the finite-dimensional likelihood ratio for testing P_α against P_0 satisfies

$$\ln f_n(x)=\frac{2\beta}{\sigma^2}\left[\alpha x(0)-\frac{\alpha^2}{2}+\alpha\sum_{\nu=0}^{n-1}\frac{y_\nu}{1+\rho_\nu}-\frac{\alpha^2}{2}\sum_{\nu=0}^{n-1}\frac{1-\rho_\nu}{1+\rho_\nu}\right] \tag{7}$$

Rewriting (7) we get

$$\ln f_n(x) = \frac{2\beta}{\sigma^2}\left[\alpha x(0) - \frac{\alpha^2}{2} + \alpha \sum_{\nu=1}^{n-1} \frac{x_\nu}{1+\rho_{\nu-1}} - \alpha \sum_{\nu=1}^{n-1} \frac{\rho_\nu x_\nu}{1+\rho_\nu} \right.$$
$$\left. + \alpha \frac{x_n}{1+\rho_{n-1}} - \alpha \frac{\rho_0 x_0}{1+\rho_0} - \frac{\alpha^2}{2} \sum_{\nu=0}^{n-1} \frac{1-\rho_\nu}{1+\rho_\nu} \right] \tag{8}$$

If $\Delta = \max(t_{\nu+1} - t_\nu)$ is made small, the right-hand side of (8) can be expressed as

$$\frac{2\beta}{\sigma^2}\left\{ \alpha x(0) - \frac{\alpha^2}{2} + \alpha \frac{x(L)}{1+\rho_{n-1}} - \alpha \frac{\rho_0 x(t_0)}{1+\rho_0} \right.$$
$$+ \frac{\alpha}{2} \sum_{\nu=1}^{n-1} x(t_\nu)\left[\frac{\beta}{2}(t_{\nu+1} - t_\nu) + \frac{\beta}{2}(t_\nu - t_{\nu-1}) + 0(\Delta^2) \right]$$
$$\left. - \frac{\alpha^2}{2} \sum_{\nu=0}^{n-1} \beta \frac{t_{\nu+1} - t_\nu + 0(\Delta^2)}{2} \right\}$$

Since the sample functions are continuous a.c. and since $\rho_\nu \to 1$, this converges to (see Note 3.2.C)

$$\ln f(x) = \frac{2\beta}{\sigma^2}\left[\frac{1}{2}\alpha x(0) + \frac{1}{2}\alpha x(L) - \frac{\alpha^2}{2} + \frac{\alpha\beta}{2}\int_0^L x(t)\,dt - \frac{1}{4}\alpha^2 \beta L \right] \tag{9}$$

Hence the likelihood ratio convergence theorem tells us that the P_α measures are equivalent, and we can use (9) for constructing tests and estimates. More about this later.

Note that $f(x)$ depends only upon the fairly simple statistic

$$x(0) + x(L) + \beta \int_0^L x(t)\,dt \tag{10}$$

Hence (10) is a sufficient statistic for the family $\{P_\alpha\}$.

With the same process, but with mean value function identically zero, a natural question is to test the hypothesis that $\beta = \beta_0$ against an alternative $\beta = \beta_1$. This is studied much more generally in Chapter 5.

Example 5 Let us consider the process from the example after Definition 2 in Chapter 1, Section 1, but with ordinates $N(\alpha, \sigma^2)$ whose sample functions are piecewise constant so that we have the sample space $X = PC(T)$.

To get the Radon-Nikodym derivative we remark that we can use the observables $t_1, t_2, \ldots, t_n, y_0, y_1, \ldots, y_n$ and get the finite-dimensional frequency function for $T=(0, L)$:

$$g_n(x^n)=\lambda^n e^{-\lambda L} \prod_{\nu=0}^{n} \frac{1}{\sqrt{2\pi}\,\sigma} e^{-(1/2\sigma^2)(y_\nu-\alpha)^2}$$

so that the likelihood ratio for testing P_α against P_0 is

$$f_n(x)=\exp\left[\frac{\alpha}{\sigma^2}\sum_0^n y_\nu - \frac{n+1}{2\sigma^2}\alpha^2\right] \tag{11}$$

But this is valid for any n, so that the limit $f(x)$ can actually be written as (11), where n is a random variable and the likelihood ratio convergence theorem tells us that we have the Radon-Nikodym derivative for the equivalent P_α measures. For example,

$$\frac{P_{\alpha_1}(dx)}{P_{\alpha_0}(dx)}=\exp\left[\frac{\alpha_1-\alpha_0}{2}\sum_0^n y_\nu + \frac{n+1}{2\sigma^2}(\alpha_0^2-\alpha_1^2)\right]$$

Hence the optimal test of H_{α_1} against H_{α_0} will be based on the test statistic Σy_ν. This is not a linear functional in $L_2(T)$, it only looks linear.

Note that this is not a Gaussian process. The one-dimensional marginals of $x(t)$ have a Gaussian distribution, but higher dimensional marginal distributions are *mixtures of Gaussian measures*.

Example 6 Let us now look at an interesting modification of the last process. Say that we make the sample functions *continuous linear splines* with the knots t_ν again generated by a homogeneous Poisson process over $\mathbb{R}$. Let the ordinates $y_\nu=x(t_\nu)$ at the knots be i.i.d. $N(\alpha, \sigma^2)$ and define the rest of the sample function by linear interpolation between contiguous knots. This is a stationary process. It is not Gaussian; not even the one-dimensional marginal distributions are Gaussian. Almost all sample functions are differentiable a.e. with

$$x'(t)=\frac{y_{\nu+1}-y_\nu}{t_{\nu+1}-t_\nu} \tag{12}$$

where $t\in(t_\nu, t_{\nu+1})$ and $y_{\nu+1}$ and y_ν are the ordinates at $t_{\nu+1}$ and t_ν.

We meet a somewhat unexpected difficulty due to the fact that the process is not differentiable in the mean. The derivative given in (12), exists both sample functionwise with probability 1 and as a limit in probability, but not as a limit in the mean. Actually the quantity does not have a variance. This sensitive distinction requires a deeper analysis than may be expected. Therefore we must exercise some caution. It would not be legitimate, although tempting, to base the last step in the following derivation on relation (3) in Section 2 of Chapter 2. It is valid only for second-order processes differentiable in the mean. Our process $x'(t)$ will turn out to have a well-defined covariance function only for nonzero lag.

To calculate the covariance function $r(0, h)=r(h)$, $h>0$, we separate cases:

I No $t_\nu \in [0, h]$.
II One $t_\nu \in [0, h]$.
III More than one $t_\nu \in [0, h]$.

In case I, which is illustrated in Figure 1(a) and which occurs with probability $e^{-\lambda h}$, the stochastic variables u, v are independent, each with frequency function $\lambda e^{-\lambda x}$. But in case I, for $h>0$,

$$z'(0)=x'(h)=\frac{y_2-y_1}{u+v+h}$$

so that

$$E[x'(0)x'(h)|\mathrm{I}, u, v]=\frac{2\sigma^2}{(u+v+h)^2} \tag{13}$$

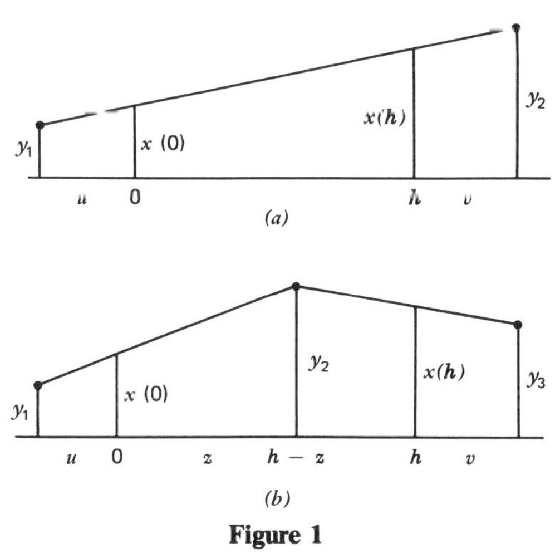

Figure 1

Integrating over u, v, (13) yields

$$E[x'(0)x'(h)|\mathrm{I}] = 2\sigma^2\lambda^2 \int_{u=0}^{\infty}\int_{v=0}^{\infty} e^{-\lambda(u+v)} \frac{du\,dv}{(u+v+h)^2} \tag{14}$$

or

$$2\sigma^2\lambda^2 \int_{x=0}^{\infty} e^{-\lambda h x} \frac{x}{(x+1)^2}\,dx$$

Case II [see Figure 1(b)] occurs with probability $\lambda h e^{-\lambda h}$. Conditioned by the event II, the point z, where the middle knot occurs, has a rectangular distribution $R(0, h)$. Also

$$x'(0) = \frac{y_2 - y_1}{u+z}$$

$$x'(h) = \frac{y_3 - y_2}{h - z + v}$$

Hence, conditioned by event II and the values u, v, and z,

$$E[x'(0)x'(h)|\mathrm{II}, u, v, z] = -\frac{\sigma^2}{(u+z)(h-z+v)} \tag{15}$$

Integrating (15) over z, we get

$$\begin{aligned} E[x'(0)x'(h)|\mathrm{II}, u, v] &= -\frac{\sigma^2}{h}\int_{z=0}^{h} \frac{\sigma^2}{(u+z)(h-z+v)} \\ &= -\frac{\sigma^2}{h(u+v+h)} \ln \frac{(u+h)(v+h)}{uv} \end{aligned}$$

and then over h, v, after some manipulations, yields

$$\begin{aligned} &E[x'(0)x'(h)|\mathrm{II}] \\ &\quad = -\frac{\sigma^2\lambda^2}{h}\int_{h=0}^{\infty}\int_{v=0}^{\infty} e^{-\lambda(u+v)} \frac{1}{u+v+h} \ln \frac{(u+h)(v+h)}{uv}\,du\,dv \\ &\quad = -2\sigma^2\lambda^2 \int_{x=0}^{\infty} e^{-\lambda h x} \frac{1}{x+1}\left[x \ln \frac{x+1}{x} + \ln(x+1)\right] dx \end{aligned} \tag{16}$$

Under case III, finally, the conditional expectation of $x'(0)x'(h)$ is zero, since $x'(0)$ and $x'(h)$ will then be conditionally independent. Hence this will not contribute to the covariance.

Combining (14) with (16) we get, for $y(t)=x'(t)$, $h>0$,

$$\rho(h)=E[y(0)y(h)]=2\sigma^2\lambda^2\int_0^\infty e^{-\lambda hx}K_h(x)\,dx \tag{17}$$

with

$$K_h(x)=e^{-\lambda h}\frac{x}{(x+1)^2}-\frac{\lambda he^{-\lambda h}}{x+1}\left[x\ln\frac{x+1}{x}+\ln(x+1)\right]$$

We can rewrite (17) into the more convenient form

$$\rho(h)=2\sigma^2\lambda^2\int_{u=1}^\infty e^{-\lambda hu}[A(u)-hB(u)]\,du;\qquad h>0 \tag{18}$$

with

$$A(u)=\frac{u-1}{u^2}$$
$$B(u)=\frac{\lambda}{u}\left[(u-1)\ln\frac{u}{u-1}+\ln u\right]$$

Integrating the B-term in (18) by parts yields, noting that $B(1)=0$,

$$\rho(h)=2\sigma^2\lambda^2\int_{u=1}^\infty e^{-\lambda hu}\left[A(u)-\frac{1}{\lambda}B'(u)\right]du;\qquad h>0 \tag{19}$$

Note that the integral in (19) does not converge for $h=0$: the variance of $x'(t)$ does not exist.

To calculate the covariance function $r(h)$ of the $x(t)$ process consider the four time points $a<0<h<b$ and the expression

$$[x(0)-x(a)][x(b)-x(h)]=\int_a^0 x'(t)\,dt\int_h^b x'(t)\,dt \tag{20}$$

The expression in (20) has the expected value

$$r(b)-r(h)-r(b-a)+r(h-a)=\int_{s=a}^0\int_{t=h}^b \rho(t-s)\,ds\,dt \tag{21}$$

Now let $a\to-\infty$, $b\to\infty$, and use the fact that $r(+\infty)=0$, which can be verified directly without difficulty. Then (21) tends to the limit, using (19),

$$-r(h)=\lim_{\substack{a\to-\infty\\ b\to+\infty}} 2\sigma^2\lambda^2\int_{s=a}^0\int_{t=h}^b\int_{u=1}^\infty e^{-\lambda(t-s)u}\left[A(u)-\frac{1}{\lambda}B'(u)\right]du$$

which can be evaluated as

$$2\sigma^2\int_{u=1}^{\infty} e^{-\lambda hu}\frac{A(u)-\dfrac{1}{\lambda}B'(u)}{u^2}\,du \tag{22}$$

Introducing the function

$$C(u)=\frac{1}{u^2}\left[\frac{B'(u)}{\lambda-A(u)}\right]=\frac{1}{u^4}\left[1-u-\ln(u-1)\right]$$

we get the desired result

$$r(h)=\int_{u=1}^{\infty} e^{-\lambda|h|u}C(u)\,du$$

which we have proved for $h>0$. Since r must be an even function, it is valid also for negative values, and since r can easily be shown to be continuous, (22) is seen to be valid for $h=0$ too (see Note 3.2.D).

Note that $C(u)$ has a logarithmic singularity at $u=1$, it is integrable, and it changes sign on $[1,\infty)$.

If we observe this process plus an unknown constant α on $T=(0, L)$, the P_α-measures will be equivalent. To test the hypothesis P_{α_1} against P_{α_0}, we could proceed as in Example 5, but using as observables the quantities $n=$number of knots, $x(0)$, $x(L)$, the abscissas t_ν, and the ordinates $x(t_\nu)$ at the knots in $(0, L)$. Note that again the number of observables is not fixed, it is a random variable n. For fixed n the two conditional measures in $\mathbb{R}^{2n+2}$ are equivalent, and it is clear that the likelihood ratio is exactly the Radon-Nikodym derivative (use the same reasoning as in Example 5). Hence the measures are equivalent.

Let us now examine two point processes on $T=(0, L)$ (see Note 3.2.E).

***Example** 7* Under the hypothesis H_i, $i=0,1$, let the P_i measure define an inhomogeneous Poisson process with intensity $\lambda_i(t)$. We only assume that the $\lambda_i(t)$ are positive a.e. and integrable over T. Our sample space X can be chosen as the family of all finite subsets of T. Other possibilities have been mentioned in Chapter 1.

A convenient choice of observables is to use $n=$number of points, and $t_\nu=$abscissa of the νth point, $\nu=1,2,\dots,n$. Then the finite-dimensional frequency function for fixed n is simply

$$\prod_{\nu=1}^{n}\lambda_i(t_\nu)\left[\int_T\lambda_i(t)\,dt\right]^{-n} \tag{23}$$

under H_i. But

$$P_i(n=k)=\frac{1}{k!}\left[\int_T \lambda_i(t)\,dt\right]^k \exp\left[-\int_T \lambda_i(t)\,dt\right] \tag{24}$$

Combining (24) with (23) we see that the P_i-measures are equivalent with the Radon-Nikodym derivative

$$\frac{P_1(dx)}{P_0(dx)}=\prod_{\nu=1}^{n}\frac{\lambda_1(t_\nu)}{\lambda_0(t_\nu)}\exp\left\{\int_T [\lambda_0(t)-\lambda_1(t)]\,dt\right\} \tag{25}$$

An alternative way of deriving (24), which can easily be extended to a general set T, is as follows. Pick a sequence of numbers $N_1, N_2, \dots,$ for example, as powers of 2. For any N in the sequence introduce the observables, for $T=(0,1)$ say,

$$x_k^{(N)}=\#\left[t_\nu' s \mid t_\nu \in \left(\frac{k-1}{N},\frac{k}{N}\right)\right]; \qquad k=1,2,\dots,N \tag{26}$$

It is clear that the knowledge of all $x_k^{(N)}$ determines the sample.

But for given N the N stochastic variables in (26) are independent of each other and $x_k^{(N)}$ is a Poisson variable with mean

$$m_{ki}^{(N)}=\int_{(k-1)/N}^{k/N}\lambda_i(t)\,dt$$

Hence the finite-dimensional measures are equivalent with the likelihood ratio

$$f_n(x)-\prod_{k=1}^{N}\left\{\left(\frac{m_{k1}^{(N)}}{m_{k0}^{(N)}}\right)^{x_k^{(N)}}\exp\left(m_{k0}^{(N)}-m_{k1}^{(N)}\right)\right\} \tag{27}$$

As N tends to infinity, all $x_k^{(N)}$ but a finite number will tend to 0, the rest will tend to 1. Using Lebesgue's theorem,

$$\lim_{\varepsilon\downarrow 0}\frac{1}{2\varepsilon}\int_{t-\varepsilon}^{t+\varepsilon}\lambda(s)\,ds=\lambda(t) \qquad \text{a.e.}$$

we see that (27) tends a.c. to (25), and the likelihood ratio convergence theorem gives us the desired result.

Once we have the Radon-Nikodym derivative, we can of course go ahead to various problems of inference. For example, the most powerful test of P_0 against P_1 is given directly by (27), together with Neyman-Pearson's lemmas, through the critical region

$$W=\left\{x \mid \prod_{\nu=1}^{n} \frac{\lambda_1(t_\nu)}{\lambda_0(t_\nu)} \geqslant k\right\} \tag{28}$$

In (28) the constant k should be chosen to make the significance level the desired one.

Example 8 We now turn to the *Pólya process* [see, e.g., Feller (1957), p. 434] with parameter β, and on the argument space $T=(0, L)$. If n events have occurred up to time t, the conditional intensity of another one is given by

$$P_n(t)=\frac{1+\beta n}{1+\beta t}$$

where β is the *contagion parameter*, $\beta \geqslant 0$. Let us test $\beta=0$, a Poisson process, against $\beta>0$. Let the observables be $n, t_1, t_2, \ldots, t_n$ as before, with the t_i ordered in increasing order. We get by the same sort of reasoning that we employed earlier for $n>0$ and $\beta>0$

$$g_\beta=\prod_{\nu=0}^{n-1} \frac{1+\beta\nu}{1+\beta t_{\nu+1}} \exp\left\{-\int_0^{t_1} \frac{dt}{1+\beta t}-\int_{t_1}^{t_2} \frac{1+\beta}{1+\beta t} dt-\cdots-\int_{t_n}^{L} \frac{1+n\beta}{1+\beta t} dt\right\}$$

or

$$\begin{aligned}
\ln g_\beta= & -\frac{1}{\beta} \ln(1+\beta t_1)-\frac{1+\beta}{\beta} \ln \frac{1+\beta t_2}{1+\beta t_1}-\cdots-\frac{1+n\beta}{\beta} \ln \frac{1+\beta L}{1+\beta t_n} \\
& +\sum_{\nu=0}^{n-1} \ln \frac{1+\beta\nu}{1+\beta t_{\nu+1}}=-\frac{1}{\beta} \ln(1+\beta L) \\
& -\sum_{\nu=1}^{n} \nu \ln \frac{1+\beta t_{\nu+1}}{1+\beta t_\nu}+\sum_{\nu=0}^{n-1} \ln \frac{1+\beta\nu}{1+\beta t_{\nu+1}}
\end{aligned}$$

with the convention $t_{n+1}=L$. Reducing the second sum, this becomes

$$\begin{aligned}
\ln g_\beta & =-\frac{1}{\beta} \ln(1+\beta L)+\sum_{\nu=1}^{n} \ln(1+\beta t_\nu)-n \ln(1+\beta L)+\sum_{\nu=0}^{n-1} \ln \frac{1+\beta\nu}{1+\beta t_{\nu+1}} \\
& =-\frac{1}{\beta} \ln(1+\beta L)+\sum_{\nu=0}^{n-1} \ln(1+\beta\nu)-n \ln(1+\beta L)
\end{aligned}$$

For $n=0$ we have simply

$$\ln g_\beta = -\frac{1}{\beta}\ln(1+\beta L)$$

For $\beta=0$ we have of course just $g_0 = e^{-L}$.

In either case the likelihood ratio convergence theorem is applicable, and the Radon-Nikodym derivative will be

$$f(x) = \frac{P_\beta(dx)}{P_0(dx)} = \frac{c(\beta)}{(1+\beta L)^n}\prod_{\nu=0}^{n-1}(1+\beta\nu)$$

where $c(\beta)$ depends upon β but not upon the sample, and the product should be evaluated as 1 when $n=0$.

The most powerful test will have the form

$$w = \left\{ x \,\Big|\, \sum_{\nu=0}^{n-1} \ln\frac{1+\beta\nu}{1+\beta L} \geqslant c \right\} \tag{29}$$

But

$$\ln\frac{1+\beta\nu}{1+\beta L} \begin{matrix} <0; & \text{if } \nu<L \\ \geqslant 0 & \text{if } \nu\geqslant L \end{matrix}$$

so that the partial sums in (29) decrease to begin with, for $\nu<L$, and then increase. Hence the critical region can be expressed as

$$W = \{x \mid n \leqslant n_1 \text{ or } n \geqslant n_2\}$$

This is intuitively reasonable since

$$\operatorname{Var}_\beta(n) = L(1+\beta L) > L = \operatorname{Var}_0(n)$$

so that positive β leads to greater variability in n than for $\beta=0$: the Poisson process. Also note that the test does not depend upon the time points t_ν of the events, only upon their total number.

Example 9 Consider the real stationary Gaussian process $x(t)$ with period p on $T=\mathbb{R}$. In complex form it can be written (see Example 1 of Chapter 2, Section 4),

$$x(t) = m + \sum_{\nu=-\infty}^{\infty} e^{2\pi i\nu t/p} z_\nu \tag{30}$$

where z_ν are Gaussian independent stochastic variables with mean zero and variance σ_ν^2 such that $\Sigma\sigma_\nu^2<\infty$. This corresponds to the bilinear representation of the covariance function

$$r_0(s,t)=r_0(s-t)=\sum_{\nu=-\infty}^{\infty}\sigma_\nu^2 e^{2\pi i\nu(s-t)/p} \tag{31}$$

Let us test H_0: $m=0$ against H_m, $m>0$. A good choice of observables is to use the Fourier coefficients

$$x_k=\frac{1}{p}\int_0^p e^{-2\pi i\nu t/p}x(t)\,dt=\begin{cases} m+z_0; & \text{for } k=0\\ z_k; & \text{otherwise}\end{cases}$$

But any z_k has the same probability measure under both hypotheses, so that when we form the finite-dimensional likelihood ratio, only x_0 will occur, and the likelihood ratio convergence theorem yields

$$f(x)=\lim_{n\to\infty}f_n(x)=e^{[\text{Re}(x_0 m)-(1/2)|m|^2]/\sigma_0^2}$$

where we have used the standard definition for a complex Gaussian stochastic variable (see Example 5 in Chapter 1, Section 1). The most powerful test will therefore be, for real positive m,

$$W=\left\{x\,\Big|\int_0^p x(t)\,dt\geqslant \text{constant}\right\}$$

If $x(t)$ is still periodic and Gaussian, but not stationary, the expansion in (30) still holds. The x_k are not longer independent, however, which complicates the analysis. Say, for example, that we observe $y(t)=x(t-\alpha)$, where the shift α is the unknown parameter. Are the corresponding P_α measures perpendicular?

We cannot answer this question by the same method, since we have lost independence for the observables. In Chapter 5 we have more to say about the general problem of testing the mean value function of Gaussian processes.

Let us now consider the same process as in (30), but with identically vanishing $m=m(t)$ and a covariance function as in (31), except for an unknown *covariance factor* α, $r(t)=\alpha r_0(t)$. Repeating the same argument, we see that now the finite-dimensional argument becomes

$$f_n(x)=\exp\left\{\frac{1}{2}\left(1-\frac{1}{\alpha}\right)\sum_{k=-n}^{n}\frac{|x_k|^2}{\sigma_k^2}\right\} \tag{32}$$

Let us say $\alpha>1$. Note that $|x_k|/\sigma_k$ all have the same distribution and are stochastically independent so that the sum in (32) tends to $+\infty$ in probability. *The two measures are perpendicular to each other* (see Note 3.2.F). A more direct way of showing this could be based on the statistic

$$\frac{1}{2n+1}\sum_{\nu=-n}^{n}\frac{|x_k|^2}{\sigma_k^2}\to\alpha \text{ in probability as } n\to\infty$$

If we modified Example 9 to a periodic Gaussian (not stationary), the expansion in (30) could be tried again. Even if it converges, it is not very useful since the z_ν would no longer be independent, and (32) would not hold. We do not yet have the analytical tools for handling this dilemma.

Example 10 With $T=S_2$, the sphere in 3-space, let the observed process be $x(t)=\alpha+y(t)$, where $y(t)$ is real, continuous in the mean, Gaussian, and isotropic in S_2. The latter means that its covariances are invariant with respect to the group of rotations.

We can then express $y(t)$ into the normalized spherical harmonics $Y_\nu^m(t)$, $\nu=0,1,2,\ldots$, $m=-\nu,\ldots,\nu$,

$$y(t)=\sum_{\nu=0}^{\infty}\sum_{m=-\nu}^{\nu}z_{\nu m}Y_\nu^m(t),\qquad t\in S_2 \tag{33}$$

The *Oboukhov expansion* in (33) converges in the mean for any t, and the isotropic property implies

$$(z_{\nu\mu},z_{\nu'\mu'})=\delta_{\nu\nu'}\delta_{\mu\mu'}f_\nu$$

where the spectral density $f_\nu\geqslant 0$ (see Note 3.2.G). The corresponding covariance function can be written as

$$r(t,t')=(y(t),y(t'))=\sum_{\nu=0}^{\infty}f_\nu\sum_{m=-\nu}^{\nu}Y_\nu^m(t)Y_\nu^m(t')$$

or, in terms of the Legendre polynomials P_ν of degree ν and the angle γ between t and t',

$$r(\gamma)=\sum_{\nu=0}^{\infty}\frac{2\nu+1}{4\pi}f_\nu P_\nu(\cos\gamma) \tag{34}$$

Multiply (34) by $P_k(\cos\gamma)$ and integrate with respect to t' over S_2, we get, using the orthogonality of the Legendre polynomials and the well-known norming constants

$$\int_{S_2} [P_\nu(\cos\gamma)]^2 \, dt = \frac{4\pi}{2\nu+1} \tag{35}$$

the Fourier coefficients

$$f_\nu = \int_{S_2} r(\gamma) P_\nu(\cos\gamma) \, dt \tag{36}$$

In (35) and (36) dt stands for surface area.

Note that when we apply the likelihood ratio convergence theorem, the finite-dimensional likelihood ratios will not depend upon any z_ν^m except for $\nu = m = 0$. Hence, as before,

$$f(x) = \frac{P_\alpha(dx)}{P_0(dx)} = \exp\left[\frac{a\sqrt{\pi}}{f_0}\int_{S_2} x(t)\, dt - \frac{2\pi^2\alpha^2}{f_0}\right]$$

Tests and estimates need therefore only depend upon the average of $x(t)$ over the sphere. All the P_α-measures are equivalent when $f_0 \neq 0$.

Example 11 Let $x(t) = \alpha + y(t)$, where $y(t)$ is the random phase model in real form (see Chapter 2, Section 1, Example 1) so that

$$y(t) = \sum_{\nu=1}^{\infty} A_\nu \cos(\lambda_\nu t + \phi_\nu)$$

with all ϕ_ν independent, and $R(0, 2\pi)$, A_ν, and λ_ν are known positive constants. We ask, of course, that $\Sigma A_\nu < \infty$; then $y \in C[(-\infty, \infty)]$.

If $T = \mathbb{R}$, it is intuitively clear that the P_α-measures are perpendicular to each other. To show this consider

$$\alpha^* = \frac{1}{2L}\int_{-L}^{L} x(t)\, dt = \alpha + \sum_{\nu=1}^{\infty} A_\nu \cos\phi_\nu \frac{\sin\lambda_\nu L}{\lambda_\nu L}$$

Hence

$$E[(\alpha^* - \alpha)^2] = \sum_{\nu=1}^{\infty} A_\nu \left(\frac{\sin\lambda_\nu L}{\lambda_\nu L}\right)^2 E(\cos^2\phi_\nu)$$

$$= \frac{1}{2}\sum_{\nu=1}^{\infty} A_\nu \left(\frac{\sin\lambda_\nu L}{\lambda_\nu L}\right)^2$$

Since $\Sigma A_\nu < \infty$ and since the quadratic factors are bounded by 1 and tend to 0 as $L \to \infty$, it follows that $\alpha^* \to \alpha$ in the mean, hence in probability, as $L \to \infty$. Perfect estimation of the parameter is possible if $T = \mathbb{R}$ so that the P_α-measures are perpendicular.

The case of a finite interval $T = [0, L]$ is harder to settle. The reader can verify that the measures can still be perpendicular, for example, when $\lambda_\nu = (2\pi\nu)/L$. We return to estimation problems of this type from the point of view of linear methods in Chapter 4, where a complete answer is given.

Example 12 Let A consist of *bounded closed convex sets* in the plane, as described in Example 2 of Chapter 1, Section 3, and consider the random sets $x = \alpha + y$, where y is a *random set* in A, and we use Minkowsky addition. Here sample space and parameter space coincide, $X = A$. The random element y will be assumed to have a known probability distribution and have a radius of curvature given by the stochastic process

$$r_y(\phi) = \sum{}' A_\nu[1 + \cos(\nu\phi - \phi_\nu)]$$

where the summation is over all integers different from 1. The angles ϕ_ν are again i.i.d. from $R(0, 2\pi)$ and the A_ν nonnegative constants such that $\{A_\nu\}$ is summable. For given constants a, b the support function of y is given as

$$s_y(\phi) = a_y \cos\phi + b_y \sin\phi + \int_0^\phi \sin(\phi - \psi) r_y(\psi)\, d\psi$$

say, with constant a_y and b_y, known to the observer.

For two given sets α_1 and α_0, are the probability measures P_{α_1} and P_{α_0} for $x = \alpha + y$ equivalent? We know that the α can be expressed uniquely in terms of their support functions

$$s_\alpha(\phi) = a_\alpha \cos\phi + b_\alpha \sin\phi + \int_0^\phi \sin(\phi - \psi) R_\alpha(d_\psi)$$

where R_α are bounded measures on $[0, 2\pi)$, satisfying the orthogonality relations

$$\int_0^{2\pi} \cos\phi R_\alpha(d\phi) = \int_0^{2\pi} \sin\phi R_\alpha(d\phi) = 0; \qquad \alpha = \alpha_1, \alpha_2$$

Observing the set $x \subset X = A$ is equivalent to observing its support function $s_x = s_\alpha + s_y$ which can also be written as

$$r_x(\phi) = a_x \cos\phi + b_x \sin\phi + \int_0^\phi \sin(\phi - \psi) R_x(d\psi)$$

Introduce

$$\xi_1 = a_x$$
$$\eta_1 = b_x$$
$$\xi_\nu = \frac{1}{2\pi}\int_0^{2\pi} \cos \nu\phi R_x(d\phi); \qquad \nu = 0, 2, 3, \ldots$$
$$\eta_\nu = \frac{1}{2\pi}\int_0^{2\pi} \sin \nu\phi R_x(d\phi); \qquad \nu = 2, 3, \ldots$$

Then

$$\xi_1 = a_\alpha + a_y$$
$$\eta_1 = b_\alpha + b_y$$
$$\xi_\nu = \frac{A_\nu}{2}\cos \nu\phi_\nu + \frac{1}{2\pi}\int_0^{2\pi} \cos \nu\phi R_\alpha(d\phi); \qquad \nu = 0, 2, 3, \ldots$$
$$\eta_\nu = \frac{A_\nu}{2}\sin \nu\phi_\nu + \frac{1}{2\pi}\int_0^{2\pi} \sin \nu\phi_\nu R_\alpha(d\phi); \qquad \nu = 2, 3, \ldots$$

Since a_y and b_y were assumed constant and known, we get perfect estimation if either $a_{\alpha_1} \neq a_{\alpha_0}$ or $b_{\alpha_1} \neq b_{\alpha_0}$. On the other hand, if we exclude this case, we need only consider the set $\{\xi_\nu, \eta_\nu;\ \nu = 2, 3, \ldots\}$ which has independent elements. Assume now that for some $\nu \geqslant 2$,

$$c_\nu^1 = \frac{1}{2\pi}\int_0^{2\pi} \cos \nu\phi R_{\alpha_1}(d\phi) \neq c_\nu^0 = \frac{1}{2\pi}\int_0^{2\pi} \cos \nu\phi_\nu R_{\alpha_0}(d\phi)$$

The sine coefficients should be handled in the same way, but we leave this out here. Then ξ_ν has a continuous and positive frequency function on the open interval $I_1 = (-A_\nu/2 + c_\nu^1,\ A_\nu/2 + c_\nu^1)$ under P_{α_1} and on $I_0 = (-A_\nu/2 + c_\nu^0,\ A_\nu/2 + c_\nu^0)$ under P_{α_0}. Since these intervals do not coincide,

$$P_1(I_1 \cap I_0^c) > 0; \qquad P_1(I_1 \cap I_0^c) = 0$$
$$P_0(I_0 \cap I_0^c) > 0; \qquad P_1(I_0 \cap I_1^c) = 0$$

Hence the measures are not equivalent.

We can claim more if

$$\sum_{\nu=1}^{\infty} A_\nu^{-2}(c_\nu^1 - c_\nu^0)^2 = +\infty \tag{37}$$

namely, that the P measures are perpendicular. To show this, introduce the statistic

$$s_n = \sum_{\nu=1}^{\infty}\left(\xi_\nu - \frac{1}{2}c_\nu^0 - \frac{1}{2}c_\nu^1\right)(c_\nu^1 - c_\nu^0)A_\nu^{-2}$$

We have

$$m_n^0 = E_0(s_n) = -\frac{1}{2}\sum_{\nu=1}^{\infty}(c'_\nu - c_\nu^0)^2 A_\nu^{-2}$$

$$m_n^1 = E_1(s_n) = +\frac{1}{2}\sum_{\nu=1}^{\infty}(c_\nu^1 - c_\nu^0)A_\nu^{-2} \tag{38}$$

and

$$\mathrm{Var}_0(s_n) = \mathrm{Var}_1(s_n) = \frac{1}{8}\sum_{\nu=1}^{\infty}A_\nu^{-2}(c_\nu^1 - c_\nu^0)^2 \tag{39}$$

Combining (38) with (39) we see that

$$\frac{s_n}{\sum_{\nu=1}^{n}A_\nu^{-2}(c_\nu^1 - c_\nu^0)^2} \to \begin{cases} -\frac{1}{2} & \text{in probability } P_0 \\ +\frac{1}{2} & \text{in probability } P_1 \end{cases}$$

because of (37). Hence perfect estimation is possible and *the measures are perpendicular*. The two given sets α_1 and α_0 can be distinguished with no error after their deformed shape $x = \alpha + y$ has been observed.

3 PERPENDICULAR MEASURES AND CONSISTENT ESTIMATION

In classical inference we are not used to encounter perpendicular measures. When they occur, they obviously present a pathological exception that can be disposed with immediately. For stochastic processes with a continuous time argument we have seen that the measures can be perpendicular quite unexpectedly, and we meet even more surprising cases later on.

If P_1 and P_2 are perpendicular, what does this signify on a more practical level? Slepian has argued that this should make us doubt the applicability of the model [see Slepian (1958)]. It is not the estimate or test that is perfect, it is *the model that lacks realism, it is too specific*. Instead of having a simple null hypothesis, we should choose a composite one with nuisance parameters that represent a lack of precise knowledge about the random phenomena we study. For composite hypotheses we are less likely to encounter the perpendicular case. Or we should incorporate measurement errors in a realistic way into the model.

If we accept this argument, we must of course still verify whether the measures are perpendicular or not, and in particular calculate the Radon-Nikodym derivative when they are equivalent.

Consistency in classical statistics means that the estimate α^*_n tends to the true value α in probability as the sample size tends to infinity. This is related to perpendicularity if we consider infinite, rather than potentially infinite, samples with $T=\mathbb{N}$.

A better way of expressing this is to say that $\alpha^*=\alpha^*(x)$ *is a consistent estimate of* α, *with the observed stochastic element* $x=x(t)$, $t\in T$, *if* $P_\alpha[\alpha^*(x)=\alpha]=1$ *for all* $\alpha\in A$. When T is $\mathbb{N}$, this statement can of course be expressed by the asymptotic relation as above, but in the present context the latter way of formulating the definition is more concise.

For stochastic processes with continuous time we meet the same problem. Say that we observe a weakly stationary stochastic process $x(t)$, continuous in the mean, and with mean α, $t\in T=(0,L)$. If we form the *time average*

$$\bar{x}_L=\frac{1}{L}\int_0^L x(t)\,dt$$

we know that it is an unbiased estimate of α and with variance

$$\operatorname{Var}(\bar{x}_L)=\frac{1}{L^2}\int_0^L\int_0^L r(s-t)\,ds\,dt=\frac{1}{L}\int_{-L}^{L} r(h)\left(1-\frac{|h|}{L}\right)dh \qquad (1)$$

Is $\bar{x}_L$ consistent? This can be answered by studying the limit of (1) as $L\to\infty$. Another way is by appealing to *the mean ergodic theorem of von Neumann*, which tells us that $\bar{x}_L$ converges to a limit in the mean.

We can be more specific. Express the process in the Cramér representation (2.4.24). Then, integrating over $(0, L)$ we get

$$\bar{x}_L=\int_{-\infty}^{\infty} K_L(\lambda)Z(d\lambda)$$

where

$$K_L(\lambda)=\frac{e^{iL\lambda}-1}{iL\lambda}$$

This function is bounded uniformly, $|K_L(\lambda)|\leqslant 1$, and $K(0)=1$. When L tends to infinity, it tends to zero everywhere except at $\lambda=0$. Hence $K_L\to K_\infty$ in the mean with respect to the F-measure associated with the Z-process and where

$$K_\infty(\lambda)=\begin{cases}1; & \text{for } \lambda=0\\ 0; & \text{otherwise}\end{cases}$$

This isometry theorem from Chapter 2, Section 3, then tells us that the limit in the mean of $\bar{x}_L$, as L tends to infinity, is the jump of the Z-process at frequency zero

$$\lim_{L\to\infty}\|\bar{x}_L-[Z(+0)-Z(-0)]\|=0$$

We also know that the variance of $\Delta z(0)$ is equal to $\Delta F(0)$, the jump of the spectral zero. In other words, *the time average is a consistent estimate of the mean if and only if* the spectrum is *continuous at zero*, contains no dc component.

While this is useful to know, it is not general enough, since it only tells us something about a particular parameter, the mean, and a particular estimate, $\bar{x}$, of it. To probe deeper into the consistency question, we must consider more general parameters.

Say that our parameter α can be expressed as a continuous function $\alpha = h(g_1, g_2, \ldots, g_p)$ of quantities g_ν that themselves are expected values of functions of the stochastic process at certain time points. For example, we may have

$$\begin{aligned} g_1 &= E_\alpha[x^2(t)] \\ g_2 &= E_\alpha[x(t)x(t+h)]; \qquad h>0 \\ g_3 &= E_\alpha[|x(t+h)-x(t)|]; \qquad h>0 \\ &\vdots \end{aligned} \tag{2}$$

If we can find consistent estimators $g_1^*, \ldots, g_p^*$ of $g_1, \ldots, g_p$, then $h(g_1^*, g_2^*, \ldots, g_p^*)$ is a consistent estimator of α.

We cannot deal with this problem using second-order properties only since we have allowed nonlinear functions in the definition of the g [see (2)]. Instead we now assume that $x(t)$ is strictly stationary, so that we can appeal to the *individual ergodic theorem* [see Hopf (1937), pp. 49–56]. It is then natural to estimate the g by time averages approximating ensemble averages:

$$\begin{aligned} g_1^* &= \frac{1}{L}\int_0^L x^2(t)\,dt \\ g_2^* &= \frac{1}{L}\int_0^{L-h} x(t)x(t+h)\,dt \\ g_3^* &= \frac{1}{L}\int_0^{L-h} |x(t+h)-x(t)|\,dt \\ &\vdots \end{aligned} \tag{3}$$

Of course we must ask that $x(t)$ be measurable and have existing moments such that the integrals in (3) exist. The individual ergodic theorem then guarantees that the averages converge with probability 1 to some limiting stochastic variables. If, moreover, $x(t)$ is ergodic, then all these stochastic variables are equal to the expected values in (2) a.c., so that we can estimate α consistently.

The question of consistent estimation therefore leads us to study ergodicity, and we mention a few results here. In the Gaussian case the following beautiful and complete answer is known (see Note 3.3.A).

THEOREM 1 ERGODICITY OF GAUSSIAN PROCESS, MARUYAMA (1949)

In order that a real and stationary Gaussian process with continuous covariance function be ergodic, it is necessary and sufficient that its spectrum is continuous.

Proof Writing the covariance function in its Bochner representation, see Theorem 5 in Chapter 2, Section 4,

$$r(t)=\int_{-\infty}^{\infty} e^{it\lambda}F(d\lambda)$$

we can assume without loss of generality that the mean is zero, also and that $r(0)=1$ so that F is an ordinary distribution function. The F-measure is symmetric since the process, and hence r, is real.

To show that the condition stated above is necessary is easy; we just consider the strictly stationary process $y(t)=x^2(t)$. The covariance function is

$$\rho(t)=E\{[x^2(s+t)-Ex^2(s+t)][x^2(s)-Ex^2(s)]\}=2r^2(t) \tag{4}$$

using the fourth-order moments for Gaussian distributions. But (4) implies that the spectral distribution F_y of ρ is twice the convolution of F with itself. We know that

$$\frac{1}{L}\int_0^L y(t)\,dt \to 1+\Delta Z_y(0) \qquad \text{with variance } \Delta F_y(0) \tag{5}$$

If $x(t)$ is ergodic, the limit in (5) should be a.c. constant, $\Delta F_y(0)=0$. But then Fourier inversion gives

$$0=\Delta F_y(0)=\lim_{L\to\infty}\frac{1}{2L}\int_{-L}^{L}\rho(t)\,dt=\lim_{L\to\infty}\frac{1}{L}\int_{-L}^{L}r^2(t)\,dt$$

The last limit is well known from classical Fourier analysis [see, e.g., Bocher (1932), p. 81]. It is twice the jump at $\lambda=0$ of F_y. But for the convolution F_y of F with itself

$$\Delta F_y(0) = \sum_\nu [\Delta F(\lambda_\nu)]^2 \tag{6}$$

summed over all points of discontinuity of F and using the fact that the F-measure is symmetric around zero. Hence (6) must be zero, which says that F must be continuous everywhere.

To prove sufficiency we recall that ergodicity is equivalent to metric transitivity. The latter means that the only sets in $\mathscr{Q}$ that are invariant with respect to translations of the time axis are the trivial ones having P measure 0 or 1. Let us assume for an indirect proof that there is an invariant set S with $P(S)=\rho$, $0<\rho<1$. We prove that F cannot be continuous.

For any given $\varepsilon>0$ we can approximate S by a cylinder I with finite-dimensional base B such that

$$\begin{aligned} &P(I)<\rho+\varepsilon \\ &P(S\cap I^c)<\varepsilon \end{aligned} \tag{7}$$

Let the time points appearing in the conditions in B be $\tau_1, \tau_2, \ldots, \tau_n$, and introduce the translate $T_t I = I_t$. Then I_t has a base associated with the time points $\tau_1+t, \tau_2+t, \ldots, \tau_n+t$. Consider now the $2n$ stochastic variables $x_i = x(\tau_i)$, $x_{n+i}=x(t+\tau_i)$, $i=1,2,\ldots,n$. They have a Gaussian probability distribution with the covariance matrix in block form:

$$\Lambda(t)=\left\{\begin{array}{cc|cc} 1 & r(\tau_1-\tau_2)\ldots & r(\tau_1-\tau_1-t) & \ldots \\ \ldots & & \ldots & \\ r(\tau_n-\tau_1) & \ldots & & \\ \hline r(\tau_1+t-\tau_1) & \ldots & 1\ r(\tau_1-\tau_2) & \ldots \\ \ldots & & \ldots & \\ r(\tau_n-t-\tau_1) & \ldots & r(\tau_n-\tau_1)\ldots & 1 \end{array}\right\}$$

$$=\left\{\begin{array}{cc} A & B(t) \\ B(t) & A \end{array}\right\}$$

Let us remind the reader that if a stationary process has a continuous spectrum, then all its finite-dimensional covariance matrices are nonsingular (see Lemma 1 in Chapter 2, Section 4).

We now complete the proof of the theorem. The covariance matrices are nonsingular for large t-values when the sets $\{t_\nu\}$ and $\{t+t_\nu\}$ do not overlap. Hence we can write

$$P(I\cap I_t)=(2\pi)^{-n}[\det\Lambda(t)]^{-1/2}\int\cdots\int\exp\left[-\frac{1}{2}Q(x)\right]dx_1dx_2\cdots dx_{2n} \tag{8}$$

integrated over $(x_1,\ldots,x_n)\in I$, $(x_{n+1},\ldots,x_{2n})\in I_t$, and where

$$Q(x)=x^T\Lambda^{-1}(t)x$$

Let us arrange all the numbers $\tau_\nu-\tau_\mu$ as $t_1,t_2,\ldots,t_N$. Direct evaluation gives

$$\frac{1}{2L}\int_{-L}^{L}\sum_1^N|r(t_k+t)|^2\,dt$$

$$=\frac{1}{2L}\int_{-L}^{L}\sum_1^N\int_{\infty}^{\infty}\int_{\infty}^{\infty}e^{i(t_k+t)\lambda-i(t_k+t)\mu}F(d\lambda)F(d\mu)\,dt$$

$$=\sum_1^N\int_{-\infty}^{\infty}\int_{-\infty}^{\infty}\frac{e^{i(t_k+L)(\lambda-\mu)}-e^{i(t_k-L)(\lambda-\mu)}}{2Li(\lambda-\mu)}F(d\lambda)F(d\mu) \tag{9}$$

Let L tend to infinity on the right-hand side in (9). For each k the integrand is bounded by 1 in absolute value and tends to 0 everywhere except on the diagonal $\lambda=\mu$ in the $(\lambda-\mu)$-plane. Note that $F\times F$ is a bounded measure and apply Lebesgue's bounded convergence theorem. The limit will be the mass of the product measure $F\times F$ on the diagonal. Since F is continuous, this mass is 0.

This implies that

$$\underline{\lim}_{t\to\infty}\sum_1^N|r(t_k+t)|^2=0$$

so that there exists a sequence $s_\nu\to+\infty$ such that the matrix introduced above $B(s_\nu)\to 0$. In other words,

$$\Lambda(s_\nu)\to\begin{Bmatrix}A & 0\\ 0 & A\end{Bmatrix} \tag{10}$$

where the limiting matrix is also nonsingular. Using (10) in (8) and again

Lebesgue's theorem on bounded convergence, we get

$$\lim_{\nu\to\infty} P(I\cap I_{s_\nu})=P(I)^2<(\rho+\varepsilon)^2 \tag{11}$$

For large values of ν (11) implies that with $\varepsilon>0$,

$$\begin{aligned}(\rho+\varepsilon)^2 &> P(I\cap I_{s_\nu})\geqslant P(S\cap I\cap I_{s_\nu})\\ &\geqslant P(S)-P(S\cap I^c)-P(S\cap I^c_{s_\nu})>\rho-2\varepsilon\end{aligned} \tag{12}$$

Here we have used the invariance of S and the stationarity of $x(t)$ to get

$$P(S\cap I^c_{s_\nu})=P\left[(I^c_{s_\nu}S)\cap I^c\right)=P(S\cap I^c)<\varepsilon$$

But (12) is only possible for arbitrarily small ε if $\rho=0$ or 1, against the assumption. Q.E.D.

For a stationary Gaussian process with continuous spectrum we can apply the estimation construction in (2) and (3) to arrive at consistent estimators. Actually, we need not limit ourselves to the case when α^* is a function of a finite number of g's. This can be extended to cover situations when α^* can be *approximated* by such functions, but we shall not go into this here.

A simple but very useful statement can be made concerning ergodicity when the process has been generated by a sequence of i.i.d. variables.

THEOREM 2

Let $\{\xi_t, t\in\mathbb{Z}\}$ form an i.i.d. sequence. Then ξ_t is ergodic.

Proof It is very easy to verify that if S is an invariant subset of X, it must have measure 0 or 1. Indeed we can approximate it by a cylinder set I as in the proof of Theorem 1 [see (7)]. If $t_1<t_2<\cdots<t_n$ are the time points associated with I, it is clear that the events I and $I_t=T^tI$ are stochastically independent as soon as $t>t_n-t_1$. Therefore (12) holds with the same implication, $P(S)=1$ or 0 (see Note 3.3.B). Q.E.D.

Example 1 The *linear processes* are defined as moving averages of i.i.d. variables with mean zero, $E[\xi_t]=0$,

$$x(t)=\sum_{\nu=-\infty}^{\infty} c_\nu\xi_{t-\nu}=f(T^t\xi) \tag{13}$$

where $\xi=(\ldots,\xi_{-1},\xi_0,\xi_1,\ldots)$. To make (13) converge a.c., we assume that

$E[\xi_t^2]<\infty$ and $\{c_\nu\}\in l_2$ [see Doob (1953), p. 108]. Such a process is automatically stationary in the strict sense. Because of Theorem 2, linear processes are ergodic so that we can estimate, for example, the covariances $r(t)$ consistently for $t\geqslant 0$ by

$$r^*(t)=\frac{1}{L}\sum_{s=1}^{L-t} x(s)x(s+t)$$

Example 2 Consider a real-valued and stationary Gaussian process with absolutely continuous spectrum. According to Theorem 6 of Section 2 it can be expressed as a moving average in uncorrelated terms. Since the terms have Gaussian distributions, uncorrelated implies independent, so that the process is linear. Then Example 1 tells us that it is ergodic, in agreement with Theorem 1.

A recurring theme in the examples of the last section was the selection of observables so that they be independent under H_0. If they are *also* independent under H_1, this results in considerable simplification of the analysis. In this case we can apply the following elegant result.

THEOREM 3 EQUIVALENCE OF PRODUCT MEASURES, KAKUTANI (1948)

Consider the two product measures $P_0=P_0^1\times P_0^2\times P_0^3\times\cdots$ *and* $P_1=P_1^1\times P_1^2\times P_1^3\times\cdots$ *defined on the same product space* $X=X_1\times X_2\times X_3\times\cdots$ *and such that* P_0^n *is equivalent with* P_1^n *for all* n. *Then* P_0 *and* P_1 *are equivalent if and only if* [*with* ρ *defined in* (1.12)],

$$\prod_{n=1}^{\infty}\rho(P_0^n, P_1^n)>0 \tag{14}$$

Proof Introduce the Radon-Nikodym derivatives

$$f^n(x)=\frac{P_1^n(dx)}{P_0^n(dx)}; \qquad \rho_n=\rho(P_0^n, P_1^n)$$

and consider

$$\prod_{n=1}^{N}\rho_n=\int_X\sqrt{R_N(x)}\,P_0(dx); \qquad R_N(x)=\prod_{n=1}^{N}f^n(x)$$

Fatou's lemma guarantees that the limit $f(x)=\lim_{N\to\infty}R_N(x)$, which exists a.c. P_0 [see the proof of necessity in Theorem 1 of Section 3), satisfies

$$E_0\left[\sqrt{R_1(x)}\,\right] \leqslant \prod_1^{\infty} \rho_n$$

If the product (14) is zero, then $R_1(x)=0$ a.c. P_0, so that P_1 and P_0 are perpendicular.

Now assume that the product does not vanish. We show that R_N converges in $L_1(X, P_0)$ norm. To do this let us consider Cauchy differences and apply Schwarz's inequality:

$$\{E_0[|R_N - R_M|]\}^2 \leqslant E_0[|R_N^{1/2} - R_M^{1/2}|^2]\,E_0[|R_N^{1/2} + R_M^{1/2}|^2] \tag{15}$$

However, if $M > N$, using the definition of R_N,

$$E_0[|R_N^{1/2} - R_M^{1/2}|^2] = E_0\left[\left(1 - \prod_{N+1}^{M} \sqrt{f^n}\right)^2 R_N\right]$$

Expanding the square we get

$$E_0\left[R_N + R_M - 2\prod_{N+1}^{M} \sqrt{f^n}\, R_N\right] = 2\left(1 - \prod_{N+1}^{M} \rho_n\right) \to 0$$

as M and $N \to \infty$. But using $(x+y)^2 \leqslant 2(x^2+y^2)$, we get

$$E_o[|R_N^{1/2} + R_M^{1/2}|^2] \leqslant 2E_o(R_N + R_M) = 4$$

so that appealing to (15) we see that R_N is a Cauchy sequence in $L_2(X, P_0)$ and

$$\int_X f(x) P_0(dx) = 1$$

Using Theorem 1 of Section 1, it follows that P_1 is absolutely continuous with respect to P_0. Q.E.D.

The question of how to choose the observables so that they are independent under both hypotheses, to be able to apply Kakutani's result, is not easy to answer in general, and we return to it several times in Chapters 4–6. Here we just look at some special cases where this question does not have to be answered.

Example 3 In several of the examples in Section 2 we exploited second-order properties. It is tempting to look at some situation when moments of order 1 and 2 do not exist.

Consider the stochastic process $x(t), t\in\mathbb{N}$, where all the $x(t)$ are independent with Cauchy distributions. Under H_0 we assume that they have the marginal frequency function

$$\frac{1}{\pi a_\nu}\frac{1}{1+\left(\dfrac{x(\nu)}{a_\nu}\right)^2} \tag{16}$$

with the scale parameter a_ν which plays a role similar to that of σ_ν in Example 2 in Section 2. Under H_1 we assume the same form as in (16), except that a_ν is replaced by b_ν.

For two Cauchy distributions with scale parameters a and b, respectively, we calculate ρ in (1.13) with $X=\mathbb{R}$ as the integral:

$$\rho(a,b)=\frac{1}{\pi\sqrt{ab}}\int_{-\infty}^{\infty}\frac{dx}{\sqrt{\left[1+\left(\dfrac{x}{a}\right)^2\right]\left[1+\left(\dfrac{x}{b}\right)^2\right]}}$$

$$=\frac{1}{\pi}\sqrt{c}\int_{-\infty}^{\infty}\frac{dx}{\sqrt{(1+x^2)(1+c^2x^2)}};\qquad c=a/b$$

To decide whether P_0 and P_1 are equivalent or not we have to study the convergence of the infinite product in Theorem 3. It is clear that the sensitive case is when $a_\nu\sim b_\nu$, as $\nu\to\infty$ making discrimination difficult, and we examine this case. In the opposite case the measures are always perpendicular.

If we write $c=1-\varepsilon$, we have

$$\rho(a,b)=\frac{\sqrt{1-\varepsilon}}{\pi}\int_{-\infty}^{\infty}\frac{dx}{(1+x^2)\sqrt{1-\varepsilon(2-\varepsilon)\dfrac{x^2}{1+x^2}}}$$

Noting that $x^2/(1+x^2)$ is bounded by 1, we can expand the square root for small ε and get

$$\rho(a,b)=\frac{1}{\pi}\left(1-\frac{1}{2}\varepsilon-\frac{1}{8}\varepsilon^2\right)\int_{-\infty}^{\infty}\frac{dx}{1+x^2}$$

$$\times\left[1+\frac{\varepsilon(2-\varepsilon)}{2}\frac{x^2}{1+x^2}+\frac{3\varepsilon^2(2-\varepsilon)^2}{8}\frac{x^4}{(1+x^2)^2}\right]+o(\varepsilon^2). \tag{17}$$

Since the linear term in ε disappears, this reduces to

$$\rho(a,b)=1-\varepsilon^2\left(\frac{5}{8}-\frac{3}{2\pi}\int_{-\infty}^{\infty}\frac{x^4}{(1+x^2)^3}dx\right)+o(\varepsilon^2) \tag{18}$$

The integral can be written as a beta integral. Evaluating it, (18) reduces to

$$\rho(a,b)=1-\frac{1}{16}\varepsilon^2+o(\varepsilon^2)$$

This means that the infinite product will diverge to zero if

$$\sum_{\nu=1}^{\infty}\varepsilon_\nu^2=\sum_{\nu=1}^{\infty}\left(\frac{b_\nu-a_\nu}{b_\nu}\right)^2=+\infty \tag{19}$$

Hence perfect discrimination between the two hypotheses is possible if and only if the above sum diverges. The occurrence of the exponent 2 in (19) is somewhat surprising since the stable law of exponent 1 could be expected to lead to the first power of ε; it is caused by the cancellation of the linear terms in ε in (17).

To use Kakutani's theorem more generally than in Example 3 when all the one-dimensional distributions are the same except for differing location or scale parameters, we develop two limit relations for the functionals that appear in (1.13), generalizing Example 3 considerably.

First, consider a frequency function f with $f'/\sqrt{f}\in L_2$ and introduce the functional

$$I(\varepsilon)=\int_{-\infty}^{\infty}\sqrt{f(x)f(x-\varepsilon)}\,dx$$

With $\phi(x)=\sqrt{f(x)}\subset L_2$ we use the L_2 Fourier transform

$$\hat{\phi}(z)=\frac{1}{\sqrt{2\pi}}\int_{-\infty}^{\infty}e^{izx}\phi(x)\,dx$$

Plancherel's theorem tells us that

$$I(\varepsilon)=\int_{-\infty}^{\infty}\phi(x)\phi(x-\varepsilon)\,dx=\int_{-\infty}^{\infty}|\hat{\phi}(z)|^2e^{iz\varepsilon}\,dz$$

so that

$$I(\varepsilon)-1=\int_{-\infty}^{\infty}\psi(z)[e^{iz\varepsilon}-1]\,dx;\qquad \psi(z)=|\hat{\phi}(z)|^2$$

since the L_2-norm of ϕ is equal to 1.

Note that ψ is in L_1 and

$$\int_{-\infty}^{\infty} z^2\psi(z)\,dz = \int_{-\infty}^{\infty} |z\hat{\phi}(z)|^2\,dz = \int_{-\infty}^{\infty} [\phi'(x)]^2\,dx$$
$$= \frac{1}{4}\int_{-\infty}^{\infty} \frac{[f'(x)]^2}{f(x)}\,dx$$

exists due to the assumption $f'/\sqrt{f} \in L_2$. Also, since ϕ is real and therefore $|\hat{\phi}|$ even,

$$\int_{-\infty}^{\infty} z|\hat{\phi}(z)|^2\,dz = 0$$

Using well-known properties for derivatives of L_1 Fourier transforms [see, e.g., Feller (1966), sec. XV.4], this tells us that

$$\lim_{\varepsilon\to 0}\frac{1}{\varepsilon^2}[1-I(\varepsilon)] = \frac{1}{2}\int_{-\infty}^{\infty} z^2\psi(z)\,dz = \frac{1}{8}\int_{-\infty}^{\infty}\frac{[f'(x)]^2}{f(x)}\,dx \tag{20}$$

so that

$$I(\varepsilon) = 1 - c\varepsilon^2 + 0(\varepsilon^2); \qquad c>0$$

Consider now the measures $P_\alpha^{(i)}$ as the product measure with marginal distributions having the frequency function $f(x-a_\nu^{(i)})$, $\nu=1,2,\ldots,\ i=1,2,$ and $a_\nu^{(2)}-a_\nu^{(1)}=\varepsilon_\nu\to 0$. Here $a_\nu^{(i)}$ plays the role of location parameter. The product appearing in Kakutani's theorem then converges or diverges together with

$$\prod_{\nu=1}^{\infty}\left[1-c\left(a_\nu^{(2)}-a_\nu^{(1)}\right)^2\right]$$

which is the same thing as convergence of divergence, respectively, for

$$\sum_{\nu=1}^{\infty}\left(a_\nu^2-a_\nu^{(1)}\right)^2 \tag{21}$$

so that we have an appealing result due to Shepp (1965).

THEOREM 4

For two product measures $P^{(1)}$ and $P^{(2)}$ with marginal measures given by the *frequency functions* $f(x-a_\nu^{(i)})$, *where* $a_\nu^{(1)}-a_\nu^{(2)}=\varepsilon_\nu\to 0$ *and* $f'/\sqrt{f}\in L_2$, *we have* $P^{(1)}$ *equivalent to* P^2 *if and only if* (21) *converges.*

A companion theorem deals with marginal measures given by frequency functions with different scale parameters:

$$\frac{1}{a_\nu^{(i)}}f\left(\frac{x}{a_\nu^{(i)}}\right); \qquad x\geqslant 0,\ \nu=1,2,\dots,\ i=1,2 \tag{22}$$

and with $a_\nu^{(1)}\sim a_\nu^{(2)}$.

We now have to study the functional

$$J(\varepsilon)\int_0^\infty \sqrt{f(x)f(xa)a}\,dx$$

With the substitution $xf(x)=g(y)$, $\ln x=y$, we have

$$J(\varepsilon)=\int_{-\infty}^\infty \sqrt{g(y)g(y+b)}\,dy; \qquad b=\ln a$$

so that the previous theorem is applicable and we have

$$\left[\frac{g'(y)}{g(y)}\right]^2=\left[1+e^y\frac{f'(e^y)}{f(e^y)}\right]^2$$

so that

$$\lim_{\varepsilon\to 0}\varepsilon^2[1-J(\varepsilon)]-\frac{1}{8}\int_{-\infty}^\infty\left[1+e^y\frac{f'(e^y)}{f(e^y)}\right]^2 e^y f(e^y)\,dy$$

$$=\frac{1}{8}\int_0^\infty\left[1+x\frac{f'(x)}{f(x)}\right]^2 f(x)\,dx$$

Integrating by parts this limit reduces to

$$\lim_{\varepsilon\to 0}\frac{1}{\varepsilon^2}[1-J(\varepsilon)]=\frac{1}{8}\left[\int_0^\infty \frac{x^2[f'(x)]^2}{f(x)}dx-1\right]$$

Proceeding in the same way as in the previous proof we get the following answer to our question about equivalence/perpendicularity. It is also due to Shepp (1965).

THEOREM 5

If $xf'/\sqrt{f} \in L_2$, the two product measures $P^{(1)}$ and $P^{(2)}$ with marginals given in (22) *are equivalent if and only if*

$$\sum_{\nu=1}^{\infty}\left(\frac{a_\nu^{(2)}-a_\nu^{(1)}}{a_\nu^{(1)}}\right)^2<\infty$$

Remark The assertions in these two theorems do not remain true if we leave out the regularity condition. In the first theorem, for example, if we do not have $f'/\sqrt{f} \in L_2$, then $I(\varepsilon)$ can behave differently from (20). For example, if f corresponds to $R(0,1)$, then

$$I(\varepsilon)=1-|\varepsilon|; \qquad |\varepsilon| \leqslant 1$$

so that (20) does not hold. In this special case the terms in (21) should instead be replaced by $|a_\nu^{(2)}-a_\nu^{(1)}|$.

CHAPTER 4

Linear Inference

1 THE FINITE-DIMENSIONAL SITUATION

We now study inference of the linear part, the mean value function $m(t)$, of the observed stochastic process. The *linear hypothesis H for stochastic processes* expressed the mean value function as a linear combination of given functions $\psi_1, \psi_2, \ldots, \psi_p$:

$$m(t) = \sum_{\nu=1}^{p} \alpha_\nu \psi_\nu(t); \qquad t \in T \tag{1}$$

where the *regression coefficients* α_ν are real or complex unknown constants. A (sub-) hypothesis restricts the α_ν linearly:

$$\alpha = \mathrm{col}(\alpha_1, \alpha_2, \ldots, \alpha_p) \in H_0$$

where H_0 is a subspace of $\mathbb{R}^p$ or $\mathbb{U}^p$. Typically we want to test H_0 against the rest of H, estimate α, or construct confidence regions for α. Here we concentrate on estimation problems, but most of the results can be applied to other sorts of inference.

To rule out degenerate cases, it will be assumed that $\{\psi_\nu\}$ is a *linearly independent* system over T.

The reduced process $y(t) = x(t) - m(t)$ has naturally its mean value function identically zero.

Consider an estimate $\alpha^* = \alpha^*(x) = (\alpha_1^*(x), \alpha_2^*(x), \ldots, \alpha_p^*(x)), \alpha_\nu^* \in L_2(x)$. Just as in the classical case, α^* is said to be *unbiased* if

$$E_\alpha[\alpha^*] \equiv \alpha$$

which vector relation should be interpreted componentwise.

To compare two unbiased estimates α_1^*, α_2^* we use their covariance matrices $M_1 = \mathrm{cov}(\alpha_1^*)$, $M_2 = \mathrm{cov}(\alpha_2^*)$ and say that α_1^* is at least as good as α_2^* if $M_1 < M_2$. Here the order relation $M_1 < M_2$ means that $M_2 - M_1$ is a nonnegative definite matrix.

Definition 1 An estimate α^* which is unbiased and at least as good as any other linear unbiased estimate is called a best linear unbiased estimate (BLUE) or the Markov estimate.

A reader who is familiar with linear estimation in finite-dimensional spaces could skip the rest of this section and go directly to Section 2.

In the classical situation $T = \{1, 2, \ldots, n\}$ one can find the BLUEs as follows for $p = 1$. We denote the covariance matrix of the process by R.

THEOREM 1 MARKOV'S THEOREM

If $\psi \in range(R) = \mathcal{R}$ a BLUE is given by

$$\alpha^* = \frac{\psi^T R_0^{-1} x}{\psi^T R_0^{-1} \psi}$$

where R_0 is the restriction of R to $\mathcal{R}$, with

$$\mathrm{Var}(\alpha^*) = \left(\psi^T R_0^{-1} \psi\right)^{-1}$$

If R is nonsingular, the coefficients of the BLUE are unique. If $\psi \notin \mathcal{R}$, the parameter is estimated consistently by

$$\alpha^* = \frac{\psi_0^T x}{\|\psi_0\|^2}$$

where $\psi_0 = P_{\mathcal{N}} \psi$, and $\mathcal{N}$ is the null space of R.

Proof To estimate $\alpha = \alpha_1$ consider a linear combination:

$$\alpha^* = \sum_{t=1}^{n} c_t x(t) = c^T x; \qquad c = \mathrm{col}[c(1), c(2), \ldots, c(n)] \tag{2}$$

In order for (2) to be unbiased we must have

$$\sum_{t=1}^{n} c_t \psi(t) = 1 = c^T \psi; \qquad \psi = \mathrm{col}[\psi(1), \psi(2), \ldots, \psi(n)] \tag{3}$$

On the other hand the variance is

$$V = \mathrm{Var}_\alpha(\alpha^*) = \mathrm{Var}(\alpha^*) = c^T R c; \qquad R = \mathrm{Cov}(x) \tag{4}$$

To minimize (4) we use a Lagrangian multiplier λ, and differentiate $\frac{1}{2}V - \lambda c^T\psi$ with respect to c:

$$Rc - \lambda\psi = 0 \tag{5}$$

If R is nonsingular, we can solve (5) and get the unique solution

$$c = \frac{R^{-1}\psi}{\psi^T R^{-1}\psi}$$

$$\alpha^* = \frac{\psi^T R^{-1} x}{\psi^T R^{-1}\psi}$$

$$\min \mathrm{Var} = (\psi^T R^{-1}\psi)^{-1} \tag{6}$$

On the other hand, *if R is singular* with the null space $\mathfrak{N}$ and the range $\mathfrak{R} = \mathbb{R}^n - \mathfrak{N}$, it follows that the support of the probability measure of y is in $\mathfrak{R}$ (see Note 4.1.A). When we consider estimates as in (2), we need therefore only consider c-vectors in $\mathfrak{R}$.

Say now that $\psi \in \mathfrak{R}$. Then we can go through the construction in (2)–(6) with the only modification that we restrict ourselves to the subspace $\mathfrak{R}$ so that R in (6) should be replaced by its restriction to $\mathfrak{R}$.

Finally if $\psi \notin \mathfrak{R}$, introduce

$$\psi_0 = P_{\mathfrak{N}}\psi \neq 0$$

and form the estimate

$$\alpha^* = \frac{\psi_0' x}{\|\psi_0\|^2}$$

We have

$$E_\alpha(\alpha^*) = \frac{\psi_0^T\psi}{\|\psi_0\|^2}\alpha = \alpha$$

since $\psi - \psi_0 \perp \psi_0$. Hence α^* is unbiased. Its variance

$$\mathrm{Var}(\alpha^*) = \frac{\psi_0^T R \psi_0}{\|\psi_0\|^4} = 0$$

since R annihilates ψ_0. Hence $P_\alpha(\alpha^*=\alpha)=1$ for all α, so that α^* is consistent; see Chapter 3, Section 3, for the definition of consistency. Q.E.D.

Remark 1 The BLUE is always unique viewed as a stochastic variable (although the form of it need not be unique viewed as a linear combination). To see this let α^* be a BLUE. Then for arbitrary ε

$$\bar{\alpha}=\alpha^*+\varepsilon[\psi(t_2)x(t_1)-\psi(t_1)x(t_2)]; \qquad t_1,t_2\in T$$

is also unbiased, now with the variance

$$\text{Var}(\bar{\alpha})=\text{Var}(\alpha^*)+2\varepsilon E\{\alpha^*[\psi(t_2)x(t_1)-\psi(t_1)x(t_2)]\}+0(\varepsilon); \qquad \varepsilon\to 0 \tag{7}$$

with real scalars and slightly modified in the complex case. In order that $\text{Var}(\bar{\alpha})\geqslant\text{Var}(\alpha^*)$ for all ε, as is required for α^* to be a BLUE, the coefficient of ε in (7) must vanish so that

$$\psi(t_2)E[\alpha^*x(t_1)]=\psi(t_1)E[\alpha^*x(t_2)]$$

Select a value t_2 for which $\psi(t_2)\neq 0$ and let $t_1=t$ be arbitrary in T, and we get the *normal equations*

$$E[\alpha^*x(t)]\equiv\text{constant}\,\psi(t)$$

where the

$$\text{constant}=\frac{1}{\psi(t_2)}E[\alpha^*x(t_2)] \tag{8}$$

But the identity (8) uniquely determines the stochastic variable α^*.

Theorem 1 immediately extends to $p>1$. In order that the vector valued estimate α^* be a BLUE it is necessary and sufficient that for any fixed vector $z\in\mathbb{R}^p$ the scalar estimate $z^T\alpha^*$ is a BLUE of the parameter $z^T\alpha$. Indeed, that $z^T\alpha^*$ is unbiased follows from α^* being unbiased. The variance is

$$\text{Var}(z^Ta^*)=z^T\text{Cov}(\alpha^*)z$$

so that if this is always minimum, it follows that α^* is as good as possible under the partial ordering of matrices introduced above. The converse conclusion is obvious.

Writing

$$x=\psi\alpha+y; \qquad y=\mathrm{col}[\,y(1), y(2),\ldots, y(n)\,] \tag{9}$$

and

$$\Psi=\begin{bmatrix} \psi_1(1) & \cdots & \psi_p(1) \\ \vdots & & \vdots \\ \psi_1(n) & & \psi_p(n) \end{bmatrix}$$

we now seek a vector valued estimate $\alpha^*=C^Tx$ with some unspecified coefficients $C=\{c_{t\nu};\ t=1,2,\ldots,n;\ \nu=1,2,\ldots,p\}$.

THEOREM 2

If all $\psi_\nu(\cdot)\in\mathcal{R}$, a BLUE is given by the matrix vector expression

$$\alpha^*=(\Psi^TR_0^{-1}\Psi)^{-1}\Psi^TR_0^{-1}x \tag{10}$$

with the covariance matrix

$$\mathrm{Cov}(\alpha^*)=(\Psi^TR_0^{-1}\Psi)^{-1} \tag{11}$$

If R is nonsingular, the form of this BLUE is unique. For singular R the parameter α can be consistently estimated if $\{P_{\mathfrak{N}}\psi_\nu\}$ are linearly independent.

Proof First let us show that $M=\Psi^TR_0^{-1}\Psi$ exists and is nonsingular when all $\psi_\nu(\cdot)\in\mathcal{R}$. Let $z=(z_1, z_2,\ldots, z_p)$ be arbitrary and consider the vector in $\mathbb{R}^n$

$$u=\Psi z=\sum_{\nu=1}^{p}\psi_\nu(\cdot)z_\nu\in\mathcal{R}$$

so that M is well defined. Since the $\psi_\nu(\cdot)$ were assumed to be linearly independent, we have $u\neq 0$. Hence

$$z^TMz=z^T\Psi^TR_0^{-1}\Psi z=u^TR_0^{-1}u>0$$

so that M is nonsingular and the expression in (10) makes sense.

It is obvious that (10) is unbiased since (9) implies that

$$E_\alpha(\alpha^*)=(\Psi^TR_0^{-1}\Psi)^{-1}\Psi^TR_0^{-1}\Psi\alpha\equiv\alpha$$

To prove that its covariance matrix is at least as small as that of any other unbiased estimate in the partial ordering used, we first establish an auxiliary result—a Schwarz-type inequality for matrices (see Note 4.1.B).

Lemma 1 Consider two matrices $A=\{a_{t\nu}\}$, $B=\{b_{t\nu}\}$ where $\nu=1,2,\ldots,p$ and $t=1,2,\ldots,n$, and A^TA is nonsingular. Then

$$B^TB \succ (A^TB)^T(A^TA)^{-1}(A^TB) \tag{12}$$

Proof of Lemma For arbitrary $\lambda,\mu\in\mathbb{R}^p$

$$(\lambda A^T+\mu B^T)(\lambda A^T+\mu B^T)^T \succ 0$$

so that

$$\lambda A^TA\lambda^T+\lambda A^TB\mu^T+\mu B^TA\lambda^T+\mu B^TB\mu^T \succ 0 \tag{13}$$

Now put $K=A^TA$, $L=A^TB$, $M=B^TB$, and rewrite (13) as

$$\begin{aligned}&\lambda K\lambda^T+\lambda L\mu^T+\mu L^T\lambda^T+\mu M\mu^T\\&=(\lambda K^{1/2}+\mu L^TK^{-1/2})(\lambda K^{1/2}+\mu L^TK^{-1/2})^T+\mu[M-L^TK^{-1}L^T]\mu^T\geqslant 0\end{aligned}$$

where we have introduced one square root of K which is possible since $K\succ 0$. Put $\lambda=-\mu L^TK^{-1}$. We get

$$M\succ L^TK^{-1}L^T$$

which can be written as (12). This proves the lemma. Q.E.D.

We can now complete the proof of Theorem 2. Assume first that R is nonsingular and put $B=R^{1/2}C$, $A=R^{-1/2}\Psi$ for some estimate $\alpha^*=Cx$. To make α^* unbiased, we must have $C^T\Psi=I$. We have, using the lemma, assuming for the moment that its assumption is satisfied,

$$\begin{aligned}\text{Cov}(\alpha^*)=CRC^T=B^TB&\succ(A^TB)(A^TA)^{-1}(A^TB)\\&=(\Psi^TC)^T(\Psi R^{-1}\Psi)^{-1}\Psi^TC\end{aligned}$$

Using $\Psi^TC=I$ this reduces to

$$(\Psi R^{-1}\Psi)^{-1}$$

as stated in (11). The lemma is applicable since $A^TA=\Psi^TR^{-1}\Psi$ is nonsingular. To see this, assume that for some vector $z\in\mathbb{R}^p$, we had $z^T\Psi^TR^{-1}\Psi z$

$=0$, or, with $y=\Psi z$, $y^TR^{-1}y=0$. Since R^{-1} is trivially nonsingular, this equation implies that $y=0$. But the ψ vectors have been assumed to be linearly independent, which implies $z=0$, and A^TA is indeed nonsingular, and the assumption in the lemma holds.

If R is singular, the above derivation still holds if the ψ-vectors belong to $\mathfrak{R}$ and if we replace R by its restriction R_0 to $\mathfrak{R}$.

Finally, if the vectors $\psi_\nu^0=P_{\mathfrak{N}}, \psi_\nu$; $\nu=1,2,\ldots,p$, are linearly independent, let us form a linear combination:

$$\alpha_\mu^* = \sum_{t=1}^{n} f(t)x(t)=f^Tx \tag{14}$$

with $f=\text{col}[f(t)]$, $x=\text{col}[x(t)]$, and

$$f(t)=\sum_{\lambda=1}^{p} c_\lambda \psi^0(t)$$

To make (14) unbiased, we need

$$E(\alpha_\mu^*)=\sum_{t=1}^{n} f(t)\sum_{\nu=1}^{n}\psi_\nu^0(t)\alpha_\nu\equiv\alpha_\mu$$

or with inner products in $\mathbb{R}^n$

$$(f,\psi_\nu^0)=\begin{cases}1; & \text{if } \nu=\mu\\ 0; & \text{otherwise}\end{cases} \tag{15}$$

Since the ψ_ν^0-vectors are linearly independent, we can find a vector f such that (15) holds. Then

$$\text{Var}(\alpha_\mu^*)=f^TRf=0; \qquad \text{since } f\in\mathfrak{N}$$

Q.E.D.

Remark 2 The theorem implies that for any fixed vector $\theta\in\mathbb{R}^p$ the estimate $\theta^T\alpha^*$ is the BLUE of the parameter $\theta^T\alpha$. Putting all components of θ equal to zero except the kth, this implies that α_k^* is the BLUE of α_k: a result which is not surprising without being entirely obvious.

The Markov estimates are optimal in the linear sense that we have used here. On the other hand, their calculation requires the inversion of the covariance matrix R, and the somewhat involved expression in (10) is certainly not intuitive.

An estimate that is easier to calculate, and perhaps to understand, is the *least-squares estimate*.

Definition 2 The estimate α^*_{LS} that solves the minimum problem

$$\|x-\Psi\alpha\|^2=\min_{\alpha} \tag{16}$$

is called the least-squares estimate of α.

THEOREM 3

In the present case (16) *is solved by*

$$\alpha^*_{LS}=(\Psi^T\Psi)^{-1}\Psi^T x \tag{17}$$

Proof We have, for any value of the vector α,

$$\begin{aligned}\|x-\Psi\alpha\|^2&=(x-\Psi\alpha)^T(x-\Psi\alpha)\\ &=x^T\left(I-\Psi(\Psi^T\Psi)^{-1}\Psi^T\right)x\\ &\quad+\left[(\Psi^T\Psi)^{1/2}\alpha-(\Psi^T\Psi)^{-1/2}\Psi^T x\right]^T\left[(\Psi^T\Psi)^{1/2}\alpha-(\Psi^T\Psi^{-1/2}\Psi^T x\right]\end{aligned} \tag{18}$$

Here $\Psi^T\Psi$ is nonsingular since the ψ-functions are linearly independent, so that (17) and (18) make sense, and nonnegative definite so that it has square roots. But (18) is then made as small as possible if we make the last term zero, which gives

$$(\psi^T\psi)^{1/2}\alpha=(\psi^T\psi)^{-1/2}\psi^T x$$

which is equivalent to (16). Note that the solution is unique, and that it is automatically unbiased. Q.E.D.

In general we cannot expect the least-squares estimate to perform as well as the BLUE, and the question arises if this can ever happen. Let us go back to the situation $p=1$ described in Theorem 1 and with nonsingular R to illustrate an important case. Can the minimum variance, given by (11), be equal to the variance of the least-squares estimate? The covariance matrix of α^*_{LS} is easily calculated to be

$$\operatorname{Var}(\alpha^*_{LS})=(\Psi^T\Psi)^{-1}\Psi^T R\Psi(\Psi^T\Psi)^{-1}$$

and for $p=1$ this reduces to

$$\frac{\psi^T R\psi}{\|\psi\|^4}$$

Equality means, therefore, if we use (11), that

$$\|\psi\|^4=(\psi^T R\psi)(\psi^T R^{-1}\psi) \tag{19}$$

Let us diagonalize R by $R=0^T D0$ via an orthogonal matrix 0, so that, with $\phi=0\psi$, (19) becomes

$$\|\psi\|^4=\|\phi\|^4=(\phi^T D\phi)(\phi^T D^{-1}\phi)=\sum_{t=1}^{n}\phi_t^2 d_t \sum_{s=1}^{n}\phi_s^2 d_t^{-1}$$

where $D=\text{diag}[d_1, d_2,\ldots, d_n]$.

Then Schwarz's inequality tells us that this is true if and only if $\phi_t d_t^{1/2}$ is proportional to $\phi_t d_t^{-1/2}$. But this means that $\phi_t=0$ except for t-values for which d_t is constant. This means that ψ_t is an eigenvector of R. Hence we have the following theorem.

THEOREM 4

For $p=1$ and nonsingular R the least-squares estimate coincides with the BLUE if and only if the ψ-vector is an eigenvector of R.

A highly special case when this occurs automatically is when $R=\sigma^2 I$, so that we have uncorrelated observations. Then all vectors are eigenvectors of R so that $\alpha_{LS}^*=\alpha_{\text{BLUE}}^*$.

Theorem 4 generalizes to the case $p>1$, but we shall not pursue this any further (see Note 4.1C).

In this chapter we do not assume any particular probability structure, except what is contained in the mean values and covariances. If, however, the probability measure happens to be Gaussian, we know that all regressions are linear. Hence *many estimation and test problems for Gaussian measure with known covariances can be reduced to the linear problems studied in this chapter.*

If the unbiased estimate α^* has a Gaussian distribution in $\mathbb{R}^p$, Cramér has introduced the *concentration ellipsoid*

$$E(\alpha^*)=\{z\,|\,z^T R^{-1}(\alpha^*)z\leqslant c\}\subset\mathbb{R}^p$$

to describe its goodness of performance [see Cramér (1945), p. 300]. One estimate α_1^* is preferred to another α_2^* if the concentration ellipsoid of α_1^* is contained in that of α_2^*: $E(\alpha_1^*)\subseteq E(\alpha_2^*)$ or $R^{-1}(\alpha_2^*)\prec R(\alpha_1^*)$ with the partial ordering of matrices. But for two positive definite matrices $A\prec B$ implies $B^{-1}\prec A^{-1}$ (see Note 4.1.D), so that we get $R(\alpha_1^*)\prec R(\alpha_2^*)$ agreeing with the ordering of preference that we have used above.

After this preparation we are ready to study infinite-dimensional linear inference.

2 CONSISTENT ESTIMATION

Let $x(t)$ be defined on the argument set $T=[a, b]$, on which we have a bounded measure denoted by dt, say Lebesgue measure, and with $V=\mathbb{R}$. We assume for the most part in this chapter that the sample space X is the Hilbert space $X=L_2(T; dt)$. It will be helpful to view the process from all three perspectives mentioned in Chapter 1, Section 1. They are equivalent, but one may be more convenient than another in a particular situation.

Let the covariance operator R be of trace type on X and known to the observer. The mean value $m=m(\cdot)$ shall be given as a linear combination of continuous functions $\psi_\nu \in X$. To begin with let us have just a single real-valued unknown parameter, $p=1$, $m(t)=\alpha\psi(t)$, where ψ is continuous on T. Actually our main interest will be processes that are continuous in the mean, which is assumed in the following. Then R will automatically be of trace type: $\mathrm{tr}(R)$ equals the integral of $r(t, t)$ over T.

Before we construct unbiased estimates we must make the notion of unbiasedness precise. In Section 1 the finite-dimensional situation was presented, and we should extend this to the present case.

Consider the set L of all finite linear combinations of the form

$$\alpha^* = \sum_{k=1}^{n} c_k x(t_k) \tag{1}$$

representing unbiased estimates, so that

$$\sum_{k=1}^{n} c_k \psi(t_k) = 1 \tag{2}$$

Then

$$\alpha^* = \alpha + \sum_{k=1}^{n} c_k y(t_k) \tag{3}$$

and we now close L (topologically) with strong convergence in X. The resulting set U will play the role of unbiased estimates.

Note that U is not linearly closed, but convex. Because of (3) the convergence properties do not depend on the values of α but only upon the y-part whose statistical structure does not depend upon α. Evaluating second-order moments of the y-part in (3), we can therefore let α be arbitrary, for example, 0.

In Section 1 consistent estimation for finite sample size could be viewed as a pathological case easily disposed with. In a Hilbert space this is no longer so, and we need a more detailed analysis.

Since R is of trace type, its spectral representation can be written as

$$R=\sum_{\nu=1}^{\infty}\lambda_\nu P_\nu$$

with the eigenvalues λ_ν forming a summable series and with projection operators P_ν down to one-dimensional subspaces. Introduce the nonnegative square root by

$$R^{1/2}=\sum_{\nu=1}^{\infty}+\sqrt{\lambda_\nu}\,P_\nu$$

which is of Hilbert-Schmidt type.

Most of the results in this chapter are based on the next theorem.

THEOREM 1 EXISTENCE OF LINEAR ESTIMATES, GRENANDER (1949, 1950)

There exists a unique BLUE α^* satisfying the equation

$$E[\alpha^* x(t)]\equiv C\psi(t);\qquad t\in T \tag{4}$$

and with variance given by

$$\operatorname{Var}[\alpha^*]=C-\alpha^2=V \tag{5}$$

Consistent estimation of α is possible if and only if $\psi\notin range(R^{1/2})$. If $\psi\in range(R^{1/2})$ with $\psi=R^{1/2}u$, $u\in range(R^{1/2})$, then

$$V=\frac{1}{\|u\|^2} \tag{6}$$

Remark 1 The question of uniqueness is not completely transparent. The BLUE considered as a stochastic variable is certainly unique, but *it can sometimes be represented in more than one way*. We have already met this ambiguity in the finite dimensional case in the previous section. The existence of inner product representation changes character now when we study an infinite-dimensional sample space. *The reason is that range* (R) *and range* ($R^{1/2}$) *do not always coincide in the infinite-dimensional case*. This question is postponed to Theorem 2.

Proof The *existence* of BLUEs can be established directly. Let $\{\alpha_n^*\}$ be a sequence from U, and with α put equal to zero in the following calculations, we have

$$\|\alpha_n^*\|^2=\operatorname{Var}(\alpha_n^*)\rightarrow\inf_{\alpha^*\in U}\operatorname{Var}(\alpha^*)=V$$

With the parallelogram law in the Hilbert space

$$\|x+y\|^2+\|x-y\|^2=2\|x\|^2+2\|y\|^2$$

we get, with $x=\alpha_n^*$, $y=\alpha_m^*$,

$$\|\alpha_n^*-\alpha_m^*\|^2=2\|\alpha_n^*\|^2+2\|\alpha_m^*\|-4\left\|\frac{\alpha_n^*+\alpha_m^*}{2}\right\|^2 \tag{7}$$

Note that $\frac{1}{2}(\alpha_n^*+\alpha_m^*)\in U$ so that it has a squared norm at least equal to V. Hence (7) tends to zero so that $\{\alpha_n^*\}$ is a Cauchy sequence converging strongly to some definite α^* which is in U and has $\mathrm{Var}(\alpha^*)=\|\alpha^*\|^2=V$.

Uniqueness follows also from convexity since if α_1^*, α_2^* are both in U and have variance V, we get with strict inequality, unless $\alpha_1^*=\alpha_2^*$,

$$\|\tfrac{1}{2}(\alpha_1^*+\alpha_2^*)\|^2=\tfrac{1}{4}\|\alpha_1^*\|^2+\tfrac{1}{4}\|\alpha_2^*\|+\tfrac{1}{2}(\alpha_1^*,\alpha_2^*)<V$$

Hence α^* is unique as a stochastic variable.

It can be characterized by a simple equation. Indeed, $\alpha^*+\varepsilon[x(t')\psi(t'')-x(t'')\psi(t')]\in U$, for any $t', t''\in T$, and has variance

$$\begin{aligned}\|\alpha^*\|^2&+2\varepsilon(\alpha^*,x(t')\psi(t'')-x(t'')\psi(t'))+0(\varepsilon^2)\\&=V+2\varepsilon(\alpha^*,x(t')\psi(t'')-x(t'')\psi(t'))+0(\varepsilon^2)\end{aligned}$$

In order that this be $\geqslant V$ for arbitrary ε, the coefficient for ε must be zero, or choosing t'' such that $\psi(t'')\neq 0$,

$$E[\alpha^*x(t)]=(\alpha^*,x(t))\equiv C\psi(t);\qquad t\in T \tag{8}$$

with some constant C. This is the *normal equation* for α^*.

If we approximate α^* as in (1) satisfying (2), the normal equation (8) implies

$$(\alpha^*,\alpha_n^*)=C$$

Let $\alpha_n^*\to\alpha^*$ strongly. Then the left-hand side of (8) tends to $V+\alpha^2$, proving (5).

Let us represent the projections P_ν introduced before the theorem by unit vectors ϕ_ν and expand the regression function

$$\psi(t)=\sum_{\nu=1}^{\infty}c_\nu\phi_\nu(t)+n(t);\qquad n\perp\{\phi_\nu\}$$

with strong convergence and the y-process

$$y(t)=\sum_{\nu=1}^{\infty}\sqrt{\lambda_\nu}\, y_\nu\phi_\nu(t);\qquad \lambda_\nu>0$$

which is just the Karhunen-Loeve representation (see Chapter 2, Section 4), and where $\{y_\nu\}$ is an orthonormal sequence of stochastic variables.

Since $\alpha^*=\alpha+a$, $a\in L_2(Y)$, let us write

$$a=\sum_{\nu=1}^{\infty} a_\nu y_\nu;\qquad \{a_\nu\}\in l_2$$

The normal equation tells us that with convergence in the mean

$$(a,y(t))=\sum_{\nu=1}^{\infty}\sqrt{\lambda_\nu}\, a_\nu\phi_\nu(t)=V\psi(t)=V\left(\sum_{\nu=1}^{\infty} c_\nu\phi_\nu(t)+n(t)\right)\tag{9}$$

If the $L_2(T)$-element $n\neq 0$, relation (9) is possibly only if $V=0$: consistent estimation. On the other hand if $n=0$, (9) gives the relation

$$a_\nu=V\frac{c_\nu}{\sqrt{\lambda_\nu}}$$

But this choice of $\{a_\nu\}$ is in l_2 for $V>0$, as required, if and only if

$$\sum_{\nu=1}^{\infty}\frac{c_\nu^2}{\lambda_\nu}<\infty\tag{10}$$

in other words, if and only if $\psi\in\text{range}(R^{1/2})$.

On the other hand, let $\psi\subset\mathfrak{R}(R^{1/2})$ with the Fourier coefficients c_ν, so that

$$\|u\|^2=\|R^{-1/2}\psi\|^2=\sum_{1}^{\infty}c_\nu^2\lambda_\nu^{-1}<\infty\tag{10a}$$

For any unbiased

$$\alpha^*=\sum_{1}^{n}a_k x(t_k)\tag{10b}$$

with variance

$$\mathrm{Var}(\alpha^*) = \sum_{\nu=1}^{\infty} \left[\sqrt{\lambda_\nu} \sum_{k=1}^{n} a_k \phi_\nu(t_k) \right]^2 \tag{10c}$$

we have

$$1 = \sum_{\nu=1}^{\infty} c_\nu \sum_{k=1}^{n} a_k \psi_\nu(t_k) = \sum_{\nu=1}^{\infty} \frac{c_\nu}{\sqrt{\lambda_\nu}} \left[\sqrt{\lambda_\nu} \sum_{k=1}^{n} a_k \psi(t_k) \right] \tag{10d}$$

and Schwarz's inequality implies, using (10a) and (10c),

$$\mathrm{Var}(\alpha^*) \geqslant \left[\sum_{1}^{\infty} c_\nu^2 \lambda_\nu^{-1} \right]^{-1} = \|u\|^{-2} \tag{10e}$$

Finally if $\psi \in \mathrm{range}(R^{1/2})$ so that $V > 0$, we can write $\psi = R^{1/2}u$, with $u \in L_2(T)$. Then from (9) we get $\sqrt{\lambda_\nu}\, a_\nu = Vc_c = V\sqrt{\lambda_\nu}\, u_\nu$ with $u_\nu = (u, \phi_\nu)$. Hence

$$V = \|a\|^2 = \sum_{\nu=1}^{\infty} a_\nu^2 = V^2 \sum_{\nu=1}^{\infty} u_\nu^2$$

and, since $V > 0$, we get

$$V = \|u\|^{-2}$$

proving (6). Q.E.D.

Can the BLUE α^* be written as an inner product (x, f) in $L_2(T)$? This question is answered by the following theorem (see Note 4.2.A).

THEOREM 2

In the regular case $V > 0$ the BLUE can be expressed as an inner product (x, f) in X if and only if $\psi \in \mathrm{range}(R)$.

Proof If $\alpha^* = (x, f)$, $f \in X$, then the normal equation can be expressed with the same notation as before:

$$V\psi(t) = E[ay(t)] = E[(y, f)y(t)] = \int_T f(s) r(s, t)\, ds = \sum_{\nu=1}^{\infty} \lambda_\nu f_\nu \phi_\nu(t)$$

where $f_\nu=(f,\phi_\nu)$. Hence $f_\nu=V(\psi,\phi_\nu)/\lambda_\nu$, and in order that $f\in L_2$, it is necessary that, since $V\neq 0$,

$$\sum_{\nu=1}^{\infty}\frac{(\psi,\phi_\nu)^2}{\lambda_\nu^2}<\infty \tag{11}$$

which is equivalent to $\psi\in\text{range}(R)$.

On the other hand if $\psi\in\text{range}(R)$, then the f defined by the f_ν just used will give an estimate (f,x) that solves the normal equation, and hence α^* can be written as the inner product. Q.E.D.

To apply these results to particular stochastic processes, we can either calculate the eigenelements of R and use the criteria in (10) and (11) or start from the normal equation (4).

Example 1 Consider the stationary periodic process in Example 1 in Chapter 2, Section 2. In complex form its eigenfunctions are

$$\frac{1}{\sqrt{2\pi p}}\exp(2\pi int/p); \qquad t\in(0,p)$$

with the eigenvalues $\lambda_n=2\pi\sigma_n^2$. Say that R is positive definite, $\sigma_n>0$ for all n, and consider a regression function $\psi\in L_2[(0,p)]$. With the Fourier coefficients

$$\psi_n=\frac{1}{\sqrt{2\pi p}}\int_0^p\exp(2\pi int/p)\psi(t)\,dt$$

we can immediately apply (10) and find that consistent estimation is possible if and only if

$$\sum_{n=-\infty}^{\infty}\psi_n^2\sigma_n^{-2}=+\infty \tag{12}$$

If (12) converges and if, moreover,

$$\sum_{n=-\infty}^{\infty}\psi_n^2\sigma_n^{-4}<+\infty$$

then the BLUE can be expressed as

$$\alpha^*=\int_0^p f(t)x(t)\,dt$$

with (see the proof of Theorem 2)

$$f(t)=\frac{1}{\sqrt{2\pi p}}\sum_{n=-\infty}^{\infty}\psi_n\sigma_n^{-2}\exp(2\pi itn/p)\left(\sum_{n=-\infty}^{\infty}(\psi_n)^2\sigma_n^{-2}\right)^{-1}$$

Example 2 Let $y(t)$ be the Wiener process on $T=(0,1)$ so that $r(s,t)=\sigma^2\min(s,t)$. To solve the normal equation let us recall that the general form of an element in $L_2(Y)$ is

$$a=\int_0^1 f(t)W(dt); \qquad f\in L_2[(0,1)]$$

Hence the normal equation takes the form

$$E[ax(s)]=\int_0^1 f(t)\min(dt,s)=\sigma^2\int_0^s f(t)\,dt\equiv c\psi(s) \tag{13}$$

If we estimate a linear regression coefficient, $\psi(t)\equiv t$, (13) gives the BLUE $\alpha^*=\alpha+a$ with $f\equiv$constant and

$$a=\text{constant}\times\int_0^1 W(dt)=\text{constant}\times W(1)$$

so that $\alpha^*=x(1)$.

Similarly, for parabolic regression, $\psi(t)=t^p$, $p>1$, we see in the same manner that f should be proportional to t^{p-1} so that

$$a=\text{constant}\times\int_0^1 t^{p-1}W(dt)=\text{constant}\left\{w(1)-(p-1)\int_0^1 t^{p-2}W(t)\,dt\right\}$$

We then find the normalizing constant $(2p-1)/p$ to make the estimate unbiased. Hence we see that the BLUE is

$$\alpha^*=\frac{2p-1}{p}\left[x(1)-(p-1)\int_0^1 t^{p-2}x(t)\,dt\right]$$

On the other hand for a constant regression function, $p=0$, the solution of (13) gives the trivial solution $f\equiv 0$, $c=0$, indicating that consistent estimation is possible. Indeed, the trivial estimate $\alpha^*=x(0)=\alpha$ is consistent. It cannot be written as an inner product in $L_2(T)$.

It is of interest to note that the last BLUE could be written as an inner product (f,x) if f is interpreted as a distribution, namely, a δ-function situated at $t=0$. Often it is possible to write down the BLUE as an inner product and interpret the ensuing normal equation in terms of distributions. Of course, this must be followed by a rigorous proof that the estimate actually coincides with the BLUE.

Example 3 With $T=\mathbb{Z}$ and $\{x_t\}$ a real and weakly stationary stochastic process, let us consider the l_2-space with norm

$$\|x\|^2=\sum_{-\infty}^{\infty} w_t x_t^2; \qquad w_t>0 \tag{14}$$

In order that the covariance operator $R=\{r_{s-t}\}$ be of trace type relative to the norm in (14), we must require that $\{w_t\}\in l_1$ (see Chapter 1, Section 2).

For a given regression sequence $\{\psi_t\}$, when is consistent estimation possible? We first examine this question in the case when $\{x_t\}$ has an absolutely continuous spectrum with a positive and continuous density from $(-\pi,\pi)$.

Then consistent estimation is possible if and only if $\{\psi_t\}\in l_2$. The proof is easy. Assuming that this condition holds, introduce the unbiased estimate

$$\alpha_n^* = K_n^{-1}\sum_{t=-n}^{n}\psi_t x_t; \qquad K_n=\sum_{t=-n}^{n}\psi_t^2$$

The variance is

$$\operatorname{Var}(\alpha_n^*)=K_n^{-2}\sum_{t=-n}^{n}\psi_s\psi_t r_{s-t}=K_n^{-2}\int_{-\pi}^{\pi}\left|P_n(e^{i\lambda})\right|^2 f(\lambda)\,d\lambda \tag{15}$$

expressed via the rational function

$$P_n(z)=\sum_{t=-n}^{n}\psi_t z^t$$

Equation (15) leads to the bound

$$\operatorname{Var}(\alpha_n^*)\leqslant K_n^{-2}\max_\lambda f(\lambda)\int_{-\pi}^{\pi}|P_n(z)|^2\,d\lambda=2\pi\max_\lambda f(\lambda)K_n^{-1}\to 0 \qquad \text{as } n\to\infty$$

and this means that consistent estimation is possible.

On the other hand assume that $\{\psi_t\}\in l_2$. Then, for any consistent estimate of form (1) satisfying (2), we have, using the same reasoning, but now together with Schwarz's inequality,

$$\operatorname{Var}(\alpha_n^*)\geqslant 2\pi\min_\lambda f(\lambda)\sum_{-n}^{n}c_t^2\geqslant 2\pi\min_\lambda f(\lambda)K_n^{-1}$$

which is uniformly bounded away from zero. Hence consistent estimation is impossible.

The assertion does not hold for a general spectrum. If, for example, the spectrum has only a finite set $S=\{\lambda_1, \lambda_2, \ldots, \lambda_n\}$ as its support, choose the vector $\{c_t\}$ to satisfy

$$\sum_{t=-n}^{n} c_t \psi_t = 1$$

$$\sum_{t=-n}^{n} c_t e^{it\lambda_\nu} = 0; \qquad \nu = 1, 2, \ldots, r$$

This is possible if the matrix of the system of linear equations is of full rank, so that there exists no nontrivial relation

$$a_0\psi_t + a_1 e^{it\lambda_1} + a_2 e^{it\lambda_2} + \cdots + a_r e^{it\lambda_r} = 0; \qquad t = -n, -n+1, \ldots, n \quad (16)$$

Since $a_0 = 0$ can be ruled out when the λ are different and $2n+1 > r$, (16) means that ψ is a trigonometric polynomial with frequencies $\lambda_1, \lambda_2, \ldots, \lambda_r$. If we exclude this regression, it follows, since the spectral measure associated with the orthogonal Z-process in the Cramér representation has no variation outside of S, that

$$\alpha_n^* = \sum_{-n}^{n} c_t x_t$$

is equal to α with probability 1 with no further condition on $\{\psi_t\}$: consistent estimation.

In the excluded case of trigonometric regression (and still with S as the support of the spectrum)

$$\psi_t = b_1 e^{it\lambda_1} + b_2 e^{it\lambda_2} + \cdots + b_r e^{it\lambda_r}$$

we have, for any unbiased α_n^* with coefficients c_t,

$$\operatorname{Var}(\alpha_n^*) = \sum_{\nu=1}^{r} F_\nu |Q(e^{i\lambda_\nu})|^2$$

with F_ν = spectral mass at the frequency λ_ν and

$$Q(z) = \sum_{t=-n}^{n} c_t z^t$$

But consistency requires that

$$1=\sum_{t=-n}^{n} c_t\psi_t=b_1Q(e^{i\lambda_1})+b_2Q(e^{i\lambda_2})+\cdots+b_rQ(e^{i\lambda_r})$$

so that using Schwarz's inequality again leads to

$$\mathrm{Var}(\alpha_n^*)\geqslant\left(\sum_{\nu=1}^{r} b_\nu^2F_\nu^{-1}\right)^{-1}>0$$

making consistent estimation impossible.

Example 4 With a finite interval of observation and $x(t)$ weakly stationary with a constant mean value function $m(t)\equiv m$, how well can we estimate m? This is one of the most studied problems in the theory of estimation in stochastic processes. We shall return to it several times in the following.

It is natural to consider unbiased averages of the form

$$m^*=\int_{-L}^{L} w(t)x(t)\,dt;\qquad \int_{-L}^{L} w(t)\,dt=1 \tag{17}$$

where we have written T as $(-L,L)$ and $w(t)$ is in L_2. The reader should be cautioned, however, that (17) is not the general form of elements in U. Among these the *equidistributed average*, which is also the least-squares estimate,

$$\bar{x}=\frac{1}{2L}\int_{-L}^{L} x(t)\,dt$$

deserves special attention as the commonsense estimate. Recalling Theorem 4 of Section 1, we are led to ask: *can the least-squares solution* $\bar{x}$ *coincide with the BLUE*?

Say that $x(t)$ has the spectral distribution F so that the normal equation can be written, if $\bar{x}=m^*$, using the Cramér representation in Chapter 2, Section 4,

$$\begin{aligned}E_m\{\bar{x}[x(t)-m]\}&=E\left\{\int_{-\infty}^{\infty}\frac{\sin L\lambda}{L\lambda}\overline{Z(d\lambda)}\int_{-\infty}^{\infty}e^{it\lambda}Z(d\lambda)\right\}\\&=\int_{-\infty}^{\infty}e^{it\lambda}\frac{\sin L\lambda}{\lambda}F(d\lambda)=\text{constant}\end{aligned}$$

Differentiate with respect to t and change the order of integration and differentiation [see, e.g., Cramér (1945), p. 68], so that

$$\int_{-\infty}^{\infty} e^{it\lambda} \sin L\lambda \, F(d\lambda) = 0; \qquad |t| \leqslant L \tag{18}$$

Evaluate (18) at $t = \pm L$ and subtract the two resulting values:

$$\int_{-\infty}^{\infty} \sin^2 L\lambda \, F(d\lambda) = 0$$

so that *a necessary condition is that F have all its mass at the frequencies $\lambda_n = n\pi/L, n \in \mathbb{Z}$.*

This condition is also sufficient since

$$E_m\{\bar{x}[x(t) - m]\} = \frac{1}{2L} \int_{-L}^{L} r(s-t)\, ds \tag{19}$$

and we can express the covariance function as

$$r(t) = \sum_{n=-\infty}^{\infty} F_n \exp i\lambda_n t$$

so that (19) reduces to

$$\frac{1}{2L} \sum_{n=-\infty}^{\infty} F_n \exp\left(\frac{-\pi i n t}{L}\right) \int_{-L}^{L} \exp\left(\frac{\pi i n s}{L}\right) ds = F_0 \equiv \text{constant} \tag{20}$$

The BLUE is unique so that it coincides with $\bar{x}$, and we have shown that the condition is also sufficient (see Note 4.2.B).

This condition is easily interpreted. It means that *the process should be periodic with period* $2L$ (see Chapter 2, Section 2).

Example 5 Consider a weakly stationary *band-limited* process $y(t)$, and $\psi(t) \equiv 1$ on $T = [0, L]$. By a band-limited process one understands one for which the support of the spectral measure F is contained in a bounded interval, $\lambda \in (-B, B)$.

For a given $A > 0$ consider the estimate

$$\alpha_A^* = \sum_{n=0}^{\infty} x^{(2n)}(0) \frac{A^{2n}}{(2n+1)!} \tag{21}$$

Since F has no mass outside of $(-B, B)$, all moments of F exist, hence all derivatives of $x(t)$ exist in the mean (see Theorem 1 in Chapter 2, Section 2, and its immediate extension to derivatives of higher order). With the usual isometry between $L_2(X)$ and $L_2[(-\infty, \infty)]$, $F(d\lambda)$, this implies that all polynomials in λ belong to the latter space. Consider the function

$$P_A(\lambda) = \sum_{n=0}^{\infty} (-1)^n \lambda^{2n} \frac{A^{2n}}{(2n+1)!} \tag{22}$$

In $(-B, B)$ the sum converges uniformly to some continuous function, so that it also belongs to the same space, and

$$\alpha_A^* = \alpha + \int_{-B}^{B} P_A(\lambda) Z_y(d\lambda)$$

where Z_y is the orthogonal set function belonging to the y-process. This also shows that the sum in (21) converges in the mean. But direct evaluation of (22) gives us the *Dirichlet kernel*

$$P_A(\lambda) = \frac{\sin A\lambda}{A\lambda}$$

so that

$$\operatorname{Var}(\alpha_A^*) = \int_{-B}^{B} \frac{\sin^2 A\lambda}{A^2\lambda^2} F(d\lambda)$$

now with the Féjer kernel. We have, as $A \to \infty$, for the uniformly bounded integrand

$$\frac{\sin^2 A\lambda}{A^2\lambda^2} \to \begin{cases} 1; & \text{if } \lambda = 0 \\ 0; & \text{otherwise} \end{cases}$$

so that if F is continuous at 0, the variance tends to zero and $\alpha_A^* \to \alpha$ in probability. Hence if $\Delta F(0) = 0$, we get *consistent estimation of the constant mean for a band-limited process, even for a finite length of observation.*

A similar but more general result can be obtained as long as the spectral energy at high frequencies is exponentially small so that, for some $a > 0$, we have

$$F(\lambda) = \begin{cases} 1 - 0(e^{-a\lambda}); & \lambda \to \infty \\ 0(e^{a\lambda}); & \lambda \to -\infty \end{cases}$$

Then $r(t)$ is analytic in a horizontal strip $|\operatorname{Im}(\lambda)| < a$ around the real axis, and the same is true for $x(t)$, with analyticity interpreted in the sense of convergence in the mean. Hence knowing $x(t)$ on an arbitrarily small interval, we can continue it to the entire real axis. Appealing to the ergodic theorem (see Chapter 3, Section 3) we get a consistent estimate if $\Delta F(0)=0$.

Theorems 1 and 2 are essentially existence theorems. With more detailed specification of the regression $\{\psi(t)\}$ we can make more informative assertions. One of the most important cases is *when $y(t)$ is stationary*, say, with spectral measure F, *and $\psi(t)$ can be expressed as a Fourier-Stieltjes integral*, say, with $T=\mathbb{Z}$:

$$\psi(t)=\int_{-\pi}^{\pi} e^{it\lambda}\mu(d\lambda) \tag{23}$$

where μ is a complex-valued measure of bounded variation. This includes trigonometric regression sequences as well as many others. When does this lead to consistent estimation on T? A complete answer was furnished in Grenander (1951b).

THEOREM 3

In order that consistent estimation be possible for the regression sequence (23), *it is necessary and sufficient that the Hellinger integral*

$$H=\int_{-\pi}^{\pi}\frac{|\mu(d\lambda)|^2}{F(d\lambda)}$$

diverge. If it converges, the lower bound of the variance of unbiased estimates is $1/H$.

Proof Consider an arbitrary unbiased estimate α^* using the observations $x(t), t=-n,-n+1,\dots,n-1,n$,

$$\alpha^*=\sum_{\nu=-n}^{n} c_\nu x(\nu)$$

Since α^* is unbiased, we must have

$$1=\int_{-\pi}^{\pi} P_n(e^{i\lambda})\mu(d\lambda) \tag{24}$$

where we have introduced the polynomial

$$P(z)=\sum_{\nu=-n}^{n} c_\nu z^\nu$$

The variance is then, as we have seen several times before,

$$\operatorname{Var}(\alpha^*) = \int_{-\pi}^{\pi} |P(e^{i\lambda})|^2 F(d\lambda) \tag{25}$$

and we get from Schwarz's inequality, for an arbitrarily small $\varepsilon > 0$, with a sufficiently fine subdivision $-\pi = \lambda_0^{(m)} < \lambda_1^{(m)} < \lambda_2^{(m)} < \dots < \lambda_m^{(m)} < \pi = \lambda_{m+1}^{(m)}$,

$$1 \leqslant \left[\int_{-\pi}^{\pi} |P_n(e^{i\lambda})||\mu(d\lambda)|\right]^2 \leqslant \left[\sum_{\nu=0}^{m} |P(e^{i\lambda_\nu^{(m)}})||\Delta_\nu \mu|\right]^2 + \varepsilon$$

$$\leqslant \sum_{\nu=0}^{m} |P_n(e^{i\lambda_\nu^{(m)}})|^2 \Delta_\nu F \sum_{\nu=0}^{m} \frac{|\Delta_\nu \mu|^2}{\Delta_\nu F} + \varepsilon \tag{26}$$

In (26) we have written $\Delta_\nu \mu$ and $\Delta_\nu F$ for the μ- and F-measures, respectively, over the interval $(\lambda_\nu^{(m)}, \lambda_{\nu+1}^{(m)})$, and $0/0$ is interpreted as 0. But it is known that the quantities

$$H_m = \sum_{\nu=0}^{m} \frac{|\Delta_\nu \mu|^2}{\Delta_\nu F} \tag{27}$$

are nondecreasing when the subdivision is made finer, and the limit H is positive and is just the definition of the *Hellinger integral* [see, e.g., Hobson (1918)]. Symbolically this is written as

$$H = \int_{-\pi}^{\pi} \frac{|\mu(d\lambda)|^2}{F(d\lambda)} \tag{28}$$

Hence we have shown that a lower bound for the variance of unbiased estimates is given by $1/H$, since the first sum on the right-hand side of (26) is the Riemann-Stieltjes sum corresponding to the integral (25).

It remains to show that we can approach the value $1/H$ as closely as we wish. In order to prove this, consider the step function

$$s^m(\lambda) = H_m^{-1} \frac{\overline{\Delta_\nu \mu}}{\Delta_\nu F}; \qquad \lambda \in (\lambda_\nu^{(m)}, \lambda_{\nu+1}^{(m)})$$

assuming for the moment that $\Delta_\nu \mu = 0$ if $\Delta_\nu F$ is 0, in which case we interpret their ratio as 1. We choose the $\lambda_\nu^{(m)}$ so that they do not coincide with the point masses of F or μ. Then we can approximate $s^m(\lambda)$ uniformly by trigonometric polynomials, except for small neighborhoods of the $\lambda_\nu^{(m)}$. But

$$\int_{-\pi}^{\pi} |s^m(\lambda)|^2 F(d\lambda) = H_m^{-1}$$

$$\int_{-\pi}^{\pi} s^m(\lambda)\mu(d\lambda) = 1$$

Hence $H^{-1} = \lim_{m\to\infty} H_m^{-1}$ is the true lower bound.

If $\Delta_\nu \mu > 0$ for some ν with $\Delta_\nu F = 0$, we define

$$s^m(\lambda) = \begin{cases} \dfrac{1}{\Delta_\nu \mu}; & \lambda \in (\lambda_\nu^{(m)}, \lambda_{\nu+1}^{(m)}) \\ 0; & \text{otherwise} \end{cases}$$

and get (24) again with $H_m = +\infty$. Q.E.D.

A curious consequence of Theorem 3, which is pointed out only in passing, concerns interpolation of stationary processes. With μ absolutely continuous with the density $1/2\pi$ we get $\psi(t) = 0$ except for $t = 0$ when $\psi(0) = 1$. That means that our minimum variance problem can be written as

$$\operatorname{Var}(\alpha^*) = \left\| x_0 - \sum_{t\neq 0} a_t x_t \right\|^2 = \min$$

But this means that we try to interpolate $\{x_t\}$ at $t = 0$ using all the other x_t-observations. Hence we have an alternative proof of Kolmogorov's well-known result (see Note 4.2.C) on interpolation of stationary processes.

Corollary 1 The minimum interpolation error is given via the Hellinger integral

$$\min \left\| x_0 - \sum_{t\neq 0} a_t x_t \right\|^2 = \frac{4\pi^2}{\int_{-\pi}^{\pi} [(d\lambda)^2 / F(d\lambda)]} \tag{29}$$

In particular, perfect interpolation is possible if and only if the integral diverges.

It can be shown that $F(d\lambda)$ in (28) can be replaced by its absolutely continuous component $f(\lambda)\,d\lambda$, so that the minimum error in (29) can be written in terms of the *harmonic mean* $2\pi H(f)$.

We also get another way of proving an already known result on consistent estimation of the mean.

Corollary 2 The constant mean of a stationary process can be estimated consistently if and only if the spectrum has no mass at frequency 0.

Proof We just have to note that a constant regression function corresponds to μ having its support at $\lambda=0$. The corollary then follows directly from Theorem 3. Q.E.D.

So far we have dealt with a single real regression parameter, $p=1$, but much of the above goes through for $p>1$. We now show how this can be achieved, but we need only sketch that part of the proof that parallels the one-dimensional case. Recall that a vector valued BLUE α^* has the property that $\text{Cov}(\alpha^*)\prec\text{Cov}(\beta^*)$ for any other unbiased vector estimate β^* (see Section 1).

The space M spanned by all the mean value functions of the form (1.1) is p-dimensional since the regression functions ψ_k were supposed to be linearly independent. One could also allow an infinite-dimensional space M, this will not be done here, but we return to it in later sections.

THEOREM 4

For $p>1$ there exists a unique BLUE $\alpha^*=\text{col}(\alpha_1^*,\alpha_2^*,\ldots,\alpha_p^*)$. It is determined by the normal equation

$$E_\alpha[\alpha_k^*\xi]=0$$

for any ξ of the form

$$\xi=\sum_\nu b_\nu x(t_\nu)$$

$$\sum_\nu b_\nu\psi_l(t_\nu)=0$$

If all $\psi_k\in\text{range}(R^{1/2})$, consistent estimation is not possible. If some linear combination of ψ_k is not in range$(R^{1/2})$, consistent estimation is possible for some linear combination of α_k. If any nontrivial linear combination of the ψ_k is outside range$(R^{1/2})$, all components of α can be estimated consistently.

Proof Consider the set of elements of the form

$$U_k=\left\{x\,|\,x=\sum_\nu a_\nu x(t_\nu),\ \sum_\nu a_\nu\psi_l(t_\nu)=\delta_{lk}\right\} \tag{30}$$

and its closure $\overline{U}_k$. Arguing in the same way as for Theorem 1, it is clear that we can find an element $\alpha_k^* \in \overline{U}_k$ with minimum variance. A variational argument with α^* replaced by

$$\{\alpha_\nu^*\} + \varepsilon\{b_\nu\}, \ \sum_\nu b_\nu \psi_l(t_\nu) = 0; \text{ for all } l \tag{31}$$

shows that we must have

$$E_\alpha\left[\alpha_k^* \sum_\nu b_\nu x(t_\nu)\right] = 0 \tag{32}$$

and these are the normal equations [together with the constraints in (31)].

It is also clear that in $\overline{U}_k$ the normal equation determines α_k^* uniquely. If there were another α_k^{**} satisfying (32), we could form $\beta = \alpha_k^* - \alpha_k^{**}$, also satisfying (32). But β can be approximated by elements of the form $\Sigma_\nu b_\nu x(t_\nu)$ with b_ν satisfying the constraint in (31). Hence (32) implies that $E(\beta^2) = 0$, and α_k^* is uniquely determined (as a stochastic variable) by the normal equations.

Form now the vector estimate $\alpha^* = \text{col}(\alpha_1^*, \alpha_2^*, \ldots, \alpha_p^*)$ from the scalar estimates just derived. Let us compare it with some other unbiased vector estimate $\hat{\beta}$ and show that the former is at least as good as the latter.

Let us choose a sequence $t_1, t_2, \ldots$ everywhere dense in T. Let $\{\alpha_k^*(n)\}$ denote the BLUE of α based on $x(t_1), x(t_2), \ldots, x(t_n)$ as discussed for a finite-dimensional sample space in Section 1. We know from Theorem 2 of Section 1 that if $\{\hat{\beta}_k(n)\}$ is any other unbiased estimate of α based on the same x-values, we have

$$\text{Cov}\left[\alpha_k^*(n)\right] \prec \text{Cov}\left[\hat{\beta}_k(n)\right]$$

If $\hat{\beta} = \{\hat{\beta}_k\}$ is another unbiased estimate, $\beta_k^* \in \overline{U}_k$, we can approximate it by some $\beta^*(n)$ as above with an arbitrarily small error in variance. But $\alpha_k^*(n)$ has minimum variance among unbiased estimates of α_k based on the same x-values (see Remark 1 following Theorem 2 of Section 1). This implies that $\alpha_k^*(n) \to \alpha_k^*$ in the mean as n tends to infinity, which, together with the approximation argument, shows that

$$\text{Cov}(\alpha_k^*) \prec \text{Cov}\left(\hat{\beta}_k\right)$$

so that α^* is indeed the (unique) BLUE.

Now let us assume that all $\psi_k \in$ range$(R^{1/2})$, so that we can write $\psi_k = R^{1/2} u_k$. Let $\gamma \in \mathbb{R}^p$ be a fixed vector of length 1, γ_0 a free parameter, and consider the restricted estimation problem of γ_0 when $\alpha = \gamma_0 \gamma$. Then we have the case $p=1$ for

$$x(t)=\gamma_0\psi(t)+y(t); \qquad \psi(t)=\sum_{n=1}^{p}\gamma_k\psi_k(t)\in\text{range}(R^{1/2}) \tag{33}$$

Theorem 1 tells us that the variance of any unbiased estimate of γ_0 has a lower bound, say V, $\text{Var}(\gamma_0^*)\geqslant V$. The estimate

$$\hat{\gamma}=\sum_{k=1}^{p}\beta_k\alpha_k^*=\beta^T\alpha^*; \qquad \beta=\text{col}(\beta_1,\beta_2,\ldots,\beta_p) \tag{34}$$

is unbiased if $\beta^T\gamma=1$, and has the variance $\beta^T\text{Cov}(\alpha^*)\beta$. This means that $\beta^T\text{Cov}(\alpha^*)\beta\geqslant V\|\beta\|^2$ if we choose $\gamma=\beta$, and hence

$$\text{Cov}(\alpha^*)\succ V\times\text{identity matrix in } \mathbb{R}^p$$

so that consistent estimation is not possible in this case.

On the other hand if for some $v\in\mathbb{R}^p$ we have $\sum_1^n v_k\psi_k\notin\text{range}(R^{1/2})$, then we can estimate γ_0 consistently. But for arbitrary β in (34) we have, if $\text{Cov}(\alpha^*)$ is nonsingular with smallest eigenvalue λ_1,

$$\text{Var}(\hat{\gamma})\geqslant\lambda_1\|\beta\|^2\geqslant\lambda_1>0$$

where $\beta^T\gamma=1$. Then consistent estimation would be impossible, against the assumption, so that $\text{Cov}(\alpha^*)$ is singular, allowing for consistent estimation for some linear combination of α_k.

Finally if any nontrivial linear combination of ψ_k is outside range$(R^{1/2})$, then γ_0 in (33) can be estimated consistently so that there exists an estimate γ_0^* with components in $\overline{U}_k$ such that γ_0^* is equal to $\gamma^T\alpha$ with probability 1 for any α. But $\gamma^T\alpha^*$ has the minimum variance and must therefore vanish so that

$$\gamma^T\text{Cov}(\alpha^*)\gamma=0$$

Since this holds for any γ, we see that

$$\text{Cov}(\alpha^*)=0$$

and consistent estimation is possible. Q.E.D.

It is also possible to extend Theorem 3 to $p>1$. Assume that the regression sequences can be expressed as Fourier-Stieltjes integrals:

$$\psi_k(f)=\int_{-\pi}^{\pi}e^{it\lambda}\mu_k(d\lambda); \qquad t\in\mathbb{Z}, \quad k=1,2,\ldots,p$$

with bounded complex measures μ_k linearly independent of each other modulo F, in the sense that there is no linear combination of the μ_k that vanishes on every Borel set of positive F-measure. Introduce the Hellinger integrals

$$H_{kl}=\int_{-\pi}^{\pi}\frac{\mu_k(d\lambda)\bar{\mu}_l(d\lambda)}{F(d\lambda)}$$

and the positive definite $p\times p$ matrix $H=\{H_{kl};\ k,l=1,2,\dots,p\}$. Note that H is nonsingular since the μ_k were linearly independent of each other. When we let the Riemann-Stieltjes sums that define the H_{kl} tend to their limits, the corresponding matrices will have nondecreasing norms. This follows from the fact that the sums in (27) are nondecreasing. If the norms tend to infinity, we write $\|H\|=\infty$, otherwise $\|H\|<\infty$.

THEOREM 5

For the estimation problem with the regression sequences (30) *in the case* $\|H\|<\infty$ *the covariance matrix is dominated from below by the inverse of the Hellinger matrix*

$$\mathrm{Cov}(\alpha^*)\geqslant H^{-1}$$

and equality can be attained. If $\|H\|=\infty$, *some linear combination of the regression coefficients* α_k *can be estimated consistently.*

Proof We use the same technique as in the previous proof. Assume first that $\|H\|<\infty$. For a fixed vector $\gamma\in\mathbb{R}^p$ consider the scalar estimation problem. Estimate γ_0 in (33) with

$$\psi(t)=\int_{-\pi}^{\pi}e^{it\lambda}\mu(d\lambda);\qquad \mu=\sum_1^p\gamma_k\mu_k$$

We know that (34) is unbiased if $\beta^T\gamma=1$ and hence, using Theorem 3,

$$\beta^T\mathrm{Cov}(\alpha^*)\beta=\mathrm{Var}(\hat{\gamma}_0)\geqslant H_\mu^{-1};\qquad H_\mu=\int_{-\pi}^{\pi}\frac{|\mu(d\lambda)|^2}{F(d\lambda)}\tag{35}$$

or

$$\beta^T\mathrm{Cov}(\alpha^*)\beta\geqslant(\gamma^TH\gamma)^{-1}\tag{36}$$

The only condition on γ and β is $\gamma^T\beta=1$. With

$$\gamma=\frac{H^{-1}}{\beta^T H^{-1}\beta} \tag{37}$$

the inequality becomes

$$\beta^T \operatorname{Cov}(\alpha^*)\beta \geqslant \beta^T H^{-1}\beta$$

or

$$\operatorname{Cov}(\alpha^*) \succ H^{-1}$$

Note, however, that in (35) equality can be obtained as we know in the scalar case.

On the other hand if $\|H\|=\infty$, then for some γ we know the right-hand side in (36) will be zero. Again this can be attained by the left-hand side so that consistent estimation is possible of some linear combination of the α_k.

Q.E.D.

If the regression sequence is simple enough, we can determine directly whether consistent estimates exist or not. Let us illustrate this for polynomial regression.

Example 6 Consider $\psi(t)=t^q, t\in\mathbb{Z}, q\in\mathbb{N}$, and with a stationary process $y(t)$. We begin with $q=0$. We have seen in Chapter 3, Section 2, that if the spectral distribution function F is continuous at frequency $\lambda=0$, then consistent linear estimation is possible having access to all $x_t, t\in\mathbb{Z}$.

But this condition can also be seen to be necessary. Indeed, consider an arbitrary unbiased estimate using the observations $x(1), x(2),\ldots, x(n)$:

$$\alpha^*=\sum_{t=1}^{n} c_t x(t); \qquad \sum_{t=1}^{n} c_t=1$$

With the usual definition we have

$$P(z)=\sum_{t=1}^{n} c_t z^t; \qquad z=e^{i\lambda}, \quad P(1)=1$$

We get a bound in terms of the saltus $\Delta F(0)$ of F at $\lambda=0$, since $P(1)=1$,

$$\operatorname{Var}(\alpha^*)=\int_{-\pi}^{\pi}|P(z)|^2F(d\lambda)\geqslant \Delta F(0)$$

Hence *consistent estimation is possible if and only if* $\Delta F(0)=0$.

Now consider $q>0$. Form the qth difference of the $x(t)$-process:

$$d(t)=x(t+q)-\binom{q}{1}x(t+q-1)+\binom{q}{2}x(t+q-2)+\cdots\pm x(t)$$

The mean value function of $d(t)$ is α times the qth difference of t^q. This difference is constant and equal to $q!$. The remainder of the $d(t)$-process is stationary and we have, in terms of the orthogonal process Z_x associated with the x-process,

$$d(t)=\int_{-\infty}^{\infty} e^{it\lambda}\left[e^{iq\lambda}-\binom{q}{1}e^{i(q-1)\lambda}+\cdots\pm 1\right]Z_x(d\lambda)$$

$$=\int_{-\infty}^{\infty} e^{it\lambda}(e^{i\lambda}-1)^q Z_x(d\lambda) \tag{38}$$

Equation (38) gives us the spectral distribution function F_α for the d-process as

$$F_\alpha(d\lambda)=\left|e^{i\lambda}-1\right|^{2q}F(d\lambda)$$

For any F we can infer from (37) that F_α is continuous at $\lambda=0$. But the mean value function of $d(t)$ was simply $\alpha q!$, so that we are back to the case $q=0$ and conclude that *in the present case*, $q>0$, *consistent estimation is possible for all spectra.*

Before leaving the subject of consistent linear estimation, let us consider briefly the situation when we observe i.i.d. elements $x_1, x_2, \ldots, x_n$ in a Hilbert space where the mean $\alpha=E(x)$ is completely unknown. The covariance operator R will be assumed to be known, as before. This differs from the earlier cases when α was known to belong to some finite-dimensional subspace.

We can no longer hope to do better than just use the straightforward average $\bar{x}$ (see Note 4.2.D). This is consistent in the sense that it converges strongly in probability to α as $n\to\infty$, as expressed by the following simple theorem.

THEOREM 6 WEAK LAW OF LARGE NUMBERS IN HILBERT SPACE

Let $x_1, x_2, \ldots$ be uncorrelated observations in a Hilbert space with mean α and with a common trace $E[\|x\|^2]<\infty$. Then, for any $\varepsilon>0$,

$$\lim_{n\to\infty} P\{\|\bar{x}-\alpha\|>\varepsilon\}=0$$

Proof We have with a completely straightforward argument

$$E_\alpha[\|\bar{x}-\alpha\|^2]=\frac{1}{n^2}E_\alpha\left[\left\|\sum_{t=1}^{n}x_t-\alpha\right\|^2\right]=\frac{1}{n^2}\sum_{s,t=1}^{n}E_\alpha(x_s-\alpha,x_t-\alpha)$$

$$=\frac{1}{n^2}\sum_{t=1}^{n}E_\alpha[\|x_t-\alpha\|^2]=\frac{1}{n}E_\alpha[\|x_t-\alpha\|^2]\to 0$$

The Chebyshev inequality then gives

$$P_\alpha(\|\bar{x}-\alpha\|>\varepsilon)\leqslant\frac{E_\alpha[\|x-\alpha\|^2]}{n\varepsilon^2}\to 0 \qquad \text{Q.E.D.}$$

A stronger version of the law of large numbers is discussed later when we study inference in Banach spaces (see Section 6).

The accuracy of the estimator $\bar{x}$ is given by the following extension of the central limit theorem to a Hilbert space (see Note 4.2.E).

THEOREM 7 CENTRAL LIMIT THEOREM IN HILBERT SPACE

Consider $x_1, x_2, x_3, \ldots$ a sequence of i.i.d. stochastic variables on the real separable Hilbert space X. Let m be the mean, and let R be the covariance operator of trace type. Then the normed average

$$y_n=\frac{1}{\sqrt{n}}\sum_{\nu=1}^{n}(x_\nu-m)$$

has a probability measure P_n converging weakly to the normal distribution $N(m, R)$ on X.

Proof For given $x^*\in X^*$ consider the Fourier transform $P_n(x^*)$ of the measure P_n associated with y_n (see Chapter 1, Section 2). This is the same thing as the characteristic function of the stochastic variable

$$(x^*,y_n)=\frac{1}{\sqrt{n}}\sum_{\nu=1}^{n}(x^*,x_\nu-m)$$

evaluated for the argument equal to one. But the classical form of the central limit theorem tells us that this characteristic function tends to

$$\exp\{-\tfrac{1}{2}\mathrm{Var}[(x^*,x_\nu-m)]\}=\exp[-\tfrac{1}{2}(Rx^*,x^*)]$$

Hence we know that the Fourier transforms $\hat{P}_n(x^*)$ of y_n tend to the Fourier transform of the normal distribution mentioned in the statement of the theorem.

According to Theorem 1 of Chapter 2, Section 1, the proof is complete if we can show that $\{P_n\}$ is tight. Express y_n in l_2-coordinates $\{\xi_\nu^n; \nu=1,2,\dots\}$. Then

$$\rho_N^n = E\left[\sum_{\nu=N}^{\infty} (\xi_\nu^n)^2\right] = \sum_{\nu=N}^{\infty} R_{\nu\nu} \tag{39}$$

the sum of the diagonal elements of the R-matrix for $\nu \geqslant N$. This follows immediately from the fact that y_n has the covariance matrix R, the same as for x. Now choose an arbitrary $\varepsilon > 0$ and introduce the set

$$C = \bigcap_{k=1}^{\infty} E_k; \qquad E_k = \left\{\xi \mid \sum_{\nu=N_k} \xi_\nu^2 \leqslant \frac{1}{l_k}\right\}$$

where $N_k \uparrow +\infty$, $N_1 = 1$, and $l_k \to +\infty$ slowly enough to make

$$\sum_{k=1}^{\infty} l_k \rho_{N_k}^n < \varepsilon; \qquad \text{all } n.$$

This can be done since (39) shows that ρ_N^n does not depend upon n and that $\rho_N^n \to 0$ as $N \to \infty$ since R is of the trace type.

Note that C is compact. Indeed, given any sequence $\xi_\nu \in C$, since $\|\xi_\nu\|^2 \leqslant 1/l_1$, we can select a subsequence ξ_{ν_i} such that for each component n we have convergence of $(\xi_{\nu_i})_n$ to some number η_n. As i and j tend to infinity, we have

$$\|\xi_{\nu_i} - \xi_{\nu_j}\|^2 = \sum_{\mu=1}^{n} \left[(\xi_{\nu_i})_\mu - (\xi_{\nu_j})_\mu\right]^2 + \sum_{\mu=n}^{\infty} \left[(\xi_{\nu_i})_\mu - (\xi_{\nu_j})_\mu\right]^2$$

For fixed n the first sum tends to zero as i and j grow toward infinity. The second sum can be bounded by $4/l_k$ if we choose $n = N_k$. Hence we have a Cauchy sequence, and C is compact. On the other hand

$$1 - P_n(C) \leqslant \sum_{k=1}^{\infty} P_n(E_k^c) \leqslant \sum_{k=1}^{\infty} l_k \rho_{N_k}^n < \varepsilon$$

so that $\{P_n\}$ is tight. Q.E.D.

3 BLUEs FOR STOCHASTIC PROCESSES

In Section 2 we studied the existence of BLUEs and in particular *the existence of consistent estimates*. We now turn to the problem of calculating BLUEs for certain stochastic processes, and we begin with a special case that has considerable potential for generalization in different directions, as will be seen later.

Consider the (linearly) Markov stationary covariance function (see end of Chapter 2, Section 2):

$$r(s,t)=e^{-\beta|s-t|}; \qquad T=(0,L)$$

and a constant mean $m(t)\equiv\alpha$.

THEOREM 1. ESTIMATION OF THE MEAN OF THE STATIONARY MARKOV PROCESS, GRENANDER (1950)

The BLUE is given by

$$\alpha^* = \frac{1}{2+\beta L}\left(x(0)+x(L)+\beta\int_0^T x(t)\,dt\right) \tag{1}$$

with the variance

$$\operatorname{Var}(\alpha^*)=\frac{2}{2+\beta L} \tag{2}$$

Proof Theorem 1 of Section 2 tells us that α^* exists and is uniquely determined by the normal equation (2.4), but what is the form of α^*? One could try to represent α^* as an inner product in X:

$$\alpha^* = \int_0^L f(t)x(t)\,dt$$

with a quadratically integrable weight function f. It will turn out, however, that no α^* of this form solves the normal equation, so that we have to try other ways of representing α^*. The next natural attempt is to try the form of a stochastic Stieltjes integral:

$$\alpha^* = \int_0^L x(t)F(dt) \tag{3}$$

where F is of bounded variation on T. Note that $x(t)$ is continuous in the mean so that (3) makes sense, being defined in the mean, and hence gives us an element in $L_2(X)$. Then (2.4) and (2.5) can be expressed as the integral equation in F:

$$E_\alpha\{\alpha^*[x(s)-\alpha]\}=\int_0^L r(s,t)F(dt)\equiv C-\alpha^2=V; \qquad \forall s\in T \tag{4}$$

and we must show that (4) is satisfied if we let F have point mass 1 at $t=0$ and L and a constant derivative β throughout T, as stated in the theorem. Indeed, the integral in (4) is with this choice of F:

$$e^{-\beta s}+e^{-\beta|L-s|}+\beta\int_0^s e^{-\beta(s-t)}dt+\beta\int_s^L e^{-\beta(t-s)}\,dt$$

Direct evaluation of this gives us the value 2. To make α^* unbiased we must normalize by dividing by the variation of F, which is $2+\beta L$. Then

$$E_\alpha\{\alpha^*[x(s)-m]\}=\frac{2}{2+\beta L}$$

so that (3) represents the BLUE when F is chosen as described, and gives the representation (1). From (2.5) we see that its variance is given by (2).

Q.E.D.

Remark The attractive form of the BLUE as an integral plus a linear combination of the values at the endpoint of T leads one to ask whether this form is of more general validity. To really penetrate this problem one has to understand the geometry of $L_2[x(t);t\in T]$ or, equivalently, that of the isometric $L_2[e^{it\lambda};t\in T]$. As we see in Chapter 5, the latter space consists for this process exactly of all functions of the form

$$\phi(\lambda)=c+(1+i\lambda)\int_0^L e^{it\lambda}c(t)\,dt \tag{5}$$

where the first c is a constant and the function $c(\cdot)$ is quadratically integrable over T with respect to Lebesgue measure [see (15) in Chapter 5, Section 1]. But then the isometry represents an arbitrary element x spanned by our observations as

$$x=cx(0)+\int_0^L c(t)x(t)\,dt+\int_0^L c(t)x(dt) \tag{6}$$

We therefore know that the BLUE must be of the form (6). Then the normal equation gives $c(\cdot)\equiv$constant, and we get back to (1).

Note that (5) is the general form of $\phi(\lambda)$ under a condition on power law asymptotic behavior of $f(\lambda)$ (see Chapter 5, Section 1). Hence, somewhat more generally, we will have BLUEs of the form appearing as in (6).

Theorem 1 seems at first to be a very restricted result, but it allows an immediate extension of considerable interest. Say that $x(t)=\alpha+y(t)$, where $y(t)$ satisfies the stochastic differential equation of order n with constant real coefficients as in Example 5 of Chapter 2, Section 4. Then $y(t)$ has an absolutely continuous spectrum with a spectral density

$$f(\lambda)=\frac{1}{|P(\lambda)|^2}$$

$$P(\lambda)=a_n(i\lambda)^n+a_{n-1}(i\lambda)^{n-1}+\cdots+a_0=a_n i^n\prod_{\nu=1}^{n}(\lambda-\lambda_\nu); \qquad a_0>0 \tag{7}$$

except for a multiplicative constant that we can put equal to 1 without loss of generality. The zeros λ_ν of the polynomial can be chosen in the upper half of the complex plane, without loss of generality, in analogy with the difference equation in Example 4 of Chapter 2, Section 4. We need only consider $n\geqslant 2$, since $n=1$ has been settled by Theorem 1.

THEOREM 2. ESTIMATION OF THE MEAN FOR A STATIONARY MARKOV PROCESS OF ORDER N, GRENANDER (1950)

The BLUE is given by

$$\alpha^*=\frac{1}{2a_1+a_0L}\left\{\sum_{\nu=0}^{n-1}\left[A_\nu x^{(\nu)}(0)+B_\nu x^{(\nu)}(L)\right]+a_0\int_0^L x(s)\,ds\right\} \tag{8}$$

with the coefficients

$$\begin{cases} A_\nu=(-1)^\nu a_{\nu+1} \\ B_\nu=a_{\nu+1} \end{cases}$$

and with the variance

$$\operatorname{Var}(\alpha^*)=\frac{2\pi}{a_0(2a_1+a_0L)}$$

Proof Since $f(\lambda)\sim\text{constant }\lambda^{-2n}$, $|\lambda|\to\infty$, it follows that λ^{2k} is integrable with respect to f if $2k<2n-1$. Hence (see Chapter 2, Section 2) $y(t)$

has derivatives in the mean up to order $n-1$. This makes it reasonable to try to represent the BLUE by the form (3) but in terms of $x(t)$, $x'(t)$, $x''(t),\cdots,x^{(n-1)}(t)$ rather than just $x(t)$. Let us try the form given in (8).

First note that (8) is unbiased. When we calculate its expected value, we get no contribution from the derivative terms. Therefore the expected value of the expression in brackets is

$$(A_0+B_0+a_0L)\alpha=(2a_1+a_0L)\alpha$$

as desired.

The expression in (8) definitely makes sense, since the derivatives exist and the denominator is not zero. Using (7) we have, recalling for the last step in (9) that a_1 is real, so that ia_1 is purely imaginary,

$$ia_1=(-1)^{n-1}i^n a_n\sum_1^n\frac{\lambda_1\lambda_2\cdots\lambda_n}{\lambda_\nu}=a_0\sum_1^n\frac{1}{\lambda_\nu}=-a_0\sum_1^n\frac{\bar{\lambda}_\nu}{|\lambda_\nu|^2} \tag{9}$$

We must have $a_0\neq 0$ for f to be integrable, so that (9) implies, since $\operatorname{Im}(\lambda_\nu)>0$, that

$$\operatorname{Im}\left(i\frac{a_1}{a_0}\right)>0 \qquad \text{or} \qquad a_1>0$$

Hence (6) is well defined, its denominator is nonzero.

Consider the expression appearing in the normal equation

$$E_a\{\alpha^*[x(s)-\alpha]\}=\frac{1}{2a_1+a_0L}\left\{\sum_{\nu=0}^{n-1}\left[A_\nu r^{(\nu)}(-s)+B_\nu r^{(\nu)}(L-s)\right]\right.$$
$$\left.+a_0\int_0^L r(s-t)\,dt\right\} \tag{10}$$

However, differentiating the spectral representation of the covariance function repeatedly, we get

$$a_n r^{(n)}(t)+a_{n-1}r^{(n-1)}(t)+\cdots+a_0r(0)=\int_{-\infty}^{\infty}e^{it\lambda}\frac{a_n(i\lambda)^n+\cdots+a_0}{|a_n(i\lambda)^n+\cdots+a_0|^2}d\lambda$$
$$=\int_{-\infty}^{\infty}e^{it\lambda}\frac{d\lambda}{a_n(-i)^n\prod_1^n(\lambda-\bar{\lambda}_\nu)} \tag{11}$$

For $t>0$ Cauchy's theorem implies [recall that $Im(\lambda_\nu)>0$] that (11) is zero. Similarly, for $t<0$ we can show that the alternating sum

$$a_n r^{(n)}(t)-a_{n-1}r^{(n-1)}(t)+\cdots+(-1)^n a_0 r(t)=0 \tag{12}$$

Integrating (11) and (12) between 0 and $t-\varepsilon$ and between $t+\varepsilon$ and L respectively, we can evaluate the integral

$$\begin{aligned} a_0\int_0^L r(s-t)\,ds &= \lim_{\varepsilon\downarrow 0}\left[\int_0^{t-\varepsilon} r(s-t)\,ds+\int_{t+\varepsilon}^{L} r(s-t)\,ds\right] \\ &= \lim_{\varepsilon\downarrow 0}\{a_1[r(-\varepsilon)-r(-t)]-a_2[r'(-\varepsilon)-r'(-t)] \\ &\quad +\cdots+(-1)^{n+1}a_n[r^{(n-1)}(-\varepsilon)-r^{n-1}(-t)] \\ &\quad -a_1[r(L-t)-r(\varepsilon)]-a_2[r'(L-t)-r'(\varepsilon)] \\ &\quad -\cdots-a_n[r^{(n-1)}(L-t)-r^{(n-1)}(\varepsilon)]\} \end{aligned} \tag{13}$$

But (13) can be calculated as

$$\begin{aligned} &2[a_1 r(0)+a_3 r''(0)+\cdots+a_\mu r^{(\mu-1)}(0)] \\ &\quad -a_1 r(-t)-a_1 r(L-t)+a_2 r'(-t)-a_2 r'(L-t) \\ &\quad +\cdots+(-1)^n a_n r^{(n-1)}(-t)-a_n r^{(n-1)}(L-t) \end{aligned}$$

where we have put $\mu=n$ if n is odd, otherwise $\mu=n-1$. Hence (10) reduces to the constant

$$\frac{2}{2a_1+a_0L}\{a_1 r(0)+a_2 r''(0)+\cdots+a_\mu r^{(\mu-1)}(0)\}\equiv V$$

and α^* satisfies the normal equation so that it is the BLUE. The variance is given by V or, integrating the two differential equations (11) and (12) between $-A$ and A, and making $A\to\infty$,

$$2a_1 r(0)+2a_3 r''(0)+\cdots+2a_\mu r^{\mu-1}(0)=\lim_{A\uparrow\infty} a_0\int_{-A}^{A} r(t)\,dt$$

Riemann-Lebesgue's lemma on Fourier coefficients tells us that the functions $r(t)$, $r'(t),\cdots,r^{(n-1)}(t)$ tend to zero as $|t|\to\infty$. Therefore by Fourier inversion and using the form (7) of f,

$$V=\frac{a_0}{2a_1+a_0L}\int_{-\infty}^{\infty} r(t)\,dt=\frac{a_0}{2a_1+a_0}2\pi f(0)=\frac{2\pi a_0}{2a_1+a_0}\frac{1}{a_0^2}$$

$$=\frac{2\pi}{a_0(2a_1+a_0L)}$$

Q.E.D.

Remark 2 The two cases in Theorems 1 and 2 hint at something else, in addition to what is discussed below. In Theorem 1 the process was continuous in the mean, but that was about all as far as regular functions are concerned. The BLUE α^* had a weight function—or rather *distribution* —that was constant except for δ-functions at the end points. In Theorem 2 the sample functions were more regular—continuous derivatives of order $n-1$ in the mean—but the weight distributions were less regular, involving derivatives of δ-functions at the end points. This is generally true: *the smoother the process, the less regular the weight distribution*. An extreme case is where $x(t)$ is analytic in the mean; (see Example 5 in Section 2).

The attractive representations (1) and (8) are suggestive in that *the main part of the estimate is the average* $\bar{x}$. We see in the next section that this statement is generally true in an asymptotic sense, and it will lead to important consequences. The BLUE is obtained as $\bar{x}$ modified by the observations of $x(t)$ and its derivatives at the end points of T. Is this just a coincidence, or is there a deeper reason for this particular form?

Let us return to Theorem 1 for a moment to simplify the discussion of this question. We note that for this Markovian process linear extrapolation of $x(t)$ outside the interval of observation T involves only the values of $x(0)$ and $x(L)$. The von Neumann ergodic theorem tells us that if $\Delta F(0)=0$,

$$\alpha=\lim_{L\uparrow\infty}\frac{1}{2L}\int_{-L}^{L} x(t)\,dt \tag{14}$$

We do not have access to the values $x(t)$, $t\notin T$; but we could try to approximate (14) by replacing these missing values by extrapolated ones.

We need for this a modified notion of optimal predictors and shall use a concept introduced in Dolph-Woodbury (1952). Form the set of elements starting from finite linear combinations,

$$\sum_{\nu=1}^{n} c_\nu x(t_\nu); \qquad t_\nu\in T, \quad \sum_{\nu=1}^{n} c_\nu=1 \tag{15}$$

and close it with convergence in the norm of $L_2(X)$; call the resulting set M.

Definition 1 The optimal unbiased linear predictor $\hat{x}(t)$ is defined as the element $\hat{x}(t)$ solving

$$\|\hat{x}(t)-x(t)\| = \inf_{x\in M} \|x-x(t)\|$$

The existence and uniqueness of $\hat{x}$ can be established just as in Section 2. Also it is seen that $\hat{x}(t)$ is uniquely determined by a normal equation:

$$E_\alpha\{[\hat{x}(t)-x(t)]x(s)\}=k; \qquad s\in T \tag{16}$$

and

$$k=\|\hat{x}(t)\|^2-(x(t),\hat{x}(t)) \tag{17}$$

Notice the similarity of (16) and (17) with the condition for a BLUE when the regression function is constant.

Given an arbitrary element $x\in L_2(X)$, we denote by $\hat{P}x$ the element that solves

$$\|\hat{P}x-x\| = \inf_{y\in M} \|y-x\|$$

and again it is easily seen that $\hat{P}x$ is unique and

$$E_\alpha[(x-\hat{P}x)x(s)]\equiv\text{constant}; \qquad s\in T \tag{18}$$

Of course $\hat{x}(t)=\hat{P}x(t)$, $t\in T$.

It is important to realize that $\hat{P}$ is not an ordinary orthogonal projection, but we establish two other properties needed for what will follow.

Lemma 1 For arbitrary $x, y\in L_2(X)$ we have

$$\|\hat{P}x-\hat{P}y\|\leqslant\|x-y\| \tag{19}$$

Proof We can write

$$x-y=(\hat{P}x-\hat{P}y)+(x-\hat{P}x)-(y-\hat{P}y)=\hat{P}x-\hat{P}y+u$$

But (18) implies $x-\hat{P}x\perp x(a)-x(b)$, any $a, b\in T$; so that $x-\hat{P}x\perp x$, any $x\in L_2(X)$, which implies $u\perp\hat{P}x-\hat{P}y$. This immediately establishes (19).

Q.E.D.

Note that Lemma 1 shows us that if $x(t)$ is continuous in the mean, the new process $\hat{P}x(t)$ is also continuous in the mean and can be integrated over a finite interval, as will be done shortly.

Lemma 2 For any combination we have

$$\hat{P}\left[\sum_{\nu=1}^{n} c_\nu x_\nu\right]=\sum_{\nu=1}^{n} c_\nu \hat{P}x_\nu \qquad \text{if } x_\nu \in L_2(X) \text{ and } \sum_{\nu=1}^{n} c_\nu=1$$

Proof The element on the right-hand side of (16) obviously belongs to M. If we can show that it satisfies the normal equation (18) with $x=\Sigma c_\nu x(t_\nu)$, the lemma is established. But using (16), we get

$$E_\alpha\left\{\sum_{\nu=1}^{n} c_\nu \hat{P}x_\nu-\hat{P}\left[\sum_{\nu=1}^{n} c_\nu x_\nu\right]\right\}x(t)$$

$$=\sum_{\nu=1}^{n} c_\nu E\left[(\hat{P}x_\nu)x(t)\right]-E\sum_{\nu=1}^{n} c_\nu x_\nu x(t)+\text{constant}$$

However

$$E\{[\hat{P}x_\nu)-x_\nu]x(t)\}$$

is also constant, and the assertion has been proved. Q.E.D.

Lemma 3 If $x(t)$ is continuous in the mean,

$$\hat{P}\left[\frac{1}{2A}\int_{-A}^{A} x(t)\,dt\right]=\frac{1}{2A}\int_{-A}^{A}\left[\hat{P}x(t)\right]dt$$

Proof The average $\bar{x}$ over $(-A, A)$ can be approximated by stochastic Riemann sums S of the form (15). Then applying Lemmas 1 and 2 we see that $\hat{P}\bar{x}$ can be approximated by stochastic Riemann sums associated with the average of the $\hat{P}x(t)$ process. Q.E.D.

There exists a remarkable relation between unbiased estimation of a constant mean and unbiased prediction. To show this, let $x(t)$ be weakly stationary; no restriction on the spectrum. First we show that the limit

$$z=\lim_{A\uparrow\infty}\frac{1}{2A}\int_{-A}^{A}\hat{x}(t)\,dt;\qquad \hat{x}(t)=\hat{P}x(t) \tag{20}$$

exists. Using Lemmas 1 and 3 we have

$$\left\|\frac{1}{2A}\int_{-A}^{A}\hat{x}(t)\,dt-\frac{1}{2B}\int_{-B}^{B}\hat{x}(t)\,dt\right\|\leqslant\left\|\frac{1}{2A}\int_{-A}^{A}x(t)\,dt-\frac{1}{2B}\int_{-B}^{B}x(t)\,dt\right\| \tag{21}$$

When $A, B\uparrow\infty$, we know from the von Neumann ergodic theorem that the right-hand side of (21) tends to zero so that the integrals defining z have the Cauchy property. Therefore the limit in (20) exists. We now show that it satisfies the normal equation (2.4) for a constant mean. With $s_1, s_2 \in T$,

$$E\{z[x(s_1)-x(s_2)]\} = \lim_{A\uparrow\infty} \frac{1}{2A}\int_{-A}^{A} E\{\hat{x}(t)[x(s_1)-x(s_2)]\}\,dt \tag{22}$$

Using (14), this becomes

$$\lim_{A\uparrow\infty} \frac{1}{2A}\int_{-A}^{A} E\{x(t)[x(s_1)-x(s_2)\}\,dt$$
$$= \lim_{A\uparrow\infty} \frac{1}{2A}\int_{-A-s_1}^{A-s_1} r(t)\,dt - \lim_{A\uparrow\infty} \frac{1}{2A}\int_{-A-s_2}^{A-s_2} r(t)\,dt = 0$$

since the last two integrals cancel each other except on a bounded interval. We have proved the following theorem (see Note 4.3.A).

THEOREM 3

The BLUE for the mean of a stationary stochastic process can be expressed as

$$\alpha^* = \lim_{A\uparrow\infty} \frac{1}{2A}\int_{-A}^{A} \hat{x}(t)\,dt$$

where $\hat{x}(t)$ is the best linear and unbiased predictor of $x(t)$ using observations from T.

We can show more than this, but only under some regularity conditions, namely, *that the BLUE can be expressed simply in terms of the (ordinary) best linear predictor $x^*(t)$.*

Consider a purely nondeterministic stationary process so that

$$x(t) = \alpha + \int_{-\infty}^{t} g(t-u)\xi(du)$$

where g is quadratically integrable, ξ is an orthogonal set function with $\|\xi(du)\|^2 = du$, and $L_2(\xi_s; s \leqslant t) = L_2(x_s; s \leqslant t)$ (see Chapter 2, Sections 4 and 5). We need a bit more, namely, that

$$\int_0^{\infty} u^3 |g(u)|^2\,du < \infty \tag{23}$$

We first show that $x^*(t)$ is integrable over $(-\infty, \infty)$ and then define

$$I = \int_{-\infty}^{\infty} x^*(t)\, dt; \qquad x^*(t) \text{ predictor from } T=(0, L) \text{ to time } t$$

To do this consider the quantity

$$\|x^*(L+h) - x(L+h)\|^2, \qquad h>0$$

Introduce the best predictor $x^{**}(t)$ using observations from $(-\infty, L)$ so that

$$\|x^*(L+h)-x(L+h)\|^2 \geqslant \|x^{**}(L+h)-x(L+h)\|^2 = \int_0^h |g(u)|^2\, du \qquad (24)$$

since

$$x^{**}(t) = \int_{-\infty}^{L} g(t-s)\xi(ds)$$

The error $x^*(L+h)-x(L+h) \perp x^*(L+h)$ so that

$$\|x(l+h)\|^2 = \int_0^{\infty} |g(u)|^2\, du = \|x^*(L+h)\|^2 + \|x^*(L+h)-x(L+h)\|^2$$

which together with (24) yields

$$\|x^*(L+h)\|^2 \leqslant \int_h^{\infty} |g(u)|^2\, du$$

and we have a similar relation for "prediction," or rather extrapolation, backwards in time to $x(t)$, $t<0$.

Hence using (22), we get the inequality

$$\int_1^{\infty} \|x^*(L+h)\|\, dh \leqslant \int_1^{\infty} \frac{dh}{h} \left(h^2 \int_h^{\infty} |g(u)|^2\, du \right)^{1/2}$$

and $x^*(L+h)$ has been shown to be integrable. However,

$$h^2 \int_h^{\infty} |g(u)|^2\, du = \frac{1}{h} \int_h^{\infty} \left(\frac{h}{u}\right)^3 u^3 |g(u)|^2\, du$$

which is dominated by

$$\frac{1}{h} \int_h^{\infty} u^3 |g(u)|^2\, du = 0(h^{-1})$$

Hence

$$\int_1^\infty \|x^*(L+h)\|\,dh \leqslant \text{constant} \times \int_1^\infty \frac{dh}{h^{3/2}} < \infty$$

Also g is integrable over $(-\infty, \infty)$ since

$$\left[\int_{-\infty}^{\infty} |g(u)|\,du\right]^2 \leqslant \int_{-\infty}^{\infty} \frac{du}{1+u^2} \int_{-\infty}^{\infty} (1+u^2)|g(u)|^2\,du < \infty \qquad (25)$$

But

$$r(t) = \int_{-\infty}^{\infty} g(u)\overline{g(u+t)}\,du; \qquad g(u)=0 \text{ for } u<0$$

so that

$$\int_A^B |r(t)|\,dt \leqslant \int_{-\infty}^{\infty} |g(u)|\,du \int_A^B |g(u+t)|\,dt - \infty$$

so that $r(t)$ is also seen to be integrable.

Using again $x^*(t)-x(t) \perp x(s); \qquad s \in T$

$$\begin{aligned} E_\alpha\{I[x(s)-\alpha]\} &= \int_{-\infty}^{\infty} E_\alpha\{x^*(t)[x(s)-\alpha]\}\,dt \\ &= \int_{-\infty}^{\infty} r(t)\,dt \equiv \text{constant}; \qquad s \in T \end{aligned}$$

But the integral of $r(t)$ is proportional to the spectral density at $\lambda=0$. If this integral, which is proportional to $f(0)$, is not zero, we can find a constant γ such that $\gamma I = \alpha^*$. We have established the following theorem.

THEOREM 4

If $x(t)$ is a purely nondeterministic stationary process satisfying (23) *and with spectral density positive at $\lambda=0$, the BLUE α^* can be represented as proportional to the integral of the linear predictor $x^*(t)$ based on $x(s)$, $s \in [0, L]$*

$$\alpha^* = \text{constant} \int_{-\infty}^{\infty} x^*(t)\,dt$$

Example 1 Let us apply this to the stationary Markov process with covariance function $r(t)=\exp(-\beta|t|)$. It is easy to calculate the linear predictor

$$x^*(L+h)=e^{-\beta h}x(L); \qquad h>0$$
$$x^*(t)=x(t); \qquad t\in T$$
$$x^*(L-h)=e^{\beta h}x(0); \qquad h>0$$

Hence

$$I=\int_{-\infty}^{\infty} x^*(t)\,dt=\frac{1}{\beta}x(0)+\frac{1}{\beta}x(L)+\int_0^L x(t)\,dt$$

so that normalizing I by multiplication with $\beta(2+\beta L)^{-1}$ we arrive again at the solution (1).

Sometimes the form of the BLUE is unexpected, as in this example.

Example 2 With ξ as a normalized orthogonal process, let

$$y(t)=\int_{t-1}^{t} \xi(du)=\xi(t)-\xi(t-1)$$

so that $y(t)$ has the covariance function

$$r(t)=\begin{cases} 1-|t|; & |t|<1 \\ 0; & \text{otherwise} \end{cases}$$

Let $T=(0, L)$, $L\in\mathbb{N}$, and consider the unbiased estimate

$$\alpha^* = \frac{1}{L+1}\sum_{n=0}^{L} x(n) \tag{26}$$

For $s\in(n, n+1)$, integral n, the stochastic variable $x(s)$ is uncorrelated with all the terms in α^* except with $x(n)$ and $x(n+1)$, so that

$$E_\alpha\{\alpha^*[x(s)-\alpha]\}=\frac{1}{L+1}[r(s-n)+r(s-n-1)]=\frac{1}{L+1} \tag{27}$$

For $s=n$ (27) also holds, since $x(s)$ is continuous in the mean. Hence Theorem 1 tells us that α^* is the BLUE with variance $1/(L+1)$. The somewhat surprising estimate (26) is also equidistributed; however, not in the way $\bar{x}$ is equidistributed, but only over the discrete set of integers. We have not obtained the solution when L is not an integer (see Note 4.3.B).

So far we have had $\psi(t)\equiv$constant. Let us consider more general regression functions and start with an example.

Example 3 A generalization of Example 2 in Section 2 of some interest is when $r(s,t)=V[\min(s,t)]$, where V is a positive, strictly increasing C_2 function with V' bounded away from zero and $\psi(t)=t$.

Try an estimate of the form (25) where f_a and f_b are constants:

$$\alpha^* = f_a x(a) + f_b x(b) + \int_a^b f(t)x(t)\,dt$$

We want to satisfy, $s \in (a,b)$,

$$E_\alpha\big[\alpha^*[x(s)-\alpha s]\big] = f_a V(a) + f_b V(s)$$

$$+\int_a^s V(t)f(t)\,dt + V(s)\int_s^b f(t)\,dt = Cs \tag{28}$$

Differentiation yields

$$f_b V'(s) + V'(s)\int_s^b f(t)\,dt \equiv C$$

or

$$\int_s^b f(t)\,dt \equiv -f_b + \frac{C}{V'(s)}$$

or

$$f(t) = C\frac{V''(s)}{[V'(s)]^2}$$

This can actually be made a solution by choosing f_a and f_b appropriately, since substituting in the right-hand side of (28) we get

$$f_a V(a) + f_b V(s) + C\int_a^s \frac{V(t)V''(t)}{[V'(t)]^2}\,dt + CV(s)\int_s^b \frac{V''(t)}{[V'(t)]^2}\,dt$$

Integrating the first integral by part, this becomes

$$= f_a V(a) + f_b V(s) - C\frac{V(s)}{V'(s)}\Big|_a^s + C(s-a) - CV(s)\frac{1}{V'(s)}\Big|_s^b$$

$$= f_a V(a) + f_b V(s) + C\frac{V(a)}{V'(a)} + C(s-a) - C\frac{V(b)}{V'(b)}$$

With

$$f_b = 0$$

$$f_a = \frac{C}{V(a)}\left[\frac{V(b)}{V'(b)} - \frac{V(a)}{V'(a)} + a\right]$$

the normal equation is satisfied, and we have found the BLUE explicitly.

Say now that we have a general continuous regression function $\psi(t)$, quadratically integrable of course over any bounded interval, but not over $\mathbb{R}$. Also assume that the process has an absolutely continuous spectrum f bounded by a constant $f_{\max}$.

Consider for fixed t the class M of *ψ-unbiased predictors* from T to t, namely, the topological closure of the set of all elements of the form

$$\sum_{\nu=1}^{n} c_\nu x(t_\nu); \qquad t_\nu \in T, \quad \sum_{\nu=1}^{n} c_\nu \psi(t_\nu) = \psi(t)$$

The set M deserves its name since for any $z \in M$ the expected value of z is $m(t) = \alpha\psi(t)$ for all values of α.

Just as before we can select the unique unbiased predictor of minimum mean square error, denote it by $\hat{x}_\psi(t)$. It satisfies the normal equation

$$E_\alpha\{[\hat{x}_\psi(t) - x(t)]x(s)\} = k\psi(s); \qquad s \in T \tag{29}$$

with some constant k. We can also show, just as in Lemma 1, that $\|\hat{x}_\psi(t') - \hat{x}_\psi(t'')\| \leqslant \|x(t') - x(t'')\|$, and therefore a statement analogous to Lemma 3 holds.

Consider the stochastic variable

$$\alpha_A^* = N_A^{-1}\int_{-A}^{A} \psi(t)\hat{x}_\psi(t)\,dt; \qquad N_A = \int_{-A}^{A} \psi^2(t)\,dt$$

Then

$$\alpha_A^* = \alpha + N_A^{-1}\int_{-A}^{A} \psi(t)\hat{y}_\psi(t)\,dt; \qquad \hat{y}_\psi(t) = \hat{x}_\psi(t) - m(t)$$

so that it is an unbiased estimate of α. Also we have the inequality

$$\|N_A^{-1}\int_{-A}^{A} \psi(t)\hat{y}_\psi(t)\,dt - N_B^{-1}\int_{-B}^{B} \psi(t)\hat{y}_\psi(dt)\|^2$$

$$\leqslant \|N_A^{-1}\int_{-A}^{A} \psi(t)y(t)\,dt - N_A^{-1}\int_{-A}^{A} \psi(t)y(t)\,dt\|^2$$

just as in (21). The latter expression can be written as

$$\int_{-\infty}^{\infty} |\phi_A(\lambda)-\phi_B(\lambda)|^2 f(\lambda)\, d\lambda \leqslant f_{\max} \|\phi_A-\phi_B\|^2 \tag{30}$$

introducing the Fourier transform

$$\phi_A(\lambda)=N_A^{-1}\int_{-A}^{A} e^{it\lambda}\psi(t)\,dt=\int_{-\infty}^{\infty} e^{it\lambda}\psi_A(t)\,dt$$

Using Parseval's relation we see that the L_2-norm on the right-hand side of (30) is bounded by a constant times the L_2-norm of the difference between ψ_A and ψ_B. For $A<B$ (and similarly for $A>B$) the latter norm is

$$\begin{aligned}(N_A^{-1}-N_B^{-1})^2\int_{-A}^{A}\psi^2(t)\,dt+N_B^{-2}\int_{-B}^{-A}\psi^2(t)\,dt+N_B^{-2}\int_{A}^{B}\psi^2(t)\,dt\\ =\frac{1}{N_A}(1-N_AN_B^{-1})^2+N_B^{-2}(N_B-N_A)\end{aligned} \tag{31}$$

Of course $N_A \leqslant N_B$ so that (31) is bounded by

$$N_A^{-1}+N_B^{-1}\to 0 \text{ as } A,\, B\to\infty$$

Hence the limit of α_A^* exists (in the mean) as $A\to\infty$; denote it by α^*. It is obviously an unbiased estimate.

We need to assume a bit more about the way that N_A diverges to $+\infty$ when A increases. We assume that the limit (32) exists (it could, for example, be zero),

$$\lim_{A\to\infty} N_A^{-1}\int_{-A}^{A}\psi(t)\,dt \tag{32}$$

We then get for $s_1, s_2 \in T$, using (29),

$$\begin{aligned}E_\alpha[\alpha^*[x(s_1)-x(x_2)]] &= \lim_{A\to\infty} N_A^{-1}\int_{-A}^{A}\psi(t)\{E\hat{x}_\psi(t)[x(s_1)-x(s_2)]\}\,dt\\ &= \lim_{A\to\infty} N_A^{-1}\int_{-A}^{A}\psi(t)\{E[\hat{x}_\psi(t)-x(t)]\\ &\qquad\times[x(s_1)-x(s_2)]\\ &\qquad+Ex(t)[x(s_1)-x(s_2)]\}\,dt\\ &= \lim_{A\to\infty} N_A^{-1}k\int_{-A}^{A}\psi(t)[\psi(s_1)-\psi(s_2)]\,dt\\ &\quad+\lim_{A\to\infty} N_A^{-1}\int_{-A}^{A}\psi(t)[r(t-s_1)-r(t-s_2)]\,dt\\ &\quad+\lim_{A\to\infty} N_A^{-1}k^2\int_{-A}^{A}\psi^2(t)[\psi(s_1)-\psi(s_2)]\,dt\end{aligned}$$

The first limit exists due to (32) and is proportional to $\psi(s_1)-\psi(s_2)$. The third limit is equal to $k^2[\psi(s_1)-\psi(s_2)]$. To evaluate the third limit we rewrite it in spectral form:

$$\lim_{A\to\infty}\int_{-\infty}^{\infty} f(\lambda)\left[\phi_1(\lambda;A)-\phi_2(\lambda;A)\right]d\lambda \tag{33}$$

where we have introduced the Fourier transform

$$\phi_\nu(\lambda,A)=N_A^{-1}\int_{-A}^{A}\psi(t)\exp(i(t-s_\nu)\lambda)\,dt=\int_{-\infty}^{\infty}\exp(it\lambda)\psi_A(t+s_\nu)\,dt$$

With Schwarz's inequality, Parseval's relation, and using the boundedness of f, the integral in (33) is bounded by a constant times the L_2-norm

$$\|\psi_A(t+s_1)-\psi_A(t+s_2)\|_2 \tag{34}$$

For $0<s_1<s_2<L$ the square of (34) is seen to be bounded by $2N_A^{-1}$ and hence tends to zero. Hence we have shown that

$$E_\alpha\{\alpha^*[x(s_1)-x(s_2)]\}=\text{constant}\times[\psi(s_1)-\psi(s_2)]$$

Therefore α^* satisfies the normal equation and coincides with the BLUE so that we can state the following theorem.

THEOREM 5

If $\psi(t)$ is continuous but not quadratically integrable over $(-\infty,\infty)$, satisfying (32), *and if the process has an absolutely continuous spectrum with bounded density, the* BLUE α^* *can be expressed in terms of the optimal ψ-unbiased predictor $\hat{x}_\psi$ as*

$$\alpha^*=\lim_{A\uparrow\infty}\frac{\int_{-A}^{A}\psi(t)\hat{x}_\psi(t)\,dt}{\int_{-A}^{A}\psi^2(t)\,dt}$$

Theorems 3–5 clearly bring out the connection between the BLUE and the corresponding predictor. In particular they tell us that the BLUE can be seen as the limit of an integral of the unbiased predictor. We shall now see that this relation can be strengthened. We do this for discrete time; the continuous case has not been settled.

Consider therefore the weakly stationary process $x(t)$ with constant mean m, the BLUE α^* based on x_t, $t\in T=\{-n+1,-n+2,\ldots,0\}$, and the unbiased predictor $\hat{x}_t$ based on the same observations. The choice of the discrete interval T above is not important and is made only for convenience.

We show, under certain conditions, that

$$\alpha^* = \lim_{N\uparrow\infty} \hat{x}_N \tag{35}$$

but a moment's reflection shows that (35) cannot hold in complete generality. Indeed if x_t is, for example, periodic, say with period 2, then $\hat{x}_N$ does not converge to anything at all as N tends to infinity.

THEOREM 6

If x_t has an absolutely continuous spectrum whose density f bounded away from zero, then (35) *holds, and if $f \in C_p$, then*

$$\|\alpha_N - \alpha^*\| = o\left(\frac{n}{N^p}\right) \tag{36}$$

Proof We first establish an auxiliary result. Consider an arbitrary element $x \in L_2(x_t; t \in T)$ and let μ_N denote the correlation coefficient between x and x_N. Then $\mu_N \to 0$ as N tends to infinity. Indeed

$$\mu_N = \frac{\int_{-\pi}^{\pi} \overline{P(e^{i\lambda})e^{iN\lambda}} f(\lambda)\, d\lambda}{\left[\int_{-\pi}^{\pi} |P(e^{i\lambda})|^2 f(\lambda)\, d\lambda \int_{-\pi}^{\pi} f(\lambda)\, d\lambda\right]^{1/2}} \tag{37}$$

with the usual trigonometric polynomial

$$P(e^{i\lambda}) = \sum_1^n a_t e^{-it\lambda}$$

associated with the element x. It is no restriction to assume $\|x\| = 1$ in (37) so that

$$1 = \int_{-\pi}^{\pi} |P(e^{i\lambda})|^2 f(\lambda)\, d\lambda \geqslant 2\pi f_{\min} \sum_1^n |a_t|^2; \qquad f_{\min} = \min_{\lambda} f(\lambda)$$

Hence $|a_t| \leqslant (2\pi f_{\min})^{-1/2}$. But (37) gives, in terms of the covariances r_t,

$$\mu_N = \frac{1}{\sqrt{r_0}} \sum_1^n a_t r_{N+t}$$

The Riemann-Lebesgue lemmas tells us that $r_t = 0(t^{-p})$ for large t, so that

$$\mu_N = o\left(\frac{n}{N^p}\right) \tag{38}$$

This leads to (36). Indeed we know that $\hat{x}_N$ is the best unbiased predictor, so that since α^* is also unbiased,

$$\|\hat{x}_N - x_N\|^2 \leqslant \|\alpha^* - x_N\|^2 \tag{39}$$

But we can write

$$\|\hat{x}_N - x_N\|^2 = \|\hat{x}_N\|^2 + r_0 - 2\|\hat{x}_N\|\sqrt{r_0}\,\mu \tag{40}$$

where μ is the correlation coefficient between $\hat{x}_N$ and x_N. Similarly,

$$\|\alpha^* - x_N\|^2 = \|\alpha^*\|^2 + r_0 - 2\|\alpha^*\|\sqrt{r_0}\,\mu' \tag{41}$$

where μ' is the correlation coefficient between α^* and x_N.

Applying (40) and (41) to (39), we get the second inequality in

$$0 \leqslant \|\hat{x}_N\|^2 - \|\alpha^*\|^2 \leqslant 2\sqrt{r_0}\,(\|\hat{x}_N\||\mu| + \|\alpha^*\||\mu'|)$$

The first inequality expresses that α^* is a BLUE. But $\|\hat{x}_N\| \leqslant \|x_N\| + \|\hat{x}_N - x_N\|$, and the last term must be uniformly bounded (compare, for example, with the unbiased predictor x_0) so that $\|\hat{x}_N\| \to \|\alpha^*\|$. Furthermore if $f \in C_p$, then, from (38),

$$\|\hat{x}_n\|^2 - \|\alpha^*\|^2 = o\left(\frac{n}{N^p}\right) \tag{42}$$

Now we use the parallelogram law, as in (2.7), and get

$$\|\hat{x}_N - \alpha^*\|^2 = 2\|\alpha^*\|^2 + 2\|\hat{x}_N\|^2 - 4\left\|\frac{\hat{x}_N + \alpha^*}{2}\right\|^2$$

Since both $\hat{x}_N$ and α^* are unbiased, we get, with $\|\alpha^*\|^2 = V$ and using (42),

$$\|\hat{x}_N - \alpha^*\|^2 \leqslant 2V + 2V + o\left(\frac{n}{N^p}\right) - 4V$$

which is (36). Q.E.D.

Theorem 6 gives an answer of the type we wanted, but it is disturbing that n, the length of the interval of observation, enters into the bound. This seems intuitively not motivated, and we shall get rid of it, although at the expense of the order of decrease in N given in (36).

The crucial step in the above proof is the analysis of the correlation coefficients μ and μ'. Introduce the *two subspaces, backwards and forward*,

$$B = L_2[x_t; t \leqslant 0]$$
$$F_h = L_2[x_t; t \geqslant h]$$

and consider the extremum problem

$$\mu_h = \sup_{\substack{x \in B \\ y \in F_h}} [\text{correlation coefficient } (x, y)] \tag{43}$$

It is clear that μ_h is nonincreasing and hence tends to a limit as $h \uparrow +\infty$. Actually the sup in (43) can be shown to be attained, but we do not make use of this. Can we assert that μ_h tends to zero at a certain rate, when f is assumed to be in C_p?

This problem is more sensitive than the earlier one and has received a good deal of attention. If $\mu_h \to 0$ as $h \to \infty$, the process is said to be *completely regular* (see Note 4.3.C). Here we derive only a simple bound on μ_h; for more information in complete regularity, see Notes.

Let x_t be a real-valued and weakly stationary process with discrete time. For the moment we only assume that it is completely nondeterministic (see Chapter 2, Section 4). Later we strengthen the condition somewhat.

We can then write

$$x_t = \sum_{\nu=0}^{\infty} c_\nu \xi_{t-\nu} \tag{44}$$

with $L_2[x_s; s \leqslant t] = L_2[\xi_s; s \leqslant t]$ and with

$$\xi_t = \int_{-\pi}^{\pi} e^{it\lambda} \frac{Z(d\lambda)}{g(e^{i\lambda})}$$

where Z is the spectral process of $\{x_t\}$ and $g(e^{i\lambda})$ should be interpreted as the radial limits of

$$g(z) = \sum_{\nu=0}^{\infty} d_n z^n$$

(See equations (5) and (15) in Chapter 2, Section 5; we also need the construction of g via (4) in the same section.)

An arbitrary element $x \in B$ with norm 1 can then be written as

$$x = \sum_{n=0}^{\infty} a_n \xi_{-n}; \qquad \sum_{n=0}^{\infty} a_n^2 = 1$$

Let y be an arbitrary element in F_h with norm 1. To find a representation of it similar to (44) we consider the *time reversed process*

$$y_t = x_{-t} = \int_{-\pi}^{\pi} e^{-it\lambda} Z(d\lambda) = \int_{-\pi}^{\pi} e^{it\lambda} z(-d\lambda)$$

so that $\{y_t\}$ has $Z(-d\lambda)$ as its spectral process. Note that $\{y_t\}$ has exactly the same geometry as $\{x_t\}$, so that it has the same spectral density f, the same function $g(z)$, and so on. Therefore

$$y_t = \sum_{n=0}^{\infty} c_n \eta_{t-n}$$

where

$$\eta_t = \int_{-\pi}^{\pi} e^{it\lambda} \frac{Z(-d\lambda)}{\overline{g(e^{i\lambda})}} = \int_{-\pi}^{\pi} e^{-it\lambda} \frac{Z(d\lambda)}{g(e^{i\lambda})}$$

The element y can then be expressed as

$$y = \sum_{n=0}^{\infty} b_n \eta_{-h-n}; \qquad \sum_{n=0}^{\infty} b_n^2 = 1 \tag{45}$$

Combining (44) with (45) we get the correlation coefficient

$$(x, y) = \sum_{n,m=0}^{\infty} h_{n+m} a_n b_m$$

with

$$h_k = (\xi_{-n}, \eta_{-h-m}) = \int_{-\pi}^{\pi} e^{-ik\lambda} e^{-ih\lambda} \frac{g(e^{i\lambda})}{\overline{g(e^{i\lambda})}} d\lambda \tag{46}$$

where we have used $E[|Z(d\lambda)|^2] = f(\lambda)\,d\lambda = |g(e^{i\lambda})|^2\,d\lambda$ and have put $k = n+m$.

Our problem hence reduces to finding the norm of the *Hankel* bilinear form $a^T Hb$ with the infinite matrix, for fixed positive lag h,

$$H_h = \{h_{n+m};\ n, m = 0, 1, \ldots\}$$

which makes it possible to bring in powerful analytical techniques that go back to Hilbert and Schur in their first formulations.

In order to get sufficiently strong bounds we now assume that $f \in C_p^+$ with

$$|f^{(p)}(\lambda)| \leqslant C$$

In the construction of $g(z)$ we started with the integrable function $\ln w(\lambda)$ and constructed the analytic function $h(z)$ whose real part is equal to $\ln w(\lambda)$ on $z = e^{i\lambda}$, $w(\lambda) = 2\pi f(\lambda)$, and whose imaginary part is zero at zero. Then, with a classical representation,

$$h(z) = \frac{1}{2\pi} \int_{-\pi}^{\pi} \ln w(\theta) \frac{e^{i} + z}{e^{i\theta} - z} d\theta$$

[see, e.g., Goursat (1942), p. 143], from which we get the imaginary part

$$\operatorname{Im}[h(re^{i\phi})] = \frac{r}{\pi} \int_{-\pi}^{\pi} \ln w(\theta) \frac{\sin(\phi - \theta)}{1 - 2r\cos(\phi - \theta) + r^2} d\theta \tag{47}$$

or, in terms of the kernel

$$K_r(\psi) = \frac{r}{\pi} \frac{\sin \psi}{1 - 2r\cos\psi + r^2}$$

we can write (47) as

$$I_r(\phi) = \operatorname{Im}[h(re^{i\phi})] = \int_{-\pi}^{\pi} u(\theta) K_r(\phi - \theta)\, d\theta \tag{48}$$

with $u(\theta) = \ln w(\theta)$.

Under the differentiality conditions above on w and t we find

$$I_r' = \frac{d}{d\phi} I_r(\phi) = \int_{-\pi}^{\pi} u(\theta) \frac{\partial}{\partial \phi} K_r(\phi - \theta)\, d\theta$$

$$= -\int_{-\pi}^{\pi} u(\theta) \frac{\partial}{\partial \theta} K_r(\phi - \theta)\, d\theta$$

Integration by parts yields

$$I_r'(\phi)=\int_{-\pi}^{\pi} u'(\theta)K_r(\phi-\theta)\,d\theta$$

and similarly for higher order derivatives up to order p.

Returning to (48), it can be written as

$$I_r(\phi)=\frac{1}{2\pi}\int_{-\pi}^{\pi} u(\theta)\frac{\partial}{\partial\theta}\ln\left[1-2r\cos(\phi-\theta)+r^2\right]d\theta$$

$$=-\frac{1}{2\pi}\int_{-\pi}^{\pi} u'(\theta)\ln\left[1-2r\cos(\phi-\theta)+r^2\right]d\theta$$

As $r\uparrow 1$, and if $p\geqslant 1$, we get by dominated convergence

$$\lim_{r\uparrow 1} I_r(\phi)=-\frac{1}{2\pi}\int_{-\pi}^{\pi} u'(\theta)\ln\left[1-\cos(\phi-\theta)\right]d\theta$$

where the singular kernel has only a weak (logarithmic) singularity at $\theta=\phi$. But

$$\frac{g(z)}{\overline{g(z)}}=\exp\left\{\tfrac{1}{2}\left[h(z)-\overline{h(z)}\right]\right\}=\exp\left[I_r(\phi)\right]$$

The radial limit of $g/\bar{g}$ is then

$$\rho(\phi)=\exp\left\{-\frac{1}{2\pi}\int_{-\pi}^{\pi} u'(\theta)\ln\left[1-\cos(\phi-\theta)\right]d\theta\right\}$$

Arguing the same manner, we see that if $u(\theta)$ has p continuous derivatives, then $\rho(\phi)$ has $p-1$ continuous derivatives. But $w\in C_p^+$ so that $u=\ln w\in C_p$ as required.

Then the Fourier coefficients of $\rho(\phi)$ are bounded by

$$o(|\nu|^{-p+1}) \tag{49}$$

for large ν, according to the Riemann-Lebesgue lemma. Hence taking the trace of H_h using (46) and (49), we get the bound

$$\|H_h\|\leqslant \operatorname{tr} H_h=o\left(\sum_{n,m=0}^{\infty}\frac{1}{(h+n+m)^{p-1}}\right)$$

But the sum is bounded by

$$\text{constant} \times \sum_{n=0}^{\infty} \frac{1}{(h+n)^{p-2}} = o\left(\frac{1}{h^{p-3}}\right)$$

Hence if $p \geqslant 3$, we have the desired result

$$\mu_h = o(h^{-p+3}) \tag{50}$$

We can sum up in the following Lemma.

Lemma 4 If $f \in C_p$, $p \geqslant 3$, the maximum correlation coefficient μ_h satisfies (50).

This gives immediately the following corollary (see Note 4.3.D).

Corollary 1 For a stationary process with an absolutely continuous spectrum whose density $f \in C_p$ the unbiased predictor $\hat{x}_N$ tends to the BLUE α^* at a rate

$$\|\hat{x}_N - \alpha^*\| = o\left(\frac{1}{N^{p-3}}\right)$$

independent of the sample size n.

The smoother the spectrum is, the closer is the unbiased predictor to the BLUE. Without smoothness assumption, however, they need not be close at all.

Let us now return to the process in Theorem 1 where it was a solution of a first-order stochastic differential equation. The covariance function $r(s,t) = e^{-\beta|s-t|}$ turns out on inspection to be the Green's function of the Sturm-Liouville problem

$$y'' - \beta^2 y = 0; \qquad y(-\infty) = y(\infty) = 0$$

This is not an accident. Similar statements can be made about processes satisfying higher order stochastic differential equations.

Unfortunately *the above Sturm-Liouville problem is for the entire real line, while our main interest is for T bounded.* It is possible to deal with this case too, as has been shown by Dolph-Woodbury (1952). This important work is not as well known as it deserves to be, and we now present an outline of it.

Let $y(t)$ be a stochastic process, not necessarily stationary, satisfying the nth-order stochastic differential equation

$$L_t y = \dot{w}; \qquad \|w(dt)\|^2 = dt \tag{51}$$

where

$$L_t y(t) = a_0(t) y^{(n)}(t) + a_1(t) y^{(n-1)}(t) + \cdots + a_n(t) y(t)$$

with all $a_k \in C_{n-1}$ with $a_0(t) \neq 0$.

Introduce the formal adjoint of the differential operator L_t:

$$M_t y(t) = [a_0(t) y(t)]^{(n)} - [a_1(t) y(t)]^{(n-1)} + \cdots + (-1)^n a_0(t) y(t)$$

We need the fundamental solutions to an initial value problem for both L and M. Define the *Riemann function* $R(t, s)$ (see Note 4.3.E) for the initial value problem

$$\begin{aligned} &L_s y(s) = f(s) \\ &y^{(k)}(t_0) = 0; \qquad k = 0, 1, 2, \ldots, n-1 \end{aligned} \tag{52}$$

as the solution of

$$\begin{aligned} &M_s[R(t, s)] = 0 \\ &\left(\frac{\partial^k R(t, s)}{\partial s^k} \right)_{s=t} = \begin{cases} 0; & k = 0, 1, \ldots, n-2 \\ \dfrac{(-1)^{n-1}}{a_0(t)}; & k = n-1 \end{cases} \end{aligned} \tag{53}$$

In the same way the *adjoint Riemann function* $R^*(t_0, s)$ is the solution of

$$\begin{aligned} &L_s[R^*(t_0, s)] = 0 \\ &\left(\frac{\partial^k R^*(t_0, s)}{\partial s^k} \right)_{s=t_0} = \begin{cases} 0; & k = 0, 1, 2, \ldots, n-2 \\ \dfrac{1}{a_0(t)}; & k = n-1 \end{cases} \end{aligned} \tag{54}$$

With these two Riemann functions we prove two auxiliary results.

Lemma 5 R and R^* satisfy (55)–(60):

$$R(t,t_0)=R^*(t_0,t) \tag{55}$$

$$L_t[R(t,t_0)]=0 \tag{56}$$

$$M_{t_0}[R^*(t_0,t)]=0 \tag{57}$$

$$\left(\frac{\partial^k R(s,t)}{\partial s^k}\right)_{s=t}=\left(\frac{\partial^k R^*(s,t)}{\partial s^k}\right)_{s=t}=0; \qquad k=0,1,2,\ldots,n-2 \tag{58}$$

$$\left(\frac{\partial^{n-1} R(s,t)}{\partial s^{n-1}}\right)_{s=t}=\frac{1}{a_0(t)} \tag{59}$$

$$\left(\frac{\partial^{n-1} R^*(s,t)}{\partial s^{n-1}}\right)_{s=t}=\frac{(-1)^{n-1}}{a_0(t)} \tag{60}$$

Proof Use the classical identity of Lagrange:

$$\int_{t_0}^{t}\{z(s)L_s[y(s)]-y(s)M_s[z(s)]\}\,ds$$

$$=\sum_{j=0}^{n-1} y^{(j)}(s)\left\{\sum_{k=j}^{n-1}(-1)^{k-j}[a_{n-k-1}(s)z(s)]^{k-j}\right\}_{s=t_0}^{s=t} \tag{61}$$

In (61) put $z(s)=R(t,s)$ and $y(s)=R^*(t_0,s)$; this yields (55) in the lemma. We then get (56) and (57) from the differential equations in (53) and (54). The remaining relations follow from the initial value assumption. Q.E.D.

Lemma 6 The solution of (52) is given by

$$y(t)=\int_{t_0}^{t}R(t,s)f(s)\,ds \tag{62}$$

Proof In Lagrange's identity put $z(s)=R(t,s)$ and use $L_s y(s)=f(s)$. We also have, using (58) and (59) in Lemma 5,

$$y^{(k)}(t)=\int_{t_0}^{t}\frac{\partial^k R(t,s)}{\partial t^k}f(s)\,ds; \qquad k=0,1,2,\ldots,n-1 \tag{63}$$

as well as

$$y^{(n)}(t)=\int_{t_0}^{t}\frac{\partial^n R(t,s)}{\partial t^n}f(s)\,ds+\left(\frac{\partial^{n-1}R(t,s)}{\partial t^{n-1}}\right)_{s=t}f(t)$$

$$=\int_{t_0}^{t}\frac{\partial^n R(t,s)}{\partial t^n}f(s)\,ds+\frac{f(s)}{a_0(t)} \tag{64}$$

Multiplying the relations (62)–(64) by $a_k(t)$ and adding, we see that (62) is indeed a solution. Q.E.D.

Applied to the stochastic differential equation (51), we can write the $y(t)$-process as a one-sided moving average (choose t_0 as $-\infty$):

$$y(t)=\int_{-\infty}^{t}R(t,u)w(du) \tag{65}$$

keeping in mind the weak interpretation of the stochastic differential equation (see Chapter 2, Section 2). Hence $y(t)$ has *a covariance function that can be expressed in terms of the Riemann function* R:

$$r(t,s)=\int_{-\infty}^{\min(t,s)}R(t,u)R(s,u)\,du \tag{66}$$

THEOREM 7

The covariance function (66) *satisfies the equation* $Nr=0$ *in both* s *and* t *for* $s\neq t$, *where* N *is the formally self-adjoint operator* M_tL_t. *Also*:

(a) *r and its first* $2n-2$ *partial derivatives are continuous at* $s=t$.

(b) *At* $s=t$ *its derivative of order* $2n-1$ *with respect to* t *has a jump at* $s=t$ *equal to* $(-1)^n/a_0^2(s)$.

Proof Using (56) in Lemma 5, we see that for $t>s$

$$L_t[r(t,s)]=\int_{-\infty}^{s}L_t[R(t,u)]R(s,u)\,du=0$$

For $t<s$ we can write $r(t,s)$ as in (66), and Lemma 6 implies that $L_t[r(t,s)]=R(s,t)$ and

$$M_tL_t[r(t,s)]=M_t[R(s,t)]=0$$

Similarly for t replaced by s.

To show (a) we use (66) and Lemma 5 and get, for $t>s$,

$$\frac{\partial^k r(t,s)}{\partial t^k}=\int_{-\infty}^{s}\frac{\partial^k R(t,u)}{\partial t^k}R(s,u)\,du; \qquad k=1,2,\dots,2n-1 \tag{67}$$

The derivatives in the integrand exist since Lemma 5 and (65) tell us that

$$a_0(t)\frac{\partial^n R(t,s)}{\partial t^n}=\sum_{k=1}^{n}a_k(t)\frac{\partial^{n-k}R(t,s)}{\partial t^{n-k}}$$

and all $a_r(t)$ are in C^{n-1}.

For $t<s$ we use Lemma 5 [(58) and (59)],

$$\frac{\partial^k r(t,s)}{\partial t^k}=\int_{-\infty}^{t}\frac{\partial^k R(t,u)}{\partial t^k}R(s,u)\,du; \qquad k=0,1,\dots,n-1$$

and

$$\frac{\partial^n r(t,s)}{\partial t^n}=\int_{-\infty}^{t}\frac{\partial^n R(t,u)}{\partial t^n}R(s,u)\,du+R(s,t)\left(\frac{\partial^{n-1}R(t,s)}{\partial t^{n-1}}\right)_{s=t} \tag{68}$$

For $n<k\leqslant 2n-1$ we get extra terms of the type

$$\frac{\partial^m}{\partial t^m}\left[R(s,t)\left(\frac{\partial^p R(t,s)}{\partial t^p}\right)_{s=t}\right]$$

with $m+p<2n-1$, $m=0,1,\dots,n-1$, $p=0,1,\dots,n-1$. But (53) tells us that there will be only one contribution when $t=s$, namely, for $k=2n-1$, [see last term of (68)]. Hence

$$\lim_{\varepsilon\downarrow 0}\left(\frac{\partial^k r(t,s)}{\partial t^k}\right)_{t=s-\varepsilon}=\int_{-\infty}^{t}\frac{\partial^k R(s,u)}{\partial s^k}R(s,u)\,du$$
$$+\delta\left(\frac{\partial^{n-1}R(t,s)}{\partial t^{n-1}}\right)_{s=t}\left(\frac{\partial^{n-1}R(t,s)}{\partial s^{n-1}}\right)_{s=t} \tag{69}$$

where $\delta=0$ except for $k=2n-1$ when $\delta=1$. Put $s=t$ in (67), and subtracting it from (69), we get stated result in (b). Q.E.D.

We are now ready to deal with a bounded interval $T=(0,L)$ and treat the case $n=1$; the higher order case can be treated in a similar way.

Consider a self-adjoint differential operator

$$N_t[y(t)] = [p(t)y'(t)]' + q(t)y(t); \qquad p \neq 0$$

possessing a Green's function over an interval containing T, and such that

$$\lim_{\varepsilon \downarrow 0} \left[\left(\frac{\partial r(t,s)}{\partial t} \right)_{t=s+\varepsilon} - \left(\frac{\partial r(t,s)}{\partial t} \right)_{t=s-\varepsilon} \right] = -\frac{1}{p(s)} \tag{70}$$

We can now pick two linearly independent solutions ϕ_1 and ϕ_2 of $N_t\phi = 0$, so that the Wronskian of ϕ_1 and ϕ_2 does not vanish identically. Using Lagrange's identity, now in the form

$$0 = \phi_1 N_t(\phi_2) - \phi_2 N_t(\phi_1) = \frac{d}{dt}[p(\phi_1\phi_2' - \phi_2\phi_1')]$$

we get

$$\frac{d}{dt}[p \text{ Wronskian}(\phi_1, \phi_2)] = 0$$

Hence the Wronskian is proportional to $1/p(t)$. However, with

$$K(t,s) = \begin{cases} \phi_1(t)\phi_2(s); & \text{for } t \geqslant s \\ \phi_1(s)\phi_2(t); & \text{for } t < s \end{cases}$$

we have for the derivative of order $2n-1=1$

$$\left(\frac{\partial K}{\partial t} \right)_{t=s+0} - \left(\frac{\partial K}{\partial t} \right)_{t=s-0} = \phi_1'(t)\phi_2(t) - \phi_1(t)\phi_2'(t) \tag{71}$$

which is the Wronskian. Hence (71) is proportional to $1/p(t)$, as required in Theorem 7, and choosing a multiple of K, we have the covariance function.

We want to solve the normal equation

$$f(t) = \int_0^L r(t,s)G(ds) \tag{72}$$

where G is of bounded variation and $f \in C_2$. As we are going to see, it will be enough to allow G to be absolutely continuous with density g in $(0, L)$ and have jumps G_0 and G_L at the end points.

THEOREM 8

Let $r(t,s)$ be a covariance function satisfying $N_t r(t,s)=0$ for $t\neq s$, with the properties described above and of the form

$$r(t,s)=\begin{cases}\phi_1(t)\phi_2(s); & \text{for } t\geqslant s\\ \phi_1(s)\phi_2(t); & \text{for } t\leqslant s\end{cases} \tag{73}$$

where ϕ_1 and ϕ_2 are linearly independent and with $\phi_1(L)\neq 0$, $\phi_2(0)\neq 0$. Then (72) *is solved by*

$$\begin{cases} g(t)=-N_t[f(t)]\\ G_0=\dfrac{p(0)}{\phi_2(0)}[\phi_2'(0)f(0)-\phi_2(0)f'(0)]\\ G_1=-\dfrac{p(L)}{\phi_1(L)}[\phi_1'(L)f(L)-\phi_1(L)f'(L)]\end{cases}$$

Proof If a solution of (72) exists with a density g, it is clear that $g=-N_t[f(t)]$ in the open interval. Let us therefore try this form, so that (72) can be written

$$S=-\int_0^L r(t,s)N_s[f(s)]\,ds+G_0 r(t,0)+G_1 r(t,L)=f(s)$$

We show that (72) can be satisfied if we choose G_0 and G_1 as described.

First apply Lagrange's identity, which now takes the form

$$\int_0^L\{z(s)N_s[y(s)]-y(s)N_s[z(s)]\}\,ds=p(y'z-z'y)\,\Big|_0^L$$

This gives us

$$\begin{aligned}S=&-\int_0^L f(s)N_s[r(t,s)]\,ds+G_0 r(t,0)+G_1 r(t,L)\\ &-\left\{p(s)\left[f'(s)r(t,s)-\frac{\partial r(t,s)}{\partial s}f(s)\right]\right\}\Big|_{s=0}^{t}\\ &-\left\{p(s)\left[f'(s)r(t,s)-\frac{\partial r(t,s)}{\partial s}f(s)\right]\right\}\Big|_{s=t}^{L}\end{aligned}$$

A standard Green's function argument shows that the first term vanishes, and after some rearrangement, using (70), we get

$$S=f(t)+\phi_2(t)\{-p(L)[f'(L)\phi_1(L)-\phi_1'(L)f(L)]+G_L\phi_1(L)\}$$
$$+\phi_1(t)\{p(0)[f'(0)\phi_2(0)-\phi_2'(0)f(0)]+G_0\phi_2(0)\}$$

Choosing G_0 and G_1 as stated, $S=f(t)$. Q.E.D.

Theorem 8 makes it clear why the BLUE in Theorem 1 had the remarkably simple form: a continuous weight function over the open interval T and discrete weights at the end points. Similarly we can explain the BLUE in Theorem 2 by considering the differential operator of order n.

But we have not yet exploited the result in Theorems 1 and 2 to its full potential. Recalling the construction in Chapter 2, Section 6, of reproducing kernel Hilbert space (see Example 4 of Chapter 2, Section 6), we can notice that the representation of an arbitrary element in $L_2(x(s);\ x\in T)$ has a form that is quite reminiscent of our BLUE. This can be pursued much further as has been done especially in Parzen's work (see Note 4.3.4).

Say that the process $x(t)$, $t\in T$, has a reproducing kernel space H with inner product $(\cdot,\cdot)_R$, where $R(s,t)$ is the covariance kernel. There is an isometric map between $L_2(X)=L_2[x(t);\ t\in T]$ and H, denoted by $h\leftrightarrow (h,X)_R$,

$$\mathrm{Cov}[(h,X)_R,(g,X)_R]=(h,g)_R$$

In particular,

$$(R(\cdot,t),X)_R=x(t)$$
$$E[(h,X)_R]=(h,m)_R; \qquad m=\text{mean value function}$$

With the mean value function as in (1.1), the possible mean value functions span a p-dimensional subspace M. Say that we want to estimate the value of $\psi(m)$ of the fixed linear functional ψ unbiasedly. Then, if we use the estimate $\beta^*=(g,X)_R$, $g\in H$, we should have

$$E_\alpha(\beta^*)=E_\alpha[(g,X)_R]=(g,m)_R=\psi(m); \qquad \text{all } m\in M \tag{74}$$

where m is related to α by (1.1). But the resulting variance is $(g,g)_R$, so that to find the BLUE of ψ we just have to find the element g among those satisfying (74) with the smallest R-norm.

In a BLUE $\alpha^* = (\alpha_1^*, \alpha_2^*, \ldots, \alpha_p^*)$ with $\alpha_\nu^* = (g_\nu, X)_R$, $g_\nu \in H$, we should have

$$E_\alpha[(g_\nu, X)_R] = (g_\nu, m)_R = \sum_\nu \alpha_\mu (g_\nu, \psi_\nu)_R \equiv \alpha_\nu$$

The minimum variance condition then tells us that we should pick g's such that $\|g_\nu\|_R^2 = \text{minimum}$ for $(g_\nu, g_\mu)_R = \delta_{\nu\mu}$. Hence we should choose g's in M, and the usual argument gives us the following elegant representation.

THEOREM 9 REPRODUCING KERNEL ESTIMATES, PARZEN (1961).

If all $\psi_\nu \in H$ and are linearly independent, the BLUE is given by

$$\alpha^* = ((\psi_\nu, \psi_\mu)_R; \quad \nu, \mu = 1, 2, \ldots, p)^{-1} \operatorname{col}[(\psi_1, X)_R, \ldots, (\psi_p, X)_R]$$

with

$$\operatorname{Cov}(\alpha^*) = ((\psi_\nu, \psi_\mu)_R; \quad \nu, \mu = 1, 2, \ldots, p)^{-1}$$

Hence if we can find the reproducing kernel space H in a tractable explicit form, the linear estimation problem can be regarded as a good deal closer to its solution. As we know, this is possible in many important cases.

Whether we use the RKHS formalism or direct solutions in terms of the eigenfunctions of the covariance kernels $r(s,t)$, one would expect that closed-form solutions would be possible only rarely. This view is too pessimistic, however, and it is possible to deal successfully with quite a number of special covariance kernels (see Note 4.3.G).

Space does not allow us to cover this topic extensively, and we present only one such case in addition to the ones we have already seen. Put

$$r(s,t) = \begin{cases} 1 - k|s-t|, & |s-t| \leq 1/k \\ 0; & \text{otherwise} \end{cases} \tag{75}$$

with $T = [0, 1]$ and k is a positive integer. Actually this can be generalized a good deal with the same method as follows.

Introducing

$$\alpha(t) = \max(0, t - 1/k)$$

$$\beta(t) = \min(1, t + 1/k)$$

we can write our integral equation for the eigenfunctions of the covariance kernel in (75) as

$$\lambda\phi(t)=\int_{\alpha(t)}^{t}[1-k(t-s)]\phi(s)\,ds+\int_{t}^{\beta(t)}[1-k(s-t)]\phi(s)\,ds \tag{76}$$

Differentiating (76) twice, which is clearly admissible, we get

$$\lambda\phi''(t)=\begin{cases} k\phi(t+1/k)-2k\phi(t); & 0\leqslant t\leqslant 1/k \\ k\phi(t+1/k)-2k\phi(t)+k\phi(t-1/k); & 1/k\leqslant t\leqslant 1-1/k \\ \qquad -2k\phi(t)+k\phi(t-1/2); & 1-1/k\leqslant t\leqslant 1 \end{cases} \tag{77}$$

To solve the system (77) of difference-differential equations we introduce the intervals

$$i_r=(r/k,(r+1)/k); \qquad r=0,1,\ldots,k-1$$

with the associated functions

$$\begin{aligned} &f_{-1}(s)=0 \\ &f_r(s-r/k)=\phi(s); \qquad s\in i_r \\ &f_k(s)=0 \end{aligned} \tag{78}$$

We can then write (77) as

$$\lambda f_r''(x)=kf_{r-1}(x)-2kf_r(x)+f_{r+1}(x); \qquad 0<x<1/k \tag{79}$$

where we have put $x=t-r/k$.

But (79) is a system of second-order ordinary differential equations with constant coefficients so that we should attempt a solution of the form

$$f_r(x)=\sum_{\nu} a_{r\nu}e^{\omega_\nu x} \tag{80}$$

where we could allow the usual degeneracies for coinciding ω_ν. Substituting (80) in (79), we get

$$\lambda\omega_\nu^2 a_{r\nu}=ka_{r-1\nu}-2ka_{r\nu}+ka_{r+1\nu}$$

for $r=0,1,2,\ldots,k-1$. In order that (80) be a nontrivial solution, at least one coefficient $a_{r\nu}$ must be different from zero. But then $\lambda\omega_\nu^2/k$ must be an eigenvalue of the well-known Toeplitz matrix

$$M=\begin{Bmatrix} -2 & 1 & 0 & 0 & \dots & 0 \\ 1 & -2 & 1 & 0 & \dots & 0 \\ \dots & & & & & \\ 0 & 0 & 0 & 0 & \dots & -2 \end{Bmatrix}$$

whose eignevalues are of the form

$$-2+2\cos\frac{\pi\mu}{k+1}; \qquad \mu=1,2,\dots,k$$

[see, e.g., Grenander-Szegő (1957), p. 67].

On the other hand $\lambda>0$ so that ω_ν must be purely imaginary:

$$\omega_\nu=\pm i\sqrt{\frac{2k}{\lambda}\left(1-\cos\frac{\pi\nu}{k+1}\right)}=\pm 2i\sqrt{\frac{k}{\lambda}}\ \sin\frac{\pi\nu}{2(k+1)} \tag{81}$$

Let us enumerate these ω such that $\omega_\nu=-\omega_{-\nu}$, $\nu=\pm 1,\pm 2,\dots,\pm k$.

Let us now choose the constants $A_{r\nu}$ so that

$$\phi(s)=\sum_{\nu=-k}^{k} A_{r\nu}e^{\omega_\nu s}; \qquad s\in i_r \tag{82}$$

is an eigenfunction; $A_{-1\nu}=A_{k\nu}=A_{r0}=0$ corresponding to (79). But then we should have

$$\lambda\phi(s)=\alpha_r+\beta_r s+\sum_{\nu=-k}^{k}\gamma_{r\nu}e^{\omega_\nu s}; \qquad s\in i_r$$

with constants α_r, β_r, $\gamma_{r\nu}$ that can be calculated directly from (76). In particular,

$$\gamma_{r\nu}=\frac{k}{\omega_\nu^2}e^{-\omega_\nu/k}A_{r-1\nu}-2\frac{k}{\omega_\nu^2}A_{r\nu}+\frac{k}{\omega_\nu^2}A_{r+1\nu}$$

In order that (82) be a solution, we must have

$$\begin{cases} \alpha_r=0 \\ \beta_r=0 \\ \gamma_{r\nu}=\lambda A_{r\nu} \end{cases} \tag{83}$$

Fixing the value of λ, let $a=\operatorname{col}[A_{r\nu}e^{r\omega_\nu/k};\ r=0,1,\ldots,k-1]$ so that

$$\frac{k}{\omega_\nu^2}Ma=\lambda a \tag{84}$$

But (84) shows that a is proportional to that eigenvector of M that is associated with the eigenvalue $\lambda\omega_\nu^2/k$, so that with some b_ν,

$$r\text{th component of } a=b_\nu \sin\frac{(r+1)\nu\pi}{k+1}$$

and

$$A_{r\nu}=b_\nu e^{-r\omega_\nu/k}\sin\frac{(r+1)\nu\pi}{k+1} \tag{85}$$

However, at least one b_ν must be nonzero. The two first conditions in (83) give us a homogeneous system of $2k$ linear equations in the $2k$ unknowns b_ν. Putting its determinant equal to zero gives us a transcendental equation in λ determining the eigenvalues of our integral equation. Using, (81), (83), and (85), gives us the eigenfunctions.

4 ASYMPTOTIC OPTIMALITY; DISCRETE TIME

Assuming that we have solved an inference problem on a given argument space T, and obtained, for example, a consistent estimate as in the previous sections; what happens if our observations are only for $t\in T'\subset T$? As T' is made to tend to T, how should the estimates be constructed in order to be asymptotically optimal?

We could have $T=\mathbb{R}^+$ and $T'=[0,L]$, where we make L tend to infinity. Can we find an estimate, or test, based only on $x(t)$ for $t\in T'$, such that it is asymptotically optimal?

Or, the argument space $T=\mathbb{N}$ and $T'=\{1,2,3,\ldots,L\}$. Again we want to construct asymptotically optimal estimates or tests.

Or, with $T=[0,L]$, let $T'=\{(k/n)L;\ k=0,1,2,\ldots,n\}$ and ask for asymptotically efficient inference methods as $n\to\infty$ so that T' becomes more and more dense in T.

We consider mainly estimation problems in Sections 4 and 5, but the methods used extend after some modification also to other questions of inference.

As several times before in the previous section, our starting point is our old standby, Theorem 1 of Section 3, but we now ask for estimates α_L^* that are only asymptotically optimal as $L\to\infty$, by which we mean the following.

Definition 1 An unbiased linear estimate α_L^{**} of α is said to be asymptotically optimal if

$$\lim_{L\to\infty} \frac{\operatorname{Var}(\alpha_L^{**})}{\operatorname{Var}(\alpha_L^{*})} = 1$$

where α_L^* is a BLUE.

For the situation in Theorem 1 of Section 3, the BLUE was found to be [see (3.1)]

$$\alpha_L^* = \frac{1}{2+\beta L}\left[x(0) + x(L) + \beta\int_0^L x(t)\,dt\right] \tag{1}$$

It is striking that (1) is almost the same as $\bar{x}$, except for the two terms involving the boundary of T and for the fact that the weight function over T is not exactly but only approximately equal to $1/L$. On the other hand the least-squares estimate $\alpha_L^{**} = \bar{x}$ is the commonsense average:

$$\bar{x} = \frac{1}{L}\int_0^L x(t)\,dt$$

so that with $x(t) = \alpha + y(t)$, and using the unbiasedness of α_L^* and α_L^{**},

$$\alpha_L^* - \alpha_L^{**} = \frac{y(0)+y(L)}{2+\beta L} - \frac{2}{L(2+\beta L)}\int_0^L y(t)\,dt = \frac{y(0)+y(L)}{2+\beta L} - \frac{2}{2+\beta L}\bar{y}$$

Hence

$$\|\alpha_L^* - \alpha_L^{**}\| \leqslant \frac{2+2\|\bar{y}\|}{2+\beta L} \tag{2}$$

so that α_L^* is asymptotically close to α_L^{**} in relative L_2-error.

To find the asymptotic behavior of $\|\bar{x}\|$ as $L\to\infty$, we use the Cramér representation of the $y(t)$-process and get

$$\bar{y} = \int_{-\infty}^{\infty} \frac{e^{iL\lambda}-1}{iL\lambda} Z_y(d\lambda)$$

where the ratio in the integrand is interpreted as taking the value L at $\lambda = 0$. Hence

$$\|\bar{y}\|^2 = \frac{4}{L^2}\int_{-\infty}^{\infty} \frac{\sin^2(L/2)\lambda}{\lambda^2} F_y(d\lambda) \tag{3}$$

From now on we assume that F is absolutely continuous with a positive and continuous density f, $f \in C^{+}(-\infty,\infty)$. At the end of this section we examine to what extent this condition is essential.

We can then write (3) in the form

$$L\|\bar{y}\|^2 = \pi \int_{-\infty}^{\infty} K_L\left(\frac{\lambda}{2}\right) f(\lambda)\, d\lambda \tag{4}$$

with the *Fejer kernel*

$$K_L(\lambda) = \frac{1}{\pi L} \frac{\sin^2 L\lambda}{\lambda^2}$$

Using a well-known property of the Féjer kernel [see, e.g., Bochner (1932), p. 23], we get, since f is continuous at the singularity $\lambda = 0$ of the kernel,

$$\lim_{L\to\infty} L\|\bar{y}\|^2 = 2\pi f(0) \tag{5}$$

a result that will be needed later on. At the moment it is enough to notice that (4) implies the bound (see Note 4.4.A)

$$L\|\bar{y}\|^2 \leqslant 2\pi f_{\max}; \qquad f_{\max} = \max_{\lambda} f(\lambda) \tag{6}$$

and $\|\bar{y}\| \to 0$ as $L \to \infty$. Using (2) we get, since the numerator in (2) is uniformly bounded,

$$\|\alpha_L^{**}\| \leqslant \|\alpha^*\| + \|\alpha_L^{**} - \alpha_L^*\| \leqslant \|\alpha^*\| + 0(L^{-1})$$

But Theorem 1 of Section 3 tells us that $\|\alpha_L^*\|^2 = 2(2+\beta L)^{-1}$. Dividing both sides of (6) by $\|\alpha_L^*\|$, we get

$$\underline{\lim}_{L\to\infty} \frac{\|\alpha_L^{**}\|}{\|\alpha_L^*\|} \geqslant 1 \tag{7}$$

Since the ratio in (7) is always less than or equal to 1, we have proved a result that looks deceptively special but will be generalized later on.

THEOREM 1

For the Markovian stationary process with constant mean the least-squares estimate $\bar{x}$ is asymptotically optimal.

With the same method we can show, using Theorem 2 of Section 3, the following more general statement about asymptotic optimality for $\bar{x}$.

THEOREM 2

For the pth-order Markovian stationary process with constant mean and with spectral density (3.4) *the least-squares estimate* $\bar{x}$ *is asymptotically optimal.*

Striving for much greater generality, we shall relax the conditions on Theorems 1 and 2 in two different directions. First, *the spectral density will no longer be restricted to any particular form*, only that it be continuous and positive. Second, *we shall allow a general regression sequence* $\{\psi_t, t \in T\}$.

In the remainder of this section the argument is assumed to be discrete, $T = \mathbb{N}$, and in the next one we return to the continuous case, $T = \mathbb{R}$.

In order to ensure the existence of a consistent estimate of α we must impose a weak condition on the regression sequence $\{\psi_t; t \in \mathbb{N}\}$ (compare with Example 6 in Section 2). We assume from now on that the spectral density is positive and continuous, $f \in C^+[(-\pi, \pi)]$.

Lemma 1 Among all linear estimates of α there exists a consistent one if and only if $\Phi(+\infty) = +\infty$, where

$$\Phi(n) = \sum_{t=1}^{n} |\psi_t|^2$$

Proof Sufficiency is easy to show since the least-squares estimate (see Definition 2 in Section 1) is

$$\alpha_{LS}^* = \Phi^{-1}(n) \sum_{t=1}^{n} \bar{\psi}_t x_t \tag{8}$$

But (8) is unbiased with variance

$$\operatorname{Var}(\alpha_{LS}^*) = \Phi^{-2}(n) \sum_{t=1}^{n} \psi_t \bar{\psi}_s r(t-s) = \Phi^{-2}(n) \int_{-\infty}^{\infty} |\sum_{1}^{n} \psi_t e^{it\lambda}|^2 f(\lambda)\, d\lambda \tag{9}$$

However, (9) can be bounded by

$$\Phi^{-2}(n) f_{\max} \int_{-\infty}^{\infty} |\sum_{1}^{n}{}_t\, e^{it\lambda}|^2 d\lambda = 2\pi f_{\max} \Phi^{-1}(n) \to 0 \qquad \text{as } n \to \infty \tag{10}$$

so that α_{LS}^* is consistent.

To prove necessity, let α_n^* be a sequence tending to α in $L_2(T)$, and write it as

$$\alpha_n^* = \sum_{t=1}^{n} c_t^{(n)} x_t$$

Since

$$E_\alpha\left[(\alpha_n^* - \alpha)^2\right] = \operatorname{Var}(\alpha_n^* - \alpha) + \left[E_\alpha(\alpha_n^*) - \alpha\right]^2$$

we must have

$$\lim_{n\to\infty} \sum_{t=1}^{n} c_t^{(n)} \psi_t = 1 \tag{11}$$

and applying Schwarz's inequality (11) implies, for large n and some $\varepsilon > 0$,

$$(1-\varepsilon)^2 \leqslant \sum_{t=1}^{n} \left(c_t^{(n)}\right)^2 \Phi(n) \tag{12}$$

However, with the same argument as in (10), but in the opposite direction,

$$\operatorname{Var}(\alpha_n^*) = \int_{-\pi}^{\pi} \left|\sum c_t^{(n)} e^{it\lambda}\right|^2 f(\lambda)\, d\lambda \geqslant 2\pi f_{\min} \sum_{t=1}^{n} \left(c_t^{(n)}\right)^2 \tag{13}$$

Combining $\operatorname{Var}(\alpha_n^*) \to 0$ in (13) with (12), we conclude that $\Phi(n)$ must tend to infinity. Q.E.D.

From now on $\Phi(+\infty) = +\infty$ is always assumed.

Definition 2 The regression sequence $\{\psi_t\}$ will be called regular if the limits exist:

$$R_h = \lim_{n\to\infty} \Phi^{-1}(n) \sum_{t=1}^{n} \psi_t \bar{\psi}_{t+h}; \qquad h \in \mathbb{N} \tag{14}$$

Definition 3 If for all $k \in \mathbb{N}$ the regular sequence $\{\psi_{t+k}\}$ has the same limits,

$$R_h = \lim_{n\to\infty} \Phi^{-1}(n) \sum_{t=1}^{n} \psi_{t+k} \bar{\psi}_{t+k+h} \tag{15}$$

the sequence is said to be stationary.

Lemma 2 In order that a regular sequence be stationary, it is necessary and sufficient that $\Phi(n)$ be slowly increasing, that is,

$$\lim_{n\to\infty} \frac{\Phi(n+k)}{\Phi(n)} = 1; \qquad k \in \mathbb{N} \tag{16}$$

Proof Suppose that the regression sequence is stationary so that (15) holds. Then we can write R_0, both with $k=0$ and with some $k\neq 0$, as

$$R_0=\lim_{n\to\infty}\Phi^{-1}(n)\sum_{t=1}^{n}|\psi_t|^2=\lim_{n\to\infty}\Phi^{-1}(n)\left[\sum_{1}^{k}|\psi_t|^2+\sum_{k+1}^{n}|\psi_t|^2\right]$$

$$=\lim_{n\to\infty}\Phi^{-1}(n)\sum_{1}^{n-k}|\psi_{t+k}|^2=\lim_{n\to\infty}\frac{\Phi(n-k)}{\Phi(n)}\frac{1}{\Phi(n-k)}\sum_{t=1}^{n-k}|\psi_t|^2 \quad (17)$$

Hence $\Phi(n)$ must be slowly increasing, (16) holds.

Assume instead that $\Phi(n)$ is slowly increasing. Then using $\Phi(+\infty)=\infty$, we get

$$\lim_{n\to\infty}\Phi^{-1}(n)\sum_{t=1}^{n}\psi_{t+k}\bar{\psi}_{t+h+k}=\lim_{n\to\infty}\Phi^{-1}(n)\sum_{1}^{n-k}\psi_t\bar{\psi}_{t+h}$$

and the slowly increasing character of $\Phi(n)$ implies that (17) is equal to R_{n+h}: the relation in (14) holds. Q.E.D.

We now use the fundamental concept of a *regression spectrum* (see Note 4.4.B).

THEOREM 3

A stationary regression sequence has a spectrum in the sense that

$$R_h=\lim_{n\to\infty}\Phi^{-1}(n)\sum_{t=1}^{n}\psi_t\bar{\psi}_{t+h}=\int_{-\pi}^{\pi}e^{ih\lambda}\psi(d\lambda); \qquad h\in\mathbb{Z} \quad (18)$$

where the regression spectral measure ψ *has mass* 1 *on* $[-\pi,\pi)$.

Remark 1 In the real case the integral need only be extended over $[0,\pi]$, and $e^{ih\lambda}$ should then be replaced by $2\cos h\lambda$ in (18). The support of ψ will then be a subset of $[0,\pi]$.

Proof For $h<0$ some of the terms in the sum in (18) may be undefined, but since $\Phi(+\infty)=+\infty$, we can assign arbitrary values to them without affecting the value of the limit. It therefore makes sense to allow arbitrary integral values for h in (18). For example, let $\psi_t=0$ for $t\leqslant 0$. Then for $h<0$,

$$R_{-h}=\lim_{n\to\infty}\Phi^{-1}(n)\sum_{t=1}^{n}\psi_t\bar{\psi}_{t-h}=\lim_{n\to\infty}\Phi^{-1}(n)\sum_{t=h+1}^{n}\psi_t\bar{\psi}_{t-h}$$

$$=\lim_{n\to\infty}\frac{\Phi(n-h)}{\Phi(n)}\Phi^{-1}(n-h)\sum_{1}^{n-h}\psi_t\bar{\psi}_{t+h}=\bar{R}_h$$

so that $\{R_h\}$ is an even (in the complex domain Hermitian) sequence.

For any vector $z \in \mathbb{R}^q$ we have, with $R_q = \{R_{s-t};\ s,t = 1,2,\ldots,q\}$,

$$z^T R_q z = \lim_{n\to\infty} \Phi^{-1}(n) \sum_{t=1}^{n} \left(\sum_{\nu=1}^{q} z_\nu \psi_{\nu+h} \right)^2 \geqslant 0$$

Hence R_q is nonnegative definite, and there exists a measure ψ with variation 1 over $[-\pi, \pi)$ for which the representation (18) is valid (see Theorem 3 in Chapter 2, Section 4).

In our estimation problem the support of ψ, to be denoted by $S = \text{support}(\psi)$, is crucial, and its elements are spoken of as the *distinguished frequencies* of the *regression spectrum*.

We now calculate the asymptotic efficiency of the least-squares estimate $\bar{\alpha}_L$:

$$e(\bar{\alpha}) = \lim_{L\to\infty} \frac{\operatorname{Var}(\alpha_L^*)}{\operatorname{Var}(\bar{\alpha}_L)}$$

and then study when $\bar{\alpha}$ is optimal in the sense of being asymptotically efficient, $e(\bar{\alpha}) = 1$. The main result gives $e(\bar{\alpha})$ in an elegant form.

THEOREM 4 ASYMPTOTIC EFFICIENCY IN LINEAR ESTIMATION, GRENANDER (1954)

The least-squares estimate $\bar{\alpha}_L$, when the regression sequence is stationary, has asymptotic efficiency

$$e(\bar{\alpha}) = \frac{H_\phi(f)}{A_\phi(f)} \tag{19}$$

expressed in the harmonic and arithmetic means with respect to ϕ:

$$H_\phi(f) = 1 \Big/ \int_{-\pi}^{\pi} \frac{1}{f(\lambda)} \phi(d\lambda)$$

$$A_\phi(f) = \int_{-\pi}^{\pi} f(\lambda) \phi(d\lambda)$$

Proof Let f_1 and f_2 be two spectral densities and consider the variances calculated under these two hypotheses for some arbitrary linear combination:

$$\alpha^* = \sum_{1}^{n} c_t x_t \tag{20}$$

Since

$$\operatorname{Var}_r(\alpha^*)=\int_{-\pi}^{\pi}\left|\sum_{1}^{n} c_t e^{it\lambda}\right|^2 f_r(\lambda)\,d\lambda;\qquad r=1,2$$

we get immediately, if $f_1 \leqslant f_2$,

$$\operatorname{Var}_1(\alpha^*) \leqslant \operatorname{Var}_2(\alpha^*) \tag{21}$$

Now assume that y_t satisfies the stochastic difference equation

$$\alpha_0 y_{t+p}+\alpha_1 y_{t+p-1}+\cdots+\alpha_p y_t=\varepsilon_{t+p};\qquad t\in\mathbb{Z} \tag{22}$$

as usual with the roots of the characteristic polynomial P inside with $(\varepsilon_t, \varepsilon_s)=\delta_{t,s}$ (see Chapter 2, Section 4). Then

$$f(\lambda)=\frac{1}{2\pi}|P(e^{i\lambda})|^{-2} \tag{23}$$

and $(y_s, \varepsilon_t)=0$ when $s<t+1$. This is the discrete analogue of the case dealt with in Theorem 2. Also the y_t are not constrained by any non trivial linear relation, so that we can apply Gram-Schmidt orthogonalization (see Chapter 2, Section 1) to the sequence of stochastic variables $y_1, y_2, \ldots, y_p$. Say that the result is

$$\begin{aligned}
&c_{10}y_1 &&=\eta_1\\
&c_{21}y_2+c_{20}y_1 &&=\eta_2\\
&\vdots\\
&c_{pp-1}y_p+c_{pp-2}y_{p-1}+\cdots+c_{p0}y_0 &&=\eta_p
\end{aligned} \tag{24}$$

with $(\eta_\nu, \eta_\mu)=\delta_{\nu\mu}$, $\nu, \mu=0,1,\ldots,p-1$, and $c_{\nu 0}\neq 0$. Extend the definition by defining $\eta_\nu=\varepsilon_\nu$ for $\nu>p$; the infinite sequence $(\eta_1, \eta_2, \ldots)$ is orthonormal.

Note that any y_ν is a linear combination of $\eta_\nu, \eta_{\nu-1}, \eta_{\nu-2}, \ldots, \eta_0$. Let L be the $(n+1)\times(n+1)$ matrix that transforms $(y_1, y_2, \ldots, y_n)$ into $(\eta_1, \eta_2, \ldots, \eta_n)$. L has the rows

$$\begin{array}{llll}
(c_{10}, 0, 0, & \cdots & & 0)\\
(c_{20}, c_{21}, & \cdots & & 0)\\
\cdots & & & \\
(c_{p0}, c_{p1}, & & & 0)\\
(\alpha_p, \alpha_{p-1}, \ldots, & & \alpha_0, & 0, \ldots 0)
\end{array} \tag{25}$$

and so on, so that it is nonsingular since the diagonal elements of this subdiagonal matrix are not zero.

Define

$$L\,\mathrm{col}[x_1, x_2, \ldots, x_n] = \mathrm{col}[\eta_1, \eta_2, \ldots, \eta_n]$$
$$L\,\mathrm{col}[\psi_1, \psi_2, \ldots, \psi_n] = \mathrm{col}[\beta_1, \beta_2, \ldots, \beta_n] \tag{26}$$

As L is nonsingular, any linear estimate (20) can be decomposed into two parts:

$$\alpha^* = \alpha \sum_{t=1}^{n} \gamma_t \beta_t + \sum_{t=1}^{n} \gamma_t \eta_t \tag{27}$$

But now the η are orthonormal so that the variance of the BLUE is (see Section 1)

$$\mathrm{Var}(\alpha^*) = \left(\sum_{t=1}^{n} |\beta_t|^2 \right)^{-1} \tag{28}$$

We now derive the asymptotic behavior of (28) when n becomes large. The contribution to the sum in (28) of the first p terms is bounded and can be neglected asymptotically. For the rest we use

$$\Phi^{-1}(n) \sum_{t=p+1}^{n} |\beta_t|^2 = \sum_{u,v=0}^{P} \alpha_u \bar{\alpha}_v \Phi^{-1}(n) \sum_{p+1}^{n} \psi_{t-u} \psi_{t-v} \tag{29}$$

which implies, together with (28) and (15),

$$\lim_{n\to\infty} \Phi^{-1}(n)\,\mathrm{Var}^{-1}(\alpha^*) = \int_{-\pi}^{\pi} \left| \sum_{0}^{p} \alpha_t e^{it\lambda} \right|^2 \phi(d\lambda) \tag{30}$$

But (23) allows us to rewrite (30) as

$$\lim_{n\to\infty} \Phi(n)\,\mathrm{Var}(\alpha^*) = \frac{2\pi}{\displaystyle\int_{-\pi}^{\pi} \frac{1}{f(\lambda)} \phi(d\lambda)} = 2\pi H_\phi(f) \tag{31}$$

Let us now derive the asymptotic behavior of the least-squares estimate $\bar{\alpha}$, which is easier since we have its closed-form expression. We get immediately

$$\mathrm{Var}(\bar{\alpha}) = \phi^{-2}(n) \sum_{s,t=1}^{n} \psi_s \bar{\psi}_t r_{s-t} \tag{32}$$

or

$$\phi(n)\operatorname{Var}(\bar{\alpha})=\sum_{0}^{n-1} r_\nu\Phi^{-1}(n)\sum_{t=1}^{n-\nu}\bar{\psi}_t\psi_{t+\nu}+\sum_{-1}^{-n+1} r_\nu\Phi^{-1}(n)\sum_{t=-1}^{n}\bar{\psi}_t\psi_{t+\nu} \tag{33}$$

In the special case (23) the covariances decrease geometrically (see Note 4.4.C) so that $r_\nu=0(a^{|\nu|})$, $0<a<1$. We also have

$$\phi^{-2}(n)\left(\sum_{t=1}^{n-\nu}\bar{\psi}_t\psi_{t+\nu}\right)^2\leqslant\Phi^{-1}(n)\sum_{1}^{n-\nu}|\psi_t|^2\Phi^{-1}(n)\sum_{\nu}^{n}|\psi_t|^2\leqslant 1 \tag{34}$$

We can therefore carry out the limit operation termwise in (33) and get by dominated convergence

$$\lim_{n\to\infty}\Phi(n)\operatorname{Var}(\bar{\alpha})=\sum_{-\infty}^{\infty} r_\nu R_{-\nu}=\sum_{-\infty}^{\infty}\int_{-\pi}^{\pi} e^{-i\nu\lambda}\phi(d\lambda) \tag{35}$$

But in the case when the spectral density is of the form in (23), the usual inversion formula

$$f(\lambda)=\frac{1}{2\pi}\sum_{-\infty}^{\infty} r_\nu e^{-i\nu\lambda} \tag{36}$$

holds with uniform convergence. Therefore (35) can be written as

$$\lim_{n\to\infty}\Phi(n)\operatorname{Var}(\bar{\alpha})=2\pi A(f) \tag{37}$$

Combining (37) with (31) we have established (19) in the case when the spectral density is the inverse absolute square of a trigonometric polynomial.

To complete the proof we must extend our result to a general density $f\in C^+$. But $1/f$ is continuous and can hence be approximated uniformly from above and below by trigonometric polynomials. In other words, for any $\varepsilon>0$ we can find densities f_1 and f_2 of the form (23) with

$$f_1(\lambda)\leqslant f(\lambda)\leqslant f_2(\lambda);\qquad \lambda\in[-\pi,\pi] \tag{38}$$

and with

$$f_2(\lambda)-f_1(\lambda)\leqslant\varepsilon,\qquad \lambda\in[-\pi,\pi] \tag{39}$$

We now use (21), both for the BLUE and for the least-squares estimate.

We get, for either estimate,

$$\mathrm{Var}_{f_1}(\alpha^*) \leqslant \mathrm{Var}_f(\alpha^*) \leqslant \mathrm{Var}_{f_2}(\alpha^*) \tag{40}$$

where the ratio between the first and second terms can be made arbitrarily close to zero. Applying (31) and (37) to (40), the stated result follows.

Q.E.D.

Theorem 4 helps us answer the basic question on asymptotic optimality in the discrete case with a single parameter.

THEOREM 5

For discrete time and a single parameter, in order that the least-squares estimate for stationary regression be asymptotically optimal for all $f \in C^+$, it is necessary and sufficient that the regression spectrum have only one distinguished frequency.

Proof We use the form of the asymptotic efficiency of $\bar{\alpha}$ in (19). We know that a harmonic mean is strictly less than the arithmetic mean if and only if the integrand is not constant. The latter statement means that there should be a set $N \subseteq [-\pi, \pi]$ with positive ψ-measure on which f is not constant.

Therefore, for a given $f \in C^+$, the asymptotic efficiency is 1 only if

$$f(\lambda) = \text{constant}, \ \lambda \in \text{support}(\psi)$$

If support (ψ) contains two distinguished frequencies λ_1 and λ_2, we can always find an $f \in C^+$ with $f(\lambda_1) \neq f(\lambda_2)$, making asymptotic optimality impossible.

On the other hand, if support (ψ) reduces to a single point, the expression in (19) takes the value 1 for all $f \in C^+$. Q.E.D.

We are now ready to consider special types of regression and shall deal especially with polynomial and trigonometric regression.

Consider *polynomial regression*, $\psi_t = t^q$, $q > 0$. We then have

$$\Phi(n) = \sum_{t=1}^{n} t^{2q} \sim \frac{n^{2q+1}}{2q+1}$$

from which it follows immediately that $\Phi(n)$ is slowly increasing. Also

$$R_h = \lim_{n\to\infty} \frac{2q+1}{n^{2q+1}} \sum_{t=1}^{n} t^q (t+h)^q \sim 1$$

so that the sequence $\{t^q;\ t\in\mathbb{N}\}$ is stationary with all the mass of its spectrum concentrated at the single distinguished frequency $\lambda=0$. Hence we have the following corollary.

Corollary 1 For polynomial regression $\psi_t=t^q$, $q\in\mathbb{N}$, the least-squares estimate,

$$\bar{\alpha}=\Phi^{-1}(n)\sum_{t=1}^{n} t_t^{q_x}$$

is asymptotically optimal.

For *trigonometric regression* $\psi_t=e^{i\gamma t}$, $\gamma\in[-\pi,\pi)$. We get

$$\Phi(n)=\sum_{t=1}^{n}|e^{i\gamma t}|^2=n$$

so that $\Phi(n)$ is slowly increasing. Also

$$R_n=\lim_{n\to\infty}\frac{1}{n}\sum_{t=1}^{n}e^{-ih\gamma}=e^{-ih\gamma}$$

so that the sequence $\{e^{i\gamma t};\ t\in\mathbb{Z}\}$ is stationary with all the mass of the spectrum concentrated at $\lambda=-\gamma$. Since there is only one distinguished frequency, Theorem 4 tells us that the maximum likelihood estimate has asymptotic efficiency 1.

Corollary 2 For trigonometric regression, $\psi_t=e^{i\gamma t}$, the least-squares estimate

$$\bar{\alpha}=\frac{1}{n}\sum_{t=1}^{n}e^{-i\gamma t}x_t$$

is asymptotically optimal.

We study more general forms of trigonometric-polynomial regression at the end of this section. Now we look only at a few special cases.

Example 1 Let $\psi_t=A_1e^{i\lambda_1 t}+A_2e^{i\lambda_2 t}$ with $\lambda_1\neq\lambda_2$ so that the regression spectrum has two distinguished frequencies λ_1 and λ_2. We cannot claim that the least-squares estimate is asymptotically efficient. If our only knowledge about the spectral density $f(\lambda)$ is that it does not deviate too must from a flat spectrum in the sense that $\max f(\lambda)/\min f(\lambda)=1+d$, where d is fairly small, then Theorem 4 gives us a bound on the asymptotic efficiency of $\bar{\alpha}$, since

$$\int_{-\pi}^{\pi}\frac{1}{f(\lambda)}\psi(d\lambda)\int_{-\pi}^{\pi}f(\lambda)\psi(d\lambda)=\left[\frac{\Delta\psi(\lambda_1)}{f(\lambda_1)}+\frac{\Delta\psi(\lambda_2)}{f(\lambda_2)}\right]$$
$$\times\left[f(\lambda_1)\Delta\psi(\lambda_1)+f(\lambda_2)\Delta\psi(\lambda_2)\right]$$
$$=\left[\Delta\psi(\lambda_1)\right]^2+\left[\Delta\psi(\lambda_2)\right]^2+\Delta\psi(\lambda_1)\Delta\psi(\lambda_2)\left[\frac{f(\lambda_2)}{f(\lambda_1)}+\frac{f(\lambda_1)}{f(\lambda_2)}\right] \tag{41}$$

The total mass of ψ is 1, so that (41) is

$$1+\left[\frac{f(\lambda_1)}{f(\lambda_2)}+\frac{f(\lambda_2)}{f(\lambda_1)}-2\right]\Delta_1\Delta_2$$

The expression in brackets is at most equal to d^2 so that $e(\bar{\alpha})\geqslant(1+d^2\Delta_1\Delta_2)^{-1}$.

Example 2 Consider an ergodic strongly stationary process ψ_t with mean m_μ and covariance spectrum F_ψ. We pick the regression sequence ψ_t as one realization of the ψ-process—this is the so-called *fixed variate* model from econometrics.

Since ψ_t is ergodic, $\Phi(n)/n\to E(\psi_t)$ a.c. as $n\to\infty$, so that $\Phi(n)$ is a.c. slowly increasing. Also the limits R_h in (18) exist a.c. and are equal to $E(\psi_t\psi_{t+h})$, corresponding to a spectral ψ-measure equal to F_ψ except for $\lambda=0$, where the discrete mass m_μ^2 has to be added.

Hence we know exactly what the regression spectrum is in the fixed variate model and can apply Theorems 4 and 5 to find the asymptotic properties of the least-squares estimate.

We now turn to a situation that differs radically from the ones we have encountered so far.

Example 3 Let $\psi_t=a^t$, $a>1$, *exponential regression*. We then have

$$\Phi(n)=\sum_{t=1}^{n}a^{2n}\sim\frac{a^{2(n+1)}}{a^2-1}\qquad\text{as } n\to\infty$$

Hence

$$\lim_{n\to\infty}\frac{\Phi(n+h)}{\Phi(n)}=a^{2h}\neq 1$$

so that Φ is not slowly increasing and the regression spectrum is not defined.

Since ψ_t increases at a rapid rate when t tends to infinity, it seems intuitively that $\hat{a}_n = x_n/\psi_n$ should be a good estimate. Its variance is, of course,

$$\mathrm{Var}(\hat{\alpha}_n) = a^{-2n}\int_{-\pi}^{\pi} f(\lambda)\, d\lambda$$

The least-squares estimate, on the other hand, has a variance

$$\mathrm{Var}(\bar{\alpha}_n) = \left(\sum_1^n a^{2t}\right)^{-2} a^2 \int_{-\pi}^{\pi} \left|\frac{1-a^n e^{in\lambda}}{1-ae^{i\lambda}}\right|^2 f(\lambda)\, d\lambda$$

$$= a^{2(n+1)}\left(\sum_1^n a^{2t}\right)^{-2} \int_{-\pi}^{\pi} \left|\frac{a^{-n}-e^{in\lambda}}{1-ae^{i\lambda}}\right|^2 f(\lambda)\, d\lambda \tag{42}$$

Letting n tend to infinity in (42), we get asymptotically

$$\mathrm{Var}(\bar{\alpha}_n) \sim (a^2-1)^2 a^{-2(n+1)} \int_{-\pi}^{\pi} \frac{f(\lambda)}{|1-ae^{i\lambda}|^2} f(\lambda)\, d\lambda$$

The asymptotic relative efficiency of the least-squares estimate $\bar{\alpha}_n$ compared with $\hat{\alpha}_n$ is then

$$\lim_{n\to\infty} \frac{\mathrm{Var}(\hat{\alpha}_n)}{\mathrm{Var}(\bar{\alpha}_n)} = \frac{a^2}{(a^2-1)^2} \frac{\int_{-\pi}^{\pi} f(\lambda)\, d\lambda}{\int_{-\pi}^{\pi} \frac{f(\lambda)}{|1-ae^{i\lambda}|^2}} \tag{43}$$

Say that f has most of its mass close to $\lambda=0$. Then the expression in (43) is close to $a^2/(a+1)^2$, so that $\hat{\alpha}_n$ is asymptotically better than $\bar{\alpha}_n$. *In this case, when the regression spectrum is not defined, the least-squares estimate is not asymptotically optimal for all f.*

We turn now to the case with several regression parameters $\alpha_1, \alpha_2, \ldots, \alpha_p$; $p>1$. The approach is quite similar to $p=1$ so that we can leave out some of the details.

Generalizing what we have learned about the single-parameter case, we assume

$$f \in C^{+}([-\pi, \pi]) \tag{44a}$$

$$\Psi_r(n) = \sum_{t=1}^{n} |\psi_r(t)|^2 \text{ slowly increasing toward } +\infty; \qquad r = 1, 2, \ldots, p \tag{44b}$$

$$\lim_{n\to\infty} \Psi_r^{-1/2}(n)\Psi_s^{-1/2}(n) \sum_{t=1}^{n} \psi_r(t)\overline{\psi_s(t+h)} = R_{rs}(h) \tag{44c}$$

exist. With the same proof as for Theorem 3 we can now state directly the following theory.

THEOREM 6

For all h the matrix valued sequence $R(h)$, $h \in \mathbb{Z}$, can be written as a Fourier-Stieltjes integral

$$R(h) = \int_{-\pi}^{\pi} e^{ih\lambda}\psi(d\lambda) \tag{45}$$

where ψ is a matrix valued measure on $[-\pi, \pi]$, the regression spectral measure.

In addition to (44) we need in the p-parameter case an additional condition that does not appear in the one-parameter situation:

$$R(0) = \psi(\pi - 0) - \psi(-\pi) \text{ is nonsingular} \tag{46}$$

The role of this condition is to prevent the regression sequences from being asymptotically linearly related.

Let us recall from Chapter 4, Section 1, that for finite sample size n the BLUE is given by

$$\alpha^* = (\psi^T R^{-1}\psi)^{-1}\psi^T R^{-1}x$$

where R is the $n \times n$ covariance matrix of the x-process, and should not be confused with the $R(h)$-matrices for the regression sequences. Also

$$\text{Cov}(\alpha^*) = (\psi^T R^{-1}\psi)^{-1}$$

On the other hand the least-squares estimate $\bar{\alpha}$ is given as

$$\bar{\alpha}=(\psi^T\psi)^{-1}\psi x$$

with

$$\operatorname{Cov}(\bar{\alpha})=(\psi^T\psi)^{-1}\psi^T R\psi(\psi^T\psi)^{-1} \tag{47}$$

Of course $\bar{\alpha}$ is easier to compute and does not require knowledge of the covariance matrix R. On the other hand it is in general less efficient than α^*.

Introduce the diagonal matrix

$$D_n=\operatorname{diag}\left[\Psi_1^{1/2}(n),\Psi_2^{1/2}(n),\ldots,\Psi_p^{1/2}(n)\right]\in\mathbb{R}^p\times\mathbb{R}^p$$

We shall now find the asymptotic properties of α^* and $\bar{\alpha}$, generalizing the proof of Theorem 4.

The crucial step is to establish the generalization of relation (37). For an arbitrary linear combination $z^Tx, z\in\mathbb{R}^n$, we have, with variances computed for f_1, f, f_2, relation (38), and the approximation argument used there carries over to the present case $p>1$.

First, consider the matrix

$$D_n\operatorname{Cov}(\bar{\alpha})D_n=D_n(\psi^T\psi)^{-1}\psi^T R\psi(\psi^T\psi)^{-1}D_n$$

We could now use a generalization of the argument in (32)–(37), but it will be instructive to look at a variation of the technique. Assume to begin with that our process is a finite moving average,

$$y_t=\sum_{\nu=1}^{\alpha}a_\nu\eta_{t-\nu};\qquad \{\eta_t\}\text{ orthonormal}$$

so that it has a spectral density given by the trigonometric polynomial

$$f(\lambda)=\sum_{\nu=-\alpha+1}^{\alpha-1}f_\nu e^{i\nu\lambda} \tag{48}$$

with covariances

$$r_\nu=\begin{cases}2\pi f_\nu; & |\nu|<\alpha\\ 0; & \text{otherwise}\end{cases}$$

We get directly

$$(D_n^{-1}\psi^T R\psi D_n^{-1})_{\nu\mu} = \sum_{u,v=1}^{n} \frac{\psi_\nu(u) r_{u-v}\bar{\psi}_\mu(v)}{\sqrt{\Psi_\nu(n)\Psi_\mu(n)}}$$

This can be evaluated as

$$2\pi \sum_{m=0}^{\alpha-1} f_m \sum_{v=1}^{n-m} \frac{\psi_\mu(v)\bar{\psi}_\nu(m+v)}{\sqrt{\Psi_\nu(n)\Psi_\mu(n)}} + 2\pi \sum_{m=-1}^{-\alpha+1} f_m \sum_{u=1}^{n+m} \frac{\psi_\mu(u-m)\overline{\psi_\nu(u)}}{\sqrt{\Psi_\nu(n)\Psi_\nu(n)}} \tag{49}$$

Using the definition in (44c), we get the limit of (49):

$$2\pi \sum_{-\alpha+1}^{\alpha-1} f_m R_{\mu\nu}(m)$$

or, with (45),

$$\lim_{n\to\infty} D_n^{-1}\psi^T R\psi D_n^{-1} = 2\pi \int_{-\pi}^{\pi} f(\lambda)\psi(d\lambda)$$

On the other hand

$$\lim_{n\to\infty} D_n^{-1}\psi^T\psi D_n^{-1} = R(0)$$

so that (47) yields

$$\lim_{n\to\infty} D_n^{-1}\operatorname{Cov}(\bar{\alpha})D_n^{-1} = 2\pi R^{-1}(0)\int_{-\pi}^{\pi} f(\lambda)\psi(d\lambda)R^{-1}(0) \tag{50}$$

for the particular spectral density in (48). The approximation argument employing (38)–(40) shows that (50) holds for any $f \in C^+$.

The treatment of $\operatorname{Cov}(\alpha^*)$ can be modeled on (22)–(31), and we get, leaving out the details (see Note 4.4.D),

$$\lim_{n\to\infty} D_n^{-1}\operatorname{Cov}(\alpha^*)D_n^{-1} = 2\pi\left[\int_{-\pi}^{\pi} \frac{1}{f(\lambda)}\psi(d\lambda)\right]^{-1} \tag{51}$$

Combining (50) with (51), we have the following theorem.

THEOREM 7 ASYMPTOTIC EFFICIENCY IN LINEAR ESTIMATION, $p > 1$, GRENANDER-ROSENBLATT (1956)

If $f \in C^+[(-\pi, \pi)]$ *and if the regression sequences satisfy* (44) *and* (46), *the asymptotic behavior of the covariance matrices of the BLUE* α^* *and the least-squares estimate* $\bar{\alpha}$ *are given by* (50) *and* (51).

Consider the regression spectrum $S = \text{support } (\psi)$ consisting of all the distinguished frequencies ψ for which $\psi(\lambda+0) \neq \psi(\lambda-0)$. We now encounter *a major difference with the single-parameter case, and we have to introduce the elements of the spectrum* in order to overcome the additional difficulty.

Introduce the matrix valued function

$$N(\lambda) = R^{-1/2}(0)\psi(\lambda)R^{-1/2}(0) \tag{52}$$

which is nondecreasing with the usual partial order, bounded, and hence defines a matrix valued measure on the Borel sets of S. Note that $N(S) = I$.

Suppose now that there exist no disjoint Borel sets E_1, E_2, $E_1 \cup E_2 = S$, such that

$$N(E_i) = \int_{E_i} N(d\lambda); \qquad i = 1,2$$

with $N(E_1)N(E_2) = 0$, but $N(E_1)$ and $N(E_2)$ both nonzero. We then say that S has a single element, itself.

Otherwise we can try to decompose E_1 or E_2 further. Since the dimension of the space in which N operates is p, this procedure can lead to at most a partition with p elements $E_1, E_2, \ldots, E_p$:

$$\begin{gathered} N(E_i) > 0 \\ \sum_i N(E_i) = I \\ N(E_i)N(E_j) = 0; \quad i \neq j \end{gathered} \tag{53}$$

We now show that such *indecomposable elements* satisfying (53) are uniquely determined.

Lemma 3 A set of indecomposable elements with properties (53) is unique.

Proof Assume that we have two partitions $\{E_i\}$, $\{E_j'\}$ into elements satisfying (53). Multiplying the second relation in (53) by $N(E_j)$ and use the third relation. It shows that $N^2(E_i)=N(E_i)$. Since the N are Hermitian, it follows that all the $N(E_i)$ are projections.

Since the partitions do not coincide, we can find sets E_k, E_i', E_j' such that

$$N(E_k\cap E_i')>0$$
$$N(E_k\cap E_j')>0$$

Let us write $E_k=(E_i'\cap E_k)\cup F=D\cup F$. Since $N(E_k)=N(D)+N(F)$ is orthogonal to every $N(E_l)$, $l\neq k$, it follows that the same is true for $N(D)$ and $N(F)$.

To see this, let

$$M=M_1+M_2$$
$$N=N_1+N_2$$

be Hermitian nonnegative matrices with $MN=0$. Diagonalize M by a unitary transformation with the first m diagonal elements$\neq 0$. Then the first m diagonal elements of N must be zero (after the same transformation), and the same then holds for N_1 and N_2. Use the nonnegative property of the matrices; it follows that the only nonzero elements of N_1 and N_2 are those with subscripts (r,s) with $m+1\leqslant r,s\leqslant p$. This implies that $M_iN_j=0$.

We now apply this to $N(D)$, $N(F)$, $N(E_l)$. Since

$$N(E_i')N\left[(E_i')^c\right]=0$$

it follows for the subsets $E_i'\cap E_k\subset E_i'$ and $F\subset(E_i')^c$,

$$N(E_i'\cap E_k)N(F)=0$$

which says that E_k can be decomposed further into $E_i'\cap E_k$ and $F=E_k\cap(E_i')^c>E_k\cap E_j'>0$. This is a contradiction. Q.E.D.

With this partition of the regression spectrum we are ready to announce the main result on asymptotic optimality for $p>1$.

THEOREM 8

With stationary regression sequences with a nonsingular $R(0)$ the least-squares estimate is asymptotically optimal for all spectral densities $f\in C^+$ if and only if the spectrum has q elements consisting of distinct points $\lambda_1,\lambda_2,\ldots,\lambda_q$, $q\leqslant p$.

Proof Assume that the spectrum S consists of $\lambda_1, \lambda_2, \ldots, \lambda_q$ as elements. Evaluating the right-hand side of (50) pre- and postmultiplied by $R^{1/2}(0)$, with the elements $E_\nu = \{\lambda_\nu\}$, we get

$$2\pi \int_{-\pi}^{\pi} f(\lambda) N(d\lambda) = 2\pi \sum_{\nu=1}^{q} f(\lambda_\nu) N(E_\nu) \tag{54}$$

Similarly for (51) we get, after the same pre- and postmultiplication,

$$2\pi \left[\int_{-\pi}^{\pi} \frac{1}{f(\lambda)} N(d\lambda) \right]^{-1} = 2\pi \left[\sum_{\nu=1}^{q} \frac{1}{f(\lambda_\nu)} N(E_\nu) \right]^{-1} \tag{55}$$

Using the fact that the $N(E_\nu)$ are projections, we can evaluate the matrix inverse in (55). The result is equal to the right-hand side of (54), so that $\bar{\alpha}$ is asymptotically optimal for all $f \in C^+$.

Assume now instead that asymptotic optimality holds for all $f \in C^+$. Treating (50) and (51) in the same way as we just did shows that we then must have the matrix relation.

$$\int_{-\pi}^{\pi} f(\lambda) N(d\lambda) \int_{-\pi}^{\pi} \frac{1}{f(\lambda)} N(d\lambda) = I \tag{56}$$

But the integrals in (56) can be written in terms of the elements E_ν of the regression spectrum

$$\sum_{\nu=1}^{q} \int_{E_\nu} f(\lambda) N(d\lambda) \sum_{\nu=1}^{q} \int_{E_\nu} \frac{1}{f(\lambda)} N(d\lambda) = I$$

What does this imply if it holds for any $f \in C^+$?

Diagonalize the two integrals in (56) simultaneously by a unitary transformation U so that

$$\int_{-\pi}^{\pi} f(\lambda) [UN(d\lambda)U^T]_{ii} \int_{-\pi}^{\pi} \frac{1}{f(\lambda)} [UN(d\lambda)U^T]_{ii} = 1; \qquad i = 1, 2, \ldots, p \tag{57}$$

Schwarz's inequality tells us that equality in (57) holds only if the integrands f and $1/f$ are proportional on the support of the real-valued measure $N_i = (UNU^T)_{ii}$. f has to be some constant c_i on each of these sets A_i, $i = 1, 2, \ldots, p$.

Consider now any chain of sets $A_1, A_{i_1}, \ldots, A_{i_k}$ such that

$$N_1(A_1 \cap A_{i_1}) > 0$$
$$N_{i_1}(A_{i_1} \cap A_{i_2}) > 0$$
$$\vdots$$
$$N_{i_k}(A_{i_{k-1}} \cap A_{i_k}) > 0$$

If there exists such a chain, clearly f must be equal to one and the same constant c_1 on all sets in the chain. Let B_1 be the union of all A_i which appear in such chains starting with A_1; then $f(\lambda) \equiv \gamma_1$ for $\lambda \in B_1$. Similarly start with the other A_i, leading to disjoint sets B_i, so that $f(\lambda) = \gamma_i$, $\lambda \in B_1$, $i = 1, 2, \ldots, r \leqslant p$. Here we have left out those sets B_i that occur more than once. Note that S is the union of the B_i.

We can now write (56) as

$$\sum_{i=1}^{r} N^2(B_i) = I \tag{58}$$

But $N(B_i) \geqslant 0$, and

$$\sum_{i=1}^{r} N(B_i) = I$$

so that $N(B_i) \leqslant I$ and hence $N^2(B_i) \leqslant I$. But then (58) shows that we have $N^2(B_i) = N(B_i)$ so that the $N(B_i)$ are projections orthogonal to each other and any B_i is the union of some elements of the regression spectrum S, and f must be constant on each element.

If any element consists of more than a single point, we can find an $f \in C^+$ which is not constant on the element so that (56) does not hold and we have arrived at a contradiction. Q.E.D.

We now apply Theorems 7 and 8 to regression sequences of the form

$$\begin{aligned}
&\psi_\nu(t) = t^\nu e^{it\lambda_1}; \qquad \nu = 0, 1, \ldots, p_1\\
&\psi_{p_1+1+\nu}(t) = t^\nu e^{it\lambda_2}; \qquad \nu = 0, 1, \ldots, p_2\\
&\qquad\vdots\\
&\psi_{p_1+p_2+\cdots p_{m-1}+m-1+\nu}(t) = t^\nu e^{it\lambda_m}; \qquad \nu = 0, 1, \ldots, p_m
\end{aligned} \tag{59}$$

combining trigonometric and polynomial regression from Corollaries 1 and 2. The subscript k corresponds to a pair (ν, s), where ν is the power of t

and s numbers the frequencies λ_s. All λ_s are assumed different. It is clear that

$$\psi_k(n) \sim \frac{n^{2\nu+1}}{2\nu+1}$$

and is slowly increasing. We get by direct evaluation

$$R_{kk'}(h) = \lim_{n\to\infty} \frac{1}{\sqrt{\Psi_k(n)\Psi_{k'}(n)}} \sum_{t=1}^{n} t^{\nu+\nu'} e^{it(\lambda_s-\lambda_{s'})-ih\lambda_{s'}}$$

$$= \frac{\sqrt{(2\nu+1)(2\nu'+1)}}{\nu+\nu'+1} \delta_{ss'} e^{-ih\lambda_{s'}}$$

Hence the spectral matrix measure ψ has points of increase only at the frequencies $-\lambda_s$ where the discrete matrix mass is

$$\psi(\Delta\lambda_s) = \begin{Bmatrix} 0 & 0 & 0 \\ 0 & \psi_s & 0 \\ 0 & 0 & 0 \end{Bmatrix}$$

and where the first $p_1 + p_2 + \cdots + p_{s-1} + s$ columns are zero, the next $p_s + 1$ have the matrix

$$\psi_s = \left\{ \frac{\sqrt{(2\nu+1)(2\nu'+1)}}{\nu+\nu'+1}; \qquad \nu, \nu' = 0, 1, \ldots, p_s \right\}$$

in it, and the remaining columns are zero. The matrix ψ_s is the Hilbert matrix

$$\left\{ \frac{1}{\nu+\nu'+1}; \qquad \nu, \nu' = 0, 1, \ldots, p_s \right\}$$

pre- and postmultiplied by the diagonal matrix

$$D = \operatorname{diag}\left[\sqrt{2\nu+1}\ ; \nu = 0, 1, \ldots, p_s\right]$$

Since the Hilbert matrix is known to be nonsingular, the same is true of ψ_s. Hence $R(0)$ is also nonsingular.

Since the $\psi(\Delta\lambda_s)$ matrices are orthogonal to each other, the points λ_s represent the elements of S and we have proved the following corollaries.

Corollary 3 For mixed trigonometric-polynomial regression as in (59) with distinct frequencies λ_s the least-squares estimate is asymptotically optimal for all spectral densities $f \in C^+$.

To generalize to even more general trigonometric integrals, let us consider $p=2$ with the two regression sequences

$$\begin{aligned}\psi_1(t)&=P(t)\int_{-\pi}^{\pi} e^{it\lambda}\alpha(d\lambda)\\ \psi_2(t)&=Q(t)\int_{-\pi}^{\pi} e^{it\lambda}\beta(d\lambda)\end{aligned}\tag{60}$$

where P and Q are polynomials of degree u and v, respectively. The measures α and β should be bounded signed measures with atoms. Actually we shall show that the jumps constitute the important part of them for the following analysis.

Let us introduce the discrete parts α_d, β_d, and the continuous parts $\alpha_c=\alpha-\alpha_d$, $\beta_c=\beta-\beta_d$. Calculate the Ψ-functions and begin with the relation

$$\begin{aligned}\sum_{t=1}^{n}\left|\int_{-\pi}^{\pi} e^{it\lambda}\alpha_c(d\lambda)\right|^2&=\int_{-\pi}^{\pi}\int_{-\pi}^{\pi} e^{i(\lambda-\lambda')}\frac{e^{in(\lambda-\lambda')}-1}{e^{i(\lambda-\lambda')}-1}\alpha_c(d\lambda)\overline{\alpha_c(d\lambda')}\\ &\leqslant n\int\int_{|\lambda-\lambda'|<\varepsilon}|\alpha_c(d\lambda)||\alpha_c(d\lambda')|+\frac{1}{\varepsilon}\{(|\alpha_c|)([-\pi,\pi)]\}^2\end{aligned}\tag{61}$$

Since α_c is continuous, the first term on the right-hand side of (61) can be made arbitrarily small by choosing ε small enough. That means that the sum on the left-hand side in (61) is $o(n)$. Similarly for β_c.

Consider now

$$\begin{aligned}\Psi_1(n)=\sum_{t=1}^{n}|P(t)|^2\Bigg[\left|\int_{-\pi}^{\pi} e^{it\lambda}\alpha_c(d\lambda)\right|^2&+2\operatorname{Re}\int_{-\pi}^{\pi} e^{it\lambda}\alpha_c(d\lambda)\\ \times\int_{-\pi}^{\pi} e^{-it\lambda}\overline{\alpha_d}(d\lambda)&+\left|\int_{-\pi}^{\pi} e^{it\lambda}\alpha_d(d\lambda)\right|^2\Bigg]\end{aligned}\tag{62}$$

The first term is, with Abelian rearrangement,

$$\begin{aligned}\sum_{t=1}^{n}|P(t)|^2\left|\int_{-\pi}^{\pi} e^{it\lambda}\alpha_c(d\lambda)\right|^2&=|P(n)|^2\sum_{t=1}^{n}\left|\int_{-\pi}^{\pi} e^{it\lambda}\alpha_c(d\lambda)\right|^2\\ &\quad-\sum_{t=1}^{n-1}(|P(t+1)|^2-|P(t)|^2)\sum_{k=1}^{t}\left|\int_{-\pi}^{\pi} e^{ik\lambda}\alpha_c(d\lambda)\right|^2\end{aligned}$$

However, $|P(t+1)|^2-|P(t)|^2=O(t^{2u-1})$, and if we appeal to the $o(n)$ relation just established, we get

$$\sum_{t=1}^{n}|P(t)|^2\left|\int_{-\pi}^{\pi}e^{it\lambda}\alpha_c(d\lambda)\right|^2=o(n^{2u+1}) \tag{63}$$

But

$$\sum_{1}^{n}|P(t)|^2\left|\int_{-\pi}^{\pi}e^{it\lambda}\alpha_d(d\lambda)\right|^2\sim|p_u|^2\frac{n^{2u+1}}{2u+1}\sum|a_s|^2 \tag{64}$$

where p_u is the leading coefficient of P, and the a_s are the point masses of α_d.

On the other hand Schwarz's inequality gives

$$\sum_{1}^{n}|P(t)|^2 2\,\mathrm{Re}\left(\int_{-\pi}^{\pi}e^{it\lambda}\alpha_c(d\lambda)\int_{-\pi}^{\pi}e^{-it\lambda}\bar{\alpha}_d(d\lambda)\right)|$$

$$\leqslant 2\left[\sum_{1}^{n}|P(t)|^2\left|\int_{-\pi}^{\pi}e^{it\lambda}\alpha_c(d\lambda)\right|^2\sum_{1}^{n}|P(t)|^2\left|\int_{-\pi}^{\pi}e^{it\lambda}\alpha_d(d\lambda)\right|^2\right]^{1/2}$$

Using (63) and (64) we see that this contribution to (62) is $o(n^{2u+1})$. Hence asymptotically

$$\Psi_1(n)\sim\frac{|p_u|^2}{2u+1}\sum|a_s|^2$$

and a similar relation for $\Psi_2(n)$.

With the same sort of reasoning we get

$$\sum_{t=1}^{n}\psi_1(t)\bar{\psi}_2(t+h)\sim\frac{p_u q_v}{u+v+1}n^{u+v+1}\sum a_s b_s e^{-ih\lambda_s}$$

where $\{\lambda_s\}$ denote *all* the atoms of α *and* β, and $b_s=\beta(\{\lambda_s\})$; q_v is the leading coefficient of Q.

Hence our 2×2-valued measure Ψ is atomic with point masses

$$D^{-1}\begin{Bmatrix}\dfrac{|p_u|^2}{2u+1}|a_s|^2 & \dfrac{p_u\bar{q}_v}{u+v+1}a_s\bar{b}_s\\[2ex] \dfrac{\bar{p}_u q_v}{u+v+1}\bar{a}_s b_s & \dfrac{|q_v|^2}{2v+1}|b_s|^2\end{Bmatrix}D^{-1} \tag{65}$$

at the frequencies $-\lambda_s$ and with

$$D=\operatorname{diag}\left[\frac{|p_u|}{\sqrt{2u+1}}\left(\sum|a_s|^2\right)^{1/2};\qquad \frac{|q_v|}{\sqrt{2v+1}}\left(\sum|b_s|^2\right)^{1/2}\right]$$

To avoid that $R(0)$ is singular, we must assume that

$$\begin{vmatrix}\sum|a_s|^2 & \sum a_s\bar{b}_s\\ \sum\bar{a}_s b_s & \sum|b_s|^2\end{vmatrix}\neq 0$$

In other words, we allow no linear relation between the vectors $\{a_s\}$ and $\{b_s\}$.

Of course the regression spectrum $S=\{-\lambda_s\}$, but what are its elements, at most two in number? Assume first that there are two elements E_1, E_2 so that

$$R(0)=\Psi_1+\Psi_2=\int_{E_1}\psi(d\lambda)+\int_{E_2}\psi(d\lambda)$$

with [see (52) and (53)]

$$\Psi_i R^{-1}(0)\Psi_j=\delta_{ij}\Psi_i;\qquad i,j=1,2$$

Of course Ψ_1 and Ψ_2 must have rank 1, and their ranges have only the zero-vector in common. Hence the matrices ψ_1 and ψ_2, obtained by summing (65) over E_1 and E_2, respectively, must be singular, which implies that the corresponding determinants

$$\frac{1}{(2u+1)(2v+1)}\sum_{E_i}|a_s|^2\sum_{E_i}|b_s|^2-\frac{1}{(u+v+1)^2}\left|\sum_{E_i}a_s\bar{b}_s\right|^2=0;\qquad i=1,2 \tag{66}$$

When does (66) occur? We could of course have $a_s\equiv 0$, $s\in E_1$ and $b_s\equiv 0$, $s\in E_2$, or the other way around. Otherwise there will be some $a_s=0$ in E_1 and $a_t=0$ in E_2 so that both terms in (66) are positive. The difference can be zero only if $u=v$ and $\{a_s,\ s\in E_1\}$, $\{b_s,\ s\in E_1\}$ are linearly related. Similarly for E_2. Of course these two linear relationships must be different from each other. We excluded the opposite to make $R(0)$ nonsingular.

Let us sum up and see when asymptotic optimality holds for all $f\in C^+$. If there is only a single element, S must reduce to a single frequency, say μ, and $\psi(\Delta\mu)$ should be nonsingular, so that $u\neq v$.

If there are two elements, S will consist of two frequencies, say μ_1 and μ_2. Then we have just seen that there are two possibilities. Either α_d has an atom at μ_1 and β_d an atom at μ_2, or both have atoms at μ_1 and μ_2, with $u=v$ and $a_1b_2-a_2b_1\neq 0$. Hence we have proved the following corollary.

Corollary 4 In order to have asymptotic optimality for the least-squares estimate for all $f\in C^+$, when the regression sequences are given by (60), it is necessary and sufficient that we have one of three cases:

Case I

$$\psi_1(t)=P(t)\left[e^{it\lambda}+\int_{-\pi}^{\pi}e^{it\lambda}\alpha_c(d\lambda)\right]$$

$$\psi_2(t)=Q(t)\left[e^{it\lambda}+\int_{-\pi}^{\pi}e^{it\lambda}\beta_c(d\lambda)\right]$$

with $u\neq v$.

Case II

$$\psi_1(t)=P(t)\left[e^{it\lambda_1}+\int_{-\pi}^{\pi}e^{it\lambda}\alpha_c(d\lambda)\right]$$

$$\psi_2(t)=Q(t)\left[e^{it\lambda_2}+\int_{-\pi}^{\pi}e^{it\lambda}\beta_c(d\lambda)\right]$$

with $\lambda_1\neq\lambda_2$.

Case III

$$\psi_1(t)=P(t)\left[a_1e^{it\lambda_1}+a_2e^{it\lambda_2}+\int_{-\pi}^{\pi}e^{it\lambda}\alpha_c(d\lambda)\right]$$

$$\psi_2(t)=Q(t)\left[b_1e^{it\lambda_1}+b_2e^{it\lambda_2}+\int_{-\pi}^{\pi}e^{it\lambda}\beta_c(d\lambda)\right]$$

with $u=v$, $\lambda_1\neq\lambda_2$, $a_1b_2-a_2b_1\neq 0$.

The case when the process is vector valued can be studied by a similar method (see Note 4.4.E), but we do not go into this here.

The technique used in this section is powerful enough to answer the main questions of asymptotic optimality. It should be noted, however, that the results assume that the spectral density $f\in C^+([-\pi,\pi])$.

How essential is the condition? Let us study this for $p=1$ and for a constant mean as the unknown parameter. The continuity of f is not really needed, as has been shown in Grenander (1951b), pp. 566–568. Very weak regularity conditions on f are sufficient.

The requirement that f does not vanish on the regression spectrum, in this case $S=\{0\}$, is a different story, however. Let us first look at a simple example with considerable potential for generalization (see Note 4.4.F).

Example 4 Consider the spectral density $f(\lambda)=\pi^{-1}(1-\cos\lambda) =\pi^{-1}|1-e^{i\lambda}|^2$ with a second-order zero at $\lambda=0$. The x_t-process can then be written as the moving average

$$x_t=\varepsilon_t-\varepsilon_{t-1}; \qquad \{\varepsilon_t\} \text{ orthonormal}$$

Let us find the BLUE $\alpha^*=c^Tx$. The normal equations are $Rc=\mu\psi$, with some constant μ and the regression sequence $\psi=(1,1,\ldots,1)$ so that they can be written

$$\begin{aligned} 2c_1-c_2&=\mu \\ 2c_k-(c_{k-1}+c_{k+1})&=\mu; \qquad k=2,3,\ldots,n-1 \\ 2c_n-c_{n-1}&=\mu \end{aligned} \tag{67}$$

Solving (67) and adjusting the value of μ to make α^* unbiased we get

$$c_k=\frac{6}{n+2}\,\frac{k}{n}\left(1-\frac{k}{n+1}\right)$$

and we can directly calculate its variance

$$\operatorname{Var}(\alpha^*)=\frac{12}{n(n+1)(n+2)}\sim 12n^{-3}$$

contrasted to the decrease constant$\times n^{-1}$ that we are accustomed to.

On the other hand the least-squares estimate $\hat{\alpha}=\bar{x}$ has

$$\operatorname{Var}(\bar{x})=2n^{-2}$$

We therefore see that the asymptotic efficiency of the least-squares estimate is 0, not 1 as when $f\in C^+([-\pi,\pi])$.

This result is of more general scope than one would suspect at first glance. Let us introduce the *parabolic estimate of the mean*

$$\alpha_p^*=\frac{6n}{n^2-1}\sum_{t=1}^{n}\frac{t}{n}\left(1-\frac{t}{n}\right)x_t \tag{68}$$

where the name emphasizes that instead of the equidistributed estimate $\bar{x}$ we now use weights with the parabolic graph $u(1-u)$, $u=t/n$. The answer to our problem was given in Vitale (1973) (See also Notes 4.4.B and 4.4.C).

THEOREM 9

Let the spectral density f be positive except for a second-order zero at $\lambda=0$ and be continuous. Then the parabolic estimate (68) *is asymptotically optimal.*

Proof To get the asymptotic variance of the parabolic estimate we factor the spectral density into

$$f(\lambda)=\frac{1}{2\pi}|1-e^{i\lambda}|^2 g(\lambda) \tag{69}$$

with $g\in C^+([-\pi,\pi])$. Then we show that $n^3 \operatorname{Var}(\alpha_p^*)\sim 12g(0)$. We have

$$\frac{n^3}{12}\operatorname{Var}(\alpha_p^*)=\frac{n^3}{12}\frac{1}{2\pi}\int_{-\pi}^{\pi}|P_n(e^{i\lambda})|^2|1-e^{i\lambda}|^2 g(\lambda)\,d\lambda$$

expressed as usual in a polynomial P_n, in this case

$$P_n(z)=\frac{6n}{n^2-1}\sum_{t=1}^{n}\frac{t}{n}\left(1-\frac{t}{n}\right)z^t; \qquad z=e^{i\lambda} \tag{70}$$

Introduce

$$h_n(\lambda)=\frac{n^3}{12}\frac{1}{2\pi}|1-e^{i\lambda}|^2|P_n(e^{i\lambda})|^2$$

and observe that Example 4 implies that

$$\lim_{n\to\infty}\int_{-\pi}^{\pi}h_n(\lambda)\,d\lambda=1$$

We now need a singular-kernel argument, showing that for any $\varepsilon>0$ the function $h_n(\lambda)\to 0$ uniformly outside $(-\varepsilon,\varepsilon)$ as $n\to\infty$.

But

$$\left(\frac{n^3}{24\pi}\right)^{1/2}P_n(e^{i\lambda})=\frac{6n^{3/2}}{\sqrt{24\pi}\,(n^2-1)}\sum_{t=1}^{n}\frac{t}{n}\left(1-\frac{t}{n}\right)z^n \tag{71}$$

Direct calculation of the finite geometric series and their derivatives gives us the value of (71) as

$$\frac{6n^{3/2}}{\sqrt{24\pi}\,(n^2-1)}\,\frac{1}{n(1-e^{i\lambda})^3}\Big[(n-1)e^{i\lambda}+(n+1)e^{i(n+1)\lambda}-(n+1)e^{2i\lambda}-(n-1)e^{i(n+2)\lambda}\Big]$$

leading to the bound

$$[h_n(\lambda)]^{1/2}\leqslant\frac{6n^{3/2}}{\sqrt{24\pi}\,(n^2-1)}\,\frac{4n}{n|1-e^{i\lambda}|^2}$$

For $|\lambda|>\varepsilon$ we get, with some $\delta>0$,

$$[h_n(\lambda)]^{1/2}\leqslant\frac{6n^{3/2}}{\sqrt{24\pi}\,(n^2-1)}\,\frac{4}{\delta}\to 0\qquad\text{as } n\to\infty$$

Hence the singular-kernel argument applies, leading to

$$\lim_{n\to\infty}\frac{n^3}{12}\operatorname{Var}(\alpha_p^*)=12g(0)\tag{72}$$

It is more difficult to find the asymptotic variance of $\operatorname{Var}(\alpha^*)$. Let us find two spectral densities $f_\nu(\lambda)=(2\pi)^{-1}|1-e^{i\lambda}|^2 g_\nu(\lambda)$ such that for given $\delta>0$,

$$0\leqslant g(\lambda)-g_1(\lambda)<\delta$$
$$0\leqslant g_2(\lambda)-g(\lambda)<\delta$$

and where $1/g_\nu$ are positive trigonometric polynomials. We are going to show that $\operatorname{Var}(\alpha^*)$ behaves asymptotically as in (72) for f_1 and f_2. Then the approximation argument in the proof of Theorem 4 will complete the proof.

A spectral density of the form

$$\frac{1}{2\pi}\,\frac{|1-e^{i\lambda}|^2}{|B(e^{i\lambda})|^2};\qquad B(z)=b_pz^p+b_{p-1}z^{p-1}+\cdots+b_0\tag{73}$$

corresponds to the solution of the stochastic differences equation

$$b_py_{t+p}+b_{p-1}y_{t+p-1}+\cdots+b_0y_t=\varepsilon_{t+p}-\varepsilon_{t+p-1}$$

with $\{\varepsilon_t\}$ orthonormal and the roots of $B(z)$ inside the unit circle.

Now we can express the y_t in terms of the corresponding ε and carry out the Gram-Schmidt orthogonalization of the first $p-1$ y's. Introduce $\eta=\Delta y$, where Δ is the matrix of the form

$$\begin{array}{lcl}
d_{11}y_1+d_{12}y_2+ & \cdots & +d_{1p-1}y_{p-1}=\eta_1 \\
d_{21}y_1+d_{22}y_2+ & \cdots & +d_{2p-2}y_{p-2}=\eta_2 \\
\vdots & & -b_py_p=\eta_p \\
b_0y_1+b_1y_2+ & \cdots & +b_py_{p+1}=\eta_{p+1}=\varepsilon_{p+1}-\varepsilon_p \\
b_0y_2+ & \cdots & +b_py_{p+2}==\eta_{p+2}=\varepsilon_{p+2}-\varepsilon_{p+1} \\
\vdots & & \\
b_0y_{n-p}+b_1y_{n-p+1}+ & \cdots & +b_py_n=\eta_n=\varepsilon_n-\varepsilon_{n-1}
\end{array} \tag{74}$$

Note that $\{y_1, y_2, \ldots, y_{p-1}\}$ is orthogonal to $\{\varepsilon_{p+1}-\varepsilon_p, \varepsilon_{p+2}-\varepsilon_{p+1}, \ldots, \varepsilon_n-\varepsilon_{n-1}\}$. We can choose the entries in Δ such that $E(\eta_i\eta_j)=0$ for $i, j\leqslant p$, $i\neq j$, and $E(\eta_i^2)=1$, $i<p$. Call $E(\eta_p^2)=\beta$. Also for $i, j>p$ we have $E(\eta_i^2)=1$ and $E(\eta_i\,\eta_{i+1})=-1$. We also have

$$E(\eta_p\eta_{p+1})=E\left[-b_py_p(\varepsilon_{p+1}-\varepsilon_p)\right]=-b_pE(y_p\varepsilon_{p+1})$$

Using

$$b_0y_0+\cdots+b_py_p=\varepsilon_p-\varepsilon_{p-1}$$

we get $E(y_p\varepsilon_p)=1/b_p$ so that

$$E(\eta_p\eta_{p+1})=-1$$

The new stochastic vector η then has some covariance matrix

$$Q=E(\eta\eta^T)=\begin{Bmatrix} A & 0 \\ 0 & B \end{Bmatrix}=\{q_{ij}; \qquad i, j=1,2,\ldots, n\}$$

where A is the $(p-1)\times(p-1)$ identity matrix and B is of dimension $(n-p+1)\times(n-p+1)$ and has the form

$$B=\begin{Bmatrix} \beta & -1 & 0 & 0 & \cdots \\ -1 & 2 & 1 & 0 & \cdots \\ 0 & -1 & 2 & -1 & \cdots \end{Bmatrix}$$

We know that α^* has the variance $1/\psi^T R^{-1}\psi$. But $\eta=\Delta y$ implies $R=\Delta^{-1}Q(\Delta^{-1})^T$, or $R^{-1}=\Delta^T Q^{-1}\Delta$ and

$$\operatorname{Var}(\alpha^*)=\frac{1}{(\Delta\psi)^T Q^{-1}\Delta\psi}=\frac{1}{\gamma^T Q^{-1}\gamma}; \qquad \gamma=\Delta\psi$$

Going back to (74) and recalling that $\psi_t\equiv 1$, we see that

$$\gamma_i=\sum_{j=1}^{p-i} d_{ij}; \qquad i<p$$

$$\gamma_p=-b_p$$

$$\gamma_i=\sum_{j=1}^{p} b_k=B_0; \qquad i>p$$

Hence

$$\operatorname{Var}^{-1}(\alpha^*)=\sum_{i,j=1}^{n}\bar{\gamma}_i q_{ij}^{(-1)}\gamma_j=S_1+S_2+S_3+S_4$$

and we treat the four sums separately. The first one

$$S_1=\sum_{i,j=p}^{n}\bar{\gamma}_i q_{ij}^{(-1)}\gamma_j=|B_0|^2\sum_{i,j=p}^{n} q_{ij}^{(-1)}+\varepsilon$$

$$=|B_0|^2\sum_{i,j=p}^{n}\bar{\psi}_i q_{ij}^{(-1)}\psi_j+\varepsilon=|B_0|^2\phi^T B^{-1}\phi+\varepsilon;$$

$$\phi=\operatorname{col}[\psi_t;\ t=1,2,\ldots,n-p+1]$$

where one can show $\varepsilon=0(n^3)$. Introduce $a=\operatorname{col}[a_1,a_2,\ldots,a_{n-p+1}]$ with

$$a_j=\frac{1}{2}j^2+j\left[\frac{1}{2}+\frac{(\beta-1)(l^2+1)}{2[l(\beta-1)+1]}\right]+\frac{(2-\beta)(l^2+l)}{2[l(\beta-1)+1]}; \qquad l=n-p+1$$

Then $Ba=\phi$ so that

$$\phi^T B^{-1}\phi=\phi^T a=\sum_{i=1}^{l} a_i=-\frac{1}{2}\frac{l(l+1)(2l+1)}{6}$$

$$+\frac{l(l+1)}{2}\left\{\frac{1}{2}+\frac{(\beta-1)(l^2+l)}{2[l(\beta-1)+1]}\right\}+\frac{(2-\beta)(l^3+l^2)}{2[l(\beta-1)+1]}$$

Hence

$$\lim_{n\to\infty} l^{-3}\phi^T B^{-1}\phi = \frac{1}{12}; \qquad \lim_{n\to\infty} n^{-3}S_1 = \frac{|B_0|^2}{12}$$

For the second term

$$S_2 = \sum_{i,j=1}^{p-1} \bar{\gamma}_i q_{ij}^{(-1)} \gamma_j = \sum_{i=1}^{p-1} |\gamma_i|^2$$

so that

$$\lim_{n\to\infty} n^{-3}S_2 = 0$$

In the third term

$$S_3 = \sum_{i=1}^{p-1} \sum_{j=p}^{n} \bar{\gamma}_i q_{ij}^{(-1)} \gamma_j = 0$$

since $q_{ij}^{(-1)} = 0$ for $i \leqslant p-1, j \geqslant p$. Similarly $S_4 = 0$.

To get B we note that (69) and (73) give, with $\lambda = 0$,

$$g(0) = \frac{1}{|B_0|^2}$$

Hence

$$\lim_{n\to\infty} n^{-3}\operatorname{Var}(\alpha^*) = \frac{12}{|B_0|^2} = 12g(0) \qquad \text{Q.E.D.}$$

What happens if we use the parabolic estimate when $f \in C^+([-\pi,\pi])$ so that the least-squares estimate is actually asymptotically optimal? How much do we lose in efficiency?

Not much; this is implied by the following surprising result from Vitale (1973).

THEOREM 10

If $f \in C^+([-\pi,\pi])$, the asymptotic efficiency of the parabolic estimate is $5/6$.

Proof With the notation in (70) we have, for the parabolic estimate,

$$\frac{5n}{12\pi}\operatorname{Var}(\alpha_p^*) = \frac{5n}{12\pi}\int_{-\pi}^{\pi} |P_n(e^{i\lambda})|^2 f(\lambda)\,d\lambda = \int_{-\pi}^{\pi} l_n(\lambda) f(\lambda)\,d\lambda$$

With calculations similar to the above we get the bound

$$|P_n(e^{i\lambda})| \leqslant \frac{24n}{(n^2-1)|1-e^{i\lambda}|^3}; \qquad l_n(\lambda) = \frac{5n}{12\pi}|P_n(e^{i\lambda})|^2 \tag{75}$$

Thus

$$|l_n(\lambda)|^{1/2} \leqslant \left(\frac{5}{12\pi}\right)^{1/2} \frac{24n^{3/2}}{(n^2-1)|1-e^{i\lambda}|^3} \to 0$$

uniformly for $|\lambda| > \varepsilon$ as $n \to \infty$. But, using (70) and (75),

$$\int_{-\pi}^{\pi} l_n(\lambda)\, d\lambda = \frac{30n^3}{(n^2-1)^2} \sum_{k=1}^{n} \left(\frac{k}{n}\right)^2 \left(1-\frac{k}{n}\right)^2 \sim 30\left[\frac{1}{n}\sum_{k=1}^{n}\left(\frac{k}{n}\right)^2\left(1-\frac{k}{n}\right)^2\right]$$

The quantity in brackets is a Riemann sum tending to

$$\int_0^1 x^2(1-x)^2\, dx = \tfrac{1}{30}$$

so that the integral of l_n tends to 1. Then the singular integral argument shows that

$$\lim_{n\to\infty} \frac{5n}{12\pi} \operatorname{Var}(\alpha_p^*) = f(0)$$

Combining this with $\operatorname{Var}(\alpha^*) \sim 2\pi f(0) n^{-1}$ when $f \in C^+([-\pi, \pi])$, we get the asymptotic efficiency $\frac{5}{6}$ for the parabolic estimate. Q.E.D.

5 ASYMPTOTIC OPTIMALITY; CONTINUOUS CASE

We now turn to the continuous case, $T = (0, L) \subset \mathbb{R}$, and then encounter a difficulty which was absent for discrete time t when the asymptotic optimality of least-squares estimates was studied. In Section 4 the basic results were obtained by approximating the spectral density $f(\lambda)$ by the inverse of a positive trigonometric polynomial uniformly, both from above and from below. The same sort of approximation could be applied in the continuous case, and then appeal made to Theorem 2 of Section 3, but since λ now varies over the whole real line, the spectral densities that can be thus approximated form a rather narrow class.

Instead we base our study on a *direct approach*. We show how this can be done in the case of $p = 1$ and $\alpha =$ a constant mean. Once this approach has been made clear, it can be extended to other cases, but we shall not go into this here since the rest of the argument follows the discrete case closely.

Before we examine the basic result (see Theorem 2 below), let us solve a simpler problem. Say that we want to estimate the constant mean α by a weighted average with the *shape function* w for the weight function appearing in the integral

$$\hat{\alpha} = \frac{1}{L}\int_0^L x(t)w\left(\frac{t}{L}\right)dt$$

$$\int_0^1 w(t)\,dt = 1 \tag{1}$$

We let w be an arbitrary element of a set W of equicontinuous functions in $[0,1]$. Then there exists a constant M such that

$$-M \leqslant w(t) - w(s) \leqslant M \tag{2}$$

Integrating (2) over $s \in (0,1)$ and using (1), we see that

$$-M \leqslant w(t) - 1 \leqslant M$$

so that the w are also uniformly bounded. The $\hat{\alpha}$ in (1) is clearly unbiased, and it is natural to ask how to choose the shape function w in order to minimize the variance. The solution is given by the following theorem from Grenander (1950).

THEOREM 1

Assume that the spectral density $f(\lambda)$ is continuous and positive at the origin as well as bounded over $\mathbb{R}$. Then

$$\bar{\alpha} = \frac{1}{L}\int_0^L x(t)\,dt$$

is asymptotically efficient relative to all estimates as in (1), $w \in W$.

Proof The least-squares estimate $\bar{\alpha}$ is easily dealt with since

$$\operatorname{Var}(\bar{\alpha}) = 4\int_{-\infty}^{\infty} \frac{\sin^2(L_\lambda/2)}{\lambda^2 L^2} f(\lambda)\,d\lambda = \frac{2}{L}\int_{-\infty}^{\infty} \frac{\sin^2{}'\mu}{\mu^2} f\left(\frac{\mu}{L}\right)d\mu$$

Using the well-known property of the Féjer kernel we get asymptotically

$$\operatorname{Var}(\bar{\alpha}) \sim 2\pi L^{-1} f(0) \tag{3}$$

On the other hand, for $w \in W$, we have

$$\operatorname{Var}(\hat{\alpha}) = \frac{1}{L} \int_{-\infty}^{\infty} |\gamma(\mu)|^2 f\left(\frac{\mu}{L}\right) d\mu \tag{4}$$

with the Fourier transform of the weight function

$$\hat{w}(\lambda) = \int_0^1 e^{it\lambda} w(t)\, dt \tag{5}$$

Consider for a sequence $L_\nu \to \infty$ the sequence associated with $\hat{\alpha}$-estimates $w_1, w_2, \ldots$ from W. Then it is possible to select a subsequence (use the equicontinuity of W) such that it converges pointwise to some $w \in W$. Supposing that this has already been done, we have, because of the uniform boundedness in W,

$$\int_0^1 |w(t) - w_\nu(t)|^2 dt \to 0 \qquad \text{as } \nu \to \infty$$

Plancherel's theorem then implies for the corresponding Fourier transforms $\hat{w}_\nu$ of the w_ν that

$$\int_{-\infty}^{\infty} |\hat{w}(\mu) - \hat{w}_\nu(\mu)|^2 \, d\mu \to \infty \qquad \text{as } \nu \to \infty$$

so that

$$\int_{-\infty}^{\infty} \left| \hat{w}_\nu(\mu) f^{1/2}\left(\frac{\mu}{L_\nu}\right) - \hat{w}(\mu) f^{1/2}\left(\frac{\mu}{L_\nu}\right) \right|^2 d\mu \leqslant f_{\max} \int_{-\infty}^{\infty} |\hat{w}_\nu(\mu) - \hat{w}(\mu)|^2 \, d\mu \to 0$$

Therefore

$$\int_{-\infty}^{\infty} |\hat{w}_\nu(\mu)|^2 f\left(\frac{\mu}{L_\nu}\right) d\mu - \int_{-\infty}^{\infty} |\hat{w}(\mu)|^2 f\left(\frac{\mu}{L_\nu}\right) d\mu \to 0 \tag{6}$$

Now we can get the bound

$$\left|\int_{-\infty}^{\infty}|\hat{w}(\mu)|^2 f\left(\frac{\mu}{L_\nu}\right)d\mu - f(0)\int_{-\infty}^{\infty}|\hat{w}(\mu)|^2 d\mu\right|$$

$$\leqslant \left|\int_{|\mu|<\varepsilon L_\nu}|\hat{w}(\mu)|^2\left[f\left(\frac{\mu}{L_\nu}\right)-f(0)\right]d\mu\right| + \left|\int_{|\mu|>\varepsilon L_\nu}|\hat{w}(\mu)|^2\left[f\left(\frac{\mu}{L_\nu}\right)-f(0)\right]d\mu\right| \tag{7}$$

and choose ε so small that for a given $\delta>0$ we have $|f(\mu)-f(0)|\leqslant\delta$ for $|\mu|\leqslant\varepsilon$, and then L_ν so large that

$$\int_{|\mu|>\varepsilon L_\nu}|\hat{w}(\mu)|^2 d\mu<\delta \tag{8}$$

Using the boundedness of f together with (8), we get from (6) and (7)

$$\lim_{\nu\to\infty}\int_{-\infty}^{\infty}|\hat{w}_\nu(\mu)|^2 f\left(\frac{\mu}{L_\nu}\right)d\mu = f(0)\int_{-\infty}^{\infty}|\hat{w}(\mu)|^2 d\mu$$

Plancherel's theorem tells us that

$$\int_{-\infty}^{\infty}|\hat{w}(\mu)|^2 = 2\pi\int_0^1 |w(t)|^2 dt$$

so that Schwarz's inequality applied to (4) and using (3) yields the relation

$$\lim_{\nu\to\infty}\frac{\operatorname{Var}(\hat{\alpha}_\nu)}{\operatorname{Var}(\bar{\alpha}_\nu)} = \int_0^1|w(t)|^2 dt \geqslant \left(\int_0^1 w(t)\,dt\right)^2 = 1 \tag{9}$$

Since only equality is possible in (9), we see that α is asymptotically optimal. Q.E.D.

Corollary 1 If $\hat{\alpha}$ is of the form in (1) with a fixed shape function w, it is asymptotically efficient if and only if $w(t)\equiv 1$.

Remark 1 The condition on w means that we restrict the variability of the weights in the linear estimate. Without any such restriction Theorem 1 is not valid. Indeed, consider the band-limited process in Example 5 in Section 2. We saw that then consistent estimation was possible for any finite, positive L. On the other hand the least-squares estimate has positive variance for any finite L so that we can make no assertion about asymptotic optimality. The explanation is that the estimate $\alpha_0^* = \lim_{A\to\infty}\alpha_A^*$ (see (2.21)) cannot be approximated by estimates as in (1) with the given equicontinuity condition on the weight functions w.

Still, the asymptotic optimality of $\bar{\alpha}$ in Theorems 1 and 2 of Section 3 held without any such regularity constraints. It is true that this was so only for a fairly special class of spectral densities [see (3.5)], and it is natural to ask if we can extend this much further. The answer is yes.

THEOREM 2 ASYMPTOTIC OPTIMALITY OF THE EQUIDISTRIBUTED AVERAGE, GRENANDER (1950)

Assume that $y(t)$ is stationary and continuous in the mean and:

(a) *$y(t)$ is completely nondeterministic (see Chapter 2, Section 4) so that it can be written as a one-sided moving average*

$$y(t) = \int_{-\infty}^{t} g(t-s)\xi(ds).$$

(b)

$$\int_0^\infty u^3|g(u)|^2\,du < \infty. \tag{10}$$

(c) *The spectral density at zero $f(0) > 0$. Then the least-squares estimate $\bar{\alpha} = \bar{x}_L$ is asymptotically optimal.*

Proof (See Note 4.5.A.) Using Theorem 4 of Section 3, we can write the BLUE α^* in terms of the best linear predictor $x^*(t)$ using the observations $x(s), s\in T=[0, L]$, as

$$\alpha^* = \gamma_L\left\{\int_{-\infty}^{0} x^*(t)\,dt + \int_0^L x(t)\,dt + \int_L^\infty x^*(t)\,dt\right\} = \gamma_L\{I_- + I + I_+\}$$

with some constant γ_L. We shall determine the asymptotic behavior of γ_L. We have, for $t > L$, since $\|x(t)-x^*(t)\|^2$ is at least equal to $\|x(t)-x^{**}(t)\|^2$ and using the fact that prediction errors are orthogonal to $x^*(t)$ and $x^{**}(t)$, respectively,

$$\|x^*(t)\|^2 \leqslant \|x^{**}(t)\|^2 = \int_{-\infty}^{L} |g(t-u)|^2\,du \tag{11}$$

where $x^{**}(t)$ is the best linear predictor using the observations $x(s), s\in$

$(-\infty, L)$. But the right-hand side of (11) is

$$\int_{-\infty}^{L} \frac{1}{1+(t-u)^3}\left[1+(t-u)^3\right]|g(t-u)|^2\, du$$

which is seen to be $0\{[1+(t-L)^3]^{-1}\}$ using (10b). Hence

$$\|x^*(t)\| = 0\left\{\left[1+(t-L)^3\right]^{-1}\right\}$$

so that

$$\left\|\int_L^\infty x^*(t)\, dt\right\| = \|I_+\| = 0(1)$$

Similarly for I_-. Therefore

$$\|\alpha^*\|^2 = |\gamma_L|\left[\sqrt{2\pi f(0)L} + 0(1)\right]$$

since

$$\operatorname{Var}(I) \sim 2\pi f(0) L^{-1}$$

But using $x^*(s) - x(s) \perp x(t)$, $t \in T$, we get

$$E\{\alpha^*[x(t)-\alpha]\} = \gamma_L \int_{-\pi}^{\pi} r(t)\, dt = 2\pi f(0)\gamma_L; \qquad t \in T$$

so that Theorem 1 of Section 2 [(4) and (5)] tells us that $\gamma_L \sim L^{-1}$, and hence

$$\|\alpha^*\|^2 \sim 2\pi f(0) L^{-1} \sim \|\bar{\alpha}\|^2$$

as asserted. Q.E.D.

When observing processes with a continuous time parameter it may happen that our instrumentation does not allow us to observe the process instantaneously. We mention three possibilities [see Note 4.5.B].

Say that $x(t)$ is stationary with absolutely continuous spectrum and has the spectral density $f(\lambda)$, $-\infty < \lambda < \infty$. If we can only observe the values $y_\nu = x(\nu h)$, $\nu \in \mathbb{Z}$, it is clear that $\{y_\nu\}$ is stationary with covariances

$$\operatorname{Cov}(y_t, y_{t+\nu}) = r(\nu h) = \int_{-\infty}^{\infty} e^{i\nu h\lambda} f(\lambda)\, d\lambda = \int_{-\pi}^{\pi} e^{i\nu\lambda} f_y(\lambda)\, d\lambda \qquad (12)$$

where we have introduced

$$f_y(\lambda)=\frac{1}{h}\sum_{k=-\infty}^{\infty} f\left(\frac{\lambda+2k\pi}{h}\right) \tag{13}$$

The series in (13) converges a.e. since $f\in L_1[(-\infty,\infty)]$ so that f_y is well defined and in $L_1[(-\pi,\pi)]$. Hence if we sample the x-process equidistant with the time step h, the new, discrete process has a spectral density f_y given by (13), and all inferences will be for f_y, not f.

On the other hand if what we can observe are the average values

$$z_\nu=\frac{1}{h}\int_{(\nu-\frac{1}{2})h}^{(\nu+\frac{1}{2})h} x(t)\,dt$$

the new, discrete process has covariances given by

$$\operatorname{Cov}(z_t, z_{t+\nu})=\frac{1}{h^2}\iint r(u-v)\,du\,dv \tag{14}$$

integrated over $(t-\frac{1}{2})h\leqslant u\leqslant(t+\frac{1}{2})h$ and $(t+\nu-\frac{1}{2})h\leqslant v\leqslant(t+\nu+\frac{1}{2})h$. Expressing the covariance kernel r in (14) by its spectral representation, we see after exchanging the order of integrations that (14) equals

$$\frac{4}{h^2}\int_{-\infty}^{\infty} e^{i\nu h\lambda}\frac{\sin^2(h\lambda/2)}{2}f(\lambda)\,d\lambda \tag{15}$$

Combining (15) with (13), it follows that the z-process has the spectral density

$$f_z(\lambda)=\frac{4}{h}\sin^2\frac{\lambda}{2}\sum_{k=-\infty}^{\infty}\frac{f((\lambda+2\pi k)/h)}{(\lambda+2\pi k)^2} \tag{16}$$

Finally if we observe $w_\nu=x(h\nu+\Delta_\nu)$, where $\{\Delta_\nu;\nu\in\mathbb{Z}\}$ are i.i.d. stochastic variables, we get the covariances

$$\operatorname{Cov}(w_t, w_{t+\nu})=E\left[\int_{-\infty}^{\infty} e^{ih\nu\lambda+ih(\Delta_{t+\nu}-\Delta_t)\lambda}f(\lambda)\,d\lambda\right] \tag{17}$$

Denoting the characteristic function of Δ_ν by $\phi(z)$, (17) gives us, for $\nu\neq 0$,

$$\int_{-\infty}^{\infty} e^{i\nu h\lambda}|\phi(h\lambda)|^2 f(\lambda)\,d\lambda$$

while for $\nu=0$, we get simply the value

$$\int_{-\infty}^{\infty} f(\lambda)\,d\lambda = r(0)$$

Introducing the constant

$$c=\int_{-\infty}^{\infty} f(\lambda)\left[1-|\phi(h\lambda)|^2\right]d\lambda \tag{18}$$

we see that the new, discrete w-process has the spectral density [use (13) again]

$$f_w(\lambda)=\frac{1}{h}\sum_{k=-\infty}^{\infty}|\phi(\lambda+2\pi k)|^2 f\left(\frac{\lambda+2\pi k}{h}\right)+\frac{c}{2\pi} \tag{19}$$

The discretization relations (13), (16), and (19) enable us to reduce the inference problems to discrete time inference of the sort we have already studied.

6 LINEAR INFERENCE IN BANACH SPACES

Consider a probability measure P in a real and separable Banach space X, and assume that $E[\|x\|]<\infty$. Introduce the mean $\alpha=E(x)$ which then exists (see Chapter 1, Section 2), and treat it as the unknown parameter to be estimated.

Say that we have n i.i.d. observations x_ν, $\nu=1,2,\ldots,n$, so that

$$x_\nu=\alpha+y_\nu; \qquad x_\nu\in X$$

and form the commonsense estimate

$$\alpha^*=\frac{1}{n}\sum_{\nu=1}^{n}x_\nu=\bar{x}$$

It is a consistent estimate of α. We first prove a simple result.

THEOREM 1

If $E[\|x\|]<\infty$, the estimate $\bar{x}$ converges in probability in norm to α.

$$\lim_{n\to\infty} E\left[\|\bar{x}-\alpha\|\right]=0 \tag{1}$$

Proof We show (1) by reducing it to the classical (finite-dimensional) law of large numbers. Define a random element y with values in X as follows. If $\|x\|>R$, define $y=0$. Here R is large enough that

$$\int_{B(0,R)}\|x\|P(dx)\leqslant\varepsilon;\qquad B(x,R)=\{y\mid\|y-x\|\leqslant R\}$$

where ε is some given positive number. For given $\delta>0$ we can find (see Note 4.6.A) a compact set $K\in B(0,R)$ with mass greater than $P(K)-\delta$. Cover K with $B(x,\delta)$, $x\in K$. We can then find a finite covering $B(\eta_i,\delta)$, $i=1,2,\ldots,N$ with $\eta_i\in K$. Define the map

$$y(x)=\begin{cases}0; & \text{if } \|x\|>R\\ \eta_i; & \text{if } \|x\|\leqslant R \text{ and } \|x-\eta_i\|=\min\limits_{k=1,2,\ldots,N}\|x-\eta_k\|\end{cases}$$

If i is not uniquely determined in the last line, pick one of the possible values.

We then have $E[\|x-y\|]\leqslant\varepsilon+(1+R)\delta$ and $\|\alpha-E(y)\|\leqslant\varepsilon+(1+R)\delta$ as well as

$$E\|\bar{x}-\bar{y}\|=E\left\|\frac{1}{n}\sum_{1}^{n}(x_\nu-y_\nu)\right\|\leqslant E[\,\|x-y\|\,]\leqslant\varepsilon+(1+R)\delta\tag{2}$$

But

$$\bar{y}=\sum_{k=1}^{N}\frac{f_k}{n}\eta_k$$

where f_k is the number of times the y take the value η_k. The classical law of large numbers implies that $E[\|\bar{y}-E(y)\|]$ tends to zero. Combining this with (2) and the fact that α is close to $E(y)$, we get (1). Q.E.D.

This can be strengthened considerably.

THEOREM 2 STRONG LAW OF LARGE NUMBERS IN A BANACH SPACE, MOURIER (1953)

If $E[\|x\|]<\infty$, the estimate $\bar{x}$ tends almost certainly to α in norm.

Proof We deal first with the case when x takes only denumerably many values $\eta_1,\eta_2,\ldots$. Choose an integer N and define

$$x_t^N=\begin{cases}\eta_k; & \text{if } x_t \text{ equals one of the } \eta_1,\eta_2,\ldots,\eta_N\\ 0; & \text{otherwise}\end{cases}$$

and write $x_t = x_t^N + r_t^N$. This is quite similar to the previous proof. Then

$$\bar{x} = \frac{1}{n} \sum_{t=1}^{n} x_t^N + R_n^N; \qquad R_n^N = \frac{1}{n} \sum_{t=1}^{n} r_t^N \tag{3}$$

and we see, as above, that the first sum in (3) converges almost certainly to $E(x^N)$. But we also have with probability 1

$$\lim_{n\to\infty} \frac{1}{n} \sum_{t=1}^{n} \| r_t^N \| = E\left[\| r_t^N \| \right]$$

which does not depend upon t, and which tends to zero as $N \to \infty$. This shows with a standard argument that $\bar{x} \to \alpha$ almost certainly in norm.

Now let us deal with the general case. Since X is separable, we can choose a sequence $\eta_1, \eta_2, \ldots$ everywhere dense in X. Introduce the balls

$$B_k(\varepsilon) = \{ x \mid \| x - \eta_k \| \leqslant \varepsilon \}$$

so that X is covered by a family of overlapping spheres. We want to deal with disjoint sets. Thus let us consider

$$\begin{aligned} E_1(\varepsilon) &= B_1(\varepsilon) \\ E_2(\varepsilon) &= B_2(\varepsilon) \cap B_1^c(\varepsilon) \\ E_3(\varepsilon) &= B_3(\varepsilon) \cap B_1^c(\varepsilon) \cap B_2^c(\varepsilon) \\ &\vdots \end{aligned}$$

Define a mapping T_ε by $T_\varepsilon x = \eta_k$ if $x \in E_k(\varepsilon)$. Then the stochastic elements $T_\varepsilon x_t$ are i.i.d. and take at most denumerably many elements. The first part of the proof shows that the first term on the right-hand side in (4) converges in norm almost certainly to $E[T_\varepsilon x]$

$$x = \frac{1}{n} \sum_{t=1}^{n} x_t = \frac{1}{n} \sum_{t=1}^{n} T_\varepsilon x_t + \frac{1}{n} \sum_{t=1}^{n} (T_\varepsilon x_t - x_t) \tag{4}$$

On the other hand the norm of the last term in (4) is at most ε. Choosing a sequence $\varepsilon_1, \varepsilon_2, \ldots$ tending to zero, this implies that $\bar{x}$ tends to α in norm with probability 1. Q.E.D.

Consider now *bounded compact convex sets* in $\mathbb{R}^2$ with the radius of curvature function $R = R(\phi) \in C^+([0, 2\pi)]$. Here $R(\phi)$ is the radius of curvature at the point of the boundary where the tangent makes the angle ϕ with some fixed direction.

We have seen in Chapter 1, Section 3, how to describe and analyze such sets. Our sample space X will be the Banach space $C([0,2\pi)]$ with sup norm in which x, expressed via R, takes values. We now suppose that we observe a random compact closed set $x=R(\phi)=\alpha\psi(\phi)z(\phi)$, where $\psi(\phi)$ is the radius of curvature function of a known compact convex set, $\psi\in C^{+}([0,2\pi)]$, α is an unknown *positive* parameter, and $z(\phi)$ is a stochastic process with $E[z(\phi)]\equiv 1$ and with sample function in $C^{+}([0,2\pi)]$ with probability 1. We assume as before that $z(\phi)$ is continuous in the mean with covariance function $r(\phi_1,\phi_2)$.

How can we estimate α linearly, and how can we do it well? Consider a bounded linear functional $x^*\in X^*$ and try the estimate $\alpha^*=x^*(x)$. To make it unbiased we must have [see (1.2.6)]

$$E_\alpha(\alpha^*)=E_\alpha[x^*(x)]=x^*[E_\alpha(x)]=\alpha x^*(\psi) \tag{5}$$

so that we get the condition $x^*(\psi)=1$.

On the other hand the Riesz representation theorem tells us that we can write

$$\alpha^*=x^*(R)=\int_0^{2\pi}R(u)A(du) \tag{6}$$

where A is a bounded signed measure on $[0,2\pi)$. Since we estimate a positive parameter, it is reasonable to require that $A(du)\geqslant 0$; A should be a measure.

The variance of such an estimate is then

$$\mathrm{Var}(\alpha^*)=\alpha^2\int_0^{2\pi}\int_0^{2\pi}\psi(u)\psi(v)r(u,v)A(du)A(dv) \tag{7}$$

and we now ask for a BLUE. Let v be the greatest lower bound for the variance of all the estimates satisfying (5) and (6) and consider a sequence α_n^*, in terms of A_n, such that their variances tend to v.

Since

$$x^*(\psi)=\int_0^{2\pi}\psi(u)A_n(du)=1$$

the measures A_n are uniformly bounded so that we can extract a subsequence converging in the weak convergence topology to some bounded measure A. Then, since r is continuous, we have $\mathrm{Var}(\alpha^*)=v$, with α^* corresponding to A.

This guarantees existence of the BLUE, but we can also show uniqueness of α^* by essentially the same proof as the one we used for Theorem 1 in Section 2. The measure A need not be unique, however.

Let us derive the normal equation. Say that $u_0 \in \text{support}(A)$ and that we vary A by adding ε times the A-measure restricted to a small interval around u_0. To keep the total measure unchanged, we subtract by the same procedure the same total mass at another point $u \in \text{support}(A)$. The usual variation argument applied to the double integral in (7) then gives us

$$\int_0^{2\pi} \psi(v) r(u,v) A(dv) = C; \qquad u \in \text{support}(A) \tag{8}$$

If u is not in support(A), we can only do this for $\varepsilon > 0$, since otherwise the varied A would have negative variation. We therefore get only

$$\int_0^{2\pi} \psi(v) r(u,v) A(dv) \leqslant C; \qquad u \notin \text{support}(A) \tag{9}$$

Relations (8) and (9) are the normal equations, familiar to some readers from the theory of infinite programming.

We now specialize further by assuming that the deformation mechanism that operates on the convex set with $\alpha\psi \in X$ and leads to $x = \alpha\psi z$ is *isotropic* so that $r(u,v) = r(u-v)$. Then we can use, since r is automatically periodic, the Egervary expansion as the Fourier series for $r(t)$ associated with the second relation in (2.4.10), and we get

$$\int_0^{2\pi}\int_0^{2\pi} r(u,v)\psi(u)\psi(v)A(du)A(dv) = \frac{1}{2\pi} \sum_{\nu=-\infty}^{\infty} \hat{r}_\nu |B_\nu|^2$$

where $\hat{r}_\nu \geqslant 0$ is the νth Fourier coefficient of r, and

$$B_\nu = \int_0^{2\pi} \psi(u) e^{i\nu u} A(du)$$

Note, however, that the unbiasedness condition tells us that $B_0 = 1$. Therefore if we choose A absolutely continuous with density

$$\frac{A(du)}{du} = \frac{1}{2\pi}\frac{1}{\psi(u)}$$

all the other B_k vanish, and we have minimized the variance by this choice. We sum it up as the following theorem.

THEOREM 3

For the convex sets and the deformation mechanism described above the BLUE exists in the form (6) *and is unique. It satisfies the normal equations*

(8) and (9). *If in particular r is isotropic, the BLUE can be expressed in closed form as*

$$\alpha^* = \frac{1}{2\pi}\int_0^{2\pi} \psi^{-1}(u)R(u)\,du \tag{10}$$

For example, if the original convex set is a circle, $\psi(u)\equiv\psi_0$, we get in terms of the arc elements ds of the observed set

$$\alpha^* = \frac{1}{2\pi\psi_0}\int_0^{2\pi} R(\phi)\,d\phi = \frac{1}{2\pi\psi_0}\int_0^{2\pi}\frac{ds}{d\phi}\,d\phi = \frac{L}{L_0}$$

where L_0 is the perimeter of the original set and L is the observed perimeter. For other shapes α^* is a weighted average of ds.

If the covariance function r is not that of an isotropic deformation mechanism, the estimate (10) is no longer BLUE. Since r is still continuous, we can choose a separable measurable version $z(t,\omega)$, and it follows from Fubini's theorem that the integrals

$$I = \int_0^{2\pi} |z(t,\omega)|^2\,dt$$

$$I_k = \frac{1}{\sqrt{2\pi}}\int_0^{2\pi} z(t,\omega)e^{ikt}\,dt$$

exist a.c. [see Chapter 1, Section 1]. Note that $E(I)<\infty$. Then with probability 1 Parseval's relation holds:

$$I = \|z(\cdot,\omega)\|^2 = \sum_{-\infty}^{\infty} |I_k|^2 \tag{11}$$

The partial sums of the right-hand side in (11) tend to the left-hand side in a nondecreasing manner so that $E(I)<\infty$ implies that

$$\sum_{-\infty}^{\infty} E\left[|I_k|^2\right] = E(I) < \infty$$

[see, e.g., Titchmarsh (1939), p. 346].

With $\hat{r}_k = E(|I_k|^2)$ we see that the *z-process is harmonizable* and can be written as

$$z(t) = \sum_{-\infty}^{\infty} z_k e^{-ikt};\quad t\in[0,2\pi) \tag{12}$$

with convergence in the mean [see Loève (1948), pp. 324–326]. Since T here is a finite interval, the representation (12) takes the form of a sum rather than an integral due to the nature of the group characters on $\mathbb{T}^1$ (see Table 1 in Chapter 1, Section 3).

It is important to realize that (12) is no longer the Egervary representation so that the stochastic variables z_k are not necessarily orthogonal to each other.

Example 1 Let the undeformed set be a circle, $\psi \equiv 1$, and consider the deformation mechanism (12) with nonzero terms only for $k=0, \pm 2$, so that with real z-values

$$z(u)=z_0+z_1e^{2iu}+z_{-1}e^{-2iu}; \qquad E(z_0)=1, \qquad E(z_1)=E(z_{-1})=0$$

To illustrate what may happen, we assume in addition that $z_1=z_{-1}$ so that $z_0 \geqslant 0$ and

$$E[z(u)z(v)]=1+V_0+2\rho\sqrt{V_0V_1}\,(\cos 2u+\cos 2v)+4V_1\cos 2u\cos 2v$$

with

$$\begin{aligned} V_0&=\mathrm{Var}(z_0)\\ V_1&=\mathrm{Var}(z_1)\\ \rho&=\text{correlation coefficient between } z_0 \text{ and } z_1 \end{aligned}$$

To minimize the variance with the constraint

$$\int_0^{2\pi} A(du)=1$$

we should solve

$$\begin{cases} \min\left[x\rho\sqrt{V_0V_1}+V_1x^2\right] \\ x=\displaystyle\int_0^{2\pi}\cos 2u A(du) \end{cases} \tag{13}$$

Hence if we solve (13) in x, $x=-(\rho/2)\sqrt{(V_0/V_1)}$, the BLUE is given by the weight function

$$\frac{A(du)}{du}=\frac{1}{2\pi}(1+2x\cos 2u)=\frac{1}{2\pi}\left[1-\rho\sqrt{\frac{V_0}{V_1}}\cos 2u\right]$$

if this is in $C^+([0,2\pi))$, which is the case if $|\rho|<\sqrt{(V_1/V_0)}$.

On the other hand, if $\rho>\sqrt{(V_1/V_0)}$, consider the weight function

$$\frac{A(du)}{du}=\frac{1}{2\pi}(1-\cos 2u)$$

corresponding to $x=-1$. Notice that we must have $|x|\leqslant 1$ to make $A(du)/du$ nonnegative. Hence $x=-1$ minimizes the variance in this case.

Finally if $\rho<-\sqrt{(V_1/V_0)}$, the weight function should be chosen as

$$\frac{A(du)}{du}=\frac{1}{2\pi}(1+\cos 2u)$$

7 SELECTING LINEAR FUNCTIONALS FOR ESTIMATION

Consider a probability measure in X=separable real Hilbert space with mean zero and covariance operator R of the trace type. As usual we denote the eigenelements by $\lambda_\nu, \phi_\nu, \nu=1,2,\ldots$; but in this section it is important to observe the ordering $\lambda_1\geqslant\lambda_2\geqslant\lambda_3\geqslant\cdots$. We use normalized ϕ_ν for convenience.

Even with a single observed value x we note that our information is infinite dimensional. It is therefore useful for storing the information if we can *replace x by a finite-dimensional quantity to achieve data compression*. To do this we try to select N continuous linear functionals $f_1 f_2,\ldots,f_N$, sometimes called *features*, such that knowledge of $f_1(x), f_2(x),\ldots,f_N(x)$ enables us to reconstruct x with as small an approximation error as possible. A continuous linear functional in X can be represented by a vector in the dual $X^*=X$, and it is convenient to use the same symbol for the vector as for the corresponding functional.

The best approximation to x, in the norm of X, is then simply $x^*=Px$, where P is the projection operator corresponding to the subspace spanned by our functionals (vectors) (see Chapter 2, Section 2). As we have seen several times, the squared norm of the error is then

$$E\left[\|x^*-x\|^2\right]=E\left[\|Px-x\|^2\right]=E\left[\|x\|^2\right]-E\left[\|Px\|^2\right] \tag{1}$$

which makes it natural to introduce the *efficiency of the system of features*

$$e_N=\frac{\inf_{x^*}E\left[\|x^*-x\|^2\right]}{E\left[\|x^*-x\|^2\right]} \tag{2}$$

where the infimum is taken over all sets of N linear functionals. As we shall see, the infimum is achieved, so that we can write *min* instead of *inf* in (2).

This is a linear problem so that we do not need knowledge of the probability measure, except for its second-order geometry. In Chapter 5, Section 3, we return to optimal selection of linear features, but there for another purpose: in order to test one hypothesis (measure) against another. We then need complete specification of the measures.

A general answer to the problem of optimal selection of linear features is given in the present situation by the following theorem.

THEOREM 1

Selecting as features the eigenvectors $\phi_1, \phi_2, \ldots, \phi_N$ associated with the N largest eigenvalues is efficient, $e_N = 1$.

Proof A simple proof can be based on the consideration of the projection operator Q onto the subspace spanned by $\phi_1, \phi_2, \ldots, \phi_N$. For an arbitrary projection P of dimension N we have

$$0 \leqslant \| P\phi_\nu \|^2 \leqslant 1 \tag{3}$$

while

$$\| Q\phi_\nu \|^2 = \begin{cases} 1; & \text{for } \nu = 1, 2, \ldots, N \\ 0; & \text{otherwise} \end{cases} \tag{4}$$

Since P and Q are projections, both of dimension N, we also have

$$\operatorname{tr}(P) = \sum_{\nu=1}^{\infty} \| P\phi_\nu \|^2 = N$$

$$\operatorname{tr}(Q) = \sum_{\nu=1}^{N} \| Q\phi_\nu \|^2 = N \tag{5}$$

In order to get $e_N = 1$ in (2), according to (1), we should maximize

$$\| Px \|^2 = \sum_1^{\infty} \lambda_\nu \| P\phi_\nu \|^2; \qquad \lambda_\nu = E\left[(x, \phi_\nu)^2 \right] \tag{6}$$

To maximize (6) under the conditions (3) and the first equation in (5), it is clear that we should try to make the coefficient $\| P\phi_\nu \|^2$ large for the largest λ_ν, that is, equal to 1. This is indeed possible by choosing $P = Q$ [see (4)] so that the minimum is achieved as stated. Q.E.D.

Remark 1 This result, which has been rediscovered repeatedly in the literature on pattern recognition, is an elementary Hilbert space relation that seems to go back to the beginning stage of the study of integral equations.

Remark 2 If some eigenvalues have multiplicity greater than 1, the projection with $e_N = 1$ need not be unique. When the eigenvalues are simple, the projection (and its associated subspace) is unique. The features themselves are of course never completely unique since they can be changed by multiplication by -1.

Remark 3 Theorem 1 was stated for (infinite-dimensional) Hilbert space. If the dimension of X is finite but large, the theorem can be modified to cover this case too, which is of considerable practical importance. The proof is practically the same.

While Theorem 1 is of general scope, it should be noted that we must be able to compute the eigenelements of the covariance operator in order to apply it. The same is true if X is of finite but large dimension. We then have to carry out an eigenanalysis of a big covariance matrix. In the important case of a stationary covariance it has been shown in McClure (1970) that for large N an asymptotically efficient way can be found of selecting the features involving no large computational effort.

Following McClure's elegant analysis, we derive the basic result through a series of three lemmas and two theorems. Let the process x_t, $t \in \mathbb{N}$, have mean zero, covariance function r, with an absolutely continuous spectrum normalized to $r(0)=1$. To arrive at an attractive formulation of the answer it will be convenient to write the Fourier transform with a factor $(2\pi)^{-1}$:

$$r(t) = \frac{1}{2\pi}\int_{-\pi}^{\pi} e^{it\lambda} f(\lambda)\, d\lambda$$

Introduce the function G and the sets $B(b)$ by

$$G(b) = \frac{1}{2\pi} m\{\lambda \mid f(\lambda) > b\} = \frac{1}{2\pi} m[B(b)]; \qquad B(b) = \{\lambda \mid f(\lambda) > b\}$$

so that G is nonincreasing and takes values between 0 and 1.

Without striving for full generality, we assume that $N = [\theta n], 0 < \theta < 1$, and

$$\begin{aligned} &f \text{ is continuous} \\ &G(b_0) = \theta \text{ for some } b_0 \text{ and } G \text{ is continuous at } b_0 \end{aligned} \tag{7}$$

The number θ expresses the degree to which our selection procedure has compressed our data. It is called the *compression rate*.

We have to calculate, for large values of n and hence N, the numerator in (2) which we know can be expressed as

$$\sum_{N+1}^{\infty} \lambda_\nu^{(n)} \tag{8}$$

where $\lambda_\nu^{(n)}$ is the νth eigenvalue of the $n \times n$ covariance matrix R_n. The first partial result tells us how (8) behaves for large n.

THEOREM 2

If (7) *holds, we have*

$$\lim_{n\to\infty} \frac{1}{n} \sum_{N+1}^{n} \lambda_\nu^{(n)} = 1 - \frac{1}{2\pi}\int_{B(b_0)} f(\lambda)\, d\lambda \tag{9}$$

Proof With $M = \max_\lambda f(\lambda)$ and N_n = number of $\lambda_\nu^{(n)}$ greater than b_0 we have (see Note 4.7.A)

$$\limsup_{n\to\infty} \frac{1}{n}\left| \sum_{\nu=1}^{N} \lambda_\nu^{(n)} - \sum_{\nu=1}^{N_n} \lambda_\nu^{(n)} \right| \leqslant \limsup_{n\to\infty} \frac{M}{n} |N - N_n|$$

$$= M \limsup_{n\to\infty} \left| \theta - \frac{N_n}{n} \right| \tag{10}$$

We now appeal to Szegö's fundamental theorem for Toeplitz forms. It states, when restricted to the present case, that the distribution of $\{\lambda_\nu^{(n)}\}$ tends weakly to the distribution of $f(\lambda)$ with λ uniformly distributed on $(-\pi, \pi)$. For a more general statement and a proof of Szegö's theorem the reader is referred to Grenander-Szegö (1957). Since all $\lambda_\nu^{(n)} \in [0, M]$, we can restrict ourselves to this interval in the following.

To apply this result, introduce the indicator function $1(\lambda)$ of the interval (b_0, M) and approximate it from below by $1^-(\lambda)$ and from above by $1^+(\lambda)$, both of them continuous:

$$1^-(\lambda) = \begin{cases} 0; & \text{for } \lambda < b_0 \\ \dfrac{1}{\varepsilon}(\lambda - b_0); & \text{for } b_0 \leqslant \lambda < b_0 + \varepsilon \\ 1; & \text{for } b_0 + \varepsilon < \lambda \end{cases}$$

and

$$1^+(\lambda)=\begin{cases} 0; & \text{for } \lambda<b_0-\varepsilon \\ \dfrac{1}{\varepsilon}(\lambda-b_0+\varepsilon); & \text{for } b_0-\varepsilon\leqslant\lambda<b_0 \\ 1; & \text{for } b_0<\lambda \end{cases}$$

Then, using the above weak convergence, we get, since 1^- is continuous,

$$\liminf_{n\to\infty}\frac{N_n}{n}=\liminf_{n\to\infty}\frac{1}{n}\sum_{\nu=1}^{n}1(\lambda_\nu^{(n)})$$
$$\geqslant\liminf_{n\to\infty}\frac{1}{n}\sum_{\nu=1}^{n}1^-(\lambda_\nu^{(n)})=\frac{1}{2\pi}\int_{-\pi}^{\pi}1^{-1}(\lambda)f(\lambda)\,d\lambda$$

The last integral is at least equal to

$$\frac{1}{2\pi}\int_{B(b_0+\varepsilon)}d\lambda=G(b_0+\varepsilon)$$

where G and B have been introduced above. But ε is an arbitrary positive number and G was assumed to be continuous at b_0 so that

$$\liminf_{n\to\infty}\frac{N_n}{n}\geqslant G(b_0)=\theta$$

Using 1^+ instead of 1^-, we get the opposite inequality

$$\limsup_{n\to\infty}\frac{N_n}{n}\leqslant G(b_0)=\theta$$

so that we have shown that $N_n/n\to\theta$. Using (10) this implies that the left-hand side in this equation is zero.

Introduce now the function $L(\lambda)=\lambda 1(\lambda)$, approximate it from below and above as before, and use Szegö's theorem again. We then get

$$\lim_{n\to\infty}\frac{1}{n}\sum_{\nu=1}^{N_n}\lambda_\nu^{(n)}=\frac{1}{2\pi}\int_{B(b_0)}f(\lambda)\,d\lambda \tag{11}$$

$$\frac{1}{n}\sum_{1}^{n}\lambda_\nu^{(n)}=\frac{1}{n}\operatorname{tr}(R_n)=\operatorname{Var}(x_t)=1=\frac{1}{2\pi}\int_{-\pi}^{\pi}f(\lambda)\,d\lambda \tag{12}$$

and subtracting (11) from (12) we get (9). Q.E.D.

Recall that stationarity here means invariance with respect to the group of translations of the integers. The *Fourier characters* can be expected to play an important role for our selection problem. They are for the *cyclic group* (see Table 1 in Chapter 1, Section 3)

$$\psi_\nu^{(n)}(s) = \frac{1}{\sqrt{n}} \exp\left(\frac{2\pi i}{n}\omega_\nu s\right)$$

associated with the frequency ω_ν as the integers in the interval $[-n/2, n/2)$. The *ordering* of the ω_ν, that is, the correspondence $\omega_\nu \leftrightarrow \nu$, is crucial. We say that the $\lambda_\nu^{(n)}$ are *B-ordered* if, for a subset B of $[-\pi, \pi)$ with $m(B) = 2\pi\theta$, we have the implication

$$\frac{2\pi\omega_\nu}{n} \in B \quad \text{and} \quad \frac{2\pi\omega_\mu}{n} \notin B \Rightarrow \nu < \mu \tag{13}$$

The role of (13) will become more intuitive as we proceed through the analysis.

During the asymptotics we have to use Riemann rather than Lebesgue integrals. But first we need an elementary fact about Riemann integrals that we give for the sake of completeness.

Lemma 1 If the set B has an interior B^i and a closure $\bar{B}$ with $m(b^i) = m(\bar{B})$, then the indicator function 1_B is Riemann integrable.

Proof Consider partitions of $[-\pi, \pi]$ by the partition points in the set $P_m = \{x_1, x_2, \ldots, x_{m+1}\}$, $x_{i+1} - x_i = \Delta x_i$, of width $w(P_m) = \max \Delta x_i$, such that $w(P_m) \to 0$ as $m \to \infty$. Strictly speaking the x_i should carry the superscript m, but we leave it out in order to simplify the notation.

We have to show that if $x_i' \in [x_i, x_{i+1}]$,

$$\lim_{n \to \infty} \sum_{i=1}^{m} 1_B(x_i')\Delta x_i = \int_{-\pi}^{\pi} 1_B(\lambda)\, d\lambda = m(B) \tag{14}$$

Of course B is measurable, $m(B) = M(B^i) = m(\bar{B})$.

The interior B^i is open, so that it is a countable union of open intervals. Consider an increasing sequence B_n of open sets, such that each B_n is a finite union of open intervals and $B_n \to B^i$. The indicator function 1_{B_n} is Riemann integrable and at most equal to 1_B. Hence

$$m(B_n) = \lim_{m\to\infty} \sum_{i=1}^{m} 1_{B_n}(x_i')\Delta x_i \leqslant \liminf_{m\to\infty} \sum_{i=1}^{m} 1_B(x_i')\Delta x_i$$

Now let $n \to \infty$, so that $m(B_n) \to m(B)$ and

$$m(B) \leqslant \liminf_{m \to \infty} \sum_{i=1}^{m} 1_B(x_i')\Delta x_i$$

We can also approximate $\bar{B}$ by a decreasing sequence of closed sets and argue in the same manner, resulting in

$$m(B) \geqslant \limsup_{m \to \infty} \sum_{i=1}^{m} 1_B(x_i')\Delta x_i$$

Therefore (14) holds. Q.E.D.

Our next auxiliary result concerns the stochastic vector $x = (x_1, x_2, \ldots, x_n)$.

Lemma 2 With the set B as described in Lemma 1, assume the Fourier characters to be B-ordered and let N_n be the number of ν such that $2\pi\omega_\nu n^{-1} \in B$. Then

$$\lim_{n \to \infty} \frac{1}{n} E\left\{ \sum_{s=1}^{n} \left| \sum_{\nu=1}^{N} (x, \psi_\nu^{(n)})\psi_\nu^{(n)}(s) - \sum_{\mu=1}^{N_n} (x, \psi_\mu^{(n)})\psi_\mu^{(n)}(s) \right|^2 \right\} = 0 \quad (15)$$

Proof Let $N_n > N$; the opposite case is dealt with in the same manner. We have

$$\limsup_{n \to \infty} \frac{1}{n} E\left\{ \sum_{s=1}^{n} \left| \sum_{\nu=1}^{N} (x, \psi_\nu^{(n)})\psi_\nu^{(n)}(s) - \sum_{\mu=1}^{N_n} (x, \psi_\mu^{(n)})\psi_\mu^{(n)}(s) \right|^2 \right\}$$

$$= \limsup_{n \to \infty} \frac{1}{n} E \sum_{s=1}^{n} \left\{ \left| \sum_{N+1}^{N_n} (x, \psi_\nu^{(n)})\psi_\nu^{(n)}(s) \right|^2 \right\} \quad (16)$$

Recalling that the $\psi_\nu^{(n)}$ form a CON system, (16) is equal to

$$\limsup_{n \to \infty} \frac{1}{n} E\left\{ \sum_{N+1}^{N_n} \left|(x, \psi_\nu^{(n)})\right|^2 \right\}$$

$$= \limsup_{n \to \infty} \frac{1}{n} \sum_{\nu=N+1}^{N_n} \sum_{s=1}^{n} \sum_{t=1}^{n} \overline{\psi_\nu^{(n)}(s)} r(s-t)\psi_\nu(t)$$

But the norm of the covariance operator is at most M, and the ψ-vectors have norm 1, so that the above expression is

$$\leqslant \limsup_{n\to\infty} \frac{M}{n} |N_n - N|$$

We will have proved (15) if we can show that

$$\limsup_{n\to\infty} \left| \frac{N_n}{n} - \theta \right| = 0$$

To do this we use the Riemann integrability of B. We have, because of the bijection $\omega_\nu \leftrightarrow \nu$,

$$N_n = \sum_{\nu=1}^{n} 1_B\left(\frac{2\pi\omega_\nu}{n} - \pi\right) = \sum_{\nu=1}^{n} 1_B\left(\frac{2\pi\nu}{n} - \pi\right)$$

so that, since 1_B is Riemann integrable,

$$\lim_{n\to\infty} \frac{1}{n} N_n = \lim_{n\to\infty} \sum_{\nu=1}^{n} 1_B\left(-\pi + \frac{2\pi\nu}{n}\right)\frac{1}{n} = \frac{1}{2\pi}\int_{-\pi}^{\pi} 1_B(\lambda)\,d\lambda$$

$$= \frac{1}{2\pi} m(B) = \theta$$

Q.E.D.

The point is now that $B = B(b_0)$ will give an optimal ordering. We have to show first that $1_{B(b_0)}$ is Riemann integrable.

Lemma 3 With the same notation and conditions the indicator function of $B(b_0)$ is Riemann integrable.

Proof According to Lemma 1 it suffices to prove that the interior $B^i(b_0)$ has the same Lebesgue measure as the closure $\bar{B}(b_0)$. Recalling that f is continuous, it is clear that the set

$$B(b_0) = \{\lambda \mid f(\lambda) > b_0\}$$

is open. Since G is continuous at b_0, we have

$$m(\{\lambda \mid f(\lambda) = b_0\}) = 0$$

But the set

$$\{\lambda \mid f(\lambda) \geqslant b_0\}$$

is closed and contains $B(b_0)$, so that $m[\bar{B}(b_0)] = m[B(b_0)]$. Q.E.D.

Now the fundamental result is given in the following theorem.

THEOREM 3

Again let B be such that 1_B is Riemann integrable, $m(B)=2\pi\theta$, and the Fourier characters are B-ordered. Then

$$\lim_{n\to\infty}\frac{1}{n}\sum_{s=1}^{n}E\left\{\left|x(s)-\sum_{\nu=1}^{N}(x,\psi_\nu^{(n)})\psi_\nu(s)\right|^2\right\}=1-\frac{1}{2\pi}\int_B f(\lambda)\,d\lambda \quad (17)$$

Proof Again we can assume that $N\leqslant N_n$ without loss of generality. Compare the outer sum in (17) with the same expression, but N being replaced by N_n. The difference is

$$E\left\{\sum_{s=1}^{n}\left|x(s)-\sum_{\nu=1}^{N_n}(x,\psi_\nu^{(n)})\psi_\nu^{(n)}(s)+\sum_{\nu=N+1}^{N_n}(x,\psi_\nu^{(n)})\psi_\nu^{(n)}(s)\right|^2\right.$$
$$\left.-\sum_{s=1}^{n}\left|x(s)-\sum_{\nu=1}^{N_n}(x,\psi_\nu^{(n)})\psi_\nu^{(n)}(s)\right|^2\right\} \quad (18)$$

Expanding (18) we get

$$E\left\{\sum_{s=1}^{n}\left\{\sum_{\nu=1}^{N_n}(x,\psi_\nu^{(n)})\psi_\nu^{(n)}(s)-\sum_{\nu=1}^{N}(x,\psi_\nu^{(n)})\psi_\nu^{(n)}(s)|^2\right\}\right. \quad (19)$$

Lemma 2 tells us that (19) is $o(n)$. Hence we can replace N by N_n in (17) with no effect asymptotically.

We then consider

$$\frac{1}{n}E\left\{\sum_{s=1}^{n}\left|x(s)-\sum_{\nu=1}^{N_n}(x,\psi_\nu^{(n)})\psi_\nu^{(n)}(s)\right|^2\right\}$$
$$=\frac{1}{n}E\left\{\sum_{s=1}^{n}|x(s)|^2-\sum_{\nu=1}^{N_n}\left|(x,\psi_\nu^{(n)})\right|^2\right\}$$
$$=1-\frac{1}{n}\sum_{\nu=1}^{N_n}\sum_{s=1}^{n}\sum_{t=1}^{n}r(s-t)\psi_\nu^{(n)}(s)\overline{\psi_\nu^{(n)}(t)}$$

Denoting the set of integers in $[-n/2,n/2)$ by N^*, this is equal to

$$1-\frac{1}{n}\sum_{\nu\in N^*}\sum_{s=1}^{n}\sum_{t=1}^{n}\frac{1}{2\pi n}\int_{-\pi}^{\pi}e^{i(s-t)\lambda}f(\lambda)e^{[2\pi i\nu(t-s)]/n}1_B\left(\frac{2\pi\nu}{n}\right)d\lambda$$
$$=1-\frac{1}{n}\sum_{\nu\in N^*}\int_{-\pi}^{\pi}\frac{\sin^2(n/2)(\lambda-2\pi\nu/n)}{2\pi n\sin^2(1/2)(\lambda-2\pi\nu/n)}f(\lambda)1_B\left(\frac{2\pi\nu}{n}\right)d\lambda$$

In the last term change the variable $\lambda - 2\pi\nu/n$ to λ, and with the classical Féjer kernel

$$K_n(\lambda) = \frac{1}{2\pi n}\frac{\sin^2(n\lambda/2)}{\sin^2(\lambda/2)}$$

we get the last term in the form

$$\frac{1}{n}\sum_{\nu\in N^*}\sigma_n\left(\frac{2\pi\nu}{n}\right)1_B\left(\frac{2\pi\nu}{n}\right) \tag{20}$$

where we have set

$$\sigma_n(x) = \int_{-\pi}^{\pi} K_n(\lambda) f(\lambda + x)\,d\lambda$$

It is well known that, under our conditions on f, $\sigma_n(\cdot)\to f(\cdot)$ uniformly as $n\to\infty$. We now use a Riemann integrability argument on the expression in (20). Rewriting (20) as we have

$$\frac{1}{2\pi}\sum_{\nu\in N^*}\sigma_n\left(\frac{2\pi\nu}{n}\right)1_B\left(\frac{2\pi\nu}{n}\right)\frac{2\pi}{n} = \frac{1}{2\pi}\sum_{\nu\in N^*} f_n\left(\frac{2\pi\nu}{n}\right)1_B\left(\frac{2\pi\nu}{n}\right)\frac{2\pi}{n}$$
$$+\frac{1}{2\pi}\sum_{\nu\in N^*}\left[\sigma_n\left(\frac{2\pi\nu}{n}\right) - f\left(\frac{2\pi\nu}{n}\right)\right]1_B\left(\frac{2\pi\nu}{n}\right)\frac{2\pi}{n} \tag{21}$$

The last term obviously tends to zero as $n\to\infty$. On the other hand 1_B in Riemann integrable, and so is $f1_B$ as the product of two Riemann integrable functions. Hence (21) tends to

$$\frac{1}{2\pi}\int_B f(\lambda)\,d\lambda$$

as stated. Q.E.D.

Corollary 1 Under the same conditions and with B such that $m(B)=2\pi\theta$ and 1_B Riemann integrable, if the $\psi_\nu^{(n)}$ are B-ordered, then their asymptotic efficiency is

$$\frac{1-(1/2\pi)\int_{B(b_0)} f(\lambda)\,d\lambda}{1-(1/2\pi)\int_B f(\lambda)\,d\lambda}$$

Corollary 2 Moreover, if the functionals $\psi_\nu^{(n)}$ are $B(b_0)$-ordered, they are asymptotically efficient.

This means that (asymptotically) *we do not have to carry out the eigen-analysis of R_n, we just select those Fourier characters $e^{i\lambda t}$ that correspond to frequencies λ for which $f(\lambda)$ is large*, λ = multiple of $2\pi/n$.

This powerful result depends crucially upon Toeplitz theory. Here we need not just asymptotic distribution of eigenvalues of Toeplitz operators, but also asymptotic knowledge of the eigenvectors (see Note 4.7.C).

CHAPTER 5

Testing Gaussian Processes

1 TESTING MEAN VALUE FUNCTIONS

When we leave the linear inference problems it is natural to turn to the Gaussian measures first. Not only are they of great practical interest, they are also the best understood ones.

One reason for this is that Gaussian measures have linear regression. The functionals that the inference problems give rise to will therefore be linear or linear-quadratic.

If the two measures P_0 and P_1 have the same covariance operator and differ only in their mean values m_0 and m_1, *the functionals will be linear with no quadratic component*. This case is treated in the present section, while the situation with different covariance operators is left to the next section.

Consider the Gaussian real stochastic process $x(t)$ on the finite interval $T=[a,b]$ with continuous covariance function $r(st)$ and continuous mean value functions $m_i(t)$, $i=0,1$, under the two hypotheses H_0 and H_1, respectively. Introduce the eigenelements of the integral equation of the first kind:

$$\lambda\phi(s)=\int_a^b r(s,t)\phi(t)$$

namely, the eigenvalues λ_ν and the normalized eigenfunctions $\phi_\nu(t)$. In case some eigenvalue is multiple, the ϕ_ν are arbitrary to some extent, but we choose them to be orthonormal. We can then calculate the means of the observables $x_\nu=(x(\cdot),\phi_\nu(\cdot))$:

$$E(x_\nu)=m_\nu^{(i)}=\int_a^b m_i(t)\phi_\nu(t)\,dt;\qquad \nu=1,2,\ldots,\quad i=0,1$$

We now derive a necessary and sufficient criterion for P_1 and P_0 to be equivalent/perpendicular. One could calculate the finite-dimensional likelihood ratios explicitly and appeal to Theorem 1 of Chapter 3, Section 1, as in Example 1 of Chapter 3, Section 2. This is possible since the stochastic variables x_ν are independent of each other (see Note 5.1.A).

If $\{\phi_\nu\}$ is not complete in $L_2(T)$, we adjoin an orthonormal system, corresponding to $\lambda_\nu=0$. This should be compared to Chapter 4, Section 2.

As an alternative we use instead Kakutani's theorem (see Chapter 3, Section 3) so that we should calculate

$$\rho\left(P_0^{(\nu)}, P_1^{(\nu)}\right)=\int_{-\infty}^{\infty}\sqrt{P_0^{(\nu)}(dx)P_1^{(\nu)}(dx)} \tag{1}$$

where $P_i^{(\nu)}$ denotes the probability measure of x_ν on $\mathbb{R}$. But under the two hypotheses

$$x_\nu=N\left(m^{(i)},\lambda_\nu\right); \qquad i=0,1$$

so that the integral in (1) can be written

$$\rho\left(P_0^{(\nu)}, P_1^{(\nu)}\right)=\frac{1}{\sqrt{2\pi\lambda_\nu}}\int_{-\infty}^{\infty}\exp\left\{-\frac{1}{4}\left[\frac{\left(x-m_\nu^{(0)}\right)^2}{\lambda_\nu}+\frac{\left(x-m_\nu^{(1)}\right)^2}{\lambda_\nu}\right]\right\}dx$$

An elementary evaluation gives

$$\rho\left(P_0^{(\nu)}, P_1^{(\nu)}\right)=\exp\left[-\frac{1}{8\lambda_\nu}\left(m_\nu^{(0)}-m_\nu^{(1)}\right)^2\right]$$

According to Theorem 3 of Chapter 3, Section 3, P_0 and P_1 are equivalent if and only if

$$\prod_{\nu=1}^{\infty}\rho\left(P_0^{(\nu)}, P_1^{(\nu)}\right)=\exp\left[-\frac{1}{8}\sum_{\nu=1}^{\infty}\frac{\left(m_\nu^{(0)}-m_\nu^{(1)}\right)^2}{\lambda_\nu}\right]>0$$

Hence we have proved the following result.

THEOREM 1 MEAN VALUE ALTERNATIVE, GRENANDER (1950)

In order that two Gaussian measures P_0 and P_1 with means m_0 and m_1 and common covariances be equivalent, it is necessary and sufficient that

$$\sum_{\nu=1}^{\infty} \frac{\left(m_\nu^{(0)} - m_\nu^{(1)}\right)^2}{\lambda_\nu} < \infty \tag{2}$$

The only alternatives are that they are equivalent or perpendicular.

Remark 1 This theorem shows that two Gaussian measures with common covariance are either equivalent or perpendicular. Only the regular and the singular case from Theorem 1 of Chapter 3, Section 1, can occur, not any intermediate one. We see later that this statement holds more generally (see Note 5.1.B).

Remark 2 The condition for equivalence can also be expressed as $m^{(1)} - m^{(0)} \in \text{range}(r^{1/2})$, which should be compared to Chapter 4, Section 2.

Assuming that (2) holds, we now calculate the Radon-Nikodym derivative $f(x)$, starting from the finite-dimensional likelihood ratios

$$f_n(x) = \frac{\prod_{\nu=1}^{n} \frac{1}{\sqrt{2\pi\lambda_\nu}} \exp\left[-\frac{1}{2\lambda_\nu}\left(x_\nu - m_\nu^{(1)}\right)^2\right]}{\prod_{\nu=1}^{n} \frac{1}{\sqrt{2\pi\lambda_\nu}} \exp\left[-\frac{1}{2\lambda_\nu}\left(x_\nu - m_\nu^{(0)}\right)^2\right]}$$

We know from Theorem 1 of Chapter 3, Section 1, that $f_n \to f$ in probability (P_0) and (P_1). The limit can be expressed, after some rearrangement, as

$$f(x) = \exp\left[\sum_{\nu=1}^{\infty} \frac{m_\nu^{(1)} - m_\nu^{(0)}}{\lambda_\nu}(x_\nu - c_\nu)\right] \tag{3}$$

where the constants c_ν are given by the averages

$$c_\nu = \frac{m_\nu^{(1)} + m_\nu^{(0)}}{2} \tag{4}$$

so that we have the following theorem [see Grenander (1950)].

THEOREM 2

Under the conditions in Theorem 1 *the Radon-Nikodym derivative*

$$\frac{P_1(dx)}{P_0(dx)} = f$$

is given by (3) *and* (4).

This implies that the most powerful test of P_0 against P_1 has the critical region (see Theorem 2 of Chapter 3, Section 2)

$$W = \left\{ x \,\middle|\, \sum_{\nu=1}^{\infty} \frac{m_\nu^{(1)} - m_\nu^{(0)}}{\lambda_\nu} (x_\nu - c_\nu) > \text{constant} \right\} \tag{5}$$

The sum in (5) can be written as an inner product in $L_2(T)$ if and only if $m^{(1)} - m^{(0)} \in \text{range}(R)$ (compare with Theorem 2 of Chapter 4, Section 2).

The reader will recognize condition (2) and the sum in (5), from Chapter 4. They appeared when we were studying linear estimation. We can therefore use results from Chapter 4 to give us critical regions for tests for Gaussian measures when we specialize the covariance structure as we did then.

To test m_0 against m_1 amounts to the same thing as to test the null function against $m_1 - m_0$. It is therefore no restriction to assume m_0 to be identically zero, and we do so.

We now introduce a more serious restriction. Let $m = m_1$ be given a "quellenmässig" representation [see Courant-Hilbert (1937) I, p. 97],

$$m(s) = \int_a^b r(s,t) F(dt); \qquad s \in T \tag{6}$$

where F is of bounded variation on T. Using Mercer's theorem for the kernel r, we can write

$$m_\nu \equiv m_\nu^{(1)} = \lambda_\nu \int_a^b \phi_\nu(t) F(dt)$$

which is certainly well defined since all the eigenfunctions are continuous. A simple calculation shows that

$$\sum_{\nu=1}^{\infty} \frac{\left(m_\nu^{(1)}\right)^2}{\lambda_\nu} = \int_T \int_T r(s,t) F(ds) F(dt) < \infty \tag{7}$$

so that P_1 and P_0 are equivalent. Since the double integral in (7) is finite, it follows that

$$\zeta = \int_T x(t)F(dt)$$

is well defined as an element of $L_2[x(t), t \in T]$, (see Chapter 2, Section 2). We can then evaluate the exponent in (3) as

$$\sum_{\nu=1}^{\infty} \frac{m_\nu}{\lambda_\nu}(x_\nu - c_\nu) = \zeta - \frac{1}{2}\int_T\int_T r(s,t)F(ds)F(dt)$$

$$= \zeta - \frac{1}{2}\int_T m(t)F(dt)$$

leading us to the following elegant result due to Pitcher (1959).

THEOREM 3

If m has a "quellenmässig" representation with an F of bounded variation in (6), *the* P_m*-measure is equivalent with* P_0, *and the Radon-Nikodym derivative is given by*

$$\ln f(x) = \zeta - \frac{1}{2}\int_T m(t)F(dt)$$

A case of considerable interest is when $m(t) \equiv$ constant and when the process is also Markovian in addition to being a Gaussian stochastic process continuous in the mean.

We state a result under conditions that can probably be relaxed considerably. Let the variance function $V(t)$ be constant, say $V(t) \equiv 1$, and assume that $r(t, \tau)$ is differentiable with respect to both arguments for $\tau \neq t$ and that

$$f(t) = \left(\frac{\partial r(t,\tau)}{\partial \tau}\right)_{\tau = t+0} \tag{8}$$

exists and is continuous on T (see Note 5.1.C). Introduce a measure F on T which has the density $-f$ on T and the discrete mass 1 at each of the end points $t = a$ and b. We show that F satisfies the normal equation associated with our present testing problem

$$I(s) = \int_T r(s,t)F(dt) \equiv \text{constant} \tag{9}$$

To prove that (9) holds we first remind the reader that the correlation function ρ of the Gaussian-Markov process satisfies the functional equation

$$r(t_1,t_3)=r(t_1,t_2)r(t_2,t_3); \qquad t_1<t_2<t_3 \tag{10}$$

[see, e.g., Feller (1966), p. 94]. We know that the first factor on the right-hand side of (10) is differentiable with respect to t_2. The second factor is continuous in t_2 and close to 1 if t_2 is close to t_3. Hence since the left-hand side has derivative zero with respect to t_2, we see that

$$g(f)=\left(\frac{\partial r(\tau,t)}{\partial\tau}\right)_{\tau=t-0}$$

exists and is equal to $-f(t)$.

We can express our integral as

$$I(s)=r(s,a)+r(s,b)-\int_a^s r(s,t)f(t)\,dt-\int_s^b r(s,t)f(t)\,dt \tag{11}$$

In (10), differentiating with respect to t_3 and then putting $t_1=s, t_2=t, t_3=t_2+0$, we get, for $s<t$,

$$\frac{\partial r(s,t)}{\partial t}=r(s,t)\left(\frac{\partial r(t,\tau)}{\partial\tau}\right)_{\tau=t+0}=r(s,t)f(t)$$

The last integral in (11) therefore can be evaluated to be $r(s,b)-r(s,s)$.

To get the next last integral, we differentiate (10) again, but now with respect to t_1, and get, for $s>t$,

$$\begin{aligned}\frac{\partial r(s,t)}{\partial t}=\frac{\partial r(t,s)}{\partial t}&=\left(\frac{\partial r(\tau,s)}{\partial\tau}\right)_{\tau=t-0} r(t,s)=r(s,t)g(t)\\ &=-r(s,t)f(t)\end{aligned}$$

so that the integral in question equals $r(s,a)-r(s,s)$.

Combining our four terms, we get

$$I(s)=2r(s,s)=2$$

as stated, so that we have proved the following theorem.

THEOREM 4

For a Gaussian Markov process, with variance 1 *and satisfying the above differentiability condition, the mean values function* $m(t)\equiv$*constant has a "quellenmässig" representation so that Theorem* 3 *applies with F given by the above construction.*

The reader will note that this extends Theorem 1 of Chapter 4, Section 3, to the nonstationary case. It is also related to Example 2 in Chapter 4, Section 2, but not as directly.

As mentioned after Theorem 3, the solution of the test problem (for equal covariances) can be expressed in the same form as for the corresponding linear estimation problem. Therefore the results in Chapter 4, Section 3, can be carried over to the last problem. For example, consider a stationary Gaussian process with an absolutely continuous spectrum with a continuous spectral density. If this density is the reciprocal of a polynomial, the result in Theorem 2 of Chapter 4, Section 3, extends to the test case directly. Similarly one can handle a general rational density [see Hajék (1962) for a complete treatment]. For more general densities the situation is less completely understood, and we give only one result here.

Let the density be f and test $m=0$ against $m\neq 0$. Are the two measures P_m and P_0 absolutely continuous with respect to each other?

To answer this question let us recall again that, expressed as an estimation problem, this amounts to asking what is

$$V=\inf\|\Sigma c_\nu x(t_\nu)\|^2=\inf\int_{-\infty}^{\infty}|\Sigma c_\nu e^{it_\nu\lambda}|^2 f(\lambda)\,d\lambda \tag{12}$$

where the inf is taken over all finite linear combinations in (12) satisfying $\Sigma_\nu c_\nu=1$. Relation (12) just expresses the isomorphism between $L_2[x(t), t\in(0, L)]$ and $L_2[e^{it\lambda}, t\in(0, L)]$ that we have used many times.

From Chapter 4 we know that the inf in (12) is realized by a function $\phi_0\in L_2[e^{it\lambda}, t\in(0, L)]$ with $\phi(0)=1$,

$$V=\int_{-\infty}^{\infty}|\phi_0(\lambda)|^2 f(\lambda)\,d\lambda \tag{13}$$

If f is rational, one can show with methods indicated that $V>0$, but we now show that this is also true for other densities if they have the same asymptotic behavior.

Introduce the density

$$h(\lambda)=\frac{1}{(1+\lambda^2)^n} \tag{14}$$

for some natural number n and assume that f/h is bounded away from zero and infinity. Consider all the functions of the form

$$\phi(\lambda)=P(\lambda)+(1+i\lambda)^n\int_0^L e^{it\lambda}c(t)\,dt \tag{15}$$

where P is a polynomial whose degree is at most $n-1$ and the function c is quadratically integrable over $(0, L)$ with respect to Lebesgue measure. We show that the set of all such ϕ is identical with $L_2[e^{it\lambda}, t\in(0, L)]$.

To do this, consider a ϕ as in (15). It is clear that the functions $1, \lambda, \lambda^2, \dots, \lambda^{n-1}$ belong to the space. For example, the function λ belongs to it since, by dominated convergence,

$$\int_{-\infty}^{\infty}\left|i\lambda-\frac{e^{it\lambda}-1}{t}\right|^2 f(\lambda)\,d\lambda\to 0 \qquad \text{as } t\downarrow 0$$

Similarly for the other powers indicated, and hence P belongs to the space.

Now assume temporarily that c has n bounded derivatives vanishing at 0 and L. Then, integrating by parts, we get, leaving out the polynomial part that has already been dealt with,

$$\phi(\lambda)=\int_0^L e^{it\lambda}c(t)\,dt \tag{16}$$

and the relations

$$\int_0^L e^{i\lambda t}c^{(k)}(t)\,dt=(-i\lambda)^k\phi(\lambda); \qquad k=0,1,\dots,n$$

so that the functions on the right-hand side of (16) belong to the space. If c is only quadratically integrable, we can find a sequence $\{c_\nu\}$ of functions, each one with n bounded derivatives, vanishing at 0 and L, converging in the mean to c. Then, with Parseval's relation,

$$\int_{-\infty}^{\infty}\left|(-i\lambda)^n\phi_\nu(\lambda)-(-i\lambda)^n\phi(\lambda)\right|^2 f(\lambda)\,d\lambda$$
$$\leqslant A\int_{-\infty}^{\infty}|\phi_\nu(\lambda)-\phi(\lambda)|^2\,d\lambda=2\pi A\int_0^L|c_\nu(t)-c(t)|^2\,dt \tag{17}$$

where we have put

$$\phi_\nu(\lambda)=\int_0^L e^{i\lambda t}c_\nu(t)\,dt$$

The first inequality in (17) holds since f/h was assumed to be bounded. This shows that $(-i\lambda)^n\phi(\lambda)$ also belongs to the space. Hence all functions of the form (15) belong to the space.

To show that the space consists only of such functions, we start from the obvious observation that any element in the space is the limit in the mean of functions of the type

$$\sum c_k e^{i\lambda t_k}; \qquad t_k \in (0, L)$$

Introduce the closure of all such functions, for the moment, with the additional constraint that they, together with their first $n-1$ derivatives, vanish at $\lambda = i$.

Say that $\{\phi_\nu\}$ is a sequence of such functions converging (in the topology we are using) to some ϕ, so that

$$\int_{-\infty}^{\infty} |\phi_\nu(\lambda) - \phi(\lambda)|^2 f(\lambda)\, d\lambda \to 0$$

and hence

$$\int_{-\infty}^{\infty} \left| \frac{\phi_\nu(\lambda)}{(1+i\lambda)^n} - \frac{\phi(\lambda)}{(1+i\lambda)^n} \right|^2 d\lambda \to 0 \tag{18}$$

Consider now the behavior of the function

$$k_\nu(\lambda) = \frac{\phi_\nu(\lambda)}{(1+i\lambda)^n} \tag{19}$$

Since $\phi_\nu(\lambda)$ has a zero of order n at $\lambda = i$, canceling the zero in the denominator of (19), it follows that k_ν is an entire function. Further, it is of exponential type and increases not faster than $e^{|\lambda| L}$. It is obviously quadratically integrable on the real axis.

But then we can appeal to the famous theorem of Paley and Wiener [see Paley-Wiener (1934), chap. V], implying that there exists a quadratically integrable function c_ν such that

$$k_\nu(\lambda) = \int_0^L e^{i\lambda t} c_\nu(t)\, dt$$

Because of (18) it follows that $c_\nu \to c$ in the mean, and

$$\phi(\lambda) = (1+i\lambda)^n \int_0^L e^{i\lambda t} c(t)\, dt$$

To remove the condition that $\lambda = i$ be a zero of order n, we need only add a polynomial of order at most $n-1$. Hence we have shown that functions of the form (15) exhaust the space $L_2[e^{it\lambda}, t \in (0, L)]$.

We therefore know that ϕ_0 in (13) is an entire function due to its representation (15). This immediately implies that $V>0$, and hence the estimation problem is nonsingular, which is the same as P_0 and P_m being equivalent, so that we have shown the following theorem.

THEOREM 5

Consider a stationary Gaussian process with absolutely continuous spectrum and mean m. If the spectral density f is such that f/h [see (14)], is bounded away from zero and infinity, the measures P_0 and P_m are equivalent.

The proof, involving the beautiful representation (15), which is of independent interest, is due to Rozanov (see Note 5.1.D).

2 TESTING COVARIANCE FUNCTIONS

There is a rich literature dealing with the problem of equivalence/perpendicularity of two Gaussian measures P_0 and P_1 with covariance operators R_0 and R_1, $R_0 \neq R_1$ (see Note 5.2.A). We study both very general conditions and others, less general, but easier to apply.

The main difficulty arises from the fact that R_0 and R_1 cannot in general be diagonalized by an orthogonal transformation at the same time. This makes it awkward to apply the choice of observables used with a good deal of success in Chapter 4; a nontrivial modification will be needed.

We follow the treatment in Rozanov (1962) since it is both elegant and has the advantage of giving us the Radon-Nikodym derivative f which we need anyway. His approach is based on the *entropy of P_1 with respect to P_0*:

$$H(P_1|P_0) = \int_X \ln f(x) P_0(dx)$$

We need four lemmas as an introduction to the main result.

Lemma 1 Consider two σ-algebras $\mathcal{Q}$, $\mathcal{Q}'$ with $\mathcal{Q}' \subseteq \mathcal{Q}$, and P_1', P' two probability measures defined on $\mathcal{Q}'$ and coinciding on $\mathcal{Q}'$ with two other probability measures P_1, P_0, respectively, defined on $\mathcal{Q}$. Then

$$H(P_1'|P_0') \leqslant H(P_1|P_0)$$

Proof We can assume that P_0 is absolutely continuous with respect to P_1 since f is not affected if we define it as the derivative of P_1 with respect to the part of P_0 which is absolutely continuous with respect to P_1. Then

$$g(x)=\frac{P_0(dx)}{P_1(dx)}=\frac{1}{f(x)}$$

With

$$g'(x)=\frac{P_0'(dx)}{P_1'(dx)}=\frac{1}{f'(x)}$$

we have, for an arbitrary $A\in\mathcal{Q}'$,

$$P_0(A)=\int_A g(x)P_1(dx)=\int_A g'(x)P_1(dx)$$

so that

$$E_1[g(x)|\mathcal{Q}']=g'(x)$$

But $-\ln x$ is convex so that Jensen's inequality is applicable and gives

$$E_1[-\ln g(s)|\mathcal{Q}']\geqslant -\ln g'(x)$$

and the entropy of P_1 with respect to P_0

$$\begin{aligned}H(P_1|P_0)&=E_1[\ln f(x)]=-E_1[\ln g(x)]\\&=E_1\{E_1[-\ln g(x)|\mathcal{Q}']\}\geqslant -E_1[\ln g'(x)]=H(P_1'|P_0')\end{aligned}$$

Q.E.D.

Lemma 2 Consider the monotonic sequence $\{\mathcal{Q}_n;\ n-1,2,\dots\}$ of σ-algebras on X, $\mathcal{Q}^n\subseteq\mathcal{Q}^{n+1}$, with

$$\mathcal{Q}=\bigcup_{n=1}^{\infty}\mathcal{Q}^n \tag{1}$$

For given probability measures P_1 and P_0 on $\mathcal{Q}$ suppose that P_1^n, P_0^n, are defined on $\mathcal{Q}^n$ and coincide on $\mathcal{Q}^n$ with P_1, P_0, respectively. Then

$$\sup_n H(P_1^n|P_0)<\infty \tag{2}$$

implies that P_1 is absolutely continuous with respect to P_0.

Proof Let us argue indirectly, starting with the assumption that the above conclusion is false, so that there exist sets $A_k, k=1,2,\ldots,$ such that for some $\varepsilon>0$,

$$\begin{aligned}&\tfrac{1}{2}\geqslant P_1(A_k)\geqslant\varepsilon\\&P_0(A_k)\to 0\end{aligned}\tag{3}$$

Because of (1) we can actually choose the A_k from some $\mathcal{A}^{n_k}$, using a standard argument from measure theory.

Introduce the set algebra $\mathcal{A}_k$ containing the sets A_k, A_k^c, X, and ϕ. But $\mathcal{A}_k\subseteq\mathcal{A}^{n_k}$ so that Lemma 1 tells us that

$$H(P_1^{n_k}|P_0^{n_k})\geqslant P_1(A_k)\ln\frac{P_1(A_k)}{P_0(A_k)}+[1-P_1(A_k)]\ln\frac{1-P_1(A_k)}{1-P_0(A_k)}$$

and appealing to (3) we see that the last expression tends to $+\infty$ as k increases. But this contradicts (2) so that the original assumption (P_1 is not absolutely continuous with respect to P_0) must be false. Q.E.D.

Lemma 3 Consider the stochastic variables $\eta_1,\eta_2,\ldots,\eta_n$ independent for both Gaussian measures P_1 and P_0 and with the parameters

$$\begin{aligned}&E_1(\eta_k)=m_k; &&\text{Var}_1(\eta_k)=\sigma_k^2\\&E_0(\eta_k)=0; &&\text{Var}_0(\eta_k)=1\end{aligned}$$

If $\mathcal{A}'$ denotes the σ-algebra generated by the η_k, we have for the corresponding density p'

$$\ln p'(x)=-\frac{1}{2}\sum_{k=1}^{n}\ln\sigma_k^2-\frac{1}{2}\sum_{k=1}^{n}\left[\frac{(\eta_k-m_k)^2}{\sigma_k^2}-\eta_k^2\right]\tag{4}$$

and

$$\begin{aligned}E_1[\ln p'(x)]&=\frac{1}{2}\sum_{k=1}^{n}\left[\sigma_k^2-1-\ln\sigma_k^2+m_k^2\right]\\E_0[\ln p'(x)]&=\frac{1}{2}\sum_{k=1}^{n}\left[-\sigma_k^{-2}+1-\ln\sigma_k^2-m_k^2\sigma_k^{-2}\right]\\\text{Var}_0[\ln p'(x)]&=\frac{1}{2}\sum_{k=1}^{n}\left[(\sigma_k^2-1)^2\sigma_k^{-4}+2m_k^2\sigma_k^{-2}\right]\\\text{Var}_1[\ln p'(x)]&=\frac{1}{2}\sum_{k=1}^{n}\left[(\sigma_k^2-1)^2+m_k^2\sigma_k^2\right]\end{aligned}\tag{5}$$

Proof Direct calculation using the value of the fourth central moment of a Gaussian distribution in terms of the second one. Q.E.D.

Now let a set $\{\xi\}$ of Gaussian stochastic variables on X (for both P_0 and P_1) generate $\mathcal{Q}$. Also, let $\mathcal{Q}'$ be the σ-algebra generated by some subset $\{\xi_k;\ k=1,2,\ldots,n\}$. We then have the crucial lemma.

Lemma 4 Defining

$$I(P_1'|P_0')=H(P_1'|P_0')+H(P_0'|P_1') \tag{6}$$

assume that

$$I=\sup I(P_1'|P_2')=+\infty$$

where the supremum is taken over all $\mathcal{Q}'$, then P_1 and P_0 are perpendicular.

Proof We can always assume that $E_0(\xi)=0$ since otherwise we could operate on the variables $\xi-E_0(\xi)$. Introduce the covariance matrices for our variables $\xi_1, \xi_2, \ldots, \xi_n$:

$$B^1=\left\{B^1_{kj};\qquad k,j=1,2,\ldots,n\right\}$$

$$B^0=\left\{B^0_{kj};\qquad k,j=1,2,\ldots,n\right\}$$

under the two hypotheses P_1 and P_0. We can find an $n\times n$ matrix C such that $CB^iC^T, i=0,1$, are both diagonal. Of course, we cannot assert that C is orthogonal, but this will not be needed. Then

$$\eta_k=\sum_{j=1}^{n} c_{kj}\xi_j;\qquad k=1,2,\ldots,n$$

will be independent (for both P_0 and P_1). We now use Lemma 3. Relation (5) enables us to calculate

$$I(P_1'|P_0')=E_1[\ln p'(x)]-E_0[\ln p'(x)]$$

$$=\frac{1}{2}\sum_{k=1}^{n}\left[\frac{\left(\sigma_k^2-1\right)^2}{\sigma_k^2}+m_k^2\left(1+\sigma_k^{-2}\right)\right]$$

Suppose now that the variances of the ξ-variables are uniformly bounded away from 0 and $+\infty$. This condition will be removed later. We can then find positive constants a and b such that

$$a\sum_{k=1}^{n}\left[(\sigma_k^2-1)^2+m_k^2\right]\leqslant I(P_1'|P_0')\leqslant b\sum_{k=1}^{n}\left[(\sigma_k^2-1)^2+m_k^2\right] \tag{7}$$

and, see (5) again,

$$\operatorname{Var}_i[\ln p'(x)]\leqslant b\sum_{k=1}^{n}\left[(\sigma_k^2-1)^2+m_k^2\right]; \qquad i=0,1 \tag{8}$$

Introduce the event

$$A'=\left\{x\,|\ln p(x)>\tfrac{1}{2}I(P_1'|P_0')+E_0[\ln p(x)]\right\} \tag{9}$$

and apply Chebyshev's inequality:

$$P_0(A')\leqslant\frac{\operatorname{Var}_0[\ln p'(x)]}{(1/4)I^2(P_1'|P_0')}\leqslant\frac{c}{I(P_1'|P_0')}$$

where we have used (8) and (9) and c is positive. With the aid of (7) we get the bound

$$P_1(A')=1-P_1[(A')^c]\geqslant 1-\frac{\operatorname{Var}_1[\ln p'(x)]}{(1/4)I^2(P_1'|P_0')}$$

$$\geqslant 1-\frac{c}{I(P_1'|P_0')}$$

Assume now

$$\sup_{\mathscr{Q}'} I(P_1'|P_0')=+\infty$$

Then, for any $\varepsilon>0$, we can find an A_ε such that

$$P_1(A_\varepsilon)\geqslant 1-\varepsilon$$

$$P_0(A_\varepsilon)\leqslant\varepsilon$$

which implies that P_1 and P_0 are perpendicular to each other.

We now have to deal with the case, so far excluded, when very small (and analogously very large) values of σ_k^2 can occur. Introduce the event

$$A''=\left\{x\,\|\eta-m|<\sigma^{1/2}\right\}$$

where η is one of our variables with

$$E_1(\eta)=m; \qquad \mathrm{Var}_1(\eta)=\sigma^2$$
$$E_0(\eta)=0; \qquad \mathrm{Var}_0(\eta)=1$$

Then

$$P_1(A'')=\frac{1}{\sqrt{2\pi\sigma}}\int_{m-\sigma^{1/2}}^{m+\sigma^{1/2}}\exp\left[-\frac{1}{2\sigma^2}(y-m)^2\right]dy$$
$$=\frac{1}{\sqrt{2\pi}}\int_{-\sigma^{-1/2}}^{\sigma^{-1/2}}\exp\left(-\frac{1}{2}x^2\right)dx$$

When we make $\sigma\to 0$, the last integral tends to 1, but

$$P_0(A'')=\frac{1}{\sqrt{2\pi}}\int_{m-\sigma^{1/2}}^{m+\sigma^{1/2}}\exp\left(-\frac{1}{2}x^2\right)dx$$

tends to 0. Hence P_1 and P_0 are perpendicular in this case, too.

With these lemmas the main result can be obtained directly. We have the fundamental theorem.

THEOREM 1 THE COVARIANCE ALTERNATIVE, HAJÉK (1958), FELDMAN (1958)

Two Gaussian measures are either equivalent or perpendicular; no other alternative exists.

Proof If I is infinite, Lemma 4 tells us that the two measures are perpendicular. On the other hand if I is finite, Lemma 2 together with the definition in (6) tells us that they are equivalent. Q.E.D.

We now calculate the Radon-Nikodym derivative when the two measures are equivalent (see Note 3.2.B). Consider Gaussian stochastic variables x_t, $t\in T$, where T can be an arbitrary set and $E_0[x(t)]\equiv 0$, and form as usual the Hilbert space $H=L_2[x(t), t\in T]$ with the scalar product corresponding to P_0-measure.

In the space introduce the bilinear form

$$B(\eta_1,\eta_2)=E_1\{[\eta_1-E_1(\eta_1)][\eta_2-E_1(\eta_2)]\}$$

Referring to the last part of the proof of Lemma 4, we see that this bilinear form is bounded and hence leads to a bounded self-adjoint operator, to be denoted by the same symbol B, $B(\eta_1,\eta_2)=(B\eta_1,\eta_2)$ with inner product in H.

Write the nonnegative operator B in its spectral representation:

$$B=\int_{c_1}^{c_2}\lambda E(d\lambda); \qquad 0<c_1\leqslant c_2<+\infty$$

where E denotes its spectral family, a resolution of the identity. Divide $[c_1,c_2]$ into a finite number of disjoint intervals $\Delta_k=(\lambda_k\lambda_{k+1}]$. For any subspace $H(\Delta_k)=E(\Delta_k)H$, which is not just the zero element, select an element η_k, $\|\eta_k\|^2=\mathrm{Var}_0(\eta_k)=1$. Then

$$B(\eta_k,\eta_l)=\int_{c_1}^{c_2}\lambda(E(d\lambda)\eta_k,\eta_l)=0; \qquad k\neq l$$

$$B(\eta_k,\eta_k)=\int_{\Delta_k}\lambda(E(d\lambda)\eta_k,\eta_k)$$

so that

$$\lambda_k\leqslant B(\eta_k,\eta_k)=\sigma_k^2=\mathrm{Var}_1(\eta_k)\leqslant\lambda_{k+1} \tag{10}$$

In other words the stochastic variables η_k are independent under both P_0 and P_1. But (6) shows, because of (10), that $I(P_1'|P_0')$ can be made arbitrarily large by choosing the Δ_k-partition appropriately, except in the case when (a) the spectrum is discrete and (b) the eigenvalues σ_ν^2 satisfy

$$\sum_\nu(1-\sigma_\nu^2)<\infty \tag{11}$$

and (c) the associated eigenfunctions η_ν have P_1-expectation $m_\nu=E_1(\eta_\nu)$ with

$$\sum_\nu m_\nu^2<\infty \tag{12}$$

We now apply Lemma 3, in particular (4), to the η_ν-variables and obtain, after the usual limit argument, the Radon-Nikodym derivative

$$\ln f(x)=-\frac{1}{2}\sum_{\nu=1}^{\infty}\left[\ln\sigma_\nu^2+\left(\frac{\eta_\nu-m_\nu}{\sigma_\nu}\right)^2-\eta_\nu^2\right] \tag{13}$$

as well as, using (5),

$$H(P_1|P_0)=\frac{1}{2}\sum_{\nu=1}^{\infty}(\sigma_\nu^2-1-\ln\sigma_\nu^2+m_\nu^2)$$

We have thus obtained the following theorem.

THEOREM 2

If the two Gaussian measures P_0 and P_1 are equivalent, the Radon-Nikodym derivative of P_1 with respect to P_0 is given by its logarithm in (13).

Actually one can state more, namely, this theorem.

THEOREM 3

P_1 and P_0 are equivalent if and only if the operator B satisfies conditions (a), (b), *and* (c) *above.*

Proof We only have to prove the "if" part. Under (11) and (12) we have, for any finite collection of eigenfunctions with the same notation [see Lemma 1 and (6)],

$$I(P_1'|P_0')\leqslant\frac{1}{2}\sum_{\nu}\left[\frac{(\sigma_k^2-1)^2}{\sigma_k^2}+m_k^2(1+\sigma_k^{-2})\right]<+\infty$$

Then Lemma 2 ensures that P_1 and P_0 are equivalent. Q.E.D.

This condition and equivalent ones, but expressed differently, given by several authors are definitive. They are necessary and sufficient (see Note 5.2.C). They are not easy to apply, however, so that there is a need for simpler conditions valid in less general situations. One such condition was given early in Baxter (1956), and inspired a good deal of further work.

THEOREM 4 BAXTER'S THEOREM

Let $x(t)$ be a Gaussian process $t\in[0,1]$ with a mean value function $m(t)$ with bounded derivative, let the covariance function $r(s,t)$ be continuous with uniformly bounded second derivatives for $s\neq t$, and introduce $f(t)=D^-(t)-D^+(t)$ where

$$D^-(t)=\lim_{s\uparrow t}\frac{r(t,t)-r(s,t)}{t-s}$$

$$D^+(t)=\lim_{s\downarrow t}\frac{r(t,t)-r(s,t)}{t-s}$$

Then, with probability 1,

$$\lim_{n\to\infty} \sum_{k=1}^{2^n} \left[x\left(\frac{k}{2^n}\right) - x\left(\frac{k-1}{2^n}\right) \right]^2 = \int_0^1 f(t)\, dt$$

Proof First note that the regularity conditions stated imply that D^-, D^+, and f exist and are bounded and continuous in $(0,1)$ so that f is Riemann integrable (see Note 5.2.D). To begin we assume that $m(t)\equiv 0$. Introduce the quantities

$$\begin{aligned}
\Delta x_k &= x\left(\frac{k}{2^n}\right) - x\left(\frac{k-1}{2^n}\right) \\
B_n &= \sum_{k=1}^{2^n} (\Delta x_k)^2 \\
a_{jk} &= E(\Delta x_j \Delta x_k) \\
b_n &= 2 \sum_{j,k=1}^{2^n} a_{jk}^2
\end{aligned} \tag{14}$$

We can then express the mean and variance of B_n as

$$\begin{aligned}
E(B_n) &= \sum_{k=1}^{2^n} a_{kk} \\
E(B_n^2) &= 3 \sum_{k=1}^{2^n} a_{kk}^2 + 2 \sum_{j>k} \left(a_{kk}a_{jj} + 2a_{jk}^2\right)
\end{aligned} \tag{15}$$

using the well-known expressions for the fourth-order moments of a joint Gaussian distribution in terms of the second-order moments [see, e.g., Cramér (1945), p. 208]. The relations in (15) imply that $\operatorname{Var}(B_n)=b_n$, so that Chebyshev's inequality yields

$$P\left\{ |B_n - E(B_n)| > \frac{n}{\sqrt{2^n}} \right\} < \frac{2^n b_n}{n^2} \tag{16}$$

To show almost certain convergence, we need a bound on the right-hand side of (16), and we let M be a bound for the three second derivatives of r. Expanding r in Taylor series, we then get, for $j \neq k$,

$$|a_{jk}| = \left| r\left(\frac{k}{2^n}, \frac{j}{2^n}\right) + r\left(\frac{k-1}{2^n}, \frac{j-1}{2^n}\right) - r\left(\frac{k}{2^n}, \frac{j-1}{2^n}\right) - r\left(\frac{k-1}{2^n}, \frac{j}{2^n}\right) \right| \leqslant \frac{3M}{2^{2n}} \tag{17}$$

Similarly, for $j=k=1,2,\ldots,2^n-1$,

$$a_{kk}=r\left(\frac{k}{2^n},\frac{k}{2^n}\right)+r\left(\frac{k-1}{2^n},\frac{k-1}{2^n}\right)-2r\left(\frac{k-1}{2^n},\frac{k}{2^n}\right)$$

$$=\frac{1}{2^n}\left[D^-\left(\frac{k}{2^n}\right)-D^+\left(\frac{k}{2^n}\right)\right]+0\left(\frac{1}{2^{2n}}\right) \tag{18}$$

Using (17) together with (18), we see that $2^n b_n$ is bounded.

Also, from the definition of B_n in (14) we get asymptotically as $n\to\infty$, with the aid of (18), that

$$E(B_n)=\sum_{k=1}^{2^n-1}2^{-n}f\left(\frac{k}{2^n}\right)+0\left(\frac{1}{2^n}\right) \tag{19}$$

But f is Riemann integrable so that (9) leads to

$$\lim_{n\to\infty}E(B_n)=\int_0^1 f(t)\,dt \tag{20}$$

Returning to (16) and using the fact that $2^n b_n$ is bounded, Borel-Cantelli's lemma shows that with probability 1

$$\lim_{n\to\infty}\left[B_n-E(B_n)\right]=0$$

Combining this with (20), the proof is complete, still under the condition that $m(t)\equiv 0$.

To handle the remaining case, when $m(t)$ is not identically zero, let $y(t)=x(t)-m(t)$. We then know that almost certainly

$$\lim_{n\to\infty}\sum_{k=1}^{2^n}(\Delta y_k)^2=\int_0^1 f(t)\,dt \tag{21}$$

where

$$\Delta y_k=y\left(\frac{k}{2^n}\right)-y\left(\frac{k-1}{2^n}\right)$$

With

$$\Delta m_k=m\left(\frac{k}{2^n}\right)-m\left(\frac{k-1}{2^n}\right)$$

we can apply Schwarz's inequality

$$\left(\sum_{k=1}^{2^n} \Delta m_k \Delta y_k\right)^2 \leqslant \sum_{k=1}^{2^n} (\Delta m_k)^2 \sum_{k=1}^{2^n} (\Delta y_k)^2 \tag{22}$$

We know that the second factor on the right-hand side of (22) remains bounded almost certainly as $n \to \infty$, and since $m(t)$ has a uniformly bounded derivative, the 2^n terms in the first factor on the right are $0(2^{-2n})$. Hence the right-hand side tends to zero a.c. But

$$\sum_{k=1}^{2^n} (\Delta x_k)^2 = \sum_{k=1}^{2^n} (\Delta y_k)^2 + \sum_{k=1}^{2^n} (\Delta m_k)^2 + 2\sum_{k=1}^{2^n} \Delta m_k \Delta y_k$$

where the last two terms tend to zero. Together with (21) this proves the statement. Q.E.D.

Corollary 1 Let $x(t)$ be a Gaussian process on [0, 1] under two hypotheses P_0 and P_1 for which the conditions of Theorem 4 hold. If the two functions f_0, f_1, defined under respective hypotheses as above, have different integrals

$$\int_0^1 f_1(t)\,dt \neq \int_0^1 f_2(t)\,dt$$

then P_0 and P_1 are perpendicular.

The proof is immediate since the statistic B_n converges almost certainly to different limits under P_0 and P_1.

Example 1 To illustrate the use of Baxter's theorem, let us consider a covariance function of the form in (3.73) of Chapter 4. We can then calculate D^-, D^+, and f directly and get

$$D^-(t) = \phi_1(t)\phi_2'(t)$$
$$D^+(t) = \phi_1'(t)\phi_2(t)$$

so that

$$f(t) = \phi_1(t)\phi_2'(t) - \phi_1'(t)\phi_2(t) = \text{the Wronskian } (\phi_1, \phi_2)$$

Hence a.c.

$$\lim_{n\to\infty} B_n = \int_0^1 [\phi_1(t)\phi_2'(t) - \phi_1'(t)\phi_2(t)]\,dt$$

which can also be expressed in terms of the function $p(t)$; (see Chapter 4, Section 3).

In Baxter's theorem we exploit the fact that the first derivative of r does not exist at $s=t$ (otherwise f would be trivially zero). This idea can be extended to derivatives of higher order, but this will not be pursued here for general covariances. Instead we consider the stationary case, first *when the spectral density is a rational function*. An elegant result due to Slepian (1958) is this theorem.

THEOREM 5

Let $x(t)$, $t\in[0,1]$, be a real Gaussian process with mean zero and absolutely continuous spectrum. If the spectral densities f_0 and f_1 under respective null and alternative hypotheses are rational functions satisfying

$$\lim_{\lambda\to\infty}\frac{f_1(\lambda)}{f_0(\lambda)}\neq 1$$

then the two measures are perpendicular.

Proof Consider first the case when

$$f_0(\lambda)\sim\frac{a_0}{\lambda^2},\quad f_1(\lambda)\sim\frac{a_1}{\lambda^{2p}};\qquad p\geqslant 1,\quad a_0,a_1>0 \tag{23}$$

We can then appeal to Baxter's theorem. Put

$$\phi(\lambda)=f_0(\lambda)-\frac{1}{\pi}\frac{a_0\pi}{1+\lambda^2}$$

which decreases as $0(\lambda^{-4})$ and hence has a Fourier transform which has a continuous derivative. Since ϕ is an even function, this derivative must be zero at zero, so that

$$r_0'(0+)=\left[\frac{d}{dt}\frac{1}{\pi}\int_{-\infty}^{\infty}e^{it\lambda}\frac{a_0/\pi}{1+\lambda^2}d\lambda\right]_{t=0+}=\left[\frac{d}{dt}\frac{a_0}{\pi}e^{-|t|}\right]_{t=0+}=-\frac{a_0}{\pi} \tag{24}$$

Since $r_0(t)$ (and r_1) has derivatives of all orders for $t\neq0$, and because of (24), it follows that we can use Baxter's theorem for the r_0-covariance with $f(t)=2a_0/\pi$. If $p=1$, then for the covariance r_1 we get $f(t)=2a_1/\pi$ so that the limits differ and the measures are perpendicular. If $p>1$, then the even function r is differentiable at $t=0$ with $r_1'(0)=0$, and we get $f(t)=0$. The limits are again different so that the measures are still perpendicular.

To deal with the case

$$f_0(\lambda) \sim \frac{a_0}{\lambda^{2m}}, \quad f_1(\lambda) \sim \frac{a_1}{\lambda^{2(m+p)}}; \qquad a_0, a_1 > 0; \quad p \geqslant 0$$

we can refer to the discussion following Theorem 4 of Chapter 1, Section 1, and in Example 1 of Chapter 1, Section 2, to prove that $x(t)$ has a.c. continuous derivatives of order up to $m-1$. Using $x^{(m-1)}(t)$ in place of $x(t)$, we deal with stationary processes whose spectral densities behave as in (23) asymptotically for the two hypothetical spectra, a case that we have already settled. Q.E.D.

If the spectral density of the process is not rational but behaves in some sense asymptotically as a power law, for example, as described in Theorem 5 of Section 1, criteria have been given for perpendicularity/equivalence. The conditions are somewhat awkward and are not given here (see Note 5.2.E for references).

3 SELECTING LINEAR FUNCTIONALS FOR TESTING

The problem of optimal selection of linear functionals (features) was treated from the point of view of data compression for fixed covariance structure in Chapter 4, Section 7. We now study what happens when we want to test two Gaussian hypotheses H_0 and H_1 against each other, both with mean value function zero but different nonsingular covariances R_1 and R_2. We assume the time parameter to be discrete; the continuous case has not been solved.

Say that we have observed our process x_t for $t=1,2,\ldots,n$, $x=\operatorname{col}(x_t)$, and want to base the test on the N linear functions

$$\begin{gathered} y_1 = a_{11}x_1 + \cdots + a_{1n}x_n \\ \vdots \\ y_N = a_{N1}x_1 + \cdots + a_{Nn}x_n \end{gathered}$$

In matrix form $y=\operatorname{col}(y_t)=Ax$, where A is an $N\times n$ matrix of rank N. Of course N ought to be considerably smaller than n to achieve *data compression*. We measure this by the *compression ratio* $\theta=N/n$, which will be kept fixed in $(0,1)$.

Then y has a Gaussian measure with mean value vector zero and the covariance matrix

$$S_i = AR_iA^T; \qquad i=0,1$$

when H_i is either one of the hypotheses. Let W be the critical region and say that P_i has the *a priori* probability π_i. Then the error probability is

$$P_n = \pi_1 \int_W p_1(y)\,dy + \pi_2 \int_{W^c} p_2(y)\,dy = \pi_1 I_n + \pi_2 I_n'$$

$$= \pi_2 + \int_W [\pi_1 p_1(y) - \pi_2 p_2(y)]\,dy$$

where p_i is the frequency function under H_i. The W that minimizes P_n is then given by, as we have seen in Theorem 3, Chapter 3, Section 1,

$$W = \{y \mid y^T(S_1^{-1} - S_2^{-1})y > \ln\det(S_2) - \ln\det(S_1) + 2\ln\pi_1 - 2\ln\pi_2\} \quad (1)$$

The main analytical difficulty is due to the fact that to calculate P_n for W given in (1), we have to get the inverses of the matrices S_1 and S_2, which is usually hard. We present an asymptotic solution for $n \to \infty$. We first prove Theorem 1 below, which will enable us to carry out the asymptotics.

After a linear transformation we can assume (see Note 5.3.A) that R_1 is the $n\times n$ identity matrix while R_2 is diagonal with the eigenvalues $d_i^{(n)}$:

$$R_2 = \operatorname{diag}(d_1^{(n)}, d_2^{(n)}, \ldots, d_n^{(n)}) = D^{(n)}$$

Similarly, later when we are going to choose t we could assume that S_1 is the $N\times N$ identity matrix, while

$$S_2 = \operatorname{diag}(e_1^{(N)}, e_2^{(N)}, \ldots, d_N^{(N)}) = E^{(N)} \quad (2)$$

as is shown below.

Let us first consider $n=N$, $A=$identity, and $\pi_0 = \pi_1 = \frac{1}{2}$. During the proof we see that asymptotically the values of π_0 and π_1 are irrelevant as long as they are different from 0 and 1. We now prove the following theorem [Grenander (1974)].

THEOREM 1

Assuming that the eigenvalues $d_i^{(n)}$ are contained in the finite interval (m, M) and that their distribution functions converge weakly to some F, we have

$$\lim_{n\to\infty} P_n^{1/n}=\exp\left\{\frac{1}{2}\int_m^M[\ln a(x)-a(x)+1]\,F(dx)\right\} \tag{3}$$

with

$$a(x)=\frac{x}{x+\mu(x-1)} \tag{4}$$

where μ should be determined from the equation

$$\int_m^M \frac{x-1}{x+\mu(x-1)}F(dx)=\int_m^M \ln xF(dx) \tag{5}$$

The proof is organized into a series of lemmas.

Proof To begin with, let us assume that the distribution has k spikes only, so that

$$\begin{aligned}
&d_1=d_2=\cdots=d_{\rho_1 n}=\lambda_1\\
&d_{\rho_1 n+1}=\cdots=d_{(\rho_1+\rho_2)n}=\lambda_2\\
&\vdots\\
&d_{(\rho_1+\cdots+\rho_k)n+1}=\cdots=d_{(\rho_1+\cdots+\rho_k)n}=\lambda_k
\end{aligned} \tag{6}$$

with positive ρ such that $\rho_1+\rho_2+\cdots+\rho_k=1$ and positive distinct λ with at least one of them different from 1. Then the diagonal entries of $Q=I-D^{-1}$, $D=D^{(n)}$, are

$$q_i=1-\frac{1}{\lambda_j}; \qquad (\rho_1+\cdots+\rho_{j-1})n<i\leqslant(\rho_1+\cdots+\rho_j)n$$

and we have

$$\log\det D=n\sum_{i=1}^{k}\rho_i\log\lambda_1=n\cdot c$$

Therefore we can write

$$I_n = P\left\{ \sum_{i=1}^{k} q_i K_i \geqslant cn \right\}$$

where the q_i denote the k distinct values on the diagonal of Q, and where K_i is χ^2-distributed with $\rho_i n$ degrees of freedom.

Recalling the functional form of the χ^2-distribution, we have

$$I_n = C_n \int_{E_k} \prod_{i=1}^{k} e^{-y_i/2} y_i^{(n_i/2)-1}\, dy$$

with $n_i = \rho_i n$, and where the region of integration is

$$E_k = \left\{ y \mid y_1, y_2, \ldots, y_k \geqslant 0,\ \sum q_i y_i \geqslant cn \right\}$$

Also,

$$C_n^{-1} = \prod_{i=1}^{k} \Gamma\left(\frac{n_i}{2}\right) 2^{n_i/2}$$

Changing variables to $y_i/n = u_i$, we get

$$I_n = C_n n^{\Sigma(n_i/2-1)} \int_{F_k} \prod_{i=1}^{k} \exp(-nu_i/2) u_i^{(n_i/2)-1}\, du$$

with the new region of integration

$$F_k = \left\{ u \mid u_1, u_2, \ldots, u_k \geqslant 0,\ \sum q_i u_i \geqslant c \right\}$$

Putting

$$\phi(u) = \prod_{i=1}^{k} e^{-u_i/2} u_i^{\rho_i/2}$$

we can write

$$I_n = C_n n^{n/2} \int_{F_k} \phi^{n-\alpha}(u) q(u)\, du$$

where

$$q(u)=\prod_i e^{-\alpha u_i/2}u_i^{-1+\alpha\rho_i/2}$$

is integrable when $\alpha>0$. Then

$$I_n^{1/n}=\frac{n^{1/2}}{2^{1/2}\left[\prod_1^k\Gamma(n_i/2)\right]^{1/n}}\left\{\int_{F_k}\phi^{n-\alpha}q\,du\right\}^{1/n} \tag{7}$$

Using Stirling's asymptotic expression for the Γ-function,

$$\Gamma(\alpha+1)\sim\left(\frac{\alpha}{e}\right)(2\pi\alpha)^{1/2}$$

we get

$$\left\{\prod_1^k\Gamma\left(\frac{n_i}{2}\right)\right\}^{1/n}\sim\left[n\frac{\prod_1^k\rho_i^{\rho_i}}{2e}\right]^{1/2} \tag{8}$$

which takes care of the coefficient on the right-hand side of (7).

Let us consider the behavior of ϕ in the region F_k. The unconstrained maximum of ϕ occurs for $u_i=\rho_i$, $i=1,2,\dots,k$, but this point does not belong to F_k, since

$$\sum q_i\rho_i=\sum\rho_i\left(1-\frac{1}{\lambda_i}\right)<\sum\rho_i\log\lambda_i=c \tag{9}$$

The constrained maximum cannot be attained when one or more of the coordinates vanish, since then $\phi=0$, and hence we only have the possibility left that it is attained on the simplex

$$G_k=\left\{u\,|\,u_1,u_2,\dots,u_k>0,\ \sum q_iu_i=c\right\} \tag{10}$$

Maximizing $\Sigma[-u_i+\rho\log u_i]+\mu[c-\Sigma q_iu_i]$ by the method of Lagrangian multipliers, we then get

$$\alpha_i=\frac{\rho_i}{1+\mu g_i} \tag{11}$$

where μ should be chosen to satisfy the equality in (10) when α are substituted for u, so that if we introduce the function

$$\alpha(t)=\sum_i \frac{\rho_i q_i}{1+tq_i} \tag{12}$$

we should have

$$\alpha(\mu)=c \tag{13}$$

The denominators in (11) must be positive so that t is contained in an interval (a, A) with

$$a=-\frac{1}{\max_i q_i^+}$$

$$A=-\frac{1}{\min_i q_i^-}$$

where the notation indicates the set of positive and negative q, respectively. If one of these sets is empty, a or A should be replaced by $-\infty$ or $+\infty$, respectively. In the interval (a, A) the function $\alpha(t)$ defined in (12), is continuous and decreasing and an examination of the range of α shows that (13) has just one root which we denote by μ. Summing up we get our first lemma needed for the proof of the theorem.

Lemma 1 The function ϕ attains its maximum over F_k in the point P with coordinates given by (11)–(13).

Denote the maximum by M

$$M-\exp\frac{1}{2}\sum_i(\rho_i\log\alpha_i-\alpha_i)$$

We also note that the gradient of ϕ at P points outwards from F_k.

The reason we make our integral in the form (7) is that we can now apply the classical method of Laplace [see, e.g., Hardy-Littlewood-Polya (1964), pp. 134 and 143, where p corresponds to our integrable function q], (see Note 5.3.B). Combining (7)–(9), we get the second lemma

Lemma 2 For eigenvalue distributions of the type (6) we have

$$\lim_{n\to\infty} I_n^{1/n}=\exp\frac{1}{2}\sum_i \rho_i(\log a_i-a_i+1) \tag{14}$$

with

$$a_i = \frac{1}{1+\mu q_i} = \alpha_i/\rho_i$$

To treat the integral

$$I_n' = \int_{w^C} p_2\,dy$$

we transform the y-variables by $y_i/d_i^{1/2} = z_i$, so that in the z-variables the complement of the critical region W is

$$w^C = \{z \mid z^T[D-1]z < \log\det D\}$$

and z has the identity matrix as covariance. We can now proceed in the same way as before and get a similar expression:

$$\lim_{n\to\infty} (I_n')^{1/n} = \exp\frac{1}{2}\sum_i \rho_i(\log a_i' - a_i' + 1) \tag{15}$$

but with

$$a_i' = \frac{1-q_i}{1+\mu q_i}$$

where μ is given by (13).

One simplification is possible, observing that

$$\sum_i \rho_i(\log a_i - a_i + 1) - \sum \rho_i(\log a_i' - a_i' + 1) = \sum_i \rho_i\left[\log\left(\frac{a_i}{a_i'}\right) - (a_i - a_i')\right]$$

$$= \sum_i \rho_i\left[-\log(1-q_i) - \frac{q_i}{1+\mu q_i}\right]$$

Using the definition of μ in (12) and (13), the sum reduces to

$$-\sum_i \rho_i \log(1-q_i) - c = \sum_i \rho_i \log\lambda_i - c = 0$$

Hence the two limits in (14) and (15) are identical, and we have proven the following lemma.

Lemma 3 Under the given assumption the probability of error decreases geometrically as the sample size tends to infinity and with the ratio

$$\lim_{n\to\infty} \left(P_n(\text{error})\right)^{1/n} = \exp\frac{1}{2}\sum_i \rho_i(\log a_i - a_i + 1)$$

We can use this result to find the asymptotic decrease of the error probability when the distribution of the $d_k^{(n)}$, with

$$0 < m \leqslant d_1^{(n)} \leqslant d_2^{(n)} \leqslant \cdots \leqslant d_n^{(n)} \leqslant M < \infty$$

converges weakly to a distribution over (m, M) with *df* F. We approximate the probability of the event in (1) from below and from above. We do this by shifting the $d_i^{(n)}$ a little to the left such that they form k spikes inside the interval (m, M). In a similar way we shift the values to the right. The probabilities associated with the new quadratic forms obviously form lower and upper bounds for the probability we study. Both of the bounds behave asymptotically in a way that is described by (14). For fixed, but arbitrarily large values of k, this gives us asymptotic bounds expressed through the ratios of geometric progressions. Since the a depend in a continuous manner upon the q, we can make the difference between the two ratios arbitrarily small. This gives us the result stated in (3)–(5). Q.E.D.

It can be noted that the geometric ratio in (16) is strictly less than 1 unless $a(x)\equiv 1$ a.c., which is only possible if $x\equiv 1$ a.c. so that F has all its variation in $x=1$.

Remark 1 It is intuitively plausible that the role of the *a priori* probabilities π_1, π_2 is asymptotically negligible. This can be proven by noticing that if $\pi_1 \neq \pi_2$ and $\pi_1 \neq 0, 1$, the only change in (1) is the occurrence of an additional term $\log \pi_1 - \log \pi_2$ on the right-hand side. This term will not influence the asymptotic result, so that the theorems in the last section are valid in this case, too.

Remark 2 We have scarcely exploited Laplace's method to its full strength. The problem was very well suited to it, and we could apply it directly. It seems likely, however, that using a more refined analysis [see, e.g., Pólya-Szegö (1925, 1954), vol. 1, chap. 5], we could get asymptotic expressions of the form

$$P_n(\text{error}) \sim B \cdot A^n \tag{16}$$

This theorem makes it possible to construct the optimal selection of linear functionals asymptotically by choosing A well.

To verify that S_1 and S_2 can indeed be chosen as described before (2) without loss of generality, it is enough to point out that the features $y=(y_1, y_2, \ldots, y_N)$ lead to exactly the same test as the features B_y if B is an arbitrary nonsingular $N \times N$ matrix. By appropriate choice of B we can then bring S_1 into the identity matrix at the same time as S_2 is brought into diagonal form. Order the eigenvalues

$$0 < d_1^{(n)} \leqslant d_2^{(n)} \leqslant \cdots \leqslant d_n^{(n)}$$

and denote their distribution function by

$$G_n(x) = \frac{1}{n} \#\{d_k^{(n)} | d_k^{(n)} \leqslant x;\ k=1,2,\ldots,n\}$$

We also order

$$0 \leqslant e_1^{(N)} \leqslant e_2^{(N)} \leqslant \cdots \leqslant e_N^{(N)}$$

and denote their distribution function by

$$F_N(x) = \frac{1}{N} \#\{e_k^{(N)} | e_k^{(N)} \leqslant x;\ k=1,2,\ldots,N\}$$

We follow the analysis in McClure (1975). The separation principle for eigenvalues of quadratic forms implies

$$d_k^{(n)} \leqslant e_k^{(N)}$$

$$e_{n+1-k}^{(N)} \leqslant d_{n+1-k}^{(n)}$$

for $k=1,2,\ldots,N$; [see Gantmacher (1959), p. 324]. Note that the N y's are given as certain linear combinations of the n x's. Equivalently F_N is bounded from above by the distribution function of $d_k^{(n)}$, $k=1,2,\ldots,N$; and from below by the distribution function of $d_k^{(n)}$, $k=n+1-N,\ldots,n$. This gives us directly the following lemma, recalling that $N=\theta n$.

Lemma 4 If $G_n \to G$, $F_N \to F$ weakly, we have

$$G_L(x) \leqslant F(x) \leqslant G_U(x) \tag{17}$$

where

$$G_L(x)=\max\left\{0,1-\frac{1}{\theta}[1-G(x)]\right\}$$

$$G_U(x)=\min\left\{1,\frac{1}{\theta}G(x)\right\}$$

Note that G_L and G_U describe the limiting distributions of the N largest and smallest of $d_k^{(n)}$, respectively. Also that the upper bound in (17) can be attained by choosing the rows in A, the functionals, as eigenvectors of R_2 with respect to R_1 with the smallest eigenvalues. Similarly for the lower bound in (17).

Introduce now a family $\mathcal{F}$ of extremal distribution functions. This family will be seen to contain an element that minimizes the integral in (3) and hence solves our optimality problem asymptotically.

For an $\alpha\in[G_L(1), G_U(1)]$ we define $F_\alpha\in\mathcal{F}$ such that for any F satisfying (17) the F_{α_0}, $\alpha_0=F(1)$, satisfies

$$\begin{aligned} F(x)&\geqslant F_{\alpha_0}(x); \qquad x\geqslant 1\\ F(x)&\leqslant F_{\alpha_0}(x); \qquad x<1 \end{aligned} \tag{18}$$

The form of the definition depends upon whether $G_U(1-0)\geqslant G_L(1)$ or not. If the inequality holds, define

$$\beta=\min[\alpha, G_U(1-0)]$$

$$x_0(\alpha)=\inf[x|G_U(x)\geqslant\beta]$$

$$x_1(\alpha)=\sup[x|G_L(x)\leqslant\beta]$$

and

$$F_\alpha(x)=\begin{cases} G_U(x); & x<x_0(\alpha)\\ \beta; & x_0(\alpha)\leqslant x<x_1(\alpha)\\ G_L(x); & x_1(\alpha)\leqslant x \end{cases}$$

If the inequality does not hold, we define instead

$$F_\alpha(x)=\begin{cases} G_U(x); & x<1\\ G_L(x); & x\geqslant 1 \end{cases}$$

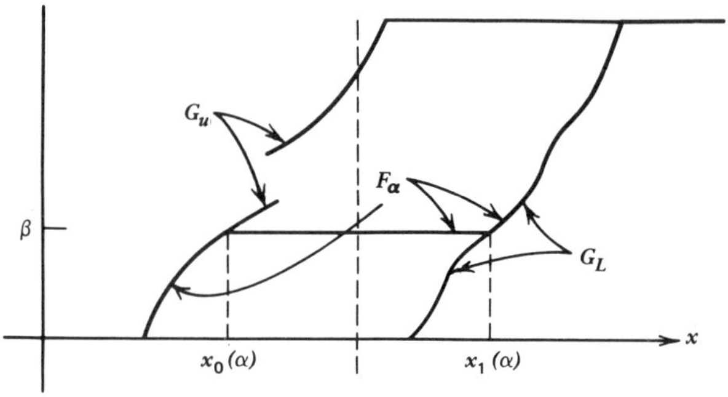

Figure 1

The family is then

$$\mathcal{F}=\{F_\alpha | G_L(1) \leqslant \alpha \leqslant G_U(1)\}$$

Figure 1 shows what the formal definition means geometrically.

Note that F_α is the weak limit of F_N if the rows of A are chosen as the eigenvectors of R_2 with respect to R_1, corresponding to the $[NF_\alpha(1)]$ smallest and $N-[NF_\alpha(1)]$ largest eigenvalues.

In the special cases $G(1-0)=0$ and $G(1)=1$ the family $\mathcal{F}$ contains only one element.

We are now ready to announce the main result. Let $F_N \to F$ where F satisfies (17). We want to maximize

$$J(F)=\int_m^M [a(x)-1-\ln a(x)]\,F(dx) \tag{19}$$

[see (3)]. We prove the following extremum property [McClure (1975)], (see Note 5.3.C).

THEOREM 2

Under the stated conditions and if $G_U(m)=0$, $G_L(M)=1$, we have

$$J(F) \leqslant J(F_{\alpha_0})$$

with $\alpha_0 = F(1)$. If $F \notin \mathcal{F}$, then the inequality is strict.

Proof Consider an arbitrary F satisfying (17), and define for $\alpha_0 = F(1)$ and $s \in [0,1]$, $t \geqslant 1$

$$F_{s,t}(x) = \begin{cases} F(x); & x<s \\ F_{\alpha_0}(x); & s \leqslant x < t \\ F(x); & t \leqslant x \end{cases}$$

Then $F_{1,1}(x) = F(x)$, $F_{m,M}(x) = F_{\alpha_0}(x)$. Differentiate $J(F_{s,t})$ with respect to s and t. We show that the derivatives are nonnegative and nonpositive, respectively, and that at least one of the derivatives is different from zero under $F = F_{\alpha_0}$.

We need the quantity $g(s,t,\mu) = J(F_{s,t})$. If we integrate (19) by parts, we obtain $g(s,t,\mu)$ equal to

$$\int_m^1 F_{s,t}(x)\left\{-\frac{1+\mu}{x+\mu(x-1)} + \frac{\mu}{[x+\mu(x-1)]^2} + \frac{1}{x}\right\}dx$$
$$+\int_1^M [1-F_{s,t}(x)]\left\{\frac{1+\mu}{x+\mu(x-1)} - \frac{\mu}{[x+\mu(x-1)]^2} - \frac{1}{x}\right\}dx$$

With

$$h(x,t,\mu) = \int_m^M \left[\frac{x-1}{x+\mu(x-1)} - \ln x\right] F_{s,t}(dx)$$

we get, after another integration by parts, that h equals

$$\int_m^1 F_{s,t}(x)\left\{\frac{1}{x} - \frac{1}{[x+\mu(x-1)]^2}\right\}dx$$
$$+\int_1^M [1-F_{s,t}(x)]\left\{-\frac{1}{x} + \frac{1}{[x+\mu(x-1)]^2}\right\}dx \qquad (20)$$

To evaluate $g(s,t,\mu)$ we determine $\mu \in (-1,0)$ uniquely from $h=0$. From (20) we see that $h(s,t,0)<0$ and $h(s,t,-1)>0$, unless $F_{s,t}$ is concentrated at $x=1$, and that $\dfrac{\partial h}{\partial \mu} < 0$ on $(-1,0)$.

We now express g using (20) and $h=0$ as

$$\int_m^1 F_{s,t}(x)\frac{(1+\mu)^2(1-x)}{[x+\mu(x-1)]^2}dx + \int_1^M [1-F_{s,t}(x)]\frac{(1+\mu)^2(x-1)}{[x+\mu(x-1)]^2}dx$$

Fix $s=1$. We evaluate

$$\frac{d}{dt}J(F_{1,t})=\frac{d}{dt}g(1,t,\mu)$$

with $h(1,t,\mu)=0$. But

$$\frac{\partial q}{\partial \mu}=-(1+\mu)\frac{\partial h}{\partial \mu}$$

and, using $h(s,t,\mu)=0$,

$$\frac{\partial}{\partial t}h(1,t,\mu)+\frac{\partial}{\partial \mu}h(1,t,\mu)\frac{d\mu}{dt}=0$$

so that

$$\frac{d}{dt}J(F_{1,t})=\frac{\partial g}{\partial t}+\frac{\partial g}{\partial \mu}\frac{d\mu}{dt}=\frac{\partial}{\partial t}\left[g(1,t,\mu)+(1+\mu)h(1,t,\mu)\right]$$

After some manipulations this yields

$$\frac{d}{dt}J(F_{1,t})=\frac{(-\mu)(1+\mu)(t-1)}{t[t+\mu(t-1)]}\left[F(t)-F_{\alpha_0}(t)\right] \tag{21}$$

For $t>1$ we see from (18) that all the factors on the right-hand side of (21) are nonnegative.

Similarly we can get

$$\frac{d}{ds}J(F_{s,M})=-\frac{(-\mu)(1+\mu)(1-s)}{s[s+\mu(s-1)]}\left[F_{\alpha_0}(s)-F(s)\right] \tag{22}$$

where the factors are nonnegative., Note that $\mu\in(-1,0)$ and recall (18).

Combining (21) with (22) we have shown, as planned, that the derivative of $J(F_{s,t})$ with respect to s is nonnegative and with respect to t is nonpositive.

If F does not belong to $\mathfrak{F}$, so that F and F_{α_0} are distinct on some interval of positive length, then one of (21) and (22) will not vanish identically. Hence from what we have proved about the signs,

$$J(F)=J(F_{1,1})\leqslant J(F_{1,M})\leqslant J(F_{mM})=J(F_{\alpha_0})$$

and one of these inequalities will be strict. Q.E.D.

The way F_α has been defined, it depends continuously upon α in the topology of weak convergence. Then μ, determined through (5), also varies continuously with α so that $J(F_\alpha)$ is also continuous. Of course the α-set is compact, so that existence of the optimal distribution function is assured.

Remark 3 Earlier work by Kadota-Shepp (1967) involved bounds for the error probabilities. These bounds may lead one to conjecture that asymptotically optimal selection of functionals will take place if we select as rows for A the eigenvectors of R_2 with respect to R_1 for which the corresponding eigenvalues λ have $\lambda + 1/\lambda$ large. Theorem 2 shows that this is not necessarily so; instead one should choose large and small λ as described in the theorem. A counterexample showing that this is so can be found in McClure (1975).

Remark 4 The recommendation to select the eigenvectors of R_2 with respect to R_1 in the way described is usually difficult to implement computationally. In the important case of stationary covariances, however, Toeplitz theory can be used to carry out the selection in an asymptotically optimal way (compare with Chapter 4, Section 7).

CHAPTER 6

Inference in Markov Processes

1 THE INFINITESIMAL GENERATOR AS PARAMETER

The processes to be studied in this chapter are defined on the argument space $T=\mathbb{R}^+$. The value space V will be $\mathbb{R}$, or a subset of it. Throughout this chapter the processes are assumed to be continuous in probability.

The measures are introduced by an initial distribution $P_0(dx)$ and a family of transition probability measures $P(x, dy; s, t)$, $s \leqslant t$. Here $P_0(dx)$ is the measure of, say $x=x(0)$, while $p(x, dy; s, t)$ denotes the measure of $y=x(t)$ conditioned by $x(s)=x$ (see Example 6 in Chapter 1, Section 1).

To make this precise we must specify the regularity properties of $P(x, dy; s, t)$, which will be done separately for each type of process that we encounter in the following sections. The basic relation governing the marginal distributions of the process is the Chapman-Kolmogorov equation, $t_1<t_2<t_3$,

$$P(x_1, dx_3; t_1, t_3)=\int_{x_2\in V} P(x_1, dx_2; t_1, t_2)P(x_2, dx_3; t_2, t_3) \tag{1}$$

[see, e.g., Feller (1966), p. 312].

We are mainly concerned with the *time homogeneous* case when $P(x, dy; s, t)$ depends on s and t only through the lag $t-s=h$. We then write the transition probability measure as $P(x, dy; h)$ instead of $P(x, dy; s, s+h)$.

In functional analytic terms the process can be studied via the probability operators T_h where

$$(T_h f)(x)=\int_{y\in V} f(y)P(x, dy; h)=E[\, f(x_{t+h})|x_t=x]$$

and the function f is, for example, bounded and continuous on V (sometimes it is natural to choose a more restricted function space for f). Equation (1) now expresses the semigroup property $T_{h_1+h_2}=T_{h_1}T_{h_2}$; $h_1, h_2>0$. [For a fuller discussion the reader is referred to Feller (1966), chap. X; see also Hille-Phillips (1957), sec. XXIII. 5.

We always assume that a separable version of the process (see Chapter 1, Section 1) has been selected so that we can state that the sample functions have desired regularity properties with probability 1.

If the process is also *space homogeneous*, so that $P(x, dy; h)$ for fixed h is a measure depending upon x only through the translation $y-x$, we write the transition probability measure simply as $P(dy; h)$. The corresponding probability operator can then be written

$$(T_h f)(x)=\int_{y\in V} f(x+y)P(dy;h)$$

Space homogeneity means that $x(t)$ is *a process with independent increments*.

If $x(t)$ is time homogeneous (whether space homogeneous or not), the semigroup $\{T_h\}$ has an *infinitesimal generator*

$$Q=\lim_{h\downarrow 0}\frac{T_h-I}{h} \tag{2}$$

where the limit must be interpreted with some care. In the extreme case when T_h is continuous in the uniform operator topology, so that $\|T_h-I\|\to 0$ as $h\downarrow 0$, then Q is defined for all bounded continuous f; otherwise its domain has to be suitably restricted.

In the inference problems to be studied here the *infinitesimal generator plays the main role as parameter*. We may ask for a test for a hypothethical Q_1 against an alternative operator Q_0. Or we may want to construct an estimator of Q given one or several sample functions of $x(t)$.

Sometimes the allowed values Q of the infinitesimal generator are indexed (in some natural way) by a parameter in a finite-dimensional Euclidean space. We can then think of the parameter space as finite dimensional. On the other hand if the values of Q are less restricted, we are dealing with *abstract inference in the double meaning of the term: both X and A may be infinite dimensional.* In this chapter we therefore anticipate to some extent the discussion of the abstract parameter case to be studied more systematically in Part III.

To indicate the way the infinitesimal operator enters the analysis of inference problems, say that the transition probability measures are absolutely continuous with respect to a fixed σ-finite measure μ, so that

$P_i(x, dy; h) = p_i(x, y; h)\mu(dy)$ and that the marginal distribution of $x(t_1)$ is also absolutely continuous with density h_i, $i=0,1$. Here $i=0,1$ and denotes the two hypotheses to be tested. Let $t_1 < t_2 < \ldots < t_n$ be points in the finite interval $T=(a,b)$ and calculate the finite-dimensional likelihood ratio f_n as in Section 1 of Chapter 3.

Due to the Markov property we get

$$f_n(x) = \frac{h_1[x(t_1)]}{h_0[x(t_1)]} \prod_{\nu=1}^{n} \frac{p_1[x(t_{\nu-1}), x(t_\nu); t_\nu - t_{\nu-1}]}{p_0[x(t_{\nu-1}), x(t_\nu); t_\nu - t_{\nu-1}]} \tag{3}$$

In order to calculate the Radon-Nikodym derivative in accordance with Theorem 1 of Chapter 3, Section 1, we should refine the partition $\{t_\nu\}$ of T so that all $h_\nu = t_{\nu+1} - t_\nu$ will be small. Then both numerator and denominator on the right-hand side of (3) will be related to the probability operator p_h for small positive values h. But *this leads us to approximations of* T_h *using* (2) *to approximate* (3), *and this is the reason why the infinitesimal operator will play the main role in our questions of inference in Markov processes.*

To calculate the limit of (3) we should express it in suitable observables. We have seen in Chapter 3 that the limit procedure can be simplified, or even eliminated completely, if we choose the observables wisely. This is the case repeatedly.

As before we are concerned with the question of equivalent/perpendicular probability measures. We also have to calculate Radon-Nikodym derivatives as a preparation for the analysis of inference problems.

The theory of Markov processes is one of the most substantial chapters in the calculus of probability. It is remarkable, however, how incomplete is our knowledge of their inference compared to what we know about linear inference and inference of stationary or of Gaussian processes.

We examine five types of Markov processes in this chapter:

1. Finite Markov chains (with a continuous time parameter).
2. Birth-and death processes.
3. Processes with independent increments; differential processes (discontinuous type).
4. Processes defined by stochastic differential equations (continuous type).
5. Stationary Markov processes.

In case 1, when $V=\{1,2,\ldots,M\}$, continuity in probability implies that T_h is continuous in the uniform operator topology so that the infinitesimal generator Q is defined everywhere on $\mathbb{R}^n$; it is just an $M \times M$ matrix.

For birth-and-death processes, case 2, we have the value space $V=\mathbb{N}$. The infinitesimal operator can again be expressed by a matrix, but now infinite dimensional, and the infinitesimal generator Q has to be restricted to a suitable everywhere dense subset of the corresponding sequence space.

The third case, both time and space homogeneous, leads to an infinitesimal generator that is traditionally expressed in Fourier analytic form: the celebrated Lévy-Khinchin representation.

In the fourth case we do not demand that the process be space homogeneous, and Q is more complicated.

For stationary Markov processes we employ the maximum likelihood method for our inference in the important situation when the process is ergodic.

2 FINITE MARKOV CHAINS

In case 1, so that $V=\{1,2,\ldots, M\}$ and μ is naturally just counting measure on V, we use the notation

$$P_{ij}(h)=P[x(t+h)=j|x(t)=i]$$

on the argument space $T=[0, L)$ with finite L, and the infinitesimal generator

$$Q=\{q_{ij};\ i, j=1,2,\ldots, M\}$$

We have

$$\begin{cases} \lim\limits_{h\downarrow 0} \dfrac{1-P_{ii}(h)}{h}=q_i=-q_{ii} \\ \lim\limits_{h\downarrow 0} \dfrac{P_{ij}(h)}{h}=q_{ij};\ i\neq j \end{cases}$$

where $q_i=\Sigma_{j\neq i}q_{ij}$ and

$$P(t)=\{P_{ij}(t);\ i, j=1,2,\ldots, M\}=\exp tQ$$

where exp stands for the exponential function with a matrix valued argument.

It is well known that, choosing a separable version, $x(t)$ has a finite number, possibly zero, of discontinuities of the first kind with probability 1. The sample functions are constant between the discontinuities.

This makes it natural to introduce as observables the number of jumps, the time points at which the discontinuities occur, and the constant values taken between jumps. The limit of (1.3) is then easy to calculate. To do this we introduce

$$\begin{aligned} \tau_0 &= 0,\ \tau_i = \text{time of } i\text{th jump} \\ t_i &= \tau_{i+1} - \tau_i \\ N(T) &= \text{largest } n \text{ for which } \tau_n < L \\ z_i &= x(\tau_i) \end{aligned} \tag{1}$$

Then the observables for $[0, L)$ can be expressed as the vector $(z_0, t_0, z_1, t_1, \ldots, z_{n-1}, t_{N-1}, z_N)$. Note, however, that since N is random, the space of observables is of infinite dimension. The space is

$$X = \bigcup_{n=0}^{\infty} X_n; \qquad X_n = \underbrace{(V\times\mathbb{R})\times\ldots\times(V\times\mathbb{R})\times V}_{n \text{ times}} \tag{2}$$

Let σ be the measure defined on the corresponding σ-algebra over X, where we use counting measure μ on V and Lebesgue measure on $\mathbb{R}$ as marginal measures to generate σ.

The density of the probability measure over X, expressed in the observables in (1), is then (with respect to σ)

$$\begin{aligned} &P[x(0)=z_0]\exp(-Lq_{z_0}) \text{ if no jump in } [0,L); \qquad x=(z_0) \\ &P[x(0)=z_0]\exp(-Lq_{z_n}) \prod_{\nu=0}^{n-1} q_{z_\nu z_{\nu+1}} \exp[-(q_{z_\nu}-q_{z_n})t_\nu] \\ &\text{with n jumps}; \qquad x=(z_0, t_0, z_1, t_1, \ldots, z_{n-1}, t_{n-1}, z_n) \end{aligned} \tag{3}$$

The first expression in (3) is obvious. In the second one the factors $q_{z_\nu z_{\nu+1}}$ correspond to the jumps from z_ν to $z_{\nu+1}$ at t_ν, where of course we only allow $z_\nu \neq z_{\nu+1}$. The remaining factor is due to the event no jump in $(0, \tau_1)$, and none in (τ_1, τ_2), and so on up to (τ_n, L), or

$$\begin{aligned} &\exp(-\tau_1 q_{z_0})\exp[-(\tau_2-\tau_1)q_{z_1}] \\ &\ldots \exp[-(\tau_n-\tau_{n-1})q_{z_{n-1}}]\exp[-(L-\tau_n)q_{z_n}] \end{aligned} \tag{4}$$

Using the definition in (1) of τ_i, (4) becomes

$$\begin{aligned} &\exp(-Lq_{z_n})\exp(-t_0 q_{z_0})\exp(-t_1 q_{z_1}) \\ &\ldots \exp(-t_{n-1}q_{z_{n-1}})\exp[(t_0+t_1+t_2+\ldots+t_{n-1})q_{z_n}] \end{aligned}$$

so that combining the factors we get (3) as stated.

Consider now two hypotheses H_0 and H_1 corresponding to the infinitesimal generators Q_0 and Q_1. To avoid unessential complications, say that both hypotheses have the same initial distribution with $x(0)=z_0$ a.c.

Introduce the set $w_i \subseteq V$ as the set of states that can be reached with positive probability starting at z_0 under hypothesis H_i. These sets can be determined directly by inspection of Q_i.

Assume first that $w_1 \neq w_0$ so that there is some state $i \notin w_0$, $i \in w_1$ (or the other way around). Then the event E that $x(t)$ takes the value i as $t \in [0, T)$ has

$$P_0(E)=0$$

$$P_1(E)>0$$

Hence P_1 is not absolutely continuous with respect to P_0 as stated.

If we exclude this case, we can apply the likelihood convergence theorem in Chapter 3, Section 1, and see that the two measures are equivalent and have the finite and nonzero Radon-Nikodym derivative given as a ratio of two expressions of the form given in (3):

$$f(x)=\frac{P_1(dx)}{P_0(dx)}=\exp\left[L\left(q^0_{z_n}-q^1_{z_n}\right)\right]\prod_{\nu=0}^{n-1}\frac{q^1_{z_\nu z_{\nu+1}}}{q^0_{z_\nu z_{\nu+1}}}\exp\left[t_\nu\left(q^1_{z_n}-q^1_{z_\nu}-q^0_{z_\nu}+q^0_{z_n}\right)\right] \tag{5}$$

if $n>0$, and a similar expression for $n=0$ [see the first expression in (3)]. We have proved the following theorem.

THEOREM 1

If $q_{ij}^{(0)}>0$, $i \neq j$, implies $q_{ij}^{(1)}>0$ and vice versa, P_0 and P_1 are equivalent, and their Radon-Nikodym derivative is given in (5).

This result immediately gives us the most powerful test of H_0 against the alternative H_1. We just apply the Neyman-Pearson lemmas in its general formulation (see Theorem 2 in Chapter 3, Section 1).

We can also find the maximum likelihood estimate of Q. Indeed, using (5) for fixed Q_0 and variable $Q_1=Q$, we get, for $n>0$,

$$\ln f(x)=-Lq_{z_n}+\sum_{\nu=0}^{n-1}\left[\ln q_{z_\nu z_{\nu+1}}+t_\nu(q_{z_n}-q_{z_\nu})\right]+\text{constant} \tag{6}$$

Introducing

$$N_{ij} = \text{number of transitions } i \to j \text{ for } x(t); \qquad 0 \leq t < L$$

$$A_i = \text{total time } x(t) \text{ spends in state } i$$

we can write (6) as

$$\sum_{i=j} N_{ij} \ln q_{ij} - \sum A_i q_i + \text{constant} \tag{7}$$

When we attempt to maximize (7) by varying the q_{ij}, we must keep in mind that q_i is the sum of q_{ij} over all subscripts j that differ from i. Rewriting (7) as

$$\sum_{i=1}^{M} \left(\sum_{j \neq i} N_{ij} \ln q_{ij} - A_i q_{ij} \right) + \text{constant}$$

we get the ML-solution

$$q_{ij}^* = \frac{N_{ij}}{A_i}; \qquad i \neq j \tag{8}$$

and q_i^* as the sum of q_{ij}^*, $j \neq i$. If some $A_i = 0$, no solution exists.

These estimates are intuitively obvious. What is more interesting, and harder to establish, is the distributional properties of q_{ij}^*, in terms of the standardized stochastic variables, at time t,

$$\xi_{ij} = (N_{ij} - q_{ij} A_i)/\sqrt{t} \tag{9}$$

It has been shown in Albert (1962) that the joint generating function of the ξ_{ij} can be calculated explicitly as follows. We use a sort of inner product for two arrays $\alpha = \{\alpha_{ij};\ i \neq j\}$ and $\beta = \{\beta_{ij};\ i \neq j\}$, where i and j range from 1 to M,

$$\langle \alpha, \beta \rangle = \sum_{i=1}^{M} \sum_{j \neq i} \alpha_{ij} \beta_{ij}$$

Let

$$\phi(\omega, t) = E\{\exp[-\langle \xi, \omega \rangle\}$$

be the generating function of the variables ξ_{ij}, $i \neq j$, in terms of the auxiliary array $\omega = \{\omega_{ij};\ i \neq j\}$. We first prove this theorem (see Note 6.2.A).

THEOREM 2

The generating function is

$$\phi(\omega) = \left(\exp(tR^T)J, e\right) \tag{10}$$

where R is the $M \times M$ *matrix with entries*

$$r_{ij} = \begin{cases} -q_i + \dfrac{1}{\sqrt{t}} \displaystyle\sum_{k \neq i} q_{ik}\omega_{ik}; & i = j \\ q_{ij}\exp\left(-\dfrac{\omega_{ij}}{\sqrt{t}}\right); & i \neq j \end{cases} \tag{11}$$

and we have introduced the two M-vectors

$$J = \operatorname{col}[\,P[\,x(0)\,] = i, i = 1, 2, \ldots, M\,]$$

and

$$e = \operatorname{col}[\,1, 1, \ldots, 1\,]$$

Proof Introduce the restricted generating function (see Note 6.2.B)

$$\phi_k(t) = E[\,\Phi(t) \text{ and } x(t) = k\,] \tag{12}$$

where, with a new array $\{\nu_{ij}\}$ of auxiliary variables,

$$\Phi(t) = \exp\left\{-\sum_{i \neq j} \nu_{ij}[\,N_{ij}(t) - q_{ij}A_i(t)\right\} \tag{13}$$

We have indicated the dependence of N_{ij} and A_i on time explicitly in (13) for clarity.

Then, for $h > 0$,

$$\phi_k(t+h) = E\{E[\,\Phi(t+h) \text{ and } x(t) \text{ fixed}\,] \text{ and } x(t+h) = k\} \tag{14}$$

In the event $x(t)=i$, $x(t+h)=k$, we have

$$\Phi(t+h)=\begin{cases}\Phi(t)\exp\left[h\sum_{j\neq i}\nu_{ij}q_{ij}\right]; & \text{if } i=k\\ \Phi(t)\exp[-\nu_{ik}+0(1)]; & \text{if } i\neq k\end{cases}$$

so that, using the asymptotic form for the transition probability $i\to k$ when h is small,

$$\phi_k(t+h)=\phi_k(t)\Big(1+h\sum_{j\neq k}\nu_{kj}q_{kj}\Big)(1-q_kh)+\sum_{i\neq k}\phi_i(t)\exp(-\nu_{ik})q_{ik}h+0(h)$$

Subtract $\phi_k(t)$ from both sides, divide by h, and let h tend to zero. We get the system of differential equations

$$\dot{\phi}_k(t)=\frac{d\phi_k(t)}{dt}=\phi_k(t)\Big(\sum_{j\neq k}\nu_{kj}q_{kj}-q_k\Big)+\sum_{i\neq k}\phi_i(t)\exp(-\nu_{ik})q_{ik} \tag{15}$$

With $\psi(t)=\text{col}[\phi_1(t),\phi_2(t),\ldots,\phi_M(t)]$ we can write (15), first, as

$$\dot{\phi}_k=\phi_k(t)\rho_{kk}+\sum_{i\neq k}\phi_i(t)\rho_{ik}$$

where ρ_{ik} is given in the same way as the r_{ik} in (11), but with $t=1$ and with ν_{ik} instead of ω_{ik}, and second, in vector form

$$\dot{\psi}(t)=P^T\psi(t) \tag{16}$$

where P is R with t set equal to 1 and ν_{ij} instead of ω_{ij} [see (11) again]. Solving (16) we get

$$\psi(t)=\exp(tP^T)\psi(0)=\exp(tP^T)\,\text{col}\{P[x(0)=k];\qquad k=1,2,\ldots,M\} \tag{17}$$

In the definition of the variables ξ_{ij} in (9) we include the factor $\sqrt{t}$ which will lead to R instead of P in (17). Hence we should put $\nu_{ij}=\sqrt{t}\,\omega_{ij}$ in (17) and then get

$$\begin{aligned}\phi(\omega,t)&=\sum_{i=1}^{M}E[\exp(-\langle\xi,\omega\rangle)\text{ and }x(t)=i]\\&=\sum_{i=1}^{M}\left(\exp(tR^T)\,\text{col}\{P[x(0)=k;k=1,2,\ldots,M\}\right)_i\end{aligned}$$

But

$$P[x(0)=i]=i\text{th component of } J$$

so that

$$\phi(\omega,t)=\left(\exp(tR^T)J,e\right)$$

as stated in (10). Q.E.D.

Theorem 2 is useful both for evaluating moments of the stochastic variables ξ_{ij}, which are not discussed here, and for deriving their asymptotic distribution. This is now done under some restriction on Q [compare with Albert (1962)].

THEOREM 3

If Q has simple eigenvalues, then

$$\lim_{t\to\infty}\phi(\omega,t)=\exp\left(\frac{1}{2}\sum_{i\neq j}K_i q_{ij}\omega_{ij}^2\right)$$

where K is the left eigenvector of Q for $\lambda=0$ normalized, so that $(K,e)=1$.

Remark 1 It is clear that $Qe=0$ so that zero is always an eigenvalue of Q. Also note that K is the (unique) equilibrium distribution.

Proof Using (11) for large values of t, we see that, for $i\neq j$,

$$r_{ij}=q_{ij}-\frac{1}{\sqrt{t}}q_{ij}\omega_{ij}+\frac{1}{2t}q_{ij}\omega_{ij}^2+o(t^{-1}) \tag{18}$$

Writing

$$a_{ij}=\begin{cases}\sum_{k\neq i}q_{ik}\omega_{ik}; & i=j\\ -q_{ij}\omega_{ij}; & i\neq j\end{cases}$$

and

$$b_{ij}=\begin{cases}0; & i=j\\ \frac{1}{2}q_{ij}\omega_{ij}^2; & i\neq j\end{cases} \tag{19}$$

we can write (18) in terms of constant matrices A and B, consisting of the elements a_{ij} and b_{ij}, respectively, as

$$R=R(t)=Q+\frac{1}{\sqrt{t}}A+\frac{1}{t}B+o\left(\frac{1}{t}\right) \qquad \text{as } t\to\infty \tag{20}$$

so that the eigenelements of $R(t)$ tend to the corresponding ones of Q. This is classical "Stőrungsrechnung." We can transform Q by the Jordan decomposition into diagonal form since the eigenvalues are simple,

$$Q=UDV$$
$$VU=I$$

where D is the diagonal matrix with some eigenvalues $\lambda_1=0, \lambda_2,\ldots, \lambda$. The U and V matrices are formed from eigenvectors (left and right) of Q associated with the λ, so that the first column of U is e, and the first row of V is K.

All the eigenvalues but one have negative real part $\text{Re}(\lambda_\nu)<0$, $\nu=2,3,\ldots, M$ (see Note 6.2.C). We can write

$$\begin{aligned} e^T\exp\left[tR^T(t)\right]J &= e^TV^T\exp\left[(V^{-1})^T tR^T(t)(U^{-1})^T\right]U^TJ \\ &= x^T\exp[S(t)]y \end{aligned} \tag{21}$$

where the first entries of the x-vector and of the y-vector are both 1, and we have set

$$S(t)=(V^{-1})^T tR^T(t)(U^{-1})^T=\left[U^{-1}tR(t)V^{-1}\right]^T$$

But using (20),

$$\begin{aligned} tU^{-1}R(t)V^{-1} &= tD+\sqrt{t}\,U^{-1}AV^{-1}+U^{-1}BV^{-1}+o(1) \\ &= tD+\sqrt{t}\,VAU+VBU+o(1) \end{aligned} \tag{22}$$

We now exploit the properties we have for D and A. The (1,1) element of $S(t)$ gets no contribution from tD since $\lambda_1=0$. The same is true for $\sqrt{t}$ VAU since $Ae=0$; remember that the first column of U is e. The only contribution is $c=(VBU)_{11}+o(1)$. In the rest of the diagonal of $S(t)$ the elements are $t\lambda_\nu+0(t)$. All the remaining elements of $S(t)$ are $o(t)$.

Therefore we have

$$S(t)=cI+tM(t)$$

where $M(t)$ has a diagonal of the form

$$(0,\lambda_2,\lambda_3,\ldots,\lambda_M)+o(1)$$

and all whose other elements are $0(1)$, so that

$$\lim_{t\to\infty} M(t)=\operatorname{diag}[0,\lambda_2,\ldots,\lambda_M]=\Delta$$

and we see that $M(t)$ has distinct eigenvalues for t big enough. Jordan decomposition gives therefore with some matrices $U(t)$ and $V(t)$ similarly to before

$$M(t)=U(t)\Delta(t)V(t);\qquad \Delta(t)\to\Delta$$

so that

$$\exp[tM(t)]=U(t)\exp[t\Delta(t)]V(t)$$

and $U(t)$ and $V(t)$ tend to the matrices U_0 and V_0 made up of eigenvectors for Δ which are orthogonal since Δ is symmetric. Hence since

$$\exp[t\Delta(t)]\to\operatorname{diag}[1,0,0,\ldots]$$

because of all $\lambda_2,\lambda_3,\ldots,\lambda_M$ having negative real parts, we see that

$$\lim_{t\to\infty}\exp[S^T(t)]=\exp(cI)U_0\operatorname{diag}[1,0,\ldots,0]V_0$$

since the identity commutes with all other matrices so that

$$\exp(cI+M)=\exp(cI)\exp(M),\qquad \text{and } M$$

We thus get, since $\exp(cI)=e^c\cdot I$,

$$\lim_{t\to\infty}\exp[S^T(t)]=e^c\left\{\begin{array}{c|c}1 & 0\\ \hline 0 & 0\end{array}\right\}$$

into block matrices of order 1×1, $1\times(M-1)$, $(M-1)\times 1$, and $(M-1)\times(M-1)$, and, using (21), since we can identify c from the $(1,1)$ element of (22),

$$\lim_{t\to\infty} e^T\exp[tR^T(t)]J=e^c=\exp[(VBU)_{11}] \tag{23}$$

But the B-matrix was given in (19), and we see that the limit in (23) is equal to

$$\exp(J^TBe)=\exp\left(\frac{1}{2}\sum_{i\neq j}K_iq_{ij}\omega_{ij}^2\right) \tag{24}$$

as stated. Q.E.D.

Remark 2 An immediate consequence of (24) is that the stochastic variables ξ_{ij} are asymptotically Gaussian, with mean zero and variances for $i\neq j$

$$\operatorname{Var}(\xi_{ij})=k_iq_{ij}$$

and, surprisingly, independent of each other.

We can use Theorem 3 to get the asymptotic distribution of the *ML*-estimates given in (8). Under the same conditions on the eigenvalues of Q and with initial distribution J equal to the equilibrium distribution K, the process is strictly stationary and ergodic. It then follows that the time averages exist a.c.

$$\lim_{L\to\infty}\frac{1}{L}\int_0^L 1_i(t)\,dt=\lim_{L\to 8be}\frac{1}{L}A_i=E[1_i(t)]=j_i \tag{25}$$

In (25) $1_i(t)$ is the indicator function of the event $x(t)=i$. Note that J is now proportional to the left eigenvector K.

In the same way a.c.

$$\lim_{N\to\infty}\frac{1}{N}\sum_{t=1}^{N}N_{ij}(t)\to E[N_{ij}(t)]=J_iq_{ij} \tag{26}$$

where $N_{ij}(t)$ is the number of transitions $i\to j$ in the interval $[t,t+1)$. Clearly (26) also holds if N is allowed to tend to infinity continuously, so that a.c.

$$\lim_{N\to\infty}\frac{1}{L}N_{ij}=J_iq_{ij}$$

But

$$\sqrt{t}\,(q_{ij}^*-q_{ij})=\frac{(N_{ij}-A_iq_{ij})/\sqrt{t}}{A_i/t} \tag{27}$$

where the numerator on the right is $N(0, K_i q_{ij})$ asymptotically, and the denominator tends to J_i. Hence the expression in (27) behaves asymptotically as Gaussian independent stochastic variables with mean zero and variances

$$\frac{K_i}{J_i^2}q_{ij}=\frac{K_i}{K_i^2}q_{ij}=\frac{q_{ij}}{K_i}$$

This can be used for asymptotic inference on Q.

Remark 3 Because of the ergodicity, the above asymptotic statement remains true if we start the process at $t=0$ with an arbitrary initial distribution.

3 BIRTH-AND-DEATH PROCESSES WITH LINEAR GROWTH

When $x(t)$ is a birth-and-death process, the infinitesimal generator can be expressed through a matrix of simple structure. Let λ and μ be the birth and mortality intensities, respectively, *constant per individual* and time unit, so that the transition probabilities satisfy, with small positive lag h, for $i>0$,

$$P(x(t+h)=j|x(t)=i)=\begin{cases} i\lambda h+o(h); & j=i+1\\ 1-i(\lambda+\mu)h+o(h); & j=i\\ i\mu h+o(h); & j=i-1\\ o(h); & \text{otherwise}\end{cases} \tag{1}$$

A small modification is needed in (1) for $i=0$. With little loss of generality we assume that we start at time zero with a single individual, $x(0)=1$.

The Kolmogorov equations for the functions $p_n(t)=P[x(t)=n]$ are

$$\frac{dp_n(t)}{dt}=(n-1)\lambda p_{n-1}(t)-n(\lambda+\mu)p_n(t)+(n+1)\mu p_{n+1}(t) \tag{2}$$

where p_{n-1} is interpreted as zero [see, e.g., Feller (1957), p. 410]. With the vector

$$p(t)=\text{col}[\,p_0(t),p_1(t),\dots\,]$$

we can write (2) as

$$\frac{dp(t)}{dt} = Qp(t)$$

so that the infinitesimal generator is

$$Q = \begin{Bmatrix} 0 & \mu & 0 & 0 & \cdots \\ 0 & -(\lambda+\mu) & 2\mu & 0 & \cdots \\ 0 & \lambda & -2(\lambda+\mu) & 3\mu & \cdots \\ 0 & 0 & 2\lambda & -3(\lambda+\mu) & \\ \cdots & & & & \end{Bmatrix} = \lambda E - (\lambda+\mu)F + \mu G$$

Here we have introduced the matrices that do not depend upon the statistical parameters λ and μ:

$$E = \begin{Bmatrix} 0 & 0 & 0 & 0 & \cdots \\ 0 & 0 & 0 & 0 & \cdots \\ 0 & 1 & 0 & 0 & \cdots \\ 0 & 0 & 2 & 0 & \cdots \\ \cdots & & & & \end{Bmatrix}; \quad F = \begin{Bmatrix} 0 & 0 & 0 & 0 & \cdots \\ 0 & 1 & 0 & 0 & \cdots \\ 0 & 0 & 2 & 0 & \cdots \\ 0 & 0 & 0 & 3 & \cdots \\ \cdots & & & & \end{Bmatrix};$$

$$G = \begin{Bmatrix} 0 & 1 & 0 & 0 & \cdots \\ 0 & 0 & 2 & 0 & \cdots \\ 0 & 0 & 0 & 3 & \cdots \\ 0 & 0 & 0 & 0 & \cdots \\ \cdots & & & & \end{Bmatrix}$$

The state space is infinite, and the domain of the infinitesimal generator has to be suitably restricted. For the present purpose it is enough to note that we can choose a separable version of the process such that $x(t)$ is constant except for a finite number of jumps in any finite interval $[0, t]$. These jumps are either births, $x(t+0) = x(t-0) + 1$, or deaths, $x(t+0) = x(t-0) - 1$. Denote by $B(t)$ the total number of births in the interval, and similarly by $D(t)$ the number of deaths. Then

$$x(t) = 1 + B(t) - D(t)$$

The *total time of exposure*, in the sense of mortality measurements, is

$$S(t) = \int_0^t x(s)\,ds$$

It is well known [see, e.g., Harris (1963), chap. V], that the mean value function for $x(t)$ is given by

$$m(t)=\exp[(\lambda-\mu)t]$$

and blows up as $t\to\infty$ in the supercritical case $\lambda>\mu$, which is the only one tat is considered here, and that a stochastic variable W then exists such that a.c.

$$\lim_{t\to\infty}\frac{x(t)}{m(t)}=W$$

Here $P(W=0)=\mu/\lambda$, and the distribution of W conditioned by $W>0$ has the exponential frequency function e^{-w}, $w>0$.

It is easy to calculate the Radon-Nikodym derivative for the hypothesis (λ,μ) against some null hypothesis (λ_0,μ_0), when $T=[0,t]$. We could use as observables $x(\tau_\nu)$, where $\{\tau_\nu\}$ is everywhere dense in T, and examine the finite-dimensional distributions of $\{x(\tau_\nu);\ \nu=1,2,\dots,N\}$ when $N\to\infty$. It is more convenient, however, to choose as observables the time points $\{t_\nu\}$, where $x(t)$ makes a jump, up or down depending upon whether t_ν is a birth and/or a death. We include for convenience among the numbered time points $t_0=0$ and $t_{n+1}=t$, where n is the total number of jumps in T. Because of the Markov property and the conditional independence that it implies, the Radon-Nikodym derivative will be a product of factors, associated with intervals $[t_\nu,t_{\nu+1})$, of the form

$$\begin{cases}\dfrac{\lambda}{\lambda_0}\exp[-(\lambda+\mu-\lambda_0-\mu_0)x(t_\nu+0)]; & \text{if } t_\nu \text{ is a birth}\\[2ex] \dfrac{\mu}{\mu_0}\exp[-(\lambda+\mu-\lambda_0-\mu_0)x(t_\nu+0)]; & \text{if } t_\nu \text{ is a death}\end{cases} \tag{3}$$

For $t_0=0$, the factor before the exponential function should be omitted.

Combining all the factors given in (3), we get

$$f(x)=\frac{P_1(dx)}{P_0(dx)}=\left(\frac{\lambda}{\lambda_0}\right)^{B(t)}\left(\frac{\mu}{\mu_0}\right)^{D(t)}\exp[-(\lambda+\mu-\lambda_0-\mu_0)S(t)] \tag{4}$$

so that we get immediately the following from the likelihood ratio convergence theorem.

THEOREM 1

For the birth-and-death process with constant intensities the Radon-Nikodym derivative is given by (4).

This gives us the following corollary.

Corollary 1 The ML-estimators of λ and μ are

$$\lambda^* = \frac{B(t)}{S(t)}$$

$$\mu^* = \frac{D(t)}{S(t)} \tag{5}$$

Proof Just differentiate (4) with respect to λ and μ, respectively, and put the derivative equal to zero. These equations are solved by (5). Q.E.D.

Following Keiding (1975) we can now find the asymptotic properties of the estimators.

THEOREM 2

When $t \to \infty$, in the supercritical case, we have a.c.

$$\lim_{t\to\infty} \exp[-(\lambda-\mu)t] \operatorname{col}[B(t), D(t), S(t)] = \frac{W}{\lambda-\mu} \operatorname{col}(\lambda, \mu, 1) \tag{6}$$

Proof We can consider $x(t)$ as a branching process with growth parameter $\lambda - \mu$ and life length distribution function $1-e^{-\mu\lambda}$. Appealing to Jagers (1974), corollaries 1, 2, and 5, the limit in (6) follows directly. Q.E.D.

Corollary 2 Conditioned by the event $x(t) \to \infty$ the ML-estimates are consistent.

Proof We just have to use (6). If $W>0$, which is the same as $x(t) \to \infty$, it is clear that

$$\operatorname{col}[B(t), S(t)] \sim \exp[(\lambda-\mu)t] \frac{W}{\lambda-\mu} \operatorname{col}(\lambda, 1)$$

so that using (5) we see that $\lambda^* \to \lambda$ a.c. Similarly for μ^*. Q.E.D.

THEOREM 3

When $t\to\infty$, in the supercritical case, we have, conditioned by $W>0$, the limit distributionwise:

$$\operatorname{col}[S^{1/2}(t)(\lambda^*-\mu)\lambda^{-1/2}, S^{1/2}(t)(\mu^*-\mu)\mu^{-1/2},$$
$$(\lambda-\mu)S(t)\exp[-(\lambda-\mu)t] \to \operatorname{col}(A, B, W) \tag{7}$$

where A, B, W are independent, and A and B are standardized normal distributions.

Proof We already know that the third component of the vector in (7) behaves as stated (see Theorem 2).

Introduce now a compound two-dimensional Poisson process $[Q(t), R(t)]$ constructed as follows. Let $U(t)$ be a Poisson process with intensity $\lambda+\mu$. At each event of $U(t)$ let $(M, N)=(1,0)$ and $(0,1)$ with probabilities $\lambda/(\lambda+\mu)$ and $\mu/(\lambda+\mu)$, respectively. If (M_i, N_i), $i=1,2,\ldots$, are independent realizations of (M, N), let

$$Q(t)=\sum_{i=1}^{U(t)} M_i$$

$$R(t)=\sum_{i=1}^{U(t)} N_i$$

Since $E(M_i)=\lambda/(\lambda+\mu)$, $E(N_i)=\mu/(\lambda+\mu)$, we have

$$E[Q(t), R(t)]=t(\lambda, \mu)$$

and also

$$\operatorname{Var}[Q(t)]=(\lambda+\mu)tE(M^2)=\lambda t$$

$$\operatorname{Var}[R(t)]=(\lambda+\mu)tE(N^2)=\mu t$$

$$\operatorname{Cov}[Q(t), R(t)]=(\lambda+\mu)tE(MN)=0$$

But $[Q(t), R(t)]$ has independent increments so that the central limit theorem implies that, distributionwise,

$$\left[\frac{Q(t)-\lambda t}{\sqrt{\lambda t}}, \frac{R(t)-\mu t}{\sqrt{\mu t}}\right]\to N \tag{8}$$

where N is a two-dimensional normal stochastic variable with means zero, standard deviations 1, and correlation coefficient 0.

To carry out a random time change of $[Q(t), R(t)]$, we introduce the sojourn times $T_1, T_2, \ldots$ of the $U(t)-$process. Define the sequence $Y_1, Y_2, \ldots$

by

$$Y_i = \begin{cases} +\infty; & \text{if } 1+\sum_{j=1}^{k}(M_j-N_j)=0 \quad \text{for some } k \leqslant i \\ \dfrac{T_i}{1+\sum_{j=1}^{i}(M_j-N_j)}; & \text{otherwise} \end{cases}$$

and the split time process $N(t)$ by

$$\{N(t)=i\} \qquad \text{for } \{Y_i \leqslant t < Y_{i+1}\}$$

In other words, $N(t)$ has intensities equal to $(\lambda+\mu)[\Sigma(M_i-N_i)+1]$, where the sum is extended from 1 to $N(t)-1$. Put

$$[B(t), D(t)] = \sum_{i=1}^{N(t)} (M_i, N_i)$$

Then $X(t)=1+B(t)-D(t)$ is a birth-and-death process with birth intensity λ and mortality intensity μ. Also $X(0)=1$, and $B(t)$ and $D(t)$ are the corresponding numbers of births and deaths, respectively. With $S(t)$ defined as the integral of $X(s)$ over $(0,t)$, we see that $[B(t), D(t)] = \{Q[S(t)], R[S(t)]\}$.

In the asymptotic relation (8) we now replace t by $S(t)$ and get

$$[Q[S(t)]-\lambda S(t)][\lambda S(t)]^{1/2}, [R[S(t)]-\mu S(t)]-S(t)][\mu S(t)]^{1/2}$$
$$= [S^{1/2}(t)(\lambda^* - \lambda)\lambda^{-1/2}, S^{1/2}(t)(\mu^* - \mu)\mu^{-1/2}]$$

One can show that the asymptotic normality still holds after the random time change [see Jagers (1974), as well as the asymptotic independence in the statement of the theorem]. Q.E.D.

Corollary 3 Under the same conditions we have in conditional distribution

$$\left[\frac{e^{(\lambda-\mu)t}-1}{\lambda-\mu}\right]^{1/2} [(\lambda^*-\lambda)\lambda^{-1/2}, (\mu^*-\mu)\mu^{-1/2}] \to W^{-1/2}(A,B) \quad (9)$$

Proof The convergence in distribution in (9) follows directly from (5) and (7). The marginal distributions on the right-hand side of (9) are Student's distributions with 2 degrees of freedom. Q.E.D.

(See Note 6.3.A for references to other results in this direction.)

4 PROCESSES WITH INDEPENDENT INCREMENTS

Let $x(t), t \in [0, \infty)$, be real valued with independent increments $x(t+h)-x(t)$ whose distributions should depend only upon h. As assumed throughout the chapter, the processes should be continuous in probability. In other words we consider time- and space-homogeneous Markov processes (differential processes). Without essential loss of generality we can also assume that $x(0)=0$ a.c.

The one-dimensional marginal distributions of $x(t)$ are then infinitely divisible so that the characteristic function $\phi(z)$ of $x(t)$ is then given by the celebrated *Lévy-Khinchin formula*. This formula can be expressed in more than one way, and we choose the form [see, e.g., Gnedenko-Kolmogorov (1954), p. 76],

$$\ln \phi(z) = i\gamma tz + t\int_{-\infty}^{\infty}\left(e^{izx}-1-\frac{izx}{1+x^2}\right)\frac{1+x^2}{x^2}G(dx) \tag{1}$$

In (1) the *trend coefficient* γ is a real constant and G is bounded and nondecreasing, normalized by $G(-\infty)=0$. The jump $\sigma^2=G(+0)-G(-0)$ of G at zero is the variance of the normal component, if any, of the process at $t=1$. The rest of the measure induced by G describes the distribution of jumps of the process due to the influence of its Poisson components.

The infinitesimal generator is in the present case expressible in terms of (1). The reader is referred to Feller (1966), chap. XVII, for the careful discussion of this. Our inference problem then has the "parameters" γ and G. Here we therefore deal with abstract inference in the double sense: both the same space and the parameter space are function spaces.

We could now proceed as in Section 2, get conditions for equivalence and an expression for the Radon-Nikodym derivative. Such knowledge is indeed available (see Note 6.4.A) so that we could then try to obtain estimates by the maximum likelihood method as in (2.8), and then prove an equivalent to Theorem 3 of Section 2.

We postpone likelihood based estimation (for continuous value space) of Markov processes to the next section where we encounter a new type of difficulty that is subjected to a deeper study in Part III.

In order to derive estimates of γ and G, having observed the process on $T=[0, L]$, we employ a technique presented in Rubin-Tucker (1959). It uses an auxiliary process $y(b, t)$ with $b \in (-\infty, \infty)$, $t \in [0, \infty)$ from which a process will be derived whose finite-dimensional marginal distributions coincide with those of $x(t)$. It is therefore natural to denote this derived process by the same symbol $x(t)$ as used above.

Let $y(b,t)$ be such that it induces a stochastic set function in $(-\infty,\infty)\times[0,\infty)$ with independent increments. The stochastic variables

$$\int_{A_\nu} y(db,dt);\qquad A_\nu=\text{Borel set}$$

should be independent when the A_ν are disjoint. We ask that the distribution of $y(b,t)$ have a characteristic function $f(z)$ with

$$\ln f(z)=itz\gamma\left[G(\infty)-\sigma^2\right]^{-1}\int\!\!\!\!\!-_{-\infty}^{b-0}G(dx)$$
$$+t\int_{-\infty}^{b-0}\left(e^{izx}-1-\frac{izx}{1+x^2}\right)\frac{1+x^2}{x^2}G(dx)\qquad(2)$$

In (2) the stroke across the integral sign indicates that the point $x=0$ should be left out from the domain of integration. It is clear that (2) defines a legitimate family of probability distributions since it is of the form stipulated by the Lévy-Khinchin formula.

Define the derived process

$$x(t)=\int_{-\infty}^{\infty}y(db,t)$$

The integral coverges in distribution due to the continuity theorem for characteristic functions and hence with probability 1, since the "summands" are independent. The resulting distribution of course coincides with the one of the original process given in (1), since the first integral in (2) tends to $G(\infty)-\sigma^2$ as b tends to infinity.

Introduce the process

$$y_0(t)=y(+0,t)-y(-0,t)\qquad(3)$$

which is Gaussian with mean zero and variance $\sigma^2 t$, and for any positive integer n and arbitrary integers k defined the processes

$$y_{n,k}(t)=\begin{cases}y\left(\frac{1}{n},t\right)-y(+0,t); & \text{for } k=1\\ y\left(\frac{k}{n},t\right)-y\left(\frac{k-1}{n},t\right) & \text{for other integer values of } k\end{cases}\qquad(4)$$

Since $y(b,t)$ had independent increments, it follows that all the y-processes in (3) and (4) are independent of each other and all have independent increments. Also

$$x(t) = y_0(t) + \sum_{k=-\infty}^{\infty} y_{n,k}(t) \tag{5}$$

converging a.c. Relation (5) is basic for the following in that it separates the Gaussian component y_0 from the Poisson components making up the rest of the process.

In the construction of estimates we use the process

$$y(t) = \sum_{k=-\infty}^{\infty} y_{n,k}(t) = x(t) - y_0(t) \tag{6}$$

and define the stochastic variables

$$Q_N^2 = \sum_{j=1}^{k_N} \left[x\left(\frac{j}{k_N}\right) - x\left(\frac{j-1}{k_N}\right) \right]^2 - L_N^2$$

$$L_N^2 = \sum_{j=1}^{k_N} \left[y\left(\frac{j}{k_N}\right) - y\left(\frac{j-1}{k_N}\right) \right]^2 \tag{7}$$

where $\{k_N\}$ is a sequence of integers growing fast enough so that

$$\sum_{N=1}^{\infty} \frac{1}{\sqrt[3]{k_N}} < \infty \tag{8}$$

and where L has been chosen as 1. Any other finite number L_0 would be treated in the same manner. The role of condition in (8) will become clear later.

The first question, constituting a substantial part of the whole solution, is to find an estimate of σ^2, to which a partial answer is given by the following lemma.

Lemma 1 If (8) holds, we have a.c.

$$\lim_{N\to\infty} Q_N^2 = \sigma^2 \tag{9}$$

Proof Expanding the square in the first relation of (7) we get, using (5) and (6),

$$Q_N^2 = \sum_{j=1}^{k_N} \left[y_0\left(\frac{j}{k_N}\right) - y_0\left(\frac{j-1}{k_N}\right) \right]^2 + 2Z_N$$

with

$$Z_N=\sum_{j=1}^{k_N}\left[y_0\left(\frac{j}{k_N}\right)-y_0\left(\frac{j-1}{k_N}\right)\right]\sum_{k=-\infty}^{\infty}\left[y_{n,k}\left(\frac{j}{k_N}\right)-y_{n,k}\left(\frac{j-1}{k_N}\right)\right] \tag{10}$$

Appealing to the independence properties of the Gaussian y_0-process, it follows that $Q_N^2-2Z_N$ has mean σ^2 and variance $2\sigma^4k_N^{-1}$ and therefore converges to σ^2 a.c. as $N\to\infty$ (see Note 6.4.B).

To show that $Z_N\to 0$ a.c., let us consider the sum with n fixed so far:

$$Z_{N,k}=\sum_{j=1}^{k_N}\left[y_0\left(\frac{j}{k_N}\right)-y_0\left(\frac{j-1}{k_N}\right)\right]\left[y_{n,k}\left(\frac{j}{k_N}\right)-y_{n,k}\left(\frac{j-1}{k_N}\right)\right] \tag{11}$$

In (11) each term is the product of two independent stochastic variables, the first of which has mean zero, so that $Z_{N,k}$ has mean zero.

With

$$U_j=y_0\left(\frac{j}{k_N}\right)-y_0\left(\frac{j-1}{k_N}\right)$$

$$V_j=y_{n,k}\left(\frac{j}{k_N}\right)-y_{n,k}\left(\frac{j-1}{k_N}\right)$$

we can express $Z_{N,k}$ as the sum of terms U_jV_j, $j=1,2,\ldots,k_N$. Hence

$$\operatorname{Var}(Z_{N,k})=\frac{\sigma^2}{k_N}\sum_{j=1}^{k_N}E(V_j^2)=\sigma^2E(V_1^2) \tag{12}$$

On the other hand (4) and (2) yield

$$E(V_1^2)=\frac{1}{k_N}\left[\int_{(k-1)/n}^{k/n}(1+x^2)G(dx)+\frac{1}{k_N}\left(\gamma\alpha_{k,n}^2+\int_{(k-1)/n}^{k/n}xG(dx)\right)^2\right]$$

where

$$\alpha_{k,n}=\frac{G(k/n)-G((k-1)/n)}{G(+\infty)-\sigma^2}$$

Thus if k is fixed, we can write

$$\operatorname{Var}(Z_{n,k})=\frac{\sigma^2}{k_N}\left(R+\frac{S}{k_N}\right) \tag{13}$$

where R and S do not depend upon N. With the aid of Chebyshev's inequality we get, for any $\varepsilon>0$, since (8) holds and $Z_{n,k}$ has mean zero,

$$\sum_{N=1}^{\infty} P(|Z_{N,k}|\geqslant\varepsilon)\leqslant\frac{1}{\varepsilon^2}\sum_{N=1}^{\infty}\operatorname{Var}(Z_{N,k})<\infty \tag{14}$$

Because of (14), Borel-Cantelli's lemma implies that $Z_{N,k}\to0$ a.c. as $N\to\infty$, k fixed.

Consider now for large values of N the behavior of

$$Z_N=\sum_{k=-\infty}^{\infty} Z_{N,k} \tag{15}$$

[see (10 and (11)]. If $k\neq0$ and 1, we see from (2) and (4) that the characteristic function of the stochastic variable $y_{n,k}(t)$ has a logarithm equal to

$$iz t\beta_{n,k}+t\int_{(k-1)/n}^{k/n}(e^{izx}-1)\frac{1+x^2}{x^2}G(dx) \tag{16}$$

where

$$\beta_{n,k}=\gamma\frac{G(k/n)-G((k-1)/n)}{G(\infty)-\sigma^2}+\int_{(k-1)/n}^{k/n}x^{-1}G(dx)$$

With $W_{n,k}(t)=y_{n,k}(t)-t\beta_{nk}$ we see that since

$$\beta=\sum_{k\neq0,1}|\beta_{n,k}|\leqslant\gamma\frac{G(+\infty)}{G(+\infty)-\sigma^2}+nG(\infty)<\infty \tag{17}$$

the series

$$\sum_{k\neq0,1}W_{n,k}(t)$$

converges in distribution. Since the terms are stochastically independent, the sum converges a.c. so that $|W_{n,k}(t)|\geqslant1/n$ occurs a.c. only for a finite number of k-values.

Consider the event A_i that i is the largest value of $|k|\geqslant2$ such that $W_{n,k}(t)\neq0$, which means that $|W_{n,k}(t)|\geqslant1/n$. See the form of the characteristic function of $W_{n,k}(t)$ using (16). Also let A_1 be the event that

$W_{n,k}(t)=0$ for all $|k|\geqslant 2$. Because of what we just saw,

$$\sum_{i=1}^{\infty} P(A_i)=P\left(\bigcup_{i=1}^{\infty} A_i\right)=1 \tag{18}$$

However Z_N can be decomposed into the four terms [see (11) and (15)],

$$Z_N=S_N+T_N+Z_{N,1}+Z_{N,0}$$

where

$$S_N=\sum_{j=1}^{k_N}\left[y_0\left(\frac{j}{k_N}\right)-y_0\left(\frac{j-1}{k_N}\right)\right]\sum_{k\neq 0,1}\left[W_{n,k}\left(\frac{j}{k_N}\right)-W_{n,k}\left(\frac{j-1}{k_N}\right)\right]$$

$$T_N=\sum_{j=1}^{k_N}\left\{\frac{1}{k_N}\sum_{k\neq 0,1}\beta_{n,k}\left[y_0\left(\frac{j}{k_N}\right)-y_0\left(\frac{j-1}{k_N}\right)\right]\right\}$$

Relation (17) together with $E(T_N)=0$ gives us, again using Chebyshev's inequality,

$$\sum_{N=1}^{\infty} P\{|T_N|\geqslant\varepsilon\}\leqslant\frac{\beta^2\sigma^2}{\varepsilon}\sum_{N=1}^{\infty} k_N^{-2}<\infty$$

Thus $T_N\to 0$ a.c. as $N\to\infty$. We know already that $Z_{N,1}$ and $Z_{N,0}\to 0$ a.c. But using indicator functions, and noting that $S_N=0$ in A_1,

$$S_N=\sum_{i=2}^{M} S_N 1_{A_i}+S_N B;\qquad B=\bigcup_{i=M+1}^{\infty} A_i$$

We know that for any fixed i, $S_N 1_{A_i}\to 0$ as $N\to\infty$ a.c. Therefore $Z_N\to 0$ a.c. in the set $\cup_{i=1}^{M} A_i$ for any M, and (18) implies that $Z_N\to 0$ a.c. Q.E.D.

This lemma is useful but not quite what we need for the construction of an estimate of σ^2. The term L_N^2 appearing in (7) and implicit in (9) is not observable. We show, instead, that a.c.

$$\lim_{N\to\infty} L_N^2=D^2=\int_T y^2(dt) \tag{19}$$

by which we mean the sum of the squares of the jumps of the y-process.

Lemma 2 Relation (19) holds and D is finite a.c.

Proof Introduce $M_N = k_N^{-1/6}$ so that condition (8) implies the two equivalent inequalities

$$\sum_1^\infty M_N^2 < \infty; \qquad \sum_1^\infty k_N^{-1} M_N^{-4} < \infty \tag{20}$$

Define two processes $U_N(t), V_N(t)$ independent of each other, both with independent increments, and with characteristic functions ϕ_U, ϕ_V such that

$$\ln f_V(z) = t\left[\int_{-\infty}^{M_N} + \int_{M_N}^{\infty} (e^{izx} - 1)\frac{1+x^2}{x^2} G(dx)\right]$$

$$\ln f_U(z) = tzt\gamma_N + t\int_{-M_N}^{M_N}\left(e^{izx} - 1 - \frac{izx}{1+x^2}\right)\frac{1+x^2}{x^2} G(dx)$$

where we have put

$$\gamma_N = \gamma - \left[\int_{-\infty}^{-M_N} + \int_{M_N}^{\infty} \frac{1}{x} G(dx)\right]$$

Of course $y(t) = U_N(t) + V_N(t)$. In this decomposition $V_N(t)$ has as sample functions step functions whose jumps are at least equal to M_N in absolute value.

Put

$$R_N = \sum_{j=1}^{k_N}\left[U_N\left(\frac{j}{k_N}\right) - U_N\left(\frac{j-1}{k_N}\right)\right]^2$$

$$S_N = \sum_{j=1}^{k_N}\left[U_N\left(\frac{j}{k_N}\right) - U_N\left(\frac{j-1}{k_N}\right)\right]\left[V_N\left(\frac{j}{k_N}\right) - V_N\left(\frac{j-1}{k_N}\right)\right]$$

$$T_N = \sum_{j=1}^{k_N}\left[V_N\left(\frac{j}{k_N}\right) - V_N\left(\frac{j-1}{k_N}\right)\right]^2$$

so that [see (7)]

$$L_N^2 = R_N + 2S_N + T_N \tag{21}$$

With arbitrary positive ε introduce the approximation G_ε to G with no mass close to $x=0$:

$$G_\varepsilon(x)=\begin{cases} G(x); & \text{for } x\leqslant -\varepsilon \\ G(-\varepsilon); & \text{for } |x|<\varepsilon \\ G(x)-G(\varepsilon)+G(-\varepsilon); & \text{for } x\geqslant \varepsilon \end{cases} \tag{22}$$

The proof of the lemma consists of three parts:

(a) For N large enough T_N is the sum of the squares of the jumps which are at least equal to M_N in absolute value.

(b) $R_N \to 0$ as $N\to\infty$.

(c) The limit of T_N exists and is finite.

When we have proved these statements, we see that $T_N \to D$ a.c. since $M_N \to 0$. Schwarz's inequality tells us that $S_N^2 \leqslant R_N T_N$ so that (b) and (c) imply that $S_N \to 0$ a.c. But then (21) shows that the statement of the lemma is true.

We begin by proving (b) and let $g(z)$ be the characteristic function of

$$U_{jN}=U_N\left(\frac{j}{k_N}\right)-U_N\left(\frac{j-1}{k_N}\right)$$

so that

$$\ln g(z)=\frac{iz}{k_N}\left[\gamma+\int_{|x|>M_N} x^{-1}G(dx)\right] + k_N^{-1}\int_{-M_N}^{M_N}\left(e^{izx}-1-\frac{izx}{1+x^2}\right)\frac{1+x^2}{x^2}G(dx)$$

Direct calculation of the first four cumulants results in

$$\kappa_1=ik_N^{-1}\left[\gamma+\int_{|x|>M_N} x^{-1}G(dx)+\int_{-M_N}^{M_N} xG(dx)\right]$$

$$\kappa_2=-k_N^{-1}\int_{-M_N}^{M_N}(1+x^2)G(dx)$$

$$\kappa_3=-ik_N^{-1}\int_{-M_N}^{M_N}x(1+x^2)G(dx)$$

$$\kappa_4=k_N^{-1}\int_{-M_N}^{M_N}x^2(1+x^2)G(dx)$$

Then

$$\operatorname{Var}(U_{jN})=\kappa_4+2\kappa_2^2+4\kappa_1\kappa_3+4\kappa_1^2\kappa_2$$

$$\operatorname{Var}(R_N)=\int_{-M_N}^{M_N} x^2(1+x^2)G(dx)+k_N^{-1}B_N$$

where $\{B_N\}$ is bounded by a finite constant B. But (20) implies that

$$\sum_N \operatorname{Var}(R_N)\leqslant[G(M_1)-G(-M_1)]\Big(\sum_N M_N^2+\sum_N M_N^4\Big)+B\sum_N k_N^{-1}<\infty$$

so that a.c., using Kolmogorov's inequality [see, e.g., Feller (1966), pp. 235-236],

$$\lim_{N\to\infty}[R_N-E(R_N)]=0 \tag{23}$$

We can calculate the expected value in (23) and get the somewhat cumbersome expression

$$\begin{aligned}E(R_N)=&-\!\!\!\!\!\!\int_{-M_N}^{M_N}(1+x^2)G(dx)\\&+k_N^{-1}\Bigg\{\gamma^2+2\gamma\int_{|x|>M_N}x^{-1}G(dx)+\left[\int_{|x|>M_N}x^{-1}G(dx)\right]^2\\&+2\gamma-\!\!\!\!\!\!\int_{-M_N}^{M_N}xG(dx)+2\int_{|x|>M_N}x^{-1}G(dx)-\!\!\!\!\!\!\int_{-M_N}^{M_N}xG(dx)\\&+\left[-\!\!\!\!\!\!\int_{-M_N}^{M_N}xG(dx)\right]^2\Bigg\}\end{aligned} \tag{24}$$

Appealing to (20) again, we see that (24) tends to zero when N tends to infinity. Together with (23) this means that $R_N\to 0$ a.c., proving (b).

Now let us prove (a). A_N now stands for the event that at least two jumps of size at least equal to M_N in absolute value occur in at least one of the k_N subintervals of length k_N^{-1},

$$P(A_N)=1-[P(X_N\leqslant 1)]^{k_N}$$

where X_N denotes the number of jumps of size at least M_N in absolute value during a given interval of length k_N^{-1}. According to Doob (1953),

p. 423, X_N has a Poisson distribution with mean

$$\lambda_N = k_N^{-1}\int_{-\infty}^{\infty}\frac{1+x^2}{x^2}G_{M_N}(dx)$$

[see also (22)]. But

$$0 \leqslant \lambda_N \leqslant k_N^{-1}M_N^{-2}(1+M_N^2)K; \qquad K = G(+\infty)-G(-\infty)$$

For simplicity say that $K=1$.

The series

$$\sum_N P(A_N) = \sum_N \left\{1-\left[e^{-\lambda_N}(1+\lambda_N)\right]^{k_N}\right\}$$

converges if and only if the product

$$\prod_0 = \prod_N \left[e^{-\lambda_N}(1+\lambda_N)\right]^{k_N} \tag{25}$$

converges. But (25) converges if and only if

$$P = \sum_N \ln\left[e^{-\lambda_N}(1+\lambda_N)\right]^{k_N} = \sum_N \left[-k_N\lambda_N + k_N\ln(1+\lambda_N)\right]$$

converges absolutely. The latter sum can be seen to equal minus

$$\sum_N \left[k_N\lambda_N - k_N\left(\lambda_N - \frac{\theta_N}{2}\lambda_N^2\right)\right] \leqslant \frac{1}{2}\sum_N k_N\lambda_N^2$$
$$\leqslant \frac{1}{2}\sum_N k_N^{-1}M_N^{-4}(1+2M_N^2+M_N^4) < \infty \tag{26}$$

if we use (20) again; in (26) $|\theta_N| \leqslant 1$. Hence

$$\sum_N P(A_N) < \infty$$

and Borel-Cantelli's lemma tells us that only a finite number of the events A_N occur a.c. This proves (a).

Finally we prove (b). The process $V_N(t)$ has a.c. only a finite number of jumps. On the other hand $U_N(t)$ has independent increments with finite variance, so that the sum of squares of its jumps has finite mean. Using (a), we get (c). Q.E.D.

Combining the lemmas we get directly the following lemma.

Lemma 3 When $N\to\infty$, under the given conditions, the quantity

$$Q_N'^2=\sum_{j=1}^{k_N}\left[x\left(\frac{j}{k_N}\right)-x\left(\frac{j-1}{k_N}\right)\right]^2-D^2$$

converges a.c. to σ^2.

Note that $Q_N'^2$ estimates σ^2 for the special interval $(0,1)$, but that any other finite interval suffices for consistent estimation.

We are now ready for the main result. Let $N_t(B)$ be the number of jumps of $x(t)$ during $[0,t]$, $B=$ Borel set on $\mathbb{R}$, with the magnitudes of the jumps in B. Define

$$J_t(x)=\int_{-\infty}^{x}\frac{b^2}{1+b^2}N_t(db)$$

and introduce the estimate, having observed the process in $(0,L)$, $L>L_0$,

$$G^*_{N,L}(u)=\begin{cases}L^{-1}J_L(u); & u<0\\ L^{-1}J_L(u)+L_0^{-1}(S_N^2-D^2); & u>0\end{cases}\tag{27}$$

where

$$S_N^2=\sum_{k=1}^{k_N}\left[x\left(\frac{kL_0}{k_N}\right)-x\left(\frac{k-1}{k_N}L_0\right)\right]^2$$

We can then announce the following theorem [see Rubin-Tucker (1959)].

THEOREM 1

The estimate $G^*_{N,L}$ is consistent (with probability 1) as $N\to\infty$ first and then $L\to\infty$. Also $G^*_{N,L}(b)-G^*_{N,L}(a)$ is an unbiased estimate of $G(b)-G(z)$ if $0\notin[a,b]$.

Proof Consider, for $x\geqslant y$,

$$J_t(x)-J_t(y)-\int_y^x G(db)=H_t(x,y)\tag{28}$$

But if $(y, x]$ does not contain zero, $N_t([y, x)]$ is finite a.c. and has expected value

$$t\int_{y+0}^{x} \frac{1+b^2}{b^2} G(db) \tag{29}$$

However, $H_t(x, y)$ is a martingale in x and in $-y$ [see Doob (1953), p. 437]. Since

$$E[|H_t(x, y)|] \leqslant t \int G(db)$$

it follows that $H_t(x, y)$ can be extended to the case when $(y, x]$ contains zero and for $y \to -\infty$. Thus $J_t(x)$ is well defined and estimates the quantity

$$t\int_{-\infty}^{x} G(db)$$

without bias. When L increases, the contributions to J_L due to disjoint subsets of $[0, L]$ are independent, with distributions only depending upon their Lebesgue measures. Then the strong law of large numbers guarantees that $G_{N,t}^*$ converges as stated. Remember that the second term on the right-hand side of (27) converges to σ^2 as shown in the lemmas. This concludes the proof of Theorem 1 (see Note 6.4.C). Q.E.D.

It remains to estimate the trend parameter γ which will be done through the estimate

$$\gamma_L^* = \frac{1}{L}\left[x(L) - \sum_n \frac{J_n^3}{1+J_n^2} \right] \tag{30}$$

where $\{J_n\}$ is the set of jumps in $(0, L)$ with arbitrary order of enumeration. Using $N_t(b)$, (30) can be expressed as the integral

$$\gamma_L^* = \frac{1}{L}\left[x(L) - \int_{-\infty}^{\infty} \frac{b^3}{1+b^2} N_L(db) \right] \tag{31}$$

We then have the following theorem.

THEOREM 2 RUBIN-TUCKER (1959)

The estimate γ_L^* is an unbiased estimate of γ and consistent: $\gamma_L^* \to \gamma$ a.c. as $L \to \infty$.

Proof Starting with $L=1$ we can again decompose $x(t)$ into a Gaussian term $u(t)$, one $v(t)$ with small jumps, and another $w(t)$ with big ones, all independent of each other and with independent increments. If f_u, f_v, f_w stand for the associated characteristic functions, they can be given in the Lévy-Khinchin form as done several times during this section:

$$\begin{aligned}
\ln f_u(z) &= -\frac{\sigma^2 z^2}{2} \\
\ln f_v(z) &= i\gamma_1 z + \int_{-K}^{K} \left(e^{izx} - 1 - \frac{izx}{1+x^2}\right)\frac{1+x^2}{x^2} G(dx) \\
\ln f_w(z) &= i\gamma_2 + \int_{|x|>K} \left(e^{izx} - 1 - \frac{izx}{1+x^2}\right)\frac{1+x^2}{x^2} G(dx)
\end{aligned} \tag{32}$$

where K is a constant such that K and $-K$ are points of continuity of G, and $\gamma_1+\gamma_2=\gamma$.

Differentiating the second equation in (32) with respect to z and putting $z=0$, we get the expected value

$$\frac{1}{i}\left(\frac{\partial \ln f_v(z)}{\partial z}\right)_{z=0} = \gamma_1 + \int_{-K}^{K} xG(dx) = E[V(1)]$$

Recalling the discussion between (28) and (29), we see that the stochastic variable

$$\bar{\gamma}_1 = V(1) - \int_{-K}^{K} \frac{x^3}{1+x^2} N_1(dx)$$

has expected value

$$E(\bar{\gamma}_1) = \gamma_1 + \int_{-K}^{K} xG(dx) - \int_{-K}^{K} \frac{x^3}{1+x^2}\frac{1+x^2}{x^2} G(dx) = \gamma_1$$

On the other hand the third equation in (32) can be given the form

$$\ln f_w(z) = iz\left[\gamma_2 - \int_{|x|>K} x^{-1} G(dx)\right] + \int_{|x|>K} (e^{izx}-1)\frac{1+x^2}{x^2} G(dx)$$

so that we can represent the stochastic variable $W(1)$ as

$$W(1) = \gamma_2 - \int_{|x|>K} x^{-1} G(dx) + \int_{|x|>K} xN_1(dx) \tag{33}$$

We know already that

$$E\left[\int_{|x|>K}\frac{x}{1+x^2}N_1(dx)\right]=\int_{|x|>K}x^{-1}G(dx)$$

and hence the new stochastic variable

$$\bar{\gamma}_2=W(1)-\int_{|x|>K}\frac{x^3}{1+x^2}N_1(dx)$$

has expectation [use (33)]

$$E(\bar{\gamma}_2)=\gamma_2-\int_{|x|>K}x^{-1}G(dx)+E\left[\int_{|x|>K}\left(x+\frac{x^3}{1+x^2}\right)N_1(dx)\right]$$

$$=\gamma_2-\int_{|x|>K}x^{-1}G(dx)+\int_{|x|>K}\frac{x}{1+x^2}\,\frac{1+x^2}{x^2}G(dx)=\gamma_2$$

Using the relation $\gamma=\gamma_1+\gamma_2$ we see, adding $\bar{\gamma}_1$ and $\bar{\gamma}_2$, that γ_L^* in (31), for $L=1$, which equals

$$x(1)-\int_{-\infty}^{\infty}\frac{x^3}{1+x^2}N_1(dx)$$

has expectation γ. Similarly for other positive values of L.

To show consistency we just apply the strong law of large numbers again. Q.E.D.

5 ESTIMATION FOR DIFFUSION PROCESSES

Let our process $x(t)$ be real valued, defined on a finite time interval, say $T=(0,1)$, and be of the diffusion type. Consider the stochastic differential equations (see Note 6.5.A) satisfied under the null and alternative hypotheses, respectively:

$$dx_t=a(x_t)\,dt+b(x_t)W(dt);\qquad \text{under } H_0$$

$$dx_t=[a(x_t)+b(x_t)h(x_t)]\,dt+b(x_t)W(dt);\qquad \text{under } H_1 \tag{1}$$

In (1) W denotes the Wiener process as usual, $a(\cdot)$ and $h(\cdot)$ are Lipschitz, $b(\cdot)$ is continuous and positive.

Are the two measures equivalent, and if so, what is their Radon-Nikodym derivative? It is not difficult to guess what form it should take if it exists.

Heuristically, for a fine partition $0=t_0<t_1<\ldots<t_n<t_{n+1}=1$ we would have, under H_0,

$$
\begin{aligned}
&\Delta x_\nu - a[x(t_\nu)]\Delta t_\nu \cong N(0, \Delta t_\nu b^2[x(t_\nu)])\\
&\Delta x_\nu = x(t_\nu) - x(t_{\nu-1})\\
&\Delta t_\nu = t_\nu - t_{\nu-1}
\end{aligned}
$$

with independence between the Δx_ν. Similarly for the alternative hypothesis H_1. Hence the finite-dimensional likelihood ratio would be approximately $\exp \psi_n$ with

$$
\begin{aligned}
\psi_n = {}& \frac{1}{2}\sum_{\nu=1}^{n+1} \{\Delta x_\nu - a[x(t_\nu)]\Delta t_\nu\}^2 \frac{1}{b^2[x(t_\nu)]\Delta t_\nu}\\
&- \frac{1}{2}\sum_{\nu=1}^{n+1} \{\Delta x_\nu - a[x(t_\nu)]\Delta t_\nu - b[x(t_\nu)]h[x(t_\nu)]\Delta t_\nu \frac{1}{b^2[x(t_\nu)]\Delta t_\nu}
\end{aligned}
\tag{2}
$$

This yields immediately

$$
\psi_n = \sum_{\nu=1}^{n+1} \left\{ \Delta x_\nu \frac{h[x(t_\nu)]}{b[x(t_\nu)]} - \frac{h[x(t_\nu)]a[x(t_\nu)]}{b[x(t_\nu)]}\Delta t_\nu - \frac{1}{2}h^2[x(t_\nu)]\Delta t_\nu \right\}
\tag{3}
$$

Making the partition finer we expect ψ_n to tend to a limit given in terms of the integrals corresponding to the Riemann sums in (3).

Indeed, it has been shown under much more general conditions in Girsanov (1960) (see also Note 6.5.B) that the two measures are absolutely continuous with the Radon-Nikodym derivative

$$
f(x) = \frac{P_1(dx)}{P_0(dx)} = \exp\left[\int_0^1 h(x_t)\frac{x(dt)}{b(x_t)} - \int_0^1 h(x_t)\frac{a(x_t)}{b(x_t)}\,dt - \frac{1}{2}\int_0^1 h^2(x_t)\,dt \right]
\tag{4}
$$

To test the null hypothesis against the alternative hypothesis, the expression in (4) gives us immediately the most powerful test of H_0 against H_1 (see Theorem 2 in Chapter 3, Section 1). The corresponding estimation problem is, however, not as easily dealt with. Then our "parameter" α is the function $h(\cdot)$, and we want to estimate it having observed a sample function $x(t)$, $t \in T$. Again both sample space and parameter space are function spaces.

Since we have the Radon-Nikodym derivative in closed form in (4), it is natural to try a likelihood based estimate. Therefore let us try to find the ML solution (see Note 6.5.C for other methods).

With some loss of generality let $a(x)\equiv 0$. The unknown function $h(x)$ shall be assumed, to begin with, to be of the form

$$h(x)=\sum_{\nu=1}^{p}\alpha_\nu\phi_\nu(x) \tag{5}$$

where the α (are unknown parameters and the ϕ_ν given linearly independent functions with the Lipschitz property. Then the ML solution of h is simply

$$h^*(x)=\sum_{\nu=1}^{p}\alpha_\nu^*\phi_\nu(x) \tag{6}$$

where the α_ν^* solve

$$\sum_{\nu=1}^{p}\alpha_\nu x_\nu-\frac{1}{2}\sum\alpha_\nu\alpha_\mu h_{\nu\mu}=\max$$

$$x_\nu=\int_0^1\phi_\nu(x)x(dt)$$

$$h_{\nu\mu}=\int_0^1\phi_\nu[x(t)]\phi_\mu[x(t)]\,dt$$

With the natural notation

$$x=\operatorname{col}(x_\nu)$$

$$H=\{h_{\nu\mu};\ \nu,\mu=1,2,\ldots,p\}$$

we get

$$\alpha^*=\operatorname{col}(\alpha_\nu^*)=H^{-1}x \tag{7}$$

Since (7) and (6) give the ML estimates in closed form, the properties of h^* can be obtained, in particular the asymptotic behavior of the estimate when the observation interval (0, 1) is replaced by (0, L), and L is made to tend to infinity.

Note that in this case the ML solution exists and is unique. It can be shown to be consistent [see Brown-Hewitt (1975)].

Arguing heuristically, treating the sample functions as smooth, we can go ahead as follows when h is not restricted to be of the form in (5), but still with $b \equiv 1$. For observed $(\cdot)$ we want to maximize the exponent in (4) as a function of $h(\cdot)$:

$$\int_0^1 h(x_t) f(x_t, x_t')\,dt - \frac{1}{2}\int_0^1 h^2(x_t)\,dt = \max_h$$

$$f(x, y) = y - a(x) \tag{8}$$

If $x \leqslant x(t) \leqslant x + \varepsilon$ in the set

$$T(x, \varepsilon) = \bigcup_1^{n(x,\varepsilon)} i_\nu$$

where i_ν is an interval with left end point t_ν, then the contribution to (2) from $T(x, \varepsilon)$ is approximately

$$h(x) \sum_1^{n(x,\varepsilon)} f(x, x_{t_\nu}')m(i_\nu) - \tfrac{1}{2}h^2(x) \sum_1^{n(x,\varepsilon)} m(i_\nu); \qquad t_\nu' \in i_\nu \tag{9}$$

But (9) is maximized by choosing [see (8)]

$$h^*(x) = \frac{\sum_1^{n(x,\varepsilon)} f(x, x_{t_\nu'})m(i_\nu)}{\sum_1^{n(x,\varepsilon)} m(i_\nu)}$$

$$= \frac{\sum_1^{n(x,\varepsilon)} x_{t_\nu'} m(i_\nu)}{\sum_1^{n(x,\varepsilon)} m(i_\nu)} - a(x) \tag{10}$$

On the other hand $m(i_\nu)|x_{t_\nu'}|$ is approximately ε if ε is small enough so that we are led to, when $\varepsilon \downarrow 0$,

$$h^*(x) = \frac{\sum_1^{n(x)} \pm 1}{\sum_1^{n(x)} 1/|x_{t_\nu'}|} - a(x) \tag{11}$$

where $n(x)$ is the number of times the sample function $x(t)$ crosses the level x. Introducing $n^+(x)$ and $n^-(x)$ as the number of crossings with positive and negative slope, respectively, we can write our estimate (11) as

$$h^*(x)=\frac{n^+(x)-n^-(x)}{n(x)}H(x'_{t_1},x'_{t_2},\ldots)-a(x) \tag{12}$$

where H represents the harmonic mean of the slope x'_t at the x-crossing abscissas.

Since $x(t)$ is not differentiable, *this reasoning is not rigorous and neglects the complicated behavior of sample functions at level crossings*. It gives us a valuable indication of how to proceed rigorously, however, which is why we have included it. The supremum of the Radon-Nikodym derivative when the parameter h varies is what we need for the ML solution. *The supremum is not attained* in the parameter space of Lipschitz functions. Therefore *we are led to conditioning the maximum by a "sieve,"* a method that is studied extensively in Part III (also see Note 6.5.B).

Also, the heuristics indicate that a promising estimate could be obtained when observing $x(t)$ over $T=(0, L)$ by replacing (12) by (10) for positive ε and where x'_t should be replaced correspondingly by a difference ratio. The important thing to keep in mind in order to make such an estimate behave well is to let ε tend to zero slowly enough when L is made to tend to infinity.

We return to many more problems with infinite-dimensional parameter space in Chapter 8 and then treat them rigorously.

6 LIKELIHOOD ESTIMATION FOR SOME GENERAL MARKOV PROCESSES

Among classical principles of estimation the maximum likelihood (ML) method is without doubt the most important. The question naturally arises what is its role in abstract inference. Most of Chapters 8 and 9 are devoted to this topic for abstract parameter spaces. Here we study the asymptotic properties of the ML estimate for real valued stochastic processes on $T=(0, L]$, but only for a real-valued parameter α.

It is well known that under appropriate regularity conditions the ML estimate is asymptotically efficient when $X=\mathbb{R}^n$, $A=\mathbb{R}$ [see Cramér (1945), pp. 501–503]. We attempt to extend this to the case when we observe a stationary process with a generalized Markov property.

For L equal to a large natural number say that we have represented the sample functions by a reference space Ω, $x(t,\omega)$ with $\omega\in\Omega$, in the following way. For each interval $(k-1,k]$, $k\in\mathbb{N}$, we use a representing parameter $\omega_k \in \Omega_0$ of the segment $x(t,\omega)$, $t\in(k-1,k]$ of the sample function. We use the same representation in all the intervals, in other words, a translation invariant one, and for ω, corresponding to T, we write $\omega = (\omega_1, \omega_2, \ldots, \omega_{L-1})$.

Let $\mathcal{A}_0$ be a σ-algebra over Ω_0 and use $\mathcal{A} = \mathcal{A}_0 \times \mathcal{A}_0 \times \ldots \times \mathcal{A}_0$ (L times) over Ω. Note that the P_α-measures should be given on the same σ-algebra for all $\alpha \in \mathbb{R}$.

We also need the σ-algebras $\mathcal{B}_k$ over Ω_0 generated by $\mathcal{A}_0 \times \mathcal{A}_0 \times \ldots \times \mathcal{A}_0$ (k times); these sub-σ-algebras are needed for the conditional distributions.

Assume now that the ω_k form a Markov process. Since $x(t) = x(t,\omega)$ is stationary, and since the ω_ν-representation is invariant with respect to translations of the time axis, it follows that the ω_k-process is also stationary. *We require for the following analysis that the process also be ergodic.*

With these assumptions as well as the obviously needed equivalence of all P_α-measures the maximum likelihood method takes the usual form

$$f(\omega;\alpha) = f(x;\alpha) = \frac{P_\alpha(dx)}{P_0(dx)} = \max_\alpha$$

and we first express this Radon-Nikodym derivative in a more suitable form.

For arbitrary sets $S_1 \in \mathcal{A}$, $S_2 \in \mathcal{B}_k$ we have, using the Markov property and obvious notation,

$$P_\alpha(S_1 \cap S_2) = \int_{S_1} P_\alpha(\omega_L \in S_2 | \omega_1, \ldots, \omega_{L-1}) P_\alpha[d(\omega_1, \omega_2, \ldots, \omega_{L-1})]$$

$$= \int_{S_1} P_\alpha(\omega_L \in S_2 | \omega_{L-1}) f(\omega_1, \omega_2, \ldots, \omega_{L-1}; \alpha)$$

$$P_0[d(\omega_1, \omega_2, \ldots, \omega_{L-1}); \alpha] \tag{1}$$

In (1) f denotes the Radon-Nikodym of P_α with respect to P_0, both measures restricted to the segment $x(t)$, $t\in(0, L-1]$.

But the probability in (1) can also be written as

$$P_\alpha(S_1 \cap S_2) = \int_{S_1 \cap S_2} f[(\omega_1, \omega_2, \ldots, \omega_L); \alpha] P_0[d(\omega_1, \omega_2, \ldots, \omega_L); \alpha]$$

so that a.c.

$$P_\alpha(\omega_L \in S_2 | \omega_{L-1}) = \int_{S_2} \frac{f[(\omega_1, \ldots, \omega_L); \alpha]}{f[(\omega_1, \ldots, \omega_{L-1}); \alpha]} P_0(d\omega_L | \omega_{L-1}) \tag{2}$$

Note that since all our measures are equivalent, we need not qualify the expression "almost certainly" by referring it to any particular α-value.

From (2) it follows that the ratio on the right under the integral sign a.c. depends only upon ω_{L-1}, ω_L, and α; we denote it by $f(\omega_L|\omega_{L-1};\alpha)$ and similarly for other values $L-1$, $L-2$, etc. Hence we can write

$$f(\omega;\alpha)=f(\omega_1,\omega_2,\dots,\omega_L;\alpha)=f(\omega_1;\alpha)f(\omega_2|\omega_1;\alpha)\dots f(\omega_L|\omega_{L-1};\alpha) \quad (3)$$

where the first factor on the right-hand side represents the absolute Radon-Nikodym derivative over (0, 1], and the others are conditional Radon-Nikodym derivatives. Since the process is stationary, we need not distinguish the last $L-1$ factors by subscripts, they are all given by the same function of the three arguments.

Taking logarithms in (3), $f>0$ a.c., we get

$$\frac{1}{L}\ln f(\omega;\alpha)=\frac{1}{L}\ln f(\omega_1;\alpha)+\frac{1}{L}\sum_2^L \ln f(\omega_k|\omega_{k-1};\alpha)$$

This makes it possible to apply the technique used in Cramér (1945) putting

$$\begin{aligned}
B_0&=\frac{1}{L}\left(\frac{\partial \ln f(\omega_1;\alpha)}{\partial\alpha}\right)_0+\frac{1}{L}\sum_2^L\left(\frac{\partial \ln f(\omega_k|\omega_{k-1};\alpha)}{\partial\alpha}\right)_0\\
B_1&=\frac{1}{L}\left(\frac{\partial^2 \ln f(\omega_1;\alpha)}{\partial\alpha^2}\right)_0+\frac{1}{L}\sum_2^L\left(\frac{\partial^2 \ln f(\omega_k|\omega_{k-1};\alpha)}{\partial\alpha^2}\right)_0 \qquad (4)\\
B_2&=\frac{1}{L}H(\omega_1)+\frac{1}{L}\sum_2^L H(\omega_k|\omega_{k-1})
\end{aligned}$$

with the analogous conditions as in the quoted reference for α in an open interval containing the true value α_0:

1. The derivatives with respect to α up to order 3 exist a.c.
2. For every α we have

$$\left|\frac{\partial f}{\partial\alpha}\right|\leqslant F_1, \qquad \left|\frac{\partial^2 f}{\partial\alpha}\right|\leqslant F_2, \qquad \left|\frac{\partial^3 \ln f}{\partial\alpha}\right|\leqslant H \qquad (5)$$

where H is a stochastic variable, where F_1 and F_2 are integrable with respect to P_0 and $E_\alpha(H)\leqslant M<\infty$, M independent of α.

3. For every α we have

$$E_\alpha\left[\left(\frac{\partial \ln f}{\partial \alpha}\right)^2\right]$$

finite and positive.

The zeros attached to the parentheses in (4) indicate that the expressions should be evaluated at $\alpha = \alpha_0 =$ the true value of α.

The ML-equation then takes the form

$$B_0 + B_1(\alpha - \alpha_0) + \frac{1}{2}\theta(\alpha - \alpha_0)^2 B_2 = 0; \qquad |\theta| \leqslant 1$$

But the stochastic variables B_0, B_1, B_2 converge in probability, as $L \to \infty$, to the limits

$$B_0 \to 0$$

$$B_1 \to -E_0\left[\left(\frac{\partial \ln f}{\partial \alpha}\right)_0^2\right]$$

$$B_2 \to E_0[H(x)] \leqslant M$$

Indeed the terms in the three sums in (4), which define our stochastic variables B_0, B_1, B_2, form ergodic stationary stochastic processes. Because of the ergodicity of the process we can appeal to the ergodic theorem, ensuring convergence as stated. The rest of the proof goes just as in Cramér (1945), so that we have the following theorem (see Note 6.6.A).

THEOREM 1

Consider a real-valued process, depending upon a real time argument and with measures dependent upon a real parameter. If the process is stationary and ergodic and has the Markov property described, and if it satisfies the conditions in (5), the the ML equation, for α in some open interval containing α_0, has a consistent and asymptotically efficient solution.

PART 3

Inference in Abstract Parameter Space

CHAPTER 7

Regularization in Abstract Parameter Space

1 PARAMETRIC METHODS FOR NONPARAMETRIC QUESTIONS

When we allow the parameter to take values in an infinite-dimensional abstract space, it will often happen that the methods of traditional inference do not apply, or that they need substantial modification in order to work under the new conditions. This problem area is at present almost *terra incognita*, but we shall try to shed some light on it in this and the following two chapters.

The general ideas in mathematical statistics can be divided into *likelihood based methods*, underlying most techniques of inference, and *direct methods*, (see the Introduction for a definition of these terms). Some of the direct methods are examined in more detail in Sections 3–5.

The oldest likelihood based method in statistics, Bayesian inference, assumes that (a) the *a priori* probabilities exist, and (b) they are known at least approximately. If (a) and (b) hold, many mathematicians, including the author, consider the Bayesian approach preferable to any other.

If A is some general topological space, not just $\mathbb{R}^n$ or a subset of it, then our knowledge is very imperfect about the behavior of Bayesian estimates and tests. We briefly examine such questions in Chapter 8, but only for certain spaces.

For the other likelihood based methods, such as maximum likelihood estimation and Neyman-Pearson theory, we show in the next chapter by studying a great number of cases that these methods have to be modified in order to work in abstract parameter cases. It can happen, however, that they work without modification in certain spaces and, in Chapter 8, where a large number of cases is treated, we search for the boundary separating their domain of applicability from the region where they have to be modified. This leads up to some general results to be presented in Chapter 9.

What is the *raison d'être* of ML estimation? Most mathematical statisticians would advocate their use of this method by pointing to various *optimality properties* which it possesses. Under certain conditions it leads to consistent estimates, under stronger conditions it is even asymptotically efficient, it can detect sufficient estimates, and so on.

According to another view, ML estimation makes sense because *it can be regarded as an instance of Bayesian estimation*, at least if one allows priors that are not legitimate probability measures. Note that seen from that perspective we are basing our procedure on the *mode* of the *a posteriori* distribution; not the mean or related quantities.

There is, however, a third way of arguing in favor of ML-estimation, namely, to take it as a *first principle* of statistical inference. After all, we know that the ML method performs well also in many cases where the conditions break down under which optimality results have been established. At the same time there exist a multitude of examples, some of them ingenious, intended to show the opposite, that these estimates do not have desirable properties; but these examples often seem contrived. Therefore *the maximum likelihood principle seems to deserve the role as a fundamental principle of estimation theory*; this will be our position in Part III. Whatever view one adopts, it is clear that ML-estimation will play an important role in inference in abstract spaces.

In order to be able to study the solution of maximum likelihood problems, we must make sure that such a solution, when it exists, is measurable. The frequency function $f=f(x;\alpha)$ with respect to a fixed σ-finite measure dx in X is therefore assumed to be a separable stochastic process when α plays the role of the time parameters (see Chapter 1, Section 1, where Theorem 1 assures that this is no essential restriction).

If we are fairly close to the traditional case of $A=\mathbb{R}^p$, for example, when A is locally compact, one would expect that the ML solution would have good properties. This is indeed the case under conditions to be stated below. If A is not locally compact the situation requires more thought and will be seen to lead to interesting consequences.

Most of the classical proofs used in the study of ML estimates assume strong smoothness properties for f. An alternative approach due to Wald (1949) (see also Note 7.1.A) avoids this and is well suited for use in abstract spaces.

Let us outline *the main idea in Wald's proof*. Let $f(x,\alpha)$ be a frequency function, x and $\alpha\in\mathbb{R}$, and consider the *formal entropy*

$$H_\alpha=-\int_{-\infty}^{\infty}\ln f(x;\alpha)f(x;\alpha)\,dx$$

as well as the related

$$H_{\alpha,\beta} = -\int_{-\infty}^{\infty} \ln f(x;\beta) f(x;\alpha)\,dx$$

It is well known that, under mild conditions, $H_{\alpha,\beta} > H_\alpha$ unless $f(x,\beta)$ is essentially equal to $f(x;\alpha)$.

Indeed, if $g(x)$ is a nonnegative measurable function with finite integral with respect to dx-measure the classical inequality between the arithmetic and geometric means with respect to P_α-measure tells us that $G_\alpha(g/f) \leqslant A_\alpha(g/f)$ or, taking logarithms,

$$\int_X \ln\left[\frac{g(x)}{f(x,\alpha)}\right] f(x,\alpha)\,dx \leqslant \ln\left[\int_X g(x)\,dx\right] \tag{1}$$

If g is also a frequency function the integral on the right of (1) equals 1, so that the left-hand side is now nonpositive. We also know that equality holds in (1) if and only if g and f are essentially proportional. Note that the left-hand side is allowed to diverge to $-\infty$.

Consider now the terms in

$$-\frac{1}{n}\ln L(\beta) = \frac{1}{n}\sum_{i=1}^{n}\left[-\ln f(x_i,\beta)\right]$$

They form an i.i.d. sample with expected value $H_{\alpha,\beta}$. If we maximize $L(\beta)$ and use the law of large numbers, we would expect that the resulting β-value should be close to the one that minimizes $H_{\alpha,\beta}$, that is, the true value α.

Naturally several things can go wrong in this heuristic argument, and we see repeatedly in Chapter 8 that modifications are needed.

If the parameter space A is compact and if the likelihood function is not too badly behaved, one would expect the ML solution to exist (uniquely or not) and the set of solutions to converge a.c. to the true value of the parameter. It is therefore natural to compactify A and see if existence and consistency can still be guaranteed. This can be done in several ways; a thought-provoking method was suggested in Bahadur (1966) based on the notion of a "*suitable compactification*" [see also Bahadur (1971)].

Definition 1 A compact metric space $\bar{A}$ is said to be a suitable compactification if:

(a) A is an everywhere dense subset of $\bar{A}$

(b) For any $\alpha_1 \in \bar{A}$, the function

$$f(x, \alpha_1, \varepsilon) = \sup\{f(x,\alpha) | \alpha \in A, d(\alpha, \alpha_1) < \varepsilon\}$$

is measurable in x for all ε small enough.

(c) Defining $f(x, \alpha_1, 0) = \lim_{\varepsilon \to 0} f(x, \alpha_1, \varepsilon)$ we have

$$\int_X f(x, \alpha_1, 0)\mu(dx) \leqslant 1; \qquad \text{any } \alpha_1 \in \bar{A} \tag{2}$$

In (2) $d(\cdot,\cdot)$ stands for the distance function in the metric space A.

The following notation will be useful. If $B \subseteq A$ and h is an extended real-valued function, we write $h(B) = \sup\{h(\alpha) | \alpha \in B\}$. Also for any set $C \subset \bar{A}$ let $C^0 = C \cap A$.

We now consider the likelihood function

$$L_n(x; \alpha) = \prod_{r=1}^{n} f(x_r; \alpha); \qquad x = (x_1, x_2, \ldots, x_n)$$

and form the set of ML-solutions, possibly empty,

$$\hat{A}_n = \{\alpha | \alpha \in A, L_n(x, \alpha) = L_n(x, A)\}$$

Bahadur's result can then be stated as the following theorem.

THEOREM 1.

Assume that:

(a) *there exists a suitable compactification* $\bar{A}$, *and*

(b) $E_\alpha\left[\ln \dfrac{f(x; A)}{f(x; \alpha)}\right] < \infty$ *for any* $\alpha \in A$.

(c) *For* $\alpha \in A$, $\alpha_1 \in \bar{A}$ *with* $\alpha \neq \alpha_1$ *we have*

$$\mu\{x | f(x, \alpha) \neq f(x, \alpha_1, 0)\} > 0.$$

(d) A *is open in* $\bar{A}$.

(e) $f(x, \alpha) = f(x, \alpha, 0)$ *for all* x, *any* $\alpha \in A$.

Under conditions (a)–(e) *it is true with probability* 1 *that* $\hat{A}_n$ *is not empty for* n *large enough, and the set* A_n *converges to the one containing the single point* $\alpha =$ *true value of the parameter.*

Proof Introduce the set in A:

$$A_n^* = \{\alpha_1 | \alpha_1 \in A, L_n(x, \alpha_1) \geqslant L_n(x, A)/2\}$$

This set certainly contains $\hat{A}_n$. Note that with probability 1 A_n^* cannot be empty since this would violate condition (b).

To show that a.c.

$$\lim_{n\to\infty} \sup\{d(\alpha, \alpha_1) | \alpha_1 \in A_n^*\} = 0 \tag{3}$$

we consider the open ball $B(\alpha_1, \varepsilon)$ in $\bar{A}$ with the center $\alpha_1 \in \bar{A}$. If $\alpha \neq \alpha_1$, we can appeal to conditions (b) and (c) together with (c) in Definition 1 to show that there exist constants $\varepsilon_1 > 0$, $c_1 > 0$ such that

$$-\infty \leqslant E_\alpha\left[\ln \frac{f(x, \alpha_1, \varepsilon_1)}{f(x, \alpha)}\right] < -c_1$$

Note the role of (c) in Definition 1 as an identification condition, and recall the basic inequality for $E_\alpha[\ln f(x; \beta)]$ that plays a fundamental role in Wald's method.

We have, since $B^0(\alpha_1, \varepsilon_1) \subseteq B(\alpha_1, \varepsilon_1)$,

$$L_n[x, B^0(\alpha_1, \varepsilon_1)] \leqslant \prod_{r=1}^{n} f(x_r, \alpha_1, \varepsilon_1)$$

so that the previous inequality yields for all n large enough:

$$\frac{1}{n} \ln \frac{L_n[x, B^0(\alpha_1, c_1)]}{L_n(x, \alpha)} < -c_1 \tag{4}$$

Choose a positive ε small enough to make A^0 nonempty with $A = \bar{A} - B(\alpha, \varepsilon)$, and apply the above argument to each $\alpha_1 \in A$. But A is compact so that we can find a finite number of elements $\alpha_1, \alpha_2, \ldots, \alpha_k$ with corresponding values $\varepsilon_1, \varepsilon_2, \ldots, \varepsilon_k$ and $c_1, c_2, \ldots, c_k$ such that

$$A \subset \bigcup_{\nu=1}^{k} B(\alpha_\nu, \varepsilon_\nu)$$

and then, of course,

$$A^0 \subset \bigcup_{\nu=1}^{k} B^0(\alpha_n, \varepsilon_\nu)$$

Hence

$$L_n(x, A^0) \leqslant L_n\left[x, \bigcup_1^k B^0(\alpha_n, \varepsilon_n)\right]$$

$$= \max_\nu L_n[x, B^0(\alpha_\nu, \varepsilon_\nu)]$$

But a.c. for n large enough (4) implies

$$\frac{1}{n}\ln\frac{L_n(x, A^0)}{L_n(x, \alpha)} \leqslant -c$$

where the new finite and positive constant $c = \min\{c_\nu\}$. Recalling the definition of A_n^* we then have a.c. for n large enough

$$A_n^* \subset B(\alpha, \varepsilon) \tag{5}$$

But ε can be made arbitrarily small; we see that (3) holds.

We must now show that $\hat{A}_n$ is nonempty for large sample size. First choose $\varepsilon > 0$ so small that the closed ball $\bar{B}(\alpha, \varepsilon) \subset A$, which is possible according to condition (d). If x and n satisfy (5), we have

$$\bar{A}_n^* \subset B^0(\alpha, \varepsilon) \tag{6}$$

We can find a sequence $\{\alpha_1, \alpha_2, \ldots, \}$ with

$$\lim_{j\to\infty} L_n(x, \alpha_j) = L_n(x, A)$$

Extract a convergent subsequence denoted by the same symbol, $\alpha_j \to \alpha_0$, $\alpha_0 \in \bar{A}_n^*$. Then $\alpha_0 \in A$ because of (6). But for any $\varepsilon_0 > 0$ we have

$$L_n(x, \alpha_j) < \prod_{r=1}^n f(x_r, \alpha_0, \varepsilon_0)$$

if j is large enough so that

$$L_n(x, A) \leqslant \prod_{r=1}^n f(x_r, \alpha_0, \varepsilon_0)$$

But ε_0 is arbitrary so that [use condition (e)]

$$L_n(x, A) \leqslant L_n(x, \alpha_0)$$

which means that $\alpha_0 \in A_n$: A_n is not empty. Q.E.D.

The choice of a suitable compactification will not be easy in general. If, for example, A is locally compact, one could try the one-point compactification.

Actually if A is locally compact, an informative result has been given in Huber (1967), compare with Roussas (1965) who also uses a variation of the technique from Wald's proof.

Assume that $\rho(x,\alpha) = -\ln f(x;\alpha)$ satisfies conditions 1–4 below.

Condition 1 The function ρ should be lower semicontinuous as a function of α:

$$\inf_{\alpha' \in N} \rho(x,\alpha') \to \rho(x,\alpha); \qquad \text{for a.e. } x$$

as the neighborhood N of α tends to $\{\alpha\}$.

Condition 2 There exists a measurable function a such that

$$E[\rho(x,\alpha) - a(x)]^- < \infty; \qquad \text{for all } \alpha$$

$$E[\rho(x,\alpha) - a(x)]^+ < \infty; \qquad \text{for some } \alpha$$

Condition 3 There exists a parameter value $\alpha_0 \in A$ such that the function

$$\gamma(\alpha) = E[\rho(x,\alpha) - a(x)]$$

satisfies $\gamma(\alpha) > \gamma(\alpha_0$ for all $\alpha \neq \alpha_0$.

Condition 4 There exists a positive continuous function b on A such that:

(a) For some integrable h

$$\inf_{\alpha} \frac{\rho(x,\alpha) - a(x)}{b(\alpha)} \geqslant h(x)$$

and

(b) $\lim_{\alpha \to \infty} \inf b(\alpha) > \gamma(\alpha_0)$
where ∞ stands for the point of infinity in the one-point compactification of A
and

(c) $E[\lim_{\alpha \to \infty} \inf \frac{\rho(x,\alpha) - a(x)}{b(\alpha)}] \geqslant 1$

Remark 1 If Condition 2 is satisfied, we can choose $a(x)$ as $\rho(x, \alpha_0)$. The term $a(x)$ is assumed to be absorbed into ρ.

Consider now a sequence of measurable estimates α_n^* with the property that a.c.

$$\frac{1}{n}\sum_{i=1}^{n}\rho(x_i, \alpha_n^*) - \inf_{\alpha}\frac{1}{n}\sum_{i=1}^{n}\rho(x_i, \alpha) \to 0 \tag{7}$$

Recalling the definition of the function ρ, the first term on the left-hand side is simply related to the logarithm of the likelihood function for $\alpha = \alpha_n^*$.

Appealing to Condition 4, we know that there exists a compact set C and a constant $\varepsilon \in (0, 1)$ such that

$$\inf_{\alpha \notin C} b(\alpha) \geqslant \frac{\gamma(\alpha_0) + \varepsilon}{1 - \varepsilon} \tag{8}$$

and we can choose C so large that

$$E\left\{\inf_{\alpha \notin C}\frac{\rho(x, \alpha)}{b(\alpha)}\right\} \geqslant 1 - \frac{\varepsilon}{2}$$

Use Condition 4, (a) and (c). Then the strong law of large numbers applied to the random variables $\rho(x_i, \alpha)/b(\alpha)$ yields a.c. for n large enough

$$\inf_{\alpha \notin C}\left[\frac{1}{n}\sum_{i=1}^{n}\frac{\rho(x_i, \alpha)}{b(\alpha)}\right] \geqslant \frac{1}{n}\sum_{i=1}^{n}\inf_{\alpha \notin C}\frac{\rho(x_i, \alpha)}{b(\alpha)} \geqslant 1 - \varepsilon \tag{9}$$

But (9) implies that for any $\alpha \notin C$ we have

$$\frac{1}{n}\sum_{i=1}^{n}\rho(x_i, \alpha) \geqslant (1 - \varepsilon) b(\alpha) \geqslant \gamma(\alpha_0) + \varepsilon$$

using (8). We have a.c. for n large enough

$$\inf_{\alpha}\frac{1}{n}\sum_{i=1}^{n}\rho(x_i, \alpha) \leqslant \frac{1}{n}\sum_{i=1}^{n}\rho(x_i, \alpha_0) \leqslant \gamma(\alpha_0) + \frac{\varepsilon}{2}$$

so that we can assert that $\alpha_n^* \in C$ ultimately a.c. since (7) was assumed to hold.

We now know that α_n^* will a.c. eventually stay in the compact set C, and we show that α_n^* tends to the true value a.c. Let U be a neighborhood of α

and consider the function γ. Because of Conditions 1,2, and 4(a) we have, using the monotone convergence,

$$E\left[\inf_{\alpha' \in U} \rho(x,\alpha')\right] \to E[\rho(x,\alpha)]$$

for neighborhoods $U \downarrow \{\alpha\}$.

The function γ is lower semicontinuous so that its infinum is attained on the compact set $C \backslash U$. This value is greater than $\gamma(\alpha_0)$ (see Condition 3) say $\geqslant \gamma(\alpha_0)+4\varepsilon$, $\varepsilon>0$. Since γ is lower semicontinuous, for any $\alpha \in C\backslash U$ there is a neighborhood U_α with

$$E\left[\inf_{\alpha' \in U_\alpha} \rho(x,\alpha')\right] \geqslant \gamma(\alpha_0)+3\varepsilon$$

We now use the compactness of $C\backslash U$ to choose a finite number of $\alpha_1,\alpha_2,\dots,\alpha_N$ such that all the U_{α_ν} cover $C\backslash U$. Again appealing to the strong law of large numbers we have a.c., if n is large enough, for any ν

$$\inf_{\alpha' \in U_\nu} \frac{1}{n}\sum_{i=1}^{n} \rho(x_i,\alpha') \geqslant \frac{1}{n}\sum_{i=1}^{n} \inf_{\alpha' \in U_\nu} \rho(x_i,\alpha') \geqslant \gamma(\alpha_0)+2\varepsilon$$

as well as the inequality

$$\frac{1}{n}\sum_{i=1}^{n} \rho(x_i,\alpha_0) \leqslant \gamma(\alpha_0)+\varepsilon$$

so that

$$\inf_{\alpha \in C/U} \frac{1}{n}\sum_{i=1}^{n} \rho(x_i,\alpha) \geqslant \inf_{\alpha \in U} \frac{1}{n}\sum_{i=1}^{n} \rho(x_i,\alpha)+\varepsilon$$

Hence with probability 1 α_n^* tends to the true value of the parameter.

Summing up we have the following theorem [see Huber (1967)].

THEOREM 2

Under the stated condition any sequence $\{\alpha_n^*\}$ of estimates satisfying (7) *will converge a.c. to the true value.*

Since our main interest is in parameter spaces that are not locally compact, we are seldom able to use these results directly. The main idea in Wald's proof is exploited repeatedly, however. To illustrate their use, let us consider the following problem which is of considerable interest on its own.

Example 1 Consider context-free languages with superimposed probability measures. More precisely, starting from a given terminal vocabulary V_T, consider context-free grammars together with the *syntax-controlled* model [see Grenander (1967), (1976), sect. 2.10, for the condition needed in order that the probability measure not be deficient in the technical sense of this term]. To each grammar corresponds a vector p in some $\mathbb{R}^n$, where n is the number of rewriting rules in the grammar and can of course vary from one grammar to another. Enumerate the grammars as

$$G_1(p), G_2(p) G_3(p), \ldots$$

and note that the dimensionality of p varies, call it n_k for G_k. Our parameter space is then

$$A = \bigcup_{k=1}^{\infty} F_k$$

where F_k denotes closed sets in R^{n_k} over which the p vector is allowed to vary, so that we can write our parameter as (k, p), where k determines the grammar and p the probability structure on it. The closed set F_k must be chosen so that the consistency conditions in Grenander (1967) hold, also on the boundary. Note that A is not the Cartesian product, but the union of the F_k. Also A is locally compact if we use the natural topology in A defined by the neighborhoods $N(\alpha)$ of $\alpha = G_k(p)$:

$$N(\alpha) = \{\alpha' | \alpha' \in F_k \text{ and } |p' - p| \lambda \varepsilon\}; \qquad \alpha' = (k, p')$$

We observe sentences $x^1, x^2, \ldots \in L(G)$, where G is the unknown grammar with $\alpha = (\beta, \kappa)$, and denote the probability distribution over the sample space $X = V_T^*$ by $P_\alpha(x) = p(x; k, p)$. In order that α be identifiable we must assume that if $(k, p) \neq (k', p')$ there exists a sentence x such that

$$P(x; k, p) \neq P(x; k', p') \tag{10}$$

We assume that for any x we have (see Note 7.1.B)

$$\sum_x p(x, \alpha) \ln p(x, \alpha) \text{ convergent}$$

$$\lim_{k \to \infty} \sup_p p(x; k, p) = 0; \qquad \forall x$$

In order to apply Theorem 1 we first define $\bar{A}$ as the one-point compactification of A. Since $\bar{A}$ is second countable and T_3, it is metrizable. Note that in the present example μ is simply a counting measure in the denumerable sample space $X = L(G)$.

Then $\bar{A}$ is a suitable compactification. Each of the three conditions (a)–(c) of Definition 1 can be verified directly. Can we apply Theorem 1 to this problem?

Condition (a) of Theorem 1 has been satisfied already, and (c) is essentially the identifiability condition (10) together with the second equation in (11). Condition (d) is automatically satisfied, and (e) follows from the continuity of $f(x,\alpha)$; as a function of p it is just a polynomial. Hence it only remains to make sure that condition (b) holds.

But this is an immediate consequence of the fact that we have a discrete sample space with μ as a counting measure. The value of $f(x,\alpha)$ is a probability, hence bounded by 1. Since the entropy $H(\alpha)$ is finite and since $f(x,A) \leqslant 1$, it follows that (b) holds.

Hence we can claim that *the ML estimate of the unknown grammar is consistent*.

To be able to apply the maximum likelihood method to the parameter spaces we encounter in Chapter 8, we introduce the concept of a *sieve in parameter space* (see Note 7.1.C).

A sieve, usually denoted by $\mathbb{S}(\mu)$, is a family of subsets of A indexed by a positive parameter μ, the *mesh size*. For any $\mu > 0$, $\mathbb{S}(\mu)$ shall be sufficiently restricted to make ML solutions exist. On the other hand as the mesh size tends to zero, the sets $\mathbb{S}(\mu)$ shall be sufficiently rich to allow the ML solution to converge to any $\alpha \in A$.

Sieves shall sometimes be introduced via regularity criteria denoted by Q, such that small values of Q mean a highly regular α. Just to give an example, say that the parameter α can be expressed as a real-valued function $\alpha: Y \to \mathbb{R}$ on some space Y, and that for any $y \in Y$ the value $(Q\alpha)(y)$ is defined for α everywhere dense in A.

Then a *local sieve* takes the form

$$(\mu) = \{\alpha \mid (Q\alpha)(y) \text{defined for all } y \text{ and of absolute value at most equal to } 1/\mu\}$$

We meet many local sieves later on.

Sometimes it is more natural, as well as computationally more convenient, to use a *global sieve*. Global sieves are of the form

$$\mathbb{S}(\mu) = \left\{ \alpha \mid (Q\alpha)(y) \text{ defined for all } y \text{ and such that } \int_Y |(Q\alpha)(y)|^p dy \leqslant 1/\mu \right\}$$

where p is some positive constant, and dy is some measure.

The choice of "good" sieves is not easy and constitutes the main problem of Chapter 8. The fact that sieves exist which make the ML estimates converge in particular cases is not difficult to establish (see Note 7.1.D), but to find sieves that are analytically tractable and computationally feasible is a more demanding task.

We postpone the detailed study of such questions to Chapter 8 and now turn to *direct methods of inference in abstract spaces*.

2 DIRECT METHODS IN ABSTRACT PARAMETER SPACE

A typical estimation procedure, not involving likelihood based methods, starts form some obvious, commonsense estimate $\hat{\alpha}: x \to \hat{A}$, where $\hat{A} \supseteq A$. The latter relation means here inclusion of *sets*, not of the topological spaces. As a matter of fact $\hat{A}$ will not always be topologized in explicit terms.

If closer examination of $\hat{\alpha}$ leads to the conclusion that it does not perform as well as expected, at least in some respect, one will try to improve it. An elementary example is the estimation of a variance by the empirical variance. Since this is not unbiased, the conventional wisdom tells us to apply a linear transformation, the $(n-1)$-correction, to remove the bias. In this one-dimensional parameter space we did not go outside of it.

In a more general setting A will be a proper subset of $\hat{A}$, and this leads to the concept of *regularization*, a notion of ancient origin in analysis. Dealing with functions in some generality, it is often convenient to replace them by better behaved ones approximating the original ones. The increase in regularity will simplify the analysis.

In abstract estimation theory the purpose is not to simplify the analysis as much as to make the regularized estimate $T\alpha = \alpha^*$ have more attractive statistical properties and, of course, take values in A.

We often let the *regularization operator* T be a member of a one-parameter semigroup $\{T_\varepsilon, \varepsilon \geqslant 0\}$, so that $T_{\varepsilon_1} T_{\varepsilon_2} = T_\varepsilon$, with some *composition function* $\varepsilon = \varepsilon(\varepsilon_1, \varepsilon_2)$. It is particularly convenient, although by no means necessary, if the composition is additive so that $\varepsilon(\varepsilon_1, \varepsilon_2) = \varepsilon_1 + \varepsilon_2$, or

$$T_{\varepsilon_1} T_{\varepsilon_2} = T_{\varepsilon_1 + \varepsilon_2}$$

$$T_\varepsilon \to I \text{ as } \varepsilon \downarrow 0$$

where the mode of convergence in the second relation will be specified in each separate case. $\{T_\varepsilon\}$ will be called the *regularization semigroup*.

Let us make this more concrete by mentioning some examples of regularization in parameter spaces. These cases are analyzed later on in the following sections of this chapter.

Example 1 Say that we want to estimate the spectral density f of a Gaussian stationary process with discrete time; $\lambda \in [-\pi, \pi)$. The obvious estimate, the periodogram $I_n(\lambda)$ [see (1) in the Section 3], has small bias but unacceptably large variance, so that it must be modified.

Let W_ε be a family of probability measures on the one-dimensional torus, parametrized as $[-\pi, \pi)$, such that the convolution equation holds:

$$W_{\varepsilon_1} * W_{\varepsilon_2} = W_{\varepsilon_1+\varepsilon_2}; \qquad \varepsilon_1, \varepsilon_2 \geqslant 0.$$

For $\varepsilon > 0$ we assume that W_ε is absolutely continuous with density $w_\varepsilon(\lambda)$. As $\varepsilon \downarrow 0$ the measures w_ε should contract in the sense of weak convergence to the distribution with all its mass at $\lambda = 0$.

Define the regularization $T_\varepsilon I_n(\lambda)$ as

$$f_\varepsilon^*(\lambda) = (I_n * w_\varepsilon)(\lambda) \tag{1}$$

Such estimates as in (1) are known as *smoothed periodograms* and are studied in section 3. It seems to be the first case of regularization in parameter spaces of the function type to be analyzed in some mathematical depth [see Grenander (1951a)]. As ε increases, the W-measures become flatter over the torus and tend (weakly) to the uniform distribution. It is worth noting that W_ε can be considered as the marginal distribution of a stochastic process with independent increments, continuous in probability, homogeneous in "time" ε, and taking values on the torus group.

Example 2 As a second case consider estimation of a frequency function f defined on $\mathbb{R}$, using a sample $x = (x_1, x_2, \ldots, x_n) \in \mathbb{R}^n$, so that A consists of nonnegative L_1-functions with integral 1. The commonsense estimate $\hat{\alpha}$ places the probability mass $1/n$ in each observed point x_i. Of course $\hat{\alpha} \notin A$, and a natural regularization is

$$\alpha^*(x) = (\hat{\alpha} * w_\varepsilon)(x) \tag{2}$$

where W_ε and w_ε are as before, except that they are now defined on $\mathbb{R}$ instead of on the torus. If we write out (2) explicitly, we get

$$\alpha^*(x) = \frac{1}{n} \sum_{i=1}^{n} w_\varepsilon(x - x_i) \tag{3}$$

the so-called *kernel estimates* of Rosenblatt (1956b) and Parzen (1962).

If ε is small, the "window" w_ε is highly concentrated, and the terms in (3) will be well separated. On the other hand, large values of ε will make them overlap. In this, as in many other similar instances of regularization, the choice of value of the regularization parameter ε is crucial for the performance of $\alpha^* = T_\varepsilon \hat{\alpha}$.

Example 3 A third case, more recent than the first two, is when α is an open bounded set, say in $\mathbb{R}^2$, and we observe a sample x consisting of points $x_i \in \mathbb{R}^2$ generated i.i.d. from a uniform distribution over α. A commonsense estimate would be to take $\hat{\alpha}=x$, the finite set observed, but this is obviously a poor choice. It does not take values in A, and, what is worse, it is not even consistent (in any acceptable sense) as $n \to \infty$.

What is wrong is clearly that $\hat{\alpha}$ is too "thin," and we should make it more solid. Consider the operation on sets

$$T_\varepsilon \alpha = \bigcup_{z \in \alpha} D(z;\varepsilon) \tag{4}$$

where $D(z;\varepsilon)$ denotes the open disk centered at x and with radius ε. For an arbitrary $\alpha \in A$ consider the two sets

$$B_1 = T_{\varepsilon_1} T_{\varepsilon_2} \alpha$$
$$B_2 = T_{\varepsilon_1+\varepsilon_2} \alpha$$

If $z \in B_1$, there are points $z_1 \in \alpha$, $\|z_2 - z_1\| < \varepsilon_2$, such that $\|z - z_2\| < \varepsilon_1$. Then the triangle inequality implies $\|z - z_1\| < \varepsilon_1 + \varepsilon_2$ so that $z \in B_2$; hence $B_1 \subseteq B_2$.

On the other hand if $z \in B_2$ there is a point $z_1 \in \alpha$ with $\|z - z_1\| < \varepsilon_1 + \varepsilon_2$. Then the point

$$z_2 = \frac{z_1 \varepsilon_1 + z \varepsilon_2}{\varepsilon_1 + \varepsilon_2}$$

satisfies

$$\|z_1 - z_2\| = \frac{\varepsilon_2}{\varepsilon_1 + \varepsilon_2} \|z - z_1\| < \frac{\varepsilon_2}{\varepsilon_1 + \varepsilon_2}(\varepsilon_1 + \varepsilon_2) = \varepsilon_2$$

and

$$\|z - z_2\| = \frac{\varepsilon_1}{\varepsilon_1 + \varepsilon_2} \|z - z_1\| < \frac{\varepsilon_1}{\varepsilon_1 + \varepsilon_2}(\varepsilon_1 + \varepsilon_2) = \varepsilon_1$$

so that $z_2 \in D(z_1, \varepsilon_2)$, $z \in D(z_2, \varepsilon_1)$, and therefore $z \in B_1$. Hence $B_2 \subseteq B_1$, implying $B_1 = B_2$ so that $\{T_\varepsilon\}$ is a semigroup with identity $T_0 = I$. This semigroup is continuous in the Hausdorff distance for sets (see Chapter 1, Section 3). Operation (4) represents a *regularization of sets*.

Example 4 Our last example has great potential for generalization. Let A consist of real-valued and sufficiently smooth functions on some interval

I, and let $\hat{A}$ have distributions (generalized functions) as element. The precise specification of A and $\hat{A}$ will have to wait until we encounter special cases. This will involve among other things appropriate boundary conditions at the end points of I.

The regularization semigroup $\{T_\varepsilon\}$ will be defined by $(T_\varepsilon \hat{a})(x)$ as the solution for $t=\varepsilon$ of the initial value problem

$$\begin{cases} \dfrac{\partial a(x,t)}{\partial t} = La(x,t) \\ a(x,0)=\hat{a}(x) \end{cases} \tag{5}$$

where L is a given linear operator on $\hat{A}$, for example, a second-order differential operator. For small values of ε the effect of regularization is weak, but as ε grows, the estimate $T_\varepsilon \hat{\alpha} = \alpha_\varepsilon^*$ will become smoother, all under appropriate conditions on L, and tend to a limit in the null space of L.

If, for example, $L=\frac{1}{2}\dfrac{\partial^2}{\partial x^2}$ and $I=(-\infty,\infty)$, so that (5) reduces to the heat equation, we get the *Weierstrass transform*

$$(T_\varepsilon a)(x) = \frac{1}{\sqrt{2\pi\varepsilon}} \int_{-\infty}^{\infty} e^{-(x-y)^2/2\varepsilon} \hat{a}(y)\,dy \tag{6}$$

and many other instances could be mentioned.

3 REGULARIZATION OF SPECTRAL ESTIMATES

Let x_t, $t \in \mathbb{N}$, be a real stationary process with Gaussian distributions, mean value zero, and an absolutely continuous spectrum with density f on the finite interval $[-\pi,\pi)$. Having been given a sample $x_1, x_2, \ldots, x_n$, we want to estimate f. Then our parameter is a nonnegative $f \in L_1$, but when we get to the asymptotic analysis of regularized estimates, it will turn out to be more natural to choose A as a subset of L_2.

The commonsense estimate of f is the time-honored *periodogram*:

$$I_n(\lambda) = \frac{1}{2\pi n}\left|\sum_{t=1}^{n} x_t e^{-it\lambda}\right|^2 = \frac{1}{2\pi n}\left\{\left(\sum_{t=1}^{n} x_t \cos t\lambda\right)^2 + \left(\sum_{t=1}^{n} x_t \sin t\lambda\right)^2\right\} \tag{1}$$

Of course $I_N(\cdot)$ belongs to the parameter space, but how does the periodogram estimate behave for large sample size? The answer is given by the following result [see Grenander (1951a)].

THEOREM 1

At any point of continuity of f the periodogram is asymptotically unbiased and with the asymptotic variance $f^2(\lambda)$ for $\lambda \neq 0$ and $2f^2(0)$ at $\lambda = 0$.

Proof We have, using the Cramér representation of $\{x_t\}$

$$E[I_n(\lambda)] = \frac{1}{2\pi n} E\left| \int_{-\pi}^{\pi} \sum_{t=1}^{n} e^{it(l-\lambda)} Z(dl) \right|^2$$

$$= \frac{1}{2\pi n} \int_{-\pi}^{\pi} \left| e^{i(l-\lambda)} \frac{1-e^{in(l-\lambda)}}{1-e^{i(l-\lambda)}} \right|^2 f(l)\, dl \tag{2}$$

But (2) can be written in terms of the Féjer kernel:

$$K_n(\mu) = \frac{1}{2\pi n} \frac{\sin^2(n/2)\mu}{\sin^2(\mu/2)}$$

as

$$E[I_n(\lambda)] = \int_{-\pi}^{\pi} K_n(l-\lambda) f(l)\, dl \tag{3}$$

Using the classical properties of the Féjer kernel it follows that (3) tends to $f(\lambda)$ as $n \to \infty$ if f is continuous at λ.

To get the variance of the periodogram, we express it as

$$I_n(\lambda) = \frac{1}{2\pi n} x^T Q(\lambda) x \tag{4}$$

where $Q(\lambda)$ is the $n \times n$ matrix with entries [see the right-hand side of (1)],

$$q_{st} = \cos s\lambda \cos t\lambda + \sin s\lambda \sin t\lambda = \cos(s-t)\lambda;\ s, t = 1, 2, \ldots, n \tag{5}$$

For a Gaussian distribution we have, however, by an elementary calculation, using the classical relation,

$$E[x_{t_1} x_{t_2} x_{t_3} x_{t_4}] = E(x_{t_1} x_{t_2}) E(x_{t_3} x_{t_4})$$
$$+ E(x_{t_1} x_{t_3}) E(x_{t_2} X_{t_4}) + E(x_{t_1} x_{t_4}) E(x_{t_2} x_{t_3}) \tag{6}$$

the formula for the variance

$$\mathrm{Var}[x^T Q(\lambda) x] = 2\,\mathrm{tr}\left[(RQ(\lambda))^2 \right] \tag{7}$$

where R is the $n \times n$ covariance matrix $\{r_{s-t};\ s, t = 1, 2, \ldots, n\}$.

Now the q_{st} are given as cosines, [see (5)] or equivalently as the sum of two complex exponentials. Combining these we get (7) in the form of four terms, which are handled in the same way. The first one is, including the factor $(2\pi n)^{-2}$ from (4),

$$\frac{1}{4}\frac{1}{2\pi^2 n^2}\Sigma r_{t-\nu}e^{i(\nu-s)\lambda}r_{s-\mu}e^{i(\mu-t)\lambda} \tag{8}$$

all indices ranging from 1 to n. Using the spectral representation for the covariances, expression (8) can be written as

$$\frac{1}{4}\frac{1}{2\pi^2 n^2}\iint_{-\pi}^{\pi}\frac{\sin^2(n/2)(u-\lambda)}{\sin^2\frac{1}{2}(u-\lambda)}\frac{\sin^2(n/2)(v-\lambda)}{\sin^2\frac{1}{2}(v-\lambda)}f(u)f(v)\,du\,dv \tag{9}$$

which tends to $f^2(\lambda)/2$. One more term leads to the same value.

Another term has the form

$$\frac{1}{4}\frac{1}{2\pi^2 n^2}\iint_{-\pi}^{\pi}\frac{\sin(n/2)(u-\lambda)}{\sin\frac{1}{2}(u-\lambda)}\frac{\sin(n/2)(v+\lambda)}{\sin\frac{1}{2}(v+\lambda)}$$

$$\frac{\sin(n/2)(u+\lambda)}{\sin\frac{1}{2}(u+\lambda)}\frac{\sin(n/2)(v-\lambda)}{\sin\frac{1}{2}(v-\lambda)}\times f(u)f(v)\,du\,dv \tag{10}$$

Here we have two separate cases. If $\lambda=0$, then (10) behaves as (9) and gives the limit $f^2(\lambda)/2$, $\lambda=0$. If $\lambda\neq 0$, this has the effect of separating the singularities from the four denominators in (10): $u=\pm\lambda$ and $v=\pm\lambda$. We can then apply dominated convergence to show that (10) tends to zero as $n\to\infty$. Similarly for the fourth term. Q.E.D.

Theorem 1 shows that the periodogram is an unacceptable estimate, and we now regularize it as described in Example 1 of Section 2. The smoothed periodogram will then take the form (2.1), or

$$f_\varepsilon^*(\lambda)=(T_\varepsilon I_n)(\lambda)=\int_{-\pi}^{\pi}w_\varepsilon(\lambda-l)I_n(l)\,dl \tag{11}$$

where the convolution semigroup $\{w_\varepsilon\}$ will be chosen to have even and continuous weight functions.

Following Grenander (1951a) pp. 521,517), in a study of *empirical spectral analysis*, we *use the expected L_2-norm of its error as a criterion*:

$$\begin{cases} \delta_n = E[\|f^* - f\|^2] = \int_{-\pi}^{\pi} E[f_\varepsilon^*(\lambda) - f(\lambda)]^2 d\lambda \\ \quad = \int_{-\pi}^{\pi} \mathrm{Var}[f_\varepsilon^*(\lambda)]d\lambda + \int_{-\pi}^{\pi} b^2(\lambda)d\lambda = \delta_n' + \delta_n'' \\ b(\lambda) = E[f^*(\lambda)] - f(\lambda) \end{cases} \tag{12}$$

In (12) we have used the variance and the bias $b(\lambda)$ of f^* and separated the total expected L_2-norm into one δ_n', expressing the variance, and another δ_n'', expressing the bias. Using Parseval's relation, we get with the covariances r_k

$$\delta_n = \frac{1}{2\pi} E\left\{ \sum_{-\infty}^{\infty} [r_k^* w_k(\varepsilon) - r_k]^2 \right\} \tag{13}$$

with

$$r_k^* = \begin{cases} \dfrac{1}{n} \sum_{s-t=k} x_s x_t; & \text{if } |k| \leqslant n-1 \\ 0; & \text{otherwise} \end{cases}$$

Relation (13) follows from the definition of the periodogram (1) and (11) combined into

$$\begin{aligned} f_\varepsilon^*(\lambda) &= \frac{1}{2\pi} \sum_k r_k^* \int_{-\pi}^{\pi} e^{ikl} w_\varepsilon(\lambda - l)dl \\ &= \frac{1}{2\pi} \sum_k r_k^* e^{ik\lambda} \int_{-\pi}^{\pi} e^{-ik\mu} w_\varepsilon(\mu)d\mu \end{aligned} \tag{14}$$

which gives (13) , introducing $w_k(\varepsilon)$ as the Fourier coefficients appearing in (14).

But

$$E[r_k^* w_k(\varepsilon) - r_k]^2 = w_k^2(\varepsilon)\mathrm{Var}(r_k^*) + r_k^2\left[w_k\left(1 - \frac{|k|}{n}\right) - 1\right]^2 \tag{15}$$

To evaluate $\mathrm{Var}(r_k^*)$ we use (6) again and get, say for $k \geqslant 0$,

$$n\,\mathrm{Var}(r_k^*) = \frac{1}{n}\Sigma r_{t-s}^2 + \frac{1}{n}\Sigma r_{t-s+k} r_{t-s-k} \tag{16}$$

summed over $1 \leqslant s, t \leqslant n-k$. Rewriting (16) we have

$$n\operatorname{Var}(r_k^*) = \Sigma r_\nu^2\left(1 - \frac{k+|\nu|}{n}\right) + \Sigma r_{\nu+k} r_{\nu-k}\left(1 - \frac{k+|\nu|}{n}\right)$$

with ν between $-n+k+1$ and $n-k-1$. As $n \to \infty$, we get the limit

$$\lim_{n\to\infty} n\operatorname{Var}(r_k^*) = \sum_{-\infty}^{\infty} \left(r^2 + r_{\nu+k} r_{\nu-k}\right) \tag{17}$$

and with the aid of Parseval's relation the sum in (17) reduces to the attractive form

$$2\pi \int_{-\pi}^{\pi} f^2(\mu)[1 + \cos 2k\mu]\, d\mu$$

We can therefore write asymptotically, using (13), (15), and (17),

$$\delta_n \cong \frac{1}{2\pi} \sum_{k=-\infty}^{\infty} \left[\frac{1}{n} w_k^2(\varepsilon) A_k + r_k^2 (1 - w_k(\varepsilon))^2\right] \tag{18}$$

putting

$$A_k = 2\pi \int_{-\pi}^{\pi} f^2(\mu)[1 + \cos 2k\mu]\, d\mu = \sum_{-\infty}^{\infty} \left(r_\nu^2 + r_{\nu+k} r_{\nu-k}\right) \tag{19}$$

Now choose $w_k(\varepsilon)$ to minimize the sum in (18). We get after some calculations

$$w_k(\varepsilon) = \frac{r_k^2 n}{A_k + r_k^2 n} \tag{20}$$

with the minimum of the sum equal to

$$m = \sum_{-\infty}^{\infty} \frac{A_k r_k^2}{A_k + r_k^2 n} \tag{21}$$

Note that (21) converges since the values of A_k are bounded away from zero, and $\{r_k\} \in l_2$ since $f \in L_2$. Also $\{w_k(\varepsilon)\} \in l_2$ so that $w_\varepsilon(\cdot) \in L_2$.

If the asymptotic properties of the $\{r_k\}$ are known and given asymptotically, one can calculate the asymptotic behavior of the minimum expected L_2-norm. This has been done for several cases in Parzen (1958), which the interested reader should consult for details (see Note 7.B.A for related results).

The A_k can be obtained as the Fourier coefficients of a measure $A(d\lambda)$ with the discrete mass at zero

$$\Delta A(0)=\sum_{-\infty}^{\infty} r_\nu^2=2\pi\int_{-\pi}^{\pi} f^2(\lambda)\,d\lambda=2\pi\|f\|^2 \tag{22}$$

and a density [use the integral representation in (19) and change variables from 2μ to λ]

$$\frac{A(d\lambda)}{d\lambda}=\pi\left[f^2\left(\frac{\lambda}{2}\right)+f^2\left(\frac{\lambda}{2}+\pi\right)\right] \tag{23}$$

The sum in (18) without the factor $1/2\pi$ can be written as $V+B$ where

$$V=\frac{1}{n}\sum_{k=-\infty}^{\infty} A_k w_k^2(\varepsilon)$$

$$B=\sum_{k=-\infty}^{\infty} r_k^2[w_k(\varepsilon)-1]^2$$

If we fix the *bias term* B and minimize the *variance term*, we get, using a Lagrange multiplier L whose value should make B take the desired value in (24),

$$\min V=\frac{1}{n}\sum_{k=-\infty}^{\infty}\frac{L^2A_kr_k^4}{(A_k/n+Lr_k^2)^2}$$

$$B=\frac{1}{n^2}\sum_{k=-\infty}^{\infty}\frac{A_kr_k^2}{(A_k/n+Lr_k^2)^2} \tag{24}$$

The terms in the first sum are increasing in L, in the second sum they are decreasing. Therefore the variance term is antagonistic to the bias term and a sort of *uncertainly principle* holds:

If we narrow the bandwidth of the weight function to lower the bias, we can expect the variance to increase, and vice versa.

Since $\{r_k^2\}$ are the Fourier coefficients of $f*f$, it follows from (20) that $w_\varepsilon(\lambda)$ is a solution of the integral equation

$$w_\varepsilon*\left[\frac{1}{n}A+f*f\right]=f*f$$

where A is the measure given by (22) and (23). This could also be shown by a direct argument from calculus of variation, expressing δ_n in terms of integrals of f and w_ε.

The semigroup property of T_ε implies that we can write the Fourier coefficients

$$w_k(\varepsilon) = e^{\varepsilon v_k}; \qquad v_k < 0 \qquad \text{for} \quad k \neq 0, \quad v_0 = 0$$

since $w_k(\lambda)$ is even. No Fourier coefficient for an absolutely continuous probability distribution can attain the value 1 in absolute value for $k \neq 0$. Then expression (18) becomes

$$\frac{1}{2\pi} \sum_{k=-\infty}^{\infty} \left[\frac{1}{n} A_k e^{2\varepsilon v_k} + r_k^2 (1 - e^{\varepsilon v_k})^2 \right] \tag{25}$$

For $\varepsilon \downarrow 0$, little smoothing, this blows up since

$$\sum_{k=-\infty}^{\infty} A_k = +\infty \tag{26}$$

On the other hand if we make $\varepsilon \to +\infty$, strong smoothing, expression (26) tends to

$$\frac{1}{2\pi n} A_0 + \frac{1}{2\pi} \sum_{k \neq 0} r_k^2 = \frac{1}{2\pi} \left[\left(\frac{2}{n} + 1 \right) \sum_{-\infty}^{\infty} r_k^2 - r_0^2 \right]$$

In general a compromise between these extremes is desirable. Expression (25) shows that if we first choose ε small to make the second term sufficiently close to zero and then n big enough, we can make $E[\|\alpha_\varepsilon^* - \alpha\|^2]$ as small as desired, and we have established this theorem.

THEOREM 2

Consider a regularized spectral estimate $\alpha^* = T_\varepsilon \hat{\alpha}$ with a continuous even weight function. When $n \to \infty$, the expected squared L_2-norm will tend to zero if we make the regularization parameter $\varepsilon \downarrow 0$ slowly enough.

Note that when $n \to \infty$, the weights w_k tend to 1, which means that the weight functions w become more peaked at $\lambda = 0$: their bandwidth will decrease to zero.

It should be mentioned that the argument in the derivation of Theorem 2 works for a particular f. But f is unknown, it is the parameter to be estimated. To make the result practically useful, a uniformity argument is needed. One way of doing this is by considering those f for which $r_k^2 = 0(s_k)$, where s_k is a given summable sequence. Then we can make the contribution of the second term uniformly small in f, for all f in the given class, and then reduce the influence of the second term.

Formula (20) for the optimal weights cannot be used directly since it assumes knowledge of $\{r_k\}$, or equivalently, of f. Nevertheless it is of considerable theoretical interest since it tells us something about the dependence of bandwidth upon sample size.

In the same problem, but with a general continuous spectrum, if we take the *spectral distribution function* F (rather than the density f) as parameter, it is more natural to embed the parameter space in $C([-\pi, \pi))$. This leads to the question of consistent estimation in this new setting.

The commonsense estimate is the *integrated periodogram*

$$F_n^*(\lambda) = \frac{1}{2\pi n} \int_{-\pi}^{\lambda} |\sum_{t=1}^{n} x_t e^{it\mu}|^2 d\mu = \frac{1}{2\pi} \sum_{k=-n+1}^{n-1} r_k^* \frac{e^{ikx} - e^{-ik\pi}}{ik} \tag{27}$$

with the natural interpretation of the term with $k=0$. Since the process is Gaussian with a continuous spectrum, it is ergodic and has moments of all orders (see Chapter 3, Section 3). It follows that with probability 1, for all k,

$$\lim_{n \to \infty} r_k^* = r_k = E(x_t x_{t+k})$$

This implies that a.c. as $n \to +\infty$,

$$\int_{-\pi}^{\pi} e^{ik\lambda} F_n^*(d\lambda) = r_k^* \to r_k = \int_{-\pi}^{\pi} e^{ik\lambda} F(d\lambda)$$

Since all the Fourier coefficients of F_n^* tend a.c. to those of F, we can claim that $F_n^* \to F$ uniformly with probability 1, which proves the following result due to Doob (1953), p. 496.

THEOREM 3

For a stationary Gaussian process with a continuous spectrum, the integrated periodogram (27) *is consistent with a.c. strong convergence in* $C([-\pi, \pi))$.

It may seem surprising that F can be estimated consistently by the commonsense estimate when we have seen that this is not true for f. The reason for this is that f, as a derivative of F, is more sensitive to random errors.

We actually know a good deal more if we impose more smoothness conditions on the spectrum, and we quote a result due to Grenander-Rosenblatt (1956):

THEOREM 4

If $\{x_t\}$ is a stationary Gaussian process with an absolutely continuous spectrum whose density f has an integrable second derivative we have with sup norm

$$\lim_{n\to\infty} P\{\sqrt{n}\,\|F_n^* - F\| \leqslant s\} = \sum_{k=-\infty}^{\infty} (-1)^k \left\{ \Phi\left[\frac{(2k+1)s}{x}\right] - \Phi\left[\frac{(2k-1)s}{x}\right]\right\} \tag{28}$$

In (28) Φ *denotes the density function of* $N(0,1)$ *and*

$$x = \sqrt{2\pi \int_0^{\pi} f^2(\lambda)\,d\lambda}$$

The proof of Theorem 4 is too complicated to be given here, and we refer the reader to Grenander-Rosenblatt (1956), P. 195. The reader can find a thorough analysis of this topic with many detailed results in Ibragimov (1963) (see also Note 7.3.B).

4 REGULARIZATION OF FREQUENCY ESTIMATES

The parameter in Section 3 described the spectral measure of a stationary process. If we let it denote a probability measure P_α instead, the new problem looks similar, but the mathematical structure is quite different.

Say that P_α on $\mathbb{R}$ is absolutely continuous with a frequency function $f = \alpha \in L_1$ with respect to Lebesgue measure on $\mathbb{R}$, and that we have observed the i.i.d. sample $x = (x_1, x_2, \ldots, x_n)$. The commonsense estimate $\hat{\alpha}$ places the mass $1/n$ at each point x_i, $i = 1, 2, \ldots, n$ (see Example 2 in Section 2). This is not an element in the current parameter space. Let the kernels w_ε be frequency functions on $\mathbb{R}$ forming a convolution semigroup say with

$$\frac{\partial w_\varepsilon(x)}{\partial \varepsilon} = L w_\varepsilon; \qquad w_\varepsilon(x) \to \delta(x) \text{ weakly as } \varepsilon \downarrow 0$$

where L is a linear operator, namely, the infinitesimal operator belonging to the regularization semigroup. Note that *the kernels $w_\varepsilon(x)$ can be viewed as the marginal frequency functions of an infinitely divisible stochastic process*, the structure of which is well understood.

The most familiar case is when

$$L=\frac{\sigma^2}{2}D^2$$

so that we get the *Weierstrass regularization kernel* [see (2.6)]:

$$w_\varepsilon(x)=\frac{1}{\sqrt{2\pi\varepsilon\sigma}}\exp\left(-\frac{x^2}{2\sigma^2\varepsilon}\right)$$

but many other continuous infinitely divisible kernels are of interest.

Let us for the moment consider w as an arbitrary weight function in L_2 to be determined later. Following Watson-Leadbetter (1963) (see Note 7.4.A), we get, if $\alpha\in L_2$, for the resulting estimate the expected L_2-norm squared by applying Parseval's relation

$$\delta_n=E[\|f_n^*-f\|^2]=\int_{-\infty}^{\infty}E\{[f_n^*(x)-f(x)]^2\}\,dx$$

$$=\frac{1}{2\pi}\int_{-\infty}^{\infty}E|\phi_n^*(z)-\phi(z)|^2\,dz \tag{1}$$

where ϕ and ϕ_n^* denote the Fourier transforms of f and f_n^*, respectively. Hence using the estimate in (2.3), we have

$$\phi_n^*(z)=v(z)\frac{1}{n}\sum_{\nu=1}^{n}e^{iz_\nu z}$$

where $v(z)$ is the Fourier transform of w_ε.

We can rewrite (1) by expanding the square and using $E[\phi_n^*(z)]=v\phi(z)$ together with

$$E[|\phi_n^*(z)|^2]=\frac{1}{n^2}\sum_{\nu,\mu=1}^{n}E[e^{i(x_\nu-x_\mu)2}]=\frac{n}{n^2}+\frac{n(n-1)}{n^2}|\phi(z)|^2$$

We then get

$$2\pi\delta_n=\int_{-\infty}^{\infty}\left\{\frac{1}{n}|v(z)|^2(1-|\phi(z)|^2)+|\phi(z)|^2[1-|v(z)|]^2\right\}dz \tag{2}$$

or

$$2\pi\delta_n = \int_{-\infty}^{\infty}\left(\frac{1}{n}+\frac{n-1}{n}|\phi(z)|^2\right)\left(\left|v(z)-\frac{|\phi(z)|^2}{1/n+(n-1)/n|\phi(z)|^2}\right|\right)^2 dz$$

$$+\int_{-\infty}^{\infty}\frac{|\phi(z)|^2(1-|\phi(z)|^2)}{1+(n-1)|\phi(z)|^2}dz \tag{3}$$

We first minimize (3) by choosing w and v appropriately. Since the last term does not depend upon v and since the first is nonnegative, the minimum is attained by

$$v(z)=\frac{|\phi(z)|^2}{1/n+(n-1)/n|\phi(z)|^2}$$

which makes the first term zero. Note that in (3) we have $v\in L_2$, since $v=0(|\phi|^2)=0(|\phi|)$ because ϕ is bounded, so that it corresponds to some $w\in L_2$.

A difficulty that does not seem to have been noticed in the literature is the question *whether the resulting estimate* α^* *is in the given parameter space*, in other words, whether $w\in L_1$. To answer this note that we can write $v=h(\psi)$, $\psi=|\phi|^2$, with $h(z)=z(1/n+[(n-1)/n]z)$. But the Fourier transform of ψ is proportional to the symmetrized convolution $f*\tilde{f}$, which is in L_1 since $f\in L_1$. Hence ψ is the absolute convergent Fourier transform of some integrable function. Note that $h(0)=0$ and that $h(z)$ is analytic in an open set surrounding the real interval $[0,1]$ of the complex plane. Then according to a result extending the classical theorem of Wiener-Lévy on absolute convergent Fourier series to other commutative Banach algebras [see Rudin (1962), chap. 6], the function $v=h(\psi)$ is also the Fourier transform of an integrable function. This shows that $w\in L_1$, and the question raised above is answered in the affirmative (see also Note 7.4.B).

With this choice of v the minimum value of δ_n becomes

$$\min\delta_n=\frac{1}{2\pi}\int_{-\infty}^{\infty}\frac{|\phi(z)|^2(1-|\phi(z)|^2)}{1+(n-1)|\phi(z)|^2}dz$$

We now return to (2) and use some regularization kernel $w=w_\varepsilon$ from an infinitely divisible process. As $\varepsilon\downarrow 0$, $v(z)\to 1$, and we choose ε so small that the integral of the second term in (2) is small, and then n large enough to make the contribution of the first term small enough; use dominated convergence. This proves the following theorem.

THEOREM 1

Consider a regularized estimate $\alpha_\varepsilon^* = T_\varepsilon \hat{\alpha}$ *as described. When* $n \to \infty$, *the expected squared* L_2*-norm of this estimate will tend to zero if* ε *is made to tend to zero slowly enough.*

Just as for Theorem 2 in Section 3, a uniformity argument is needed to make the estimate consistent for a whole class of frequency functions.

When we consider the distribution function F as the unknown parameter, rather than the frequency function, no regularization is needed for consistency in sup norm. This is a classical result.

THEOREM 2 GLIVENKO-CANTELLI'S THEOREM

For the empirical distribution function

$$F_n^*(x) = \frac{1}{n} \# \{x_\nu \leqslant x\}$$

Then with probability 1, $F_n \to F$ *in sup norm,*

$$P\left\{ \sup_x |F_n^*(x) - F(x)| \to 0 \text{ as } n \to \infty \right) = 1$$

Proof For any natural number n introduce the numbers $x_{n,k}$ defined as the largest x for which

$$F(x-0) \leqslant \frac{k}{n} \leqslant F(x)$$

Define

$$\Delta_n^{(1)} = \max_{1 \leqslant k \leqslant n} |F_n^*(x_{n,k}) - F(x_{n,k})|$$

$$\Delta_n^{(2)} = \max_{1 \leqslant k \leqslant n} |F_n^*(x_{n,k} - 0) - F(x_{n,k} - 0)|$$

so that

$$\Delta_n = \sup_x |F_n^*(x) - F(x)| \leqslant \max(\Delta_n^{(1)}, \Delta_n^{(2)}) + 1/n \tag{4}$$

For every fixed x the strong law of large numbers tells us that

$$\begin{cases} P\left[\lim_{n\to\infty} F_n^*(x) = F(x) \right] = 1 \\ P\left[\lim_{n\to\infty} F_n^*(x-0) = F(x-0) \right] = 1 \end{cases}$$

Using (4) this implies that

$$P\left\{ \limsup_{n\to\infty} \Delta_n > 1/n = 0 \right\}$$

for any n, so that a.c. $\Delta_n \to 0$ as $n\to\infty$. Q.E.D.

A more refined result, due to Kolmogorov, tells us *how fast* $\Delta_n \to 0$.

THEOREM 4

If F is continuous, we have for the sup norm of the difference $F_n^* - F$

$$\lim_{n\to\infty} P\left[\sqrt{n} \sup_x |F_n^*(x) - F(x)| \leqslant y \right] = \sum_{k=-\infty}^{\infty} (-1)^k e^{-2k^2y^2}$$

The proof is too complicated to be given here. The original proof was given in Kolmogorov (1933).

5 REGULARIZATION OF SETS

Since regularization of sets is less well understood than regularization of densities, we pay more attention to it here. The typical problem is that a set, say α, is given as a subset of a space X. We observe points in α generated i.i.d. from some probability measure P_α whose support is α, and want to estimate α well in terms of some criterion as

$$E[\mu(\alpha^* \Delta \alpha)]$$

where μ is some given σ-finite Borel measure over X. The structure of X and α will be specified from case to case.

Let us apply the regularization operator T_ε in Example 3, Section 2, so that with $\hat{\alpha} = \{z_1, z_2, \ldots, z_n\}$,

$$\alpha_\varepsilon^* = T_\varepsilon \hat{\alpha} = \bigcup_{z\in\hat{\alpha}} D(z;\varepsilon) \tag{1}$$

We show that $\alpha_\varepsilon^* \to \alpha$ [see Grenander (1978), p. 338] in the sense that

$$E_\alpha[\mu(\alpha_\varepsilon^* \Delta \alpha)] \to 0; \qquad n\to\infty \tag{2}$$

if we make ε tend to zero slowly enough. To be concrete, let $X = \mathbb{R}^2$, let P_α be the uniform distribution over α, and let μ be Lebesgue area.

THEOREM 1

If $\varepsilon_n \downarrow 0$ with $\varepsilon_n^2 n \to +\infty$, the regularized set estimate α^* is consistent in the sense of (2), *if A consists of bounded Borel sets whose boundary have Lebesgue measure zero.*

Proof Introduce the sets

$$\alpha_\varepsilon^{\text{in}} = \alpha \cap [\alpha^c + D(\varepsilon)]^c$$
$$\alpha_\varepsilon^{\text{out}} = [\alpha + D(\varepsilon)]^c \tag{3}$$

with Minkowsky addition (see Chapter 1, Section 3) and where $D(\varepsilon)$ denotes a disk centered at 0 and of radius ε. For an intuitive understanding of the geometry of (3), see Figure 1.

We have

$$E_\alpha[\mu(\alpha_\varepsilon^* \Delta \alpha)] = \int_\alpha \{1 - E_\alpha[1^*(z)]\}\, dz + \int_{\alpha^c} E_\alpha[1^*(z)]\, dz \tag{4}$$

where we have introduced the indicator function 1^* for the set α_ε^* and written dz for Lebesgue area. For any $z \in \alpha_\varepsilon^{\text{in}}$, we get

$$E_\alpha[1^*(z)] = 1 - \left[1 - \frac{\pi\varepsilon^2}{\mu(\alpha)}\right]^n \to 1$$

if $n^2\varepsilon \to +\infty$ as $n \to \infty$. Similarly for $z \in \alpha_\varepsilon^{\text{in}}$, we have

$$1^*(z) \equiv 0 \text{ a.c.}; \qquad E_\alpha[1^*(z)] = 0$$

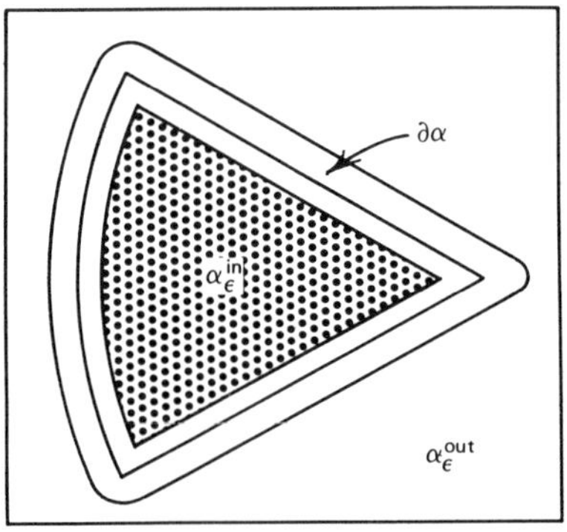

Figure 1

For the remaining parts of the integrals in (4) one should notice that the integrands are bounded and that the areas over which these parts are integrated tend to zero. The latter can be seen by writing the remaining sets as

$$\begin{aligned}(\alpha_\varepsilon^{\text{in}})^c \cap (\alpha_\varepsilon^{\text{out}})^c = \{\alpha \cap [\alpha^c + D(\varepsilon)]^c\}^c \\ \cap [\alpha + D(\varepsilon)] = \{\alpha \cup [\alpha^c + D(\varepsilon)]\} \cap [\alpha + D(\varepsilon)]\end{aligned} \tag{5}$$

This can be rewritten as

$$\{\alpha^c \cap [\alpha + D(\varepsilon)]\} \cup \{[\alpha^c + D(\varepsilon)] \cap [\alpha + D(\varepsilon)]\} = A_\varepsilon \cup B_\varepsilon$$

As $\varepsilon \downarrow 0$, the sets $\alpha + D(\varepsilon)$ decrease toward the closure of α so that A_ε decreases to

$$\alpha^c \cap \text{closure}(\alpha)$$

and hence $\mu(A_\varepsilon) \downarrow 0$. Similarly $\mu(B_\varepsilon) \downarrow 0$ so that the measure of the remaining sets in (5) tends to zero. Q.E.D.

The extension of Theorem 1 to higher dimensional Euclidean spaces is easy and is not done here. If X is a separable Hilbert space, say real, we meet a challenging analytical difficulty which we solve in the case of a Gaussian measure playing the same role as Lebesgue measure did in Theorem 1.

Say that P is a Gaussian distribution, for example, with mean zero, and with covariance operator R. We now sample from a closed set $\alpha \subset X$ according to P conditioned by α. The measure μ will be taken as P itself. If we try to use the idea of the previous proof to show that the regularized set estimate (1) satisfies the consistency relation (2), we encounter the difficulty that we do not know *what is the rate of decrease of the Gaussian measure of small balls*

$$S(\varepsilon) = \{x \mid \|x\|^2 \leqslant \varepsilon\} \subset X; \qquad \varepsilon \downarrow 0$$

To answer this question recall that R is of the trace type so that its eigenvalues $\{\lambda_\nu\}$ satisfy $\Sigma_1^\infty \lambda_\nu < \infty$. Introduce the *exponent of convergence* α as

$$\alpha = \inf\left\{ p \mid \sum_1^\infty \lambda_\nu^p < \infty \right\}$$

Of course $0 \leqslant \alpha \leqslant 1$, and we assume, slightly stronger, that $0<\alpha<1$. *The value of the exponent of convergence will determine the speed with which $P[S(\varepsilon)]$ tend to zero with ε.*

Appealing to the Karhunen-Loève expansion, we can write the random Gaussian element in the Hilbert space as (see Theorem 2 in Chapter 2, Section 4)

$$x=\sum_{\nu=1}^{\infty} \sqrt{\lambda_\nu}\, \eta_\nu \xi_\nu \tag{6}$$

where $\{\xi_\nu\}$ are the normalized eigenvectors of R and η_ν Gaussian uncorrelated stochastic variables with mean zero and variance 1, hence stochastically independent. The sum in (6) converges in the strong topology of X with probability 1. Hence a.c.

$$\|x\|^2=\sum_{\nu=1}^{\infty} \lambda_\nu \eta_\nu^2 \tag{7}$$

so that $\|x\|^2$ has a well-defined probability distribution on $\mathbb{R}^+$ given as an infinite convolution of χ^2-distribution with one degree of freedom scaled with the factors λ_ν. The characteristic function for a λ^2-distribution with one degree of freedom is, however, $(1-2iz)^{-1/2}$ [see, e.g., Cramér (1945), p. 233], so that (7) has the characteristic function

$$\phi(z)=\prod_{\nu=1}^{\infty} (1-2iz\lambda_\nu)^{-1/2}=D^{-1/2}(z) \tag{8}$$

It is clear from the continuity theorem for characteristic functions that the product for $D(z)$ defined by (8) will converge for all real z, but this can also be seen directly, since $|2iz\lambda_\nu|=2\lambda_\nu|z|$ and $\{\lambda_\nu\}$ is summable, hence the product is absolutely convergent [see, e.g., Titchmarsh (1939), p. 15]. We also have, expanding out some of the elements in the product,

$$\begin{aligned} |D(z)|^2 &= \prod_{\nu=1}^{\infty} |1-2iz\lambda_\nu|^2=\prod_{\nu=1}^{\infty}\left(1+4z^2\lambda_\nu^2\right) \\ &\geqslant 64z^6(\lambda_1\lambda_2\lambda_3)^2 \end{aligned} \tag{9}$$

so that $\phi \in L_1$. Actually we shall see that one can make much stronger statements than the bound in (9). Because of the integrability of ϕ, Fourier inversion yields the frequency function f for the random variable $\|x\|^2$;

$$f(x)=\frac{1}{2\pi}\int_{-\infty}^{\infty} e^{-ixz}D^{-1/2}(z)\,dz \tag{10}$$

Integrating from 0 to ε we get from (10)

$$P_\varepsilon = P[S(\varepsilon)] = \frac{1}{2\pi}\int_{-\infty}^{\infty} \frac{1-e^{-i\varepsilon z}}{iz} D^{-1/2}(z)\,dz \tag{11}$$

or, changing variables,

$$P_\varepsilon = \frac{1}{2\pi}\int_{-\infty}^{\infty} \frac{1-e^{-iz}}{iz} D^{-1/2}\left(\frac{z}{\varepsilon}\right) dz \tag{12}$$

To be able to exploit (12) we must examine the function $D(z)$ more carefully. So far it has been defined for real z-values, but the product converges absolutely for any z in the complex plane. It is given as a canonical product, and we thus see that *$D(z)$ is an entire function of genus zero and order α* [see Titchmarsh (1939), chap. VIII, using the convergence exponent α of the zeros].

When we continue $D(z)$ out from the real line, taking the branch of the square root that makes $D(0)=1$, $\phi(z)$ is also continued if we avoid the zeros of D:

$$z_\nu = -\frac{i}{2\lambda_\nu}$$

These zeros will give us branch point singularities for ϕ unless they are of even multiplicity. It is of fundamental importance that *the zeros are situated on a ray*, here the negative imaginary axis. This will make it possible for us to find the behavior of $|D(z)|$ and $|\phi(z)|$ for large real values of z. A classical result [see Pólya-Szegö (1925, 1954), IV, no. 61, and Titchmarsh (1939)] yields the answer, and we give its proof since it will be essential for further developments.

THEOREM 2

Let $f(z)$ be an entire function of order ρ less than 1 expressed by

$$f(z) = \prod_{n=1}^{\infty}\left(1+\frac{z}{z_n}\right)$$

where the z_n are real and positive and the number of zeros

$$n(t) = \#\{z_\nu \leqslant t\} \sim \lambda t^\rho; \qquad \lambda > 0 \tag{13}$$

Then

$$\ln f(re^{i\theta}) \sim e^{i\rho\theta} \frac{\pi\lambda}{\sin \pi\lambda} r^{\rho} \tag{14}$$

for any fixed $\theta \in (-\pi, \pi)$.

Here $\ln f$ denotes the branch of the logarithm function which is real on the positive real line.

Proof We have, beginning with real z, after Abelian rearrangement

$$\ln f(z) = \sum_{n=1}^{\infty} \ln\left(1 + \frac{z}{z_n}\right) = \sum_{n=1}^{\infty} n\left[\ln\left(1 + \frac{z}{z_n}\right) - \ln\left(1 + \frac{z}{z_{n+1}}\right)\right]$$

$$= \sum_{n=1}^{\infty} n \int_{z_n}^{z_{n+1}} \frac{z\,dt}{t(z+t)} = z\int_0^{\infty} \frac{n(t)}{t(z+t)}\,dt \tag{15}$$

But for t large enough we have, with some $\varepsilon > 0$ [see (13)]

$$(\lambda - \varepsilon)t^{\rho} < n(t) < (\lambda + \varepsilon)t^{\rho}; \qquad t > t_0$$

and (15) implies when $z \to +\infty$,

$$\ln f(z) < z\int_0^{t_0} \frac{n(t)}{t(z+t)}\,dt + z(\lambda+\varepsilon)\int_{t_0}^{\infty} \frac{t^{\rho}}{t(z+t)}\,dt$$

$$= z\int_0^{t_0} \frac{n(t) - (\lambda+\varepsilon)t}{t(z+t)} + z(\lambda+\varepsilon)\int_0^{\infty} \frac{t^{\rho}}{t(z+t)}\,dt$$

Hence, substituting $t = zn$ in the last integral,

$$\ln f(z) \leqslant 0(1) + \pi\frac{\lambda+\varepsilon}{\sin \rho\pi} z^{\rho}$$

The same proof gives an inequality in the opposite direction with $\lambda - \varepsilon$ instead of $\lambda + \varepsilon$.

Now let z move out into the complex plane along a ray $z = re^{i\theta}$, $\theta \in (-\pi, \pi)$. By analytic continuation (15) still holds, and the same reasoning shows that

$$\ln f(re^{i\theta}) \sim re^{i\theta}\lambda \int_0^{\infty} \frac{t^{\rho}}{t(re^{i\theta} + t)}\,dt$$

or, substituting $t = ue^{i\theta}$,

$$\ln f(re^{i\theta}) \sim re^{i\rho\theta}\lambda \int_0^\infty \frac{u^\rho}{u(r+u)}\,du = e^{i\rho\theta}\frac{\pi\lambda}{\sin \pi\rho} r^\rho \tag{16}$$

Q.E.D.

To apply this to our situation we just have to rotate the complex plane 90°, and we then get for real z

$$\text{Re}\{\ln|D(z)|\} \sim \cos\frac{\pi\rho}{2}\frac{\pi\lambda}{\sin \pi\rho}|z|^\rho = A|z|^\rho \tag{17}$$

Note that $\cos(\pi\rho)/2$ and $\sin \pi\rho$ are positive since $\rho \in (0,1)$. Hence $|D(z)|$ increases less than exponentially as $|z| \to \infty$ along the real line. The asymptotic relation also gives us bounds, from above and from below. If we restrict ourselves to a sector in the complex plane not containing the imaginary axis, the bounds are uniform over the angle θ.

To get an *upper bound* for $P(\varepsilon)$, we now deform the contour of integration to get as far away as possible from the zeros z_ν, as indicated in Figure 2 into the contour $\Gamma = \Gamma_1 \cup \Gamma_2 \cup \Gamma_3$, where Γ_1 is the real interval $[R, +\infty)$, Γ_2

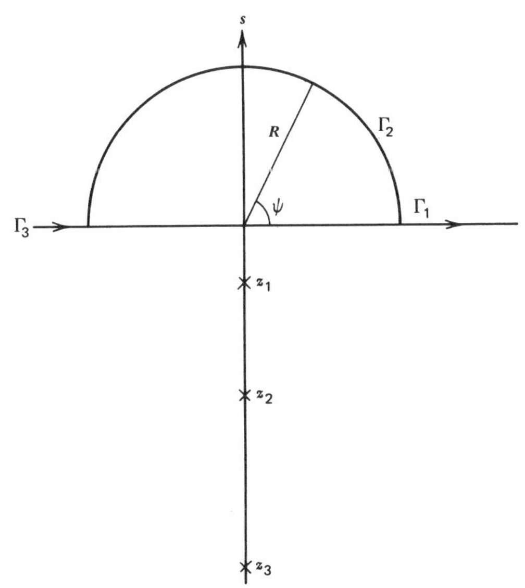

Figure 2

is a half-circle of radius R, and Γ_3 is the real interval $(-\infty, -R)$. Since the integrand is analytic inside the half-circle, $P(\varepsilon)$ can be written as in (11) but integrated along Γ.

The contribution to $P(\varepsilon)$ from $\Gamma_1 \cup \Gamma_3$ is bounded by

$$\frac{2}{\pi}\int_R^\infty \frac{|\phi(z)|}{z}\,dz = \frac{2}{\pi}\int_1^\infty |\phi(Rt)|\frac{dt}{t}$$

Using (17) this gives us the new bound

$$\frac{2}{\pi}\int_1^\infty \exp\{-AR^\rho t^\rho\}\frac{dt}{t} \tag{18}$$

where A is a number slightly smaller than

$$a = \tfrac{1}{2}\cos\frac{\pi\rho}{2}\,\frac{\pi\lambda}{\sin\pi\rho}$$

But (18) is equal to

$$\frac{2}{\pi}e^{-AR^\rho}\int_1^\infty \exp\{-AR^\rho(t^\rho-1)\}\frac{dt}{t} = 0(e^{-AR^\rho}) \tag{19}$$

Using (16) we can dominate the contribution from Γ_2 to (11) by a constant times

$$\int_0^\pi \exp\{\varepsilon R\sin\psi - AR^\rho\}\,d\psi \tag{20}$$

With the help of the Bessel function I_0 of imaginary argument, order zero, (20) can be expressed as

$$I_0(\varepsilon R)\exp\{-AR^\rho\}$$

But for large R,

$$I_0(R) = 0\left(\frac{1}{\sqrt{R}}e^R\right)$$

so that this contribution is dominated by

$$0\left[\frac{1}{\sqrt{\varepsilon R}}\exp(\varepsilon R - AR^\rho)\right]$$

Putting

$$R = B\varepsilon^{1/(\rho-1)} \tag{21}$$

we get the bound

$$0[\exp \varepsilon^{-\sigma}(B - AB^{\rho})]$$

with the crucial coefficient

$$\sigma = \frac{\rho}{1-\rho} \tag{22}$$

Since $\rho < 1$, we have

$$B - AB^{\rho} < 0$$

if B is small enough, $B > 0$. Choose B such that this is so. Then using (19) with (21), we see that $P(\varepsilon)$ tends to zero slower than

$$\exp(-\varepsilon^{-\sigma'}) \tag{23}$$

for any $\sigma^* < \sigma$.

We cannot use the same technique to derive a lower bound, since $\arg[D(z)]$ oscillates too much. Instead we use a Laplace transform argument.

To get a lower bound for $F(\varepsilon) = P(\varepsilon)$, let us put $z = is$, $s > 0$, and write the characteristic function (8) as a Laplace transform:

$$\frac{1}{\sqrt{D(is)}} = \int_0^{\infty} e^{-sx} f(x)\,dx \leqslant \int_0^{\varepsilon} f(x)\,dx + \int_{\varepsilon}^{\infty} e^{-sx} f(x)\,dx$$

Hence

$$F(\varepsilon) \geqslant \frac{1}{\sqrt{D(si)}} - e^{-s\varepsilon} \tag{24}$$

For large s the first term on the right is, using (14) with $\theta = \pi/2$, at least equal to

$$e^{-(a+\delta)s^{\rho}} \tag{25}$$

Now choose, with $\delta' > \delta$, s to satisfy

$$s\varepsilon = (a+\delta')s^{\rho}; \qquad s = \frac{(a+\delta')^{1/(1-\rho)}}{\varepsilon^{1/(1-\rho)}}$$

so that the second term on the right-hand side of (24) can be neglected asymptotically compared to the first one. The first term is now bounded from below by

$$\exp\left[-\frac{(a+\delta)(a+\delta')^{\rho/(1-\rho)}}{\varepsilon^{\sigma}}\right]$$

Hence $P(\varepsilon)$ is of larger order of magnitude than any

$$e^{-A/\varepsilon^{\sigma}}; \qquad A > a^{1/(1-\rho)}.$$

Summing up we have the following theorem (see Note 7.5.A).

THEOREM 3

The Gaussian measure in a real separable Hilbert space, mean 0 *and covariance operator* R, *for which* (13) *holds, satisfies*

$$\lim_{\varepsilon \downarrow 0} e^{\varepsilon^{1/\tau}} P\{x \,|\, \|x\|^2 \leqslant \varepsilon\} = \begin{cases} +\infty; & \text{if } \tau > \sigma \\ 0; & \text{if } \tau < \sigma \end{cases} \tag{26}$$

where σ *is given in* (22).

This settles the question for a sphere with its center located at $x=0$. What happens if the center is at $x=a$, where a now stands for an arbitrary element in X? Since we then get smaller mass, the bound in (23) is still valid, but to get an upper bound an additional argument is required. To carry this out, let us go back to one dimension for a while.

The random variable

$$\xi = (a - b\eta)^2; \qquad \eta = N(0,1); \quad b > 0$$

has the frequency function for $x > 0$

$$h(x) = \frac{1}{2b\sqrt{2\pi x}}\left\{\exp\left[-\frac{(\sqrt{x}+a)^2}{2b^2}\right] + \exp\left[-\frac{(\sqrt{x}-a)^2}{2b^2}\right]\right\} \tag{27}$$

with a weak square root singularity at $x=0$. Here a and b are just real numbers to begin with; later they will be chosen in a specific way related to the covariance operator. Its Laplace transform

$$l(y)=\int_0^\infty e^{-yx}h(x)\,dx$$

will be made up of two integrals originating from (27), the first with $+a$, the other the same except with $-a$. Let us calculate the first

$$l_+(y)=\frac{1}{2b\sqrt{2\pi}}\int_0^\infty \frac{1}{\sqrt{x}}\exp\left[-\frac{(\sqrt{x}+a)^2}{2b^2}-yx\right]dx$$

or changing to the variable $\sqrt{x}$,

$$\begin{aligned}
l_+(y)&=\frac{1}{b\sqrt{2\pi}}\int_0^\infty \exp\left[-\frac{(a+x)^2}{2b^2}-yx^2\right]dx\\
&=\frac{1}{b\sqrt{2\pi}}\exp\left(-\frac{a^2}{2b^2}\right)\int_0^\infty \exp\left[-\frac{ax}{b^2}-\frac{x^2}{2b^2}-yx^2\right]dx\\
&=\frac{1}{b\sqrt{2\pi}}\exp\left[-\frac{a^2}{2b^2}+\frac{a^2}{4b^4}\left(\frac{1}{2b^2}+y\right)^{-1}\right]\\
&\quad\times\int_0^\infty \exp-\left[x\left(\frac{1}{2b^2}+y\right)^{1/2}+\frac{a}{2b^2}\left(\frac{1}{2b^2}+y\right)^{-1/2}\right]dx \qquad (28)
\end{aligned}$$

With the substitution

$$x\left(\frac{1}{2b^2}+y\right)^{1/2}+\frac{a}{2b^2}\left(\frac{1}{2b^2}+y\right)^{-1/2}=\frac{z}{\sqrt{2}}$$

(28) becomes

$$\exp\left[-\frac{a^2y}{1+2b^2y}\right]\frac{1}{\sqrt{1+2b^2y}}\Phi\left(\frac{a}{\sqrt{2}\,b\sqrt{1+2b^2y}}\right)$$

The second term $l_-(y)$ will be similarly

$$l_-(y)=\exp\left[-\frac{a^2y}{1+2b^2y}\right]\frac{1}{\sqrt{1+2b^2y}}\Phi\left(\frac{-a}{\sqrt{2}\,b\sqrt{1+2b^2y}}\right)$$

Adding, we get the Laplace transform

$$l(y)=l_+(y)+l_-(y)=\exp\left[-\frac{a^2y}{1+2b^2y}\right]\frac{1}{\sqrt{1+2b^2y}}$$

We now put $a=a_\nu$, where a_ν are the coordinates of the element a along the coordinate axes ξ_ν, $b=\sqrt{\lambda_\nu}$, multiply the Laplace transform together, corresponding to the convolution we are concerned with, and get

$$L(y)=\exp\left[-y\sum_1^\infty \frac{a_\nu^2}{1+2\lambda_\nu y}\right]\prod_1^\infty \frac{1}{\sqrt{1+2\lambda_\nu y}} \tag{29}$$

This should be compared to (8) with $z=iy$.

Let us recall that $a_\nu=\sqrt{\lambda_\nu}\ \xi_\nu$ where $\{\xi_\nu\}$ is a set of i.i.d. variables with $N(0,1)$ as their marginal distributions. For any $\eta>0$ let us study the probability

$$P\{|\xi_\nu|\leqslant \nu^\eta t,\ \text{all } \nu\}=\prod_1^\infty[1-2\Phi(-\nu^\eta t)]$$

Using the usual asymptotic relation for Φ, we get, for large ν,

$$\Phi(-\nu^\eta t)\sim\frac{1}{\sqrt{2\pi}\ \nu^\eta t}e^{-\nu^{2\eta}t^2/2} \tag{30}$$

so that the product in (30) converges and tends to 1 as $t\to\infty$. This means that

$$\sup_\nu \nu^{-\eta}|\xi_\nu|$$

is a well-defined random variable, or that with probability 1

$$\xi_\nu=0(\nu^\eta)$$

With the eigenvalues of the order ν^{-p}, $p>1$, we can bound the sum appearing in the exponent in (29) by a constant times

$$\int_1^\infty \frac{dx}{x^{p-\eta}(1+y/x^p)}$$

with a different η, still positive, so that as $y\to\infty$, the integral is with probability 1 of order

$$y^{(1+\eta)/p-1}$$

Including the factor y from (29), the exponent in (29) is bounded from below by

$$-y^{(1+\eta)/p}$$

Hence for large y, $L(y)$ is bounded from below by [see (25)]

$$\exp-\left(y^{(1+\eta)/p}+y^{(a+\delta)s^{\rho}}\right)$$

for almost all points a.

In this case $\rho=1/p$, which implies that the reasoning behind Theorem 3 applies also as far as the lower bound is concerned; remember that η is an arbitrarily small positive number. Hence (26) with x replaced by $x-a$ still holds for almost all points $a\in X$.

Therefore let us choose $\varepsilon=\varepsilon_n$ so that

$$\lim_{n\to\infty} nP(\varepsilon_n)=+\infty \tag{31}$$

in order to be able to use the proof of Theorem 1, but now in the Hilbert space X. If we let $\varepsilon_n\downarrow 0$ as

$$\frac{1}{\ln^{1/\sigma'} n}; \qquad \sigma'>\sigma \tag{32}$$

relation (31) is satisfied. Then the proof of Theorem 1 is satisfied and we have the following theorem.

THEOREM 4

If the covariance operator R of the Gaussian measure in the Hilbert space X has eigenvalues tending to zero as ν^{-p}, $p>1$, *then the regularized set estimate in* (1) *is consistent in the sense given in Theorem* 1 *if* ε_n *is as in* (32), $\sigma=\rho/(1-\rho)=1/(p-1)$. *This constant cannot be improved.*

Note that the main part of the statement in the theorem follows already from the paragraph following it immediately. The proof after the paragraph is needed only to show that we cannot let ε_n tend to zero much faster than stated. (For alternative proofs, see Note 7.5.B).

6 GENERAL REGULARIZATION

The analysis in Sections 3–5 sheds some light on three important problems, but it does not tell us why regularization is needed in some cases but not in others. In this section we give one answer to this question, others will no doubt be given by other authors.

To fix ideas, say that our parameter takes values in a real separable Hilbert space. Say that we have access to a commonsense estimate $\hat{\alpha}$ of the parameter $\alpha \in A$ with

$$\begin{cases} E_\alpha(\hat{\alpha}) \equiv \alpha \\ \text{Cov}_\alpha(\hat{\alpha}) = \dfrac{1}{n} C \end{cases} \tag{1}$$

where the operator C must be of the trace type (see Chapter 1, Section 2). The first relation in (1) expresses that α^* is unbiased, but if this is not the case, the following analysis still goes through after appropriate modifications.

From $\hat{\alpha}$ we can also construct estimates of other parameters, say $\beta = L\alpha$, where L is some self-adjoint bounded linear operator.

THEOREM 1

The parameter $\beta = L\alpha$, with L a self-adjoint and bounded linear operator, L: $A \to A$, can be estimated by the unbiased estimate $\beta^* = L\hat{\alpha}$ whose error has the expected norm squared

$$E_\alpha[\|\beta^* - \beta\|^2] = \frac{1}{n}\,\text{tr}(LCL)$$

Proof The result is nearly obvious. First,

$$E_\alpha(\beta^*) = E_\alpha(L\hat{\alpha}) = LE_\alpha(\hat{\alpha}) = L\alpha = \beta \tag{2}$$

where we have used (1.2.7). Relation (2) shows that β^* is unbiased. Its covariance operator is [see (1.2.8)]

$$\text{Cov}_\alpha(\beta^*) = \frac{1}{n} LCL \tag{3}$$

so that, representing the squared norm as a sum of squares and using (3), we get

$$E_\alpha[\|\beta^* - \beta\|^2] = \mathrm{tr}[\mathrm{Cov}_\alpha(\beta^*)] = \frac{1}{n}\,\mathrm{tr}(LCL) \tag{4}$$

Q.E.D.

This is all very well, but the argument breaks down if L is unbounded. And this is *the reason why regularization is needed*: *the parameter in question is the result of an unbounded operator applied to the simple estimates.*

To see what happens in this more difficult case, let L be self-adjoint and defined on a domain $\mathfrak{L}$ everywhere dense in A. We express L in its spectral decomposition:

$$L = \int_{-\infty}^{\infty} \lambda E(d\lambda)$$

where $\{E(\lambda);\, \lambda \in \mathbb{R}\}$ forms a resolution of the identity [see, e.g., Riesz-Nagy (1952), sec. 120].

Since $L\hat{\alpha}$ is no longer defined everywhere, we must regularize the estimate, and we introduce the regularization operators

$$T_\varepsilon = \int_{|\lambda| < 1/\varepsilon} E(d\lambda) \tag{5}$$

For this semi group $\{T_\varepsilon\}$ the composition function (see Section 2) is $\varepsilon(\varepsilon_1, \varepsilon_2) = \max(\varepsilon_1, \varepsilon_2)$. We now prove a basic result.

THEOREM 2

The regularized estimate $\beta_\varepsilon^* = LT_\varepsilon\hat{\alpha}$ of the parameter $\beta = L\alpha$, $\alpha \in \mathfrak{L}$, will have an error whose expected squared norm tends to zero when $n \to \infty$ if ε is made to tend to zero slowly enough.

Proof We have, with $m_\varepsilon = E_\alpha(\beta_\varepsilon^*)$ and bias $b_\varepsilon = m_\varepsilon - \beta$,

$$\begin{aligned}\delta_n &= E_\alpha[\|\beta_\varepsilon^* - \beta\|^2] = E_\alpha[\|\beta_\varepsilon^* - m_\varepsilon\|^2] + \|b_\varepsilon\|^2 + 2E_\alpha[(\beta_\varepsilon^* - m_\varepsilon, b_\varepsilon)] \\ &= E_\alpha[\|\beta_\varepsilon^* - m_\varepsilon\|^2] + \|b_\varepsilon\|^2\end{aligned}$$

The first term can be rewritten as in (4) so that

$$\delta_n = \mathrm{tr}[\mathrm{Cov}(\beta_\varepsilon^*)] + \|b_\varepsilon\|^2 \tag{6}$$

We must now make both terms in (6) tend to zero as n increases.

The bias term can be expressed as

$$\|b_\varepsilon\|^2 = \|E_\alpha(LT_\varepsilon\alpha^*) - L\alpha\|^2 = \|LT_\varepsilon E_\alpha(\alpha^*) - L\alpha\|^2 = \|L(T_\varepsilon - I)\alpha\|^2$$

or, using (5),

$$\|b_\varepsilon\|^2 = \left\| \int_{|\lambda|>1/\varepsilon} \lambda E(d\lambda)\alpha \right\|^2 \tag{7}$$

With a well-known property for spectral integrals (7) can be rewritten as

$$\|b_\varepsilon\|^2 = \int_{|\lambda|>1/\varepsilon} \lambda^2 \|E(d\lambda)\alpha\|^2$$

For any *fixed* $\alpha \in \mathfrak{L}$ we can make $\|b_\varepsilon\|^2$ arbitrarily small by choosing ε small enough.

After we have done this, keeping ε fixed, the variance term in (6) is

$$\operatorname{tr}[\operatorname{Cov}(\beta_\varepsilon^*)] = \operatorname{tr}\operatorname{Cov}(LT_\varepsilon\hat{\alpha})$$

but LT_ε is bounded so that, as in (3), we get

$$\operatorname{tr}[\operatorname{Cov}(\beta_\varepsilon^*)] = \frac{1}{n}\operatorname{tr}(LT_\varepsilon CLT_\varepsilon) \to 0$$

as $n \to \infty$. Hence Δ_n can be made arbitrarily small. Q.E.D.

Remark 1 Just as we have noticed in previous sections, we can claim only that for *fixed* $\alpha \in \mathfrak{L}$, the bias term can be made small by choosing ε sufficiently small. If we want statements about the *rate* with which we should let ε tend to zero, we must introduce subsets of $\mathfrak{L}$ with prescribed regularity in order to get uniform results. The method of proof can then be used to get asymptotic expressions for the error and for how the regularization parameter ε should be made to tend to zero as n tends to infinity. The same question appears in our study of sieves in Chapters 8 and 9.

Example 1 Let $\beta = f$ be a quadratically integrable frequency function on the torus $[0, 2\pi)$, so that $f(0) = f(2\pi)$ and $\beta = L\alpha$, where α is the corresponding distribution function and $L = D$ stands for differentiation. The operator $(1/i)\,D$ on periodic functions on $[0, 2\pi]$ is self-adjoint with the spectrum $\mathbb{Z}$. To the eigenvalue $k \in \mathbb{Z}$ corresponds the normalized eigenfunction $(2\pi)^{-1/2}e^{ikx}$.

The obvious estimate of F is the empirical distribution function $F^* = \hat{\alpha}$ (see Note 7.6.A) which is unbiased and has a covariance operator given by the integral kernel

$$\frac{1}{n}\{F[\min(x,y)] - F(x)F(y)\}$$

Then, applying the regularization operator from Theorem 2, we have, for any $h \in \mathcal{L}$,

$$(LT_\varepsilon h)(x) = \frac{1}{\sqrt{2\pi}} \sum_{|k|<1/\varepsilon} e^{ikx} k \frac{1}{\sqrt{2\pi}} \int_0^{2\pi} e^{-iky} h(y)\,dy$$

Integrating by parts we get, in terms of the *Dirichlet kernel*, if $1/\varepsilon$ is an integer K,

$$D_K(z) = \frac{1}{2\pi} \frac{\sin\left(K+\frac{1}{2}\right)z}{\sin z/2}$$

the regularized estimate

$$\alpha_\varepsilon^* = \int_0^{2\pi} D_K(x-y) F^*(dy)$$

The regularized estimate can then be written as a kernel estimate in terms of the observed sample $(x_1, x_2, \ldots, x_n)$

$$\frac{1}{n} \sum_{\nu=1}^{n} D_K(x - x_\nu)$$

It is consistent in expected norm square if we let K tend slowly enough to $+\infty$ when n tends to $+\infty$.

CHAPTER 8

Sieves in Parameter Space

1 PARAMETER IN l_1-SPACE

In this chapter we study a number of parameter spaces, several of great importance in themselves, in order to develop our understanding of what can happen when inference is attempted in general parameter spaces.

In each particular parameter space we examine the ML estimates obtained by choosing suitable sieves. It is not just a matter of making the resulting estimates consistent, which is usually not too difficult, but of finding the analytic properties of the estimates and of making sure that they can be calculated conveniently. These considerations will dictate the choice of sieves.

Let us begin by letting the stochastic variable x take values $1,2,3,\ldots$ with probabilities $p_1, p_2, p_3, \ldots$ with $\alpha=(p_1, p_2, p_3, \ldots)\in A \subset l_1$. Having observed a sample $x_1, x_2, \ldots, x_n$, the likelihood function is

$$L=\prod_{i=1}^{n} \alpha_{x_i}=\prod_{k=1}^{\infty} \alpha_k^{f_k}; \qquad f_k=\#\{x_i\text{'s}=k\}, \quad \alpha_k=p_k$$

leading to the obvious ML-estimate

$$\alpha^*=\alpha^*(n)=(\alpha_1^*, \alpha_2^*, \ldots); \qquad \alpha_k^*=f_k/n$$

The maximum is attained and unique in A.

The sequence $\{\alpha^*(n);\quad n=1,2,\ldots\}$ is strongly consistent with the norm of the Banach space l_1. Indeed, with dominated convergence,

$$E(\|\alpha^*-\alpha\|)=E\left[\sum_{k=1}^{\infty}|\alpha_k^*-\alpha|\right]=\sum_{k=1}^{\infty}E|\alpha_k^*-\alpha|\to 0$$

since, for each k, we have $E|\alpha_k^* - \alpha| \to 0$, and with the domination

$$E|\alpha_k^* - \alpha| \leqslant 2\alpha_k; \qquad 2\alpha \in l_1$$

Hence *in this case the maximum likelihood method requires no sieve: it leads to a consistent estimate directly*. We can actually be more precise in evaluating the asymptotic behavior of α^* (see Note 8.1.A).

THEOREM 1

If $\Sigma\sqrt{\alpha_k} < +\infty$, we have asymptotically for the expected l_1-norm of the estimation error

$$\lim_{n\to\infty} \sqrt{n}\, E(\|\alpha^* - \alpha\|) = \sqrt{\frac{2}{\pi}} \sum_{k=1}^{\infty} \sqrt{p_k(1-p_k)}$$

Proof We have, since $f_\nu = B(n, \alpha_\nu)$, the relation

$$E|f_k - np_k| = 2\nu_k(1-p_k)b(\nu_k; n, p_k)$$

where the integer ν_k satisfies $np_k < \nu_k \leqslant np_k + 1$ [see Feller (1957), p. 226], with the usual notation for the binomial probabilities

$$b(\nu; n, p) = \binom{n}{\nu} p^\nu (1-p)^{n-\nu}$$

Hence

$$\sqrt{n}\, E(\|\alpha^* - \alpha\|) = 2\sum_{k=1}^{\infty} \frac{\nu_k}{\sqrt{n}}(1-p_k)b(\nu_k; n, p_k) \tag{1}$$

To get the asymptotic behavior of this series, we use the refined form of Stirling's formula [see Feller (1957), p. 52]:

$$\sqrt{2\pi}\, n^{n+1/2} e^{-n+1/(12n+1)} < n! < \sqrt{2\pi}\, n^{n+1/2} e^{-n+1/12n}$$

so that

$$\sqrt{n}\, b(\nu; n, p) \leqslant \sqrt{n}\, \frac{\sqrt{2\pi n}\, n^n}{e^{n-1/12n}} \times \frac{e^{\nu - 1/(12\nu+1)}}{\sqrt{2\pi\nu}\, \nu^\nu} \times \frac{e^{n-\nu-1/12(n-\nu+1)}}{\sqrt{2\pi(n-\nu)}\, (n-\nu)^{n-\nu}}$$

$$\times p^\nu (1-p)^{n-\nu} = \frac{1}{\sqrt{2\pi(\nu/n)(n-\nu)/n}} \left(\frac{pn}{\nu}\right)^\nu \left(\frac{(1-p)n}{n-\nu}\right)^{n-\nu}$$

$$\exp\left[\frac{1}{12n} - \frac{1}{12\nu+1} - \frac{1}{12(n-\nu)}\right]$$

Now use $np < \nu \leqslant np+1$. Then

$$\sqrt{n}\, b(\nu; n, p) \leqslant \frac{1}{\sqrt{2\pi}\,\sqrt{p(q-1/n)}} \times \left[\frac{(1-p)n}{(1-p)n-1}\right]^{(1-p)n} \times e^{1/12n}$$

But in the infinite series (1) we are interested in the tail, k large, and large values of n. A uniform bound is then given by

$$\sqrt{n}\, b(\nu; n, p) \leqslant \frac{\text{constant}}{\sqrt{p}} \frac{1}{(1-1/qn)^{qn}}$$

where $q = q_k > 1-\varepsilon$ if k is large enough. Hence

$$\sqrt{n}\, b(\nu; n, p) \leqslant \frac{\text{constant}}{\sqrt{p}}$$

Applying this bound together with the classical

$$\lim_{n\to\infty} \sqrt{n}\, b(\nu; n, p) = \frac{1}{\sqrt{2\pi pq}}$$

we get

$$\lim_{n\to\infty} \sqrt{n}\, E\|\alpha_n^* - \alpha\| = \sqrt{\frac{2}{\pi}} \sum_{k=1}^{\infty} \sqrt{p_k(1-q_k)}$$

since the terms in the series studied are bounded by

$$2\left(p_k + \frac{1}{n}\right) q_k b(\nu_k; n, p_k) = 0\left(\sqrt{p_k}\right)$$

and this conclude the proof by appealing to the convergence condition in the statement of the theorem. Q.E.D.

The case $\Sigma\sqrt{\alpha_k} = +\infty$ is more difficult, and we can offer only a partial answer. We start again from relation (1) with $\nu_k = [np_k]+1$ and the bounds

$$\beta(\nu; n, p)\exp\left[\frac{1}{12n+1} - \frac{1}{12\nu} - \frac{1}{12(n-\nu)}\right] \leqslant b(\nu; n, p)$$

$$\leqslant \beta(\nu; n, p)\exp\left[\frac{1}{12n} - \frac{1}{12\nu+1} - \frac{1}{12(n-\nu)+1}\right]$$

using the expression

$$\beta(\nu; n, p)=\sqrt{\frac{n}{2\pi\nu(n-\nu)}}\left(\frac{pn}{\nu}\right)^{\nu}\left(\frac{(1-p)n}{n-\nu}\right)^{n-\nu}$$

Writing

$$d_n=2\left(\sum_{p_k>1/n}+\sum_{p_k<1/n}\right)\frac{\nu_k}{n}(1-p_k)b(\nu_k; n, p_k)\triangleq d_n'+d_n''$$

we get the upper bound

$$d_n'\leqslant 2e^{1/12n}\sum_{p_k>1/n}\frac{\nu_k}{n}\beta(\nu_k; n, p_k)$$

or, using the relation between ν_k and p_k,

$$d_n'\leqslant\sqrt{\frac{2}{\pi n}}\,e^{1/12n}\sum_{p_k>1/n}\left(p_k+\frac{1}{n}\right)$$

$$\times\sqrt{\frac{1}{p_k(1-p_k-1/n)}}\left(\frac{1-p_k}{1-p_k-1/n}\right)^{n(1-p_k)}$$

Since $1/n\leqslant p_k$ in this sum, we get

$$d_n'\leqslant 2\sqrt{\frac{2}{\pi n}}\,e^{1/12n}\frac{1}{\sqrt{1-p_{\max}-1/n}}\sum_{p_k>1/n}\sqrt{p_k}\,\frac{1}{(1-1/\beta_k)^{\beta_k}}$$

with

$$\beta_k=n(1-p_k)>n(1-p_{\max})$$

This proves that

$$d_n'\leqslant\frac{\text{constant}}{\sqrt{n}}\sum_{p_k>1/n}\sqrt{p_k}=0\left[\frac{1}{\sqrt{n}}M\left(\frac{1}{n}\right)\right]$$

where

$$M(x)=\sum_{p_k>x}\sqrt{p_k}$$

With the asymptotic behavior k^{-p}, $1<p<2$, for the p_k we have, as $x\downarrow 0$,

$$M(x)=\sum_{p_k>x} k^{-p/2}=\sum_{k<x^{-1/p}} k^{-p/2}\sim \text{constant}\, x^{1/2-1/p}$$

so that $n^{-1/2}M(1/n)\sim \text{constant}\, n^{-\beta}$, $\beta=1-1/p, 0<\beta<\frac{1}{2}$.

On the other hand

$$d_n''=2\sum_{p_k<1/n}\frac{1-p_k}{n}b(1;n,p_k)=2\sum_{p_k<1/n}p_k(1-p_k)^n$$

since $\nu_k=1$ in this sum. Hence

$$d_n''\leqslant 2\sum_{p_k<1/n}p_k\sim\frac{\text{constant}}{n^\beta}$$

which proves the upper bound

$$d_n=0\left(\frac{1}{n^\beta}\right)$$

To get a lower bound, we use again the fact that $1\leqslant\nu_k\leqslant[np_{\max}]+1$, so that

$$d_n\geqslant\sqrt{\frac{2}{\pi n}}\exp\left[\frac{1}{12n+1}-\frac{1}{12}-\frac{1}{12(n-[np_{\max}]-1)}\right]$$

$$\times\sum_{p_k>1/n}p_k\sqrt{\frac{1}{(p_k+1/n)(1-p_k)}}\;\frac{1}{(1+1/np_k)^{np_k+1}}$$

bounded from below by

$$\frac{\text{constant}}{\sqrt{n}}\sum_{p_k>1/n}\sqrt{p_k}\sqrt{\frac{p_k}{p_k+1/n}}$$

But $x=np_k>1$ gives us

$$\sqrt{\frac{x}{x+1}}>\frac{1}{\sqrt{2}}$$

so that indeed d_n is at least of order

$$\frac{1}{\sqrt{n}} \sum_{p_k > 1/n} \sqrt{p_k} = \frac{1}{\sqrt{n}} M\left(\frac{1}{n}\right)$$

and we have proved the following theorem.

THEOREM 2

If $p_k \sim k^{-p}, 1<p<2$, the expected norm of the estimation error is of order $n^{-\beta}, \beta = 1-1/p$.

Remark 1 We have only obtained the *order* of d_n for large n, not an asymptotic expression, as in the earlier case. To get the leading term, we would have to take into account how the quantity

$$\nu_k - np_k = [np_n] - np_k - 1$$

varies with k *and* n, that is, the fractional part of np_k. This fractional part will show an oscillating behavior, it will not converge, and this will require an additional analysis which we have not attempted.

Remark 2 The order of the expected l_1-norm is $n^{-\beta}$. Note that β can be arbitrarily close to zero by making p close to but a little larger than 1; very slow convergence will then occur.

2 PARAMETER IN L_1-SPACE

The situation changes drastically when we go to L_1-spaces. With the sample space $X=[0,1]$ (see Note 8.2.A) consider probability measures P_α on X absolutely continuous with respect to Lebesgue measure dx. Denote the frequency function by $\alpha = \alpha(x)$, so that the parameter space can be embedded in $L_1 = L_1([0,1], dx)$.

For a given sample $x_1, x_2, \ldots, x_n$ the likelihood function

$$L(\alpha) = \prod_{\nu=1}^{n} \alpha(x_\nu)$$

can be made arbitrarily large by letting α be zero except in small neighborhoods around x_ν. Making these neighborhoods smaller, L will tend to $+\infty$ while the corresponding α do not converge to any element in L_1.

Hence *a sieve is needed*, and we first choose a local one defined by

$$\mathbb{S}(\mu)=\{\alpha\,|\,\alpha \text{ positive and absolutely continuous frequency function with } |\alpha'|\leqslant \alpha/\mu \text{ a.e.}\} \tag{1}$$

We then have the following theorem.

THEOREM 1

If the formal entropy $H(\alpha)$ is finite, a restricted ML-estimate $\alpha^*\in\mathbb{S}(\mu)$ will tend to α in the norm of L_1 if μ is made to tend to zero slowly enough as the sample size n tends to infinity.

Proof The argument will be organized into six steps. Let us first prove that the functions belonging to $\mathbb{S}(\mu)$ are uniformly bounded and equicontinuous.

For fixed positive μ, integrating the inequality in (1) from 0 to x we get

$$\alpha(0)e^{-x/\mu}\leqslant\alpha(x)\leqslant\alpha(0)e^{x/\mu} \tag{2}$$

and integrating (2) from 0 to 1, using the fact that α has integral 1, it follows that

$$\frac{1}{\mu(e^{1/\mu}-1)}\leqslant\alpha(0)\leqslant\frac{1}{\mu(1-e^{-1/\mu})} \tag{3}$$

Combining (2) with (3) we have the uniform bound in $\mathbb{S}(\mu)$

$$\frac{e^{-x/\mu}}{\mu(e^{1/\mu}-1)}\leqslant\alpha(x)\leqslant\frac{e^{x/\mu}}{\mu(1-e^{-1/\mu})} \tag{4}$$

so that the α are uniformly bounded. But this implies that $\alpha'(x)$ is also uniformly bounded [use (1) and (4)] so that the functions in $\mathbb{S}(\mu)$ are equicontinuous.

Second, we show that $\sup L$ is attained in $\mathbb{S}(\mu)$ for a given sample, and consider a sequence $\alpha_1,\alpha_2,\ldots\in\mathbb{S}(\mu)$ such that $L(\alpha_\nu)\to\sup L(\alpha),\alpha\in\mathbb{S}(\mu)$. Since α'_ν and α_ν are uniformly bounded, we can extract a subsequence $\{\alpha'_{n_\nu}\}$ from $\{\alpha'_\nu\}$ converging in L_1-norm to some bounded function β and such that $\alpha_\nu(0)$ also converges. It then follows by integration that $\alpha_{n_\nu}(x)$ converges for any x to the integral of β plus some constant; call the limit α_0. Hence $L(\alpha_{n_\nu})\to L(\alpha_0)=\sup L(\alpha),\alpha\in\mathbb{S}(\mu)$. Hence the supremum is attained in the sieve. We do not claim that the restricted ML-solution is unique.

Third, we prove that for any $\alpha \in A$ and $\delta > 0$ we can find a *regularization* α_δ in $\mathbb{S}(\mu)$, μ small enough, with small L_1-norm for the approximation error $\|\alpha_\delta - \alpha\|_1 < \delta$ and such that

$$\left| \int_0^1 \ln \alpha_\delta(x) \alpha(x)\, dx + H(\alpha) \right| < \delta$$

This is easy and can be done in many ways, one of which is to truncate the function α at the values $1/c$ and c into a modified function $\alpha_1 \in L_1$ with

$$\alpha_1(x) \qquad \text{and} \qquad \frac{1}{\alpha_1(x)} \leqslant c$$

and $\|\alpha_1 - \alpha\|_1 < \delta/2$ with

$$\left| \int_0^1 \ln \alpha_1(x) \alpha(x) + H(\alpha) \right| < \frac{\delta}{2}$$

We then choose, with an appropriate modification at the boundary points $x = 0$ and 1,

$$\alpha_2(x) = \frac{1}{2h} \int_{x-h}^{x+h} \alpha_1(y)\, dy$$

so that

$$\frac{1}{c} \leqslant \alpha_2(x) \leqslant c$$
$$-\ln c \leqslant \ln \alpha_2(x) \leqslant \ln c$$

and hence $\alpha_2 \in \mathbb{S}(\mu)$ if the mesh size μ is small enough. Since $\alpha_2(x) \to \alpha(x)$ ad $h \downarrow 0$ a.e., we can find a value h such that

$$\|\alpha_2 - \alpha_1\|_1 \leqslant \frac{\delta}{2}$$
$$\left| \int_0^1 \ln \alpha_2(x) \alpha(x)\, dx - \int_0^1 \ln \alpha_1(x)\, dx \right| \leqslant \frac{\delta}{2}$$

Then $\alpha_\delta = \alpha_2$ is the required regularization.

Fourth, we prove that $-(1/n)\ln L(\alpha_\delta)$ can be made to differ arbitrarily little from $H(\alpha)$ with a probability close to 1 when n is large enough. To see this, denote the density function corresponding to α by F and the empirical density function of the sample $x_1, x_2, \ldots, x_n$ by F^*. We have, for any absolutely continuous h,

$$\int_0^1 h(x)[F^*(dx)-F(dx)] = -\int_0^1 h'(x)[F^*(x)-F(x)]\,dx$$

With $\Delta_n(x)=\sqrt{n}\,[F^*(x)-F(x)]$, the well-known result due to Kolmogorov, see Theorem 7.4.4, tells us that $\sup_x|\Delta_n(x)|$ has a nondegenerate limiting probability distribution as n tends to infinity.

With $h(x)=\ln\alpha_\delta(x)$, this implies that

$$\frac{1}{n}\sum_{i=1}^{n}\ln\alpha_\delta(x_i)=\int_0^1 h(x)F^*(dx) \tag{5}$$

differs from

$$\int_0^1 \ln\alpha_\delta(x)\alpha(x)\,dx=\int_0^1 h(x)F(dx) \tag{6}$$

by at most

$$\frac{C_\mu}{\sqrt{n}}$$

with a probability close to 1 if C_μ is made large enough, since $|\ln\alpha_\delta|$ is uniformly bounded. Hence, with a probability arbitrarily close to 1, (5) can be made close to (6), which itself differs less than δ from $H(\alpha)$.

Fifth, introduce the auxiliary function (see Figure 1)

$$\phi(y)=y-\ln(1+y); \qquad -1<y<\infty \tag{7}$$

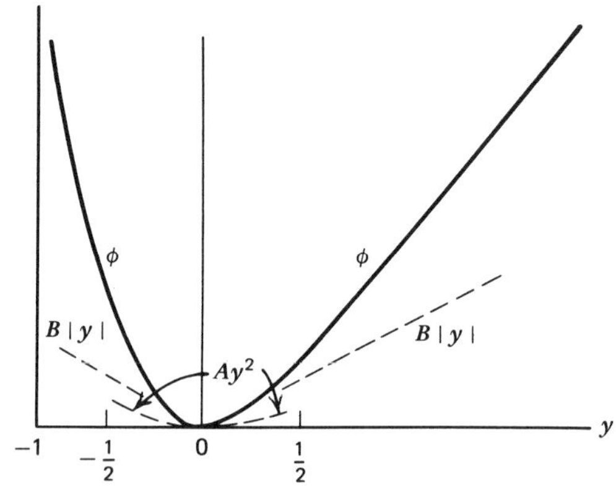

Figure 1

Since

$$\phi'(y)=\frac{y}{1+y}$$

which is positive for $y>0$ and negative for $y<0$, it is clear that $\phi(y)\geqslant 0$ with equality only for $y=0$. Also we can find positive constants A and B such that we have the inequalities

$$\begin{aligned} Ay^2 &\leqslant \phi(y); \qquad |y|<\tfrac{1}{2} \\ B|y| &\leqslant \phi(y); \qquad -1<y<-\tfrac{1}{2}, \quad y>\tfrac{1}{2} \end{aligned} \tag{8}$$

We are now ready for the sixth and final step. Substitute

$$y=\begin{cases} -1+\dfrac{\alpha^*}{\alpha}; & \text{for } x\in\{x|\alpha(x)>0\} \\ 0 & \text{otherwise} \end{cases}$$

in (7) and multiply by $\alpha(x)$. Integrating over (0, 1) we get the inequality

$$\begin{aligned} 0 &\leqslant \int_0^1 \phi\left(\frac{\alpha^*(x)}{\alpha(x)}-1\right)\alpha(x)\,dx \\ &= \int_0^1 [\alpha^*(x)-\alpha(x)]\,dx - \int_0^1 \ln \alpha^*(x)\alpha(x) - H(\alpha) \end{aligned} \tag{9}$$

Let α^* denote an ML-solution in $\mathbb{S}(\mu)$. We have shown that with probability close to 1 the last integral in (9) is greater than $H(\alpha)-\varepsilon$. The next last integral is zero since α^* and α are both frequency functions. This implies that

$$\int_0^1 \phi\left[\frac{\alpha^*(x)}{\alpha(x)}-1\right]\alpha(x)\,dx \leqslant \varepsilon \tag{10}$$

On the other hand the bounds in (8) imply together with (10), that with large probability, $\varepsilon'=\max(\varepsilon/A, \varepsilon/B)$,

$$\begin{aligned} &\int_E \left[\frac{\alpha^*(x)}{\alpha(x)}-1\right]^2 \alpha(x)\,dx \leqslant \varepsilon' \\ &\int_{E^c} |\alpha^*(x)-\alpha(x)|\,dx \leqslant \varepsilon' \\ &E=\left\{x\left|\left|\frac{\alpha^*(x)}{\alpha(x)}-1\right|\right.\leqslant \tfrac{1}{2}\right\} \end{aligned} \tag{11}$$

To exploit the first inequality in (11), we just apply Schwarz's inequality which leads to

$$\int_E \left|\frac{\alpha^*(x)}{\alpha(x)}-1\right|\alpha(x)\,dx \leqslant \left\{\int_E \left[\frac{\alpha^*(x)}{\alpha(x)}-1\right]^2 \alpha(x)\,dx \int_E a(x)\,dx\right\}^{1/2} \qquad (12)$$

On the right-hand side of (12) the first integral is at most ε' [see the first inequality in (11)] and the second is at most 1. Hence, with a probability close to 1, if n is large enough,

$$\|\alpha^*-\alpha\|_1=\int_0^1 |\alpha^*(x)-\alpha(x)|\,dx \leqslant \sqrt{\varepsilon'}\ +\varepsilon'$$

and this is the relation we wanted to establish. Q.E.D.

Several other sieves can be suggested for this combination of parameter space and sample space, two of which are mentioned here. The first one consists of the subset of A for which the frequency function α is constant on each interval $((k-1)/m,(k/m)=I_k$, $k=1,2,\ldots,m$, where the mesh size $\mu=1/m$. The ML-solution constrained by this sieve is clearly the usual *histogram*:

$$\alpha_\mu^*(x)=\frac{n_k}{n}m; \qquad x\in I_k$$

where n_k is the number of observations falling in I_k. Hence the expected L_1-norm of this estimate is

$$E\left[\|\alpha_\mu^*-\alpha\|\right]=\sum_{k=1}^m \int_{I_k} E\left[\left|\frac{n_k}{n}m-\alpha(x)\right|\right]dx$$

and $n_k=B(n,p_k)$ where the probabilities

$$p_k=\int_{I_k}\alpha(x)\,dx$$

For fixed mesh size we have a.c. as $n\to\infty$,

$$\frac{n_k}{n}m\to mp_k; \qquad k=1,2,\ldots,m$$

so that

$$\lim_{n\to\infty} E\left[\|\alpha_\mu^* - \alpha\|\right] = \sum_{k=1}^{m} \int_{I_k} |mp_k - \alpha(x)|\, dx$$

As the mesh size is made to tend to zero (and m to infinity), it is intuitively clear that the above expression tends to zero. (A proof of this is given at the end of this section for another purpose.) Hence α_μ^* is consistent in the expected norm if μ is allowed to tend to zero slowly enough for increasing the sample size.

We can be more precise. Introduce the regularization integral kernel for $\mu = 1/m$

$$K_\mu(x,y) = \begin{cases} m; & \text{in } x\in I_k \quad \text{when } y\in I_k \\ 0; & \text{otherwise} \end{cases}$$

and the regularized parameter $\hat{\alpha}_\mu$. Then $\hat{\alpha}_\mu(x) = mp_k$ for $x \in I_k$. We derive an asymptotic expression for the expected norm of $\alpha_\mu^* - \hat{\alpha}_\mu$.

We have

$$e_{n,\mu} = E\left[\|\alpha_\mu^* - \hat{\alpha}_\mu\|\right] = \sum_{k=1}^{m} E\left[\left|\frac{n_k}{n} - p_k\right|\right]$$

so that it is natural to use equation (1) in Section 1 and its consequences again. It gives us

$$e_\mu = \lim_{n\to\infty} \sqrt{n}\, c_{n,\mu} = \sqrt{\frac{2}{\pi m}} \sum_{k=1}^{m} \sqrt{mp_k(1-p_k)} \tag{13}$$

To be able to handle e_μ asymptotically for small μ, we write (13) as

$$e_\mu = \sqrt{\frac{2}{\pi} m} \int_0^1 f_m(x)\, dx$$

with the step function

$$f_m(x) = \sqrt{mp_k(1-p_k)}\ ; \qquad x \in I_k$$

We now dominate $\{f_m\}$ by a function f obtained from the Hardy-Littlewood maximal theorem in L_1 [see, e.g., Hewitt-Stromberg (1975), p. 426]. Put

$$g(x)=\sup_{\varepsilon>0}\frac{1}{|\varepsilon|}\int_x^{x+\varepsilon}\alpha(y)\,dy$$

so that $g\in L_{1/2}$ by the theorem just quoted. But $f_m\leqslant f=g^{1/2}$. Also $f_m(x)\to\sqrt{\alpha(x)}$ a.c. as $m\to\infty$, so that by dominated convergence we get the strange limit result (notice the minus sign in the exponent.)

$$\lim_{m\to\infty} m^{-1/2}e_\mu=\sqrt{\frac{2}{\pi}}\int_0^1\sqrt{\alpha(x)}\,dx$$

which should be compared to Theorem 1 in Section 1.

Still another sieve is the one consisting of α's which are continuous and piecewise linear: linear over each I_k. *This sieve is made up of linear splines* (see Note 8.2.B). We can express the regularized parameter by the ordinates y_k at $x=(k-1)/m$, $k=1,2,\ldots,m+1$, and get for the ML-solution constrained by this sieve the equations

$$y_{k-1}+y_k=2m\frac{n_k}{n};\qquad k=1,2,\ldots,m$$

This system of equations is underdetermined so that we can impose one more condition, for example, $y_0=y_m$, for periodicity if X is the torus. In spite of its considerable interest, this type of sieve will not be explored any further here.

With the same A, but *restricted by the qualitative condition that α be nonincreasing*, we get a drastically different answer, with a construction from Grenander (1956) (see Note 8.2.C).

THEOREM 2 ESTIMATING MONOTONE FREQUENCY FUNCTIONS

If A consists of nonincreasing frequency functions on [0, 1], *the maximum of the likelihood is attained and given by*

$$\alpha^*(x)=-\frac{dG^{**}}{dx}\quad \text{a.e.}$$

where G^{**} is the greatest convex minorant of $G^*(x)=1-F^*(x)$. The ML-solution is strongly L_1-consistent in probability.

Proof With the sample $x_1, x_2,\ldots, x_n$ and with the notation $\alpha_i=\alpha(x_i)$ it is clear that the maximum of $L=\alpha_1\alpha_2\cdots\alpha_n$ is attained. Indeed the α_i vary in the compact region of $\mathbb{R}^n$ bounded by the surfaces $\alpha_1=0, \alpha_2=0,\ldots,\alpha_n=0$ and $x_1\alpha_1+(x_2-x_1)\alpha_2+\cdots+(1-x_n)\alpha_n=1$, and L is a continuous function of the α. It is clear that once we have fixed α_i, in order to realize the maximum, we should make $\alpha(x)=\alpha_i$ for $x\in(x_{i-1},x_i]$. In other words the ML-*solution will be a step function continuous to the left*.

To determine this step function explicitly, we have to solve the extremum problem

$$\prod_1^n a_i=\max$$
$$\alpha_1\geqslant\alpha_2\geqslant\cdots\geqslant\alpha_n\geqslant 0$$
$$x_1\alpha_1+(x_2-x_1)\alpha_2+\cdots+(x_n-x_{n-1})\alpha_n=1$$

so that our original infinite-dimensional problem has now been reduced to a finite-dimensional one.

Denote the region in $\mathbb{R}^n$ defined by the inequalities for the α_i by R, and let us assume that the maximum is attained for

$$\alpha_1=\alpha_2=\cdots=\alpha_{i_1}=\phi_1$$
$$\alpha_{i_1+1}=\alpha_{i_2+2}=\cdots=\alpha_{i_2}=\phi_2$$
$$\vdots$$
$$\alpha_{i_p+1}=\alpha_{i_p+2}=\cdots=\alpha_n=\phi_{p+1}$$

with $\phi_1>\phi_2>\cdots>\phi_{p+1}$ and $i_1\geqslant 1, i_2\geqslant i_1+1,\ldots,n>i_p+1$. In order that this be true it is *necessary* that the gradients of L taken in R vanish, so that, with a Lagrange multiplier λ,

$$\frac{\partial}{\partial\phi_k}\Big\{\ln\big[\phi_1^{i_1}\phi_2^{i_2-i_1}\cdots\phi_{p+1}^{n-i_p}\big]-\lambda\big[\phi_1x_1+\phi_2(x_2-x_1)+\cdots+\phi_{p+1}(x_n-x_{i_p})-1\big]\Big\}=0$$

for $k=1,2,\ldots,p+1$. This results in

$$\begin{aligned}\phi_1&=\frac{1}{n}\frac{i_1}{x_1}\\ \phi_2&=\frac{1}{n}\frac{i_2-i_1}{x_2-x_1}\\ &\vdots\\ \phi_{p+1}&=\frac{1}{n}\frac{n-i_p}{x_n-x_{i_p}}\end{aligned}\tag{14}$$

This determines $\alpha^*(x_i)$ and hence the step function α for all $x \in X$. Integrating from x to 1 we get the function

$$G^{**}(x) = \int_x^1 \alpha^*(\xi)\,d\xi \tag{15}$$

which should be regarded as a provisional estimate of $G(x) = 1 - F(x)$. The function in (15) takes the value 1 at 0, is decreasing from 0 to x_n, and is convex. At the points $x = x_{i_t}$, $t = 1, 2, \ldots, p+1$, it is given by

$$G^{**}(x_{i_t}) = \frac{1}{n}\frac{i_{t+1} - i_t}{x_{i_{t+1}} - x_{i_t}}(x_{i_{t+1}} - x_{i_t}) + \cdots + \frac{1}{n}\frac{n - i_p}{x_n - x_{i_p}}(x_n - x_{i_p})$$

$$= \frac{n - i_t}{n} = G^*(x_{i_t}) = 1 - F^*(x_{i_t})$$

with the usual empirical density function F^* and its complement $1 - F^*$.

It remains to determine p and the numbers $i_1, i_2, \ldots, i_p$. From now on let G^{**} denote the greatest of all convex polygons determined by conditions of the form (14); G^{**} is the *greatest convex minorant of* G^* (see Figure 2). We show that $F^{**} = 1 - G^{**}$ corresponds to the ML-solution for α.

Call the ML-solution λ. We already know that it has the form of a G polygon with ordinates as in (14) so that $\lambda(x) = G^*(x)$ for $x = 0$, x_n, and at the points x_i, where λ has a jump. Suppose that there is a point x_i such that $G^*(x_i) < \lambda(x_i)$ (see Figure 3), where $x_{i-\nu}$ is the first corner to the left of x_i of the convex polygon λ and $x_{i+\mu}$ similarly on the right.

Let us now deform λ slightly downwards as indicated in the figure by the dashed line. To see how the likelihood changes due to this deforma-

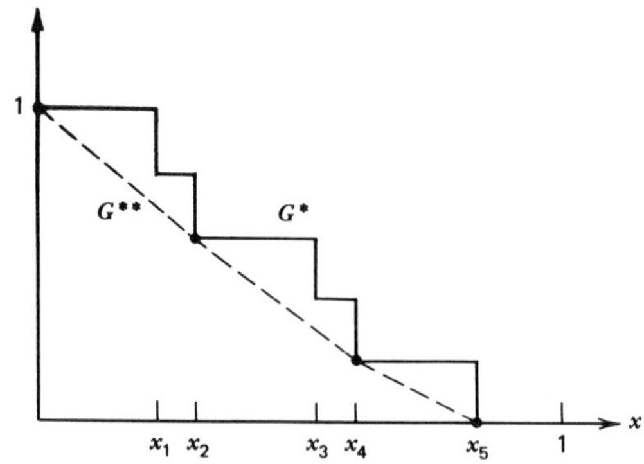

Figure 2

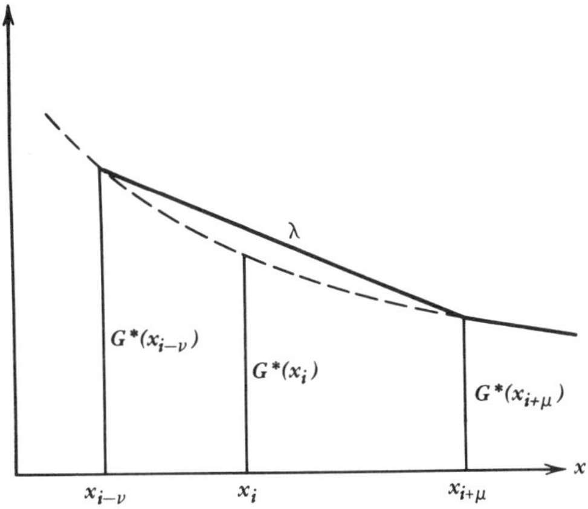

Figure 3

tion, we need only consider the values of the frequency function at observed values between $x_{i-\nu}$ and $x_{i+\mu}$. But we have the estimated value of α for $x \in (x_{i-\nu}, x_{i+\mu})$ corresponding to λ;

$$\frac{\lambda_{x_{i-\nu}} - \lambda_{x_{i+\mu}}}{\delta_1 + \delta_2} = \frac{G^*(x_{i-\nu}) - G^*(x_{i+\mu})}{\delta_1 + \delta_2}; \qquad \text{say} = f$$

where we have written $\delta_1 = x_i - x_{i-\nu}$, $\delta_2 = x_{i+\mu} - x_i$. The part of the likelihood due to the observations between $x_{i-\nu}$ and $x_{i+\mu}$ is then

$$L' = f^{\nu+\mu+1}$$

For the deformed curve we get instead

$$\begin{aligned} L'' &= \left(f + \frac{\varepsilon}{\delta_1}\right)^{\nu+1} \left(f - \frac{\varepsilon}{\delta_2}\right)^{\mu} \\ &= \left[f^{\nu+1} + \frac{\nu+1}{\delta_1} \varepsilon f^{\nu} + 0(\varepsilon^2) \right] \left[f^{\mu} - \frac{\mu}{\delta_2} \varepsilon f^{\mu-1} + 0(\varepsilon^2) \right] \\ &= f^{\nu+\mu+1} + \varepsilon \left(\frac{\nu+1}{\delta_1} - \frac{\mu}{\delta_2} \right) + 0(\varepsilon^2) \end{aligned} \tag{16}$$

We had assumed, however, that

$$\lambda(x_i) = \frac{1}{\delta_1 + \delta_2} \left[\delta_2 G^*(x_{i-\nu}) + \delta_1 G^*(x_{i+\mu}) \right] > G^*(x_i),$$

which implies that

$$\delta_2\left[G^*(x_{i-\nu})-G^*(x_i)\right]+\delta_1\left[G^*(x_{i+\mu})-G^*(x_i)\right]>0 \tag{17}$$

Note that

$$G^*(x_{i-\nu})-G^*(x_i)=\frac{\nu}{n}; \qquad G^*(x_{i+\mu})-G^*(x_i)=-\frac{\mu}{n}$$

Hence $\delta_2\nu>\delta_1\nu$ and, using (16) and (17),

$$L''=L+\frac{\varepsilon}{\delta_1\delta_2}(\delta_2\nu-\delta_1\mu+\delta_2)+0(\varepsilon^2)>L$$

if ε is small enough so that the deformed curve is still convex. This shows that the ML-solution corresponds to G^{**} as described.

To show consistency is fairly straightforward, pick a denumerable set $\{\xi_\nu\}$ everywhere dense in (0, 1). Then the strong law of large numbers tells us that

$$P\left\{\lim_{n\to\infty} G^*(\xi_\nu)=G(\xi_\nu);\quad \nu=1,2,\ldots\right\}=1$$

It follows directly that G^{**} also satisfies the limit relations $G^{**}(\xi_\nu)\to G(\xi_\nu)$ with probability 1 for all ν, so that, using the monotonic property we get, with η's stated from $\{\xi_\nu\}$, with $\eta_1<\eta_2<x<\eta_3<\eta_4$ with probability 1,

$$\begin{aligned}\lim_{n\to\infty}\frac{1}{\eta_2-\eta_1}\int_{\eta_1}^{\eta_2}\alpha^*(x)\,dx &\leqslant \underline{\lim}_{n\to\infty}\,\alpha^*(x)\leqslant \overline{\lim}_{n\to\infty}\,\alpha^*(x)\\ &\leqslant \lim_{n\to\infty}\frac{1}{\eta_4-\eta_3}\int_{\eta_3}^{\eta_4}\alpha^*(x)\,dx\end{aligned} \tag{18}$$

If x is a point of continuity, the difference between the left- and right-hand sides of (18) can be made arbitrarily small: $\alpha^*(x)$ converges at all points of continuity x to $\alpha(x)$ with probability 1.

This is not enough, however. We also have to show $\|\alpha^*-\alpha\|_1\to 0$ a.c. Note that $\alpha^*(x)\leqslant x^{-1}G^*(x)\leqslant 1/x$, so that $\alpha^*(x)$ is uniformly bounded for $x\geqslant\varepsilon>0$, and hence $E[\alpha^*(x)]\to\alpha(x)$, for any fixed $x>0$. To deal with the behavior of the estimate in a neighborhood of zero, we choose a point η from the set $\{\xi_\nu\}$ close to zero. Since $G^{**}(\eta)\to G(\eta)$ a.c. as $n\to\infty$ we have

$$\int_0^\eta \alpha^*(x)\,dx=1-G^{**}(\eta)\to 1-G(\eta)$$

which can be made as small as desired by choosing η sufficiently small. At the same time the integral of α over $(0,\eta)$ is made small. Hence combining these results we see that

$$\|\alpha^* - \alpha\|_1 \leqslant \int_0^\eta \alpha^*(x)\,dx + \int_0^\eta \alpha(x)\,dx + \int_\eta^1 |\alpha^*(x) - \alpha(x)|\,dx$$

tend to zero in probability as $n \to \infty$. Q.E.D.

The main part of Theorem 2 is the construction of the ML-estimate, not the consistency result. Nevertheless it is of interest to observe that consistency in L_1-norm holds for this ML-estimate, in contrast to what we observed for the situation in Theorem 1, where a sieve was required. And A is not locally compact in the L_1-topology. To see this, let α_0 be the frequency function of the rectangular distribution $R(0,1)$ and consider for any fixed $r \in (0,1)$ the frequency function in A,

$$\alpha_n(x) = \begin{cases} 1 - r + 2nr - 2n^2 rx; & 0 \leqslant x \leqslant \dfrac{1}{n} \\ 1 - r; & \dfrac{1}{n} \leqslant x \leqslant 1 \end{cases}$$

(see Figure 4). Let r be small and $n = 1, 2, \ldots$.

All the α_n belong to the neighborhood $\{\alpha \mid \|\alpha - \alpha_0\|_1 \leqslant 2r\}$ with $\alpha_0 = R(0,1)$, as can be shown by direct evaluation of $\|\alpha - \alpha_0\|_1$, but can have no convergent subsequence. Indeed if a subsequence $\{\beta_n\}$ converges in L_1-

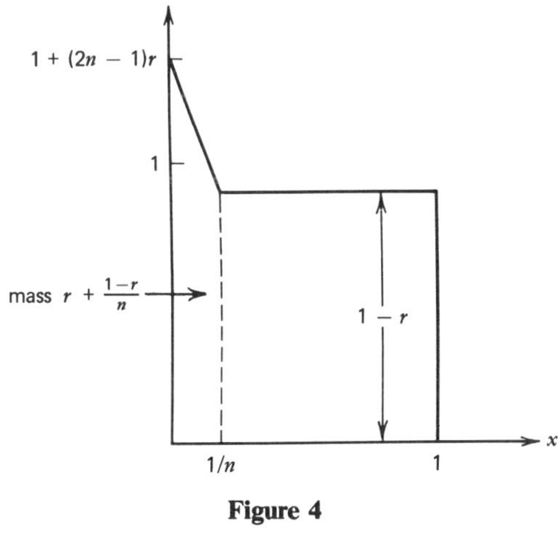

Figure 4

norm to some $\beta \in A$, then $\beta_n \to \beta$ also in the weak topology of probability measures. But this is impossible, since α_n converge in the weak topology to the measure with mass r at zero and the mass $1-r$ spread out uniformly over $(0,1)$. Since this measure is not of the type described by A, the parameter space is not locally compact. (See Note 8.2.D for further reference).

We now return to the situation in Theorem 1 (without monotonicity constraint) and ask for *constructive procedures to calculate the ML estimate*. For this purpose it is more convenient to use a global sieve of frequency functions with piecewise continuous derivatives whose variation is not too strong

$$\mathbb{S}(\mu) = \left\{ \alpha \mid \alpha \in PC_1, I(\alpha) \leqslant \frac{1}{\mu} \right\} \tag{19}$$

where

$$I(\alpha) = \int_0^1 [\alpha'(s)]^2 \, dx \tag{20}$$

Let us also assume that $\alpha(0) = \alpha(1) = 0$, which is natural if we confine our attention to a long interval, scaled down to $[0,1]$, outside of which we neglect all probability mass. Sieves built on other functionals than (20) could certainly be used, for example, one where the logarithmic derivative α'/α is used instead of α' in (20). This would correspond better to the local sieve (1). Dealing with (19) will give us the experience in handling global sieves. We need to go ahead to more difficult ones.

It is clear that if the values $\alpha(x_\nu) = \alpha_\nu$ have been fixed, then (20) will be minimized by quadratic functions between observed points. Indeed this minimum problem can be expressed in the form

$$\begin{aligned} &\alpha(x_\nu) = \alpha_\nu, \qquad \alpha(x_{\nu+1}) = \alpha_{\nu+1} \\ &\int_{x_\nu}^{x_{\nu+1}} \alpha(x) \, dx = \text{fixed constants for each } \nu \\ &\int_{x_\nu}^{x_{\nu+1}} [\alpha'(x)]^2 \, dx = \min \end{aligned}$$

With a Lagrange multiplier λ we should then solve

$$\begin{aligned} &\min_\alpha \int_{x_\nu}^{x_{\nu+1}} \{ [\alpha'(x)]^2 + \lambda \alpha(x) \} \, dx \\ &\alpha(x_\nu) = \alpha_\nu, \qquad \alpha(x_{\nu+1}) = \alpha_{\nu+1} \end{aligned}$$

which leads to the Euler equation $\alpha''+\lambda=0$. If α does not satisfy this equation between the points x_ν, we can modify it so that all α_ν are bigger but the integral of α is still 1 and $I(\alpha)$ has not changed. Hence *the ML-solution will be a quadratic spline*. It will be continuous everywhere, but its derivative may have jumps at the x_ν.

The original extremum problem had the form of the maximum problem

$$\sum_{\nu=1}^{n} \ln \alpha_\nu = \max$$

$$\int_0^1 \alpha(x)\,dx = 1$$

$$I(\alpha) = \frac{1}{\mu}$$

$$\alpha(0) = \alpha(1) = 0$$

or, with two Lagrange multipliers,

$$\max_{\alpha}\left[\int_0^1 \left\{[\alpha'(x)]^2 + \lambda_1 \alpha(x) + \lambda_2 \sum_{\nu=1}^{n} \ln \alpha(x)\delta(x-x_\nu)\right\}dx\right.$$

This leads to the Euler equation

$$\alpha''(x) + c = b\sum_{\nu=1}^{n} \frac{1}{\ln \alpha(x_\nu)}\delta(x-x_\nu) \tag{21}$$

where c and b are constants related to λ_1 and λ_2. If $G(x, y)$ denotes the Green's function for the operator d^2/dx^2 with boundary conditions $\alpha(0)=\alpha(1)=0$, we get the restricted ML-estimate expressed in the attractive form

$$\alpha^*(x) = \sum_{i=1}^{n} a_i G(x, x_i) - cg(x) \tag{22}$$

with

$$g(x) = \int_0^1 G(x, y)\,dy = \frac{1}{2}x(x-1)$$

since the Green's function G has the classical form

$$G(x, y) = \begin{cases} x(y-1); & x \leqslant y \\ y(x-1); & x \geqslant y \end{cases}$$

We see that α is a quadratic polynomial between the points x_ν at which its derivative has the jumps

$$\left(\Delta\frac{d\alpha^*(x)}{dx}\right)_{x+x_i}=a_i \tag{23}$$

Introduce the symmetric matrix

$$G=\{G(x_i,x_j);\, i,\, j=1,2,\dots,n\}$$

the vectors

$$\begin{aligned} a&=\text{col}(a_1,\dots,a_n)\\ \alpha^*&=\text{col}[\alpha^*(x_1),\dots,\alpha^*(x_n)]\\ g&=\text{col}[g(x_1),\dots,g(x_n)]\\ \gamma(x)&=\text{col}[G(x,x_1),\dots,G(x,x_n)]\\ \alpha^{-1}&=\text{col}\left[\frac{1}{\alpha^*(x_1)},\dots,\frac{1}{\alpha^*(x_n)}\right] \end{aligned}$$

and the constant

$$\gamma_0=\int_0^1 g(x)\,dx=-\frac{1}{12}$$

Integrating (21) over a small neighborhood of any x_ν and using (23), we see that we must have $a=b\alpha^{-1}$, but we also deduce from (22) that

$$\alpha^*=Ga-cg \tag{24}$$

so that

$$\alpha^*(x)=\gamma^T(x)a-cg(x); \qquad x\in[0,1)$$

Since $\alpha^*(x)$ must have integral 1, this gives us the constraint

$$1=\gamma^Ta-c\gamma_0=b\gamma^{T-1}-c\gamma_0 \tag{25}$$

with the vector γ having the entries $g(x_i)=\frac{1}{12}x_i(x_i-1)$.

The role of the parameters c and b is now becoming clear. The value of $-c$ is the second derivative between sample points. A small value means almost linear interpolation and strong smoothing, a large value allows

$\alpha^*(x)$ to vary more and leads to less smoothing. The other parameter, b, is needed to adjust $\alpha^*(x)$ to integral 1 in (25).

In the interval (x_i, x_{i+1}) we would expect $x_{i+1}-x_i$ to be roughly of the order of magnitude $(nf(x_i))^{-1}$. Using this we get a crude bound in order that $\alpha^*(x)$ be positive:

$$|c|<8n^2f^3(x)$$

When the sample size increases, we can also let $-c$ increase, less smoothing, but we had better make it $o(n^2)$ in order to have α^* positive.

If the estimate takes negative values for some argument values, we are in the uncomfortable situation that α^* does not belong to the parameter space we started with. The same dilemma occurs in many other estimation problems, also for finite-dimensional cases, for example, when estimating a positive parameter in a missing plot situation.

In itself this is not serious, but it is an indication that the estimate can be improved, sometimes by a simple truncation operation making it take values in the given parameter space (see Note 8.2.E).

To solve (24) and (25) we encounter a nonlinear problem in numerical analysis, and we shall try an iterative technique. Call the iterates $\alpha_1, \alpha_2, \alpha_3, \ldots$ for the α-vector, so that α_ν no longer stands for the νth ordinate but for one instance of the α^*-vector, and the corresponding vectors with the reciprocal values as entries $\alpha_1^{-1}, \alpha_2^{-1}, \alpha_3^{-1}, \ldots$ meaning instances of the α^{-1}-vector. Define recursively

$$\beta_{\nu+1}=\frac{1-c/12}{\gamma^T\alpha_\nu^{-1}}G\alpha_\nu^{-1}-cg$$

$$\alpha_{\nu+1}=\alpha_\nu+k(\beta_{\nu+1}-\beta_\nu) \tag{26}$$

In other words, we use a relaxation scheme with a relaxation parameter k. Indeed, in our numerical studies we have used *underrelaxation*, $k<1$, in order to stay away from the singular set where components of the α-vector vanish. In some of the simulations we also truncated α_ν from below by putting any of its components equal to a very small positive number if it became negative. This does not seem necessary, however, if k is small, for example, $k=\frac{1}{2}$.

Assuming that the true $\alpha(x)$ was a mixture of a small number of beta-distributions with integral values of p_i and q_i,

$$\alpha(x)=\sum_{i=1}^{r} B_i x^{p_i-1}(1-x)^{q_i-1}; \qquad p_i, q_i \geq 1$$

so that we could get multimodal frequency functions, we simulated the beta-distributions as ratios of quadratic forms in normal pseudo-random numbers. The mixture was then obtained by sampling with appropriate weights from the beta-distributed pseudo-random numbers.

The convergence of the iterates in (26) is so fast that only two or three iterations are needed. It becomes even faster if the value of k is increased, but this is not recommended because of the danger of approaching the singular set.

A typical result is shown in Figure 5. The true frequency function is shown full drawn in Figure 5 (a); it is bimodal with $r=2$. For small sample size, say $n=4$, as in the dashed curve in Figure 5 (b), one can clearly see the curvature of parabolic arcs of the quadratic spline. For $n=40$, as in the dashed line in Figure 5 (a), the curvature of the parabolic arcs is barely visible.

It would be of interest to carry out the numerical analysis of the iteration scheme in some detail and study the corresponding sampling properties of this restricted ML solution, but this has not been attempted.

In two dimensions the sample space X, for example, $X=T^2$ with periodic boundary conditions, will be partitioned by a finite set algebra into subsets. We assume that α is continuous on X and with grad α

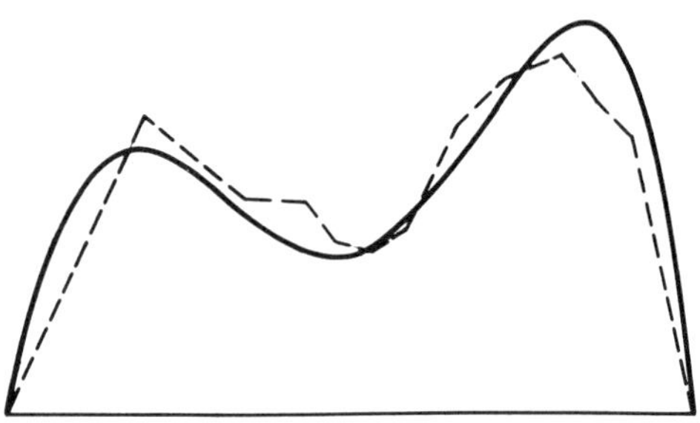

(a)

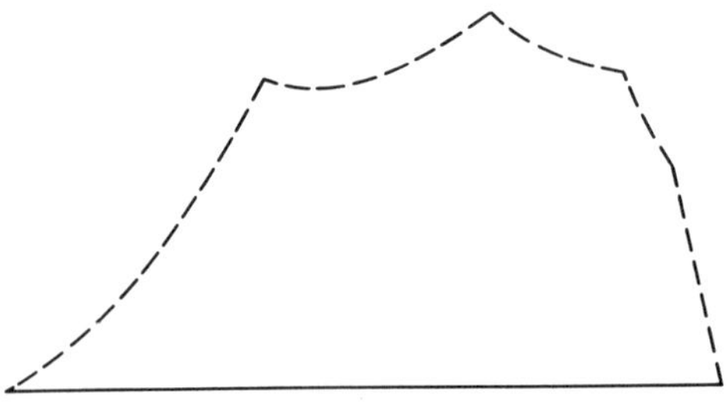

(b)

Figure 5

constant on all subsets belonging to the set algebra. This is a measurability condition via the set algebra on grad α. However, to strengthen the sieve we also assume a Dirichlet sieve on α.

To make this concrete, let us partition X, as shown in Figure 6(a), into $2m^2$ congruent triangles; $m=10$ in the figure. The reason why we have chosen this particular set algebra is that it is very convenient for computations. Geometric computations are notoriously time consuming, but the triangular partition will turn out to be very efficient and is also well suited for xy-plotters for graphical representation of α^*.

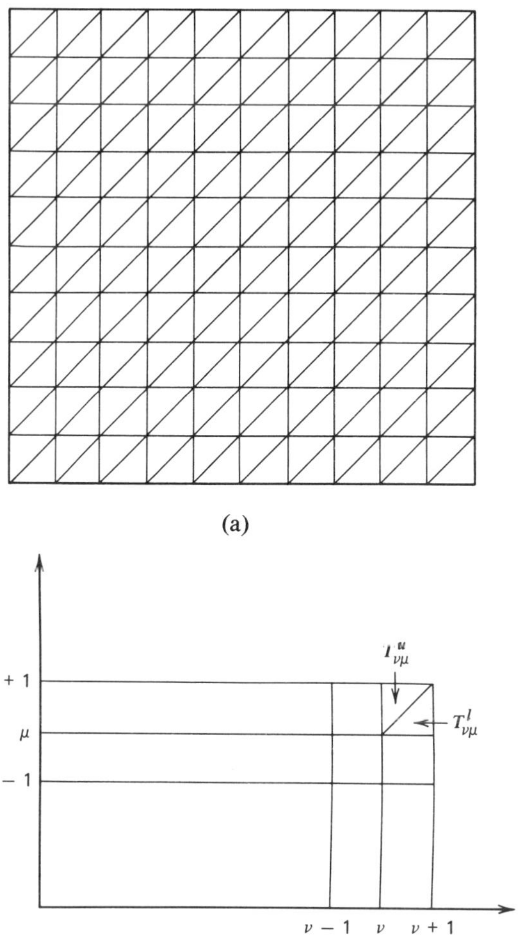

(a)

(b)

Figure 6

Denote the upper and lower triangles at the (ν, μ) corner with coordinates $(\nu/m, \mu/m)$ by $T^u_{\nu\mu}$ and $T^l_{\nu\mu}$, respectively [see Figure 6(b)]. Since grad α is constant over any of these triangles, α is a piecewise linear and continuous function of (x, y). Straightforward calculation gives, if $\alpha(\nu/m, \mu/m)$ is denoted by $\phi_{\nu\mu}$:

$$\bar{p}^u_{\nu\mu} = \bar{P}(T^u_{\nu\mu}) = \frac{1}{6m^2}\left[\phi_{\nu+1\mu+1} + \phi_{\nu\mu+1} + \phi_{\nu\mu}\right]$$

$$\bar{p}^l_{\nu\mu} = \bar{P}(T^l_{\nu\mu}) = \frac{1}{6m^2}\left[\phi_{\nu+1\mu+1} + \phi_{\nu+1\mu} + \phi_{\nu\mu}\right] \tag{27}$$

where $\bar{P}$ denotes the probability measure according to α^* with piecewise constant gradient

$$|\text{grad}\,\alpha|^2 = m^2\left[(\phi_{\nu+1\mu+1} - \phi_{\nu\mu+1})^2 + (\phi_{\nu\mu+1} - \phi_{\nu\mu})^2\right] \text{ in } T^u_{\nu\mu}$$

$$|\text{grad}\,\alpha|^2 = m^2\left[(\phi_{\nu+1\mu+1} - \phi_{\nu+1\mu})^2 + (\phi_{\nu+1\mu} - \phi_{\nu\mu})^2\right] \text{ in } T^l_{\nu\mu}$$

A natural choice of approximate likelihood function could be based on the statistics

$$\sum_{\nu,\mu=0}^{n-1}\left[\frac{(\xi^u_{\nu\mu} - p^u_{\nu\mu})^2}{p^u_{\nu\mu}} + \frac{(\xi^l_{\nu\mu} - p^l_{\nu\mu})^2}{p^l_{\nu\mu}}\right] \tag{28}$$

and

$$\xi^u_{\nu\mu} = \frac{1}{n} \text{ number of observations in } T^u_{\nu\mu}$$

$$\xi^l_{\nu\mu} = \frac{1}{n} \text{ number of observations in } T^l_{\nu\mu} \tag{29}$$

This would amount to a Poisson-normal approximation with asymptotic independence between cell frequencies. Instead of (28) we shall use the criterion

$$Q = \sum_{\nu,\mu=0}^{m-1}\left[(\xi^u_{\nu\mu} - p^u_{\nu\mu})^2 + (\xi^l_{\nu\mu} - p^l_{\nu\mu})^2\right]$$

both for computational convenience and to avoid the appearance of possibly small p-values in the denominators in (29). We do not know if this leads to any serious loss in statistical performance.

With a Dirichlet sieve [second term in (30)] and the condition that the total probability mass be zero (third term) we should then minimize the expression

$$\frac{1}{2}Q+\frac{\lambda}{2}\iint_{T^2}|\text{grad}\,\alpha(x,y)|^2dx\,dy-\kappa\iint_{T^2}\alpha(x,y)\,dx\,dy \qquad (30)$$

with the two Lagrange multipliers λ and κ. Elementary but cumbersome calculations lead us to the equations

$$\frac{1}{18m^4}L\phi_{\nu\mu}-2\lambda m^2\Delta\phi_{\nu\mu}=\frac{1}{6m^2}\eta_{\nu\mu}+\kappa \qquad (31)$$

with the partial difference operator L given by

$$L\phi_{\nu\mu}=\phi_{\nu-1\mu-1}+\phi_{\nu-1}\phi_{\mu}+\phi_{\nu\mu-1}+3\phi_{\nu\mu}+\phi_{\nu\mu+1}+\phi_{\nu+1,\mu}+\phi_{\nu+1\mu+1} \qquad (32)$$

and with the discrete Laplacian

$$\Delta\phi_{\nu\mu}=\phi_{\nu-1\mu}+\phi_{\nu\mu-1}-4\phi_{\nu\mu}+\phi_{\nu\mu+1}+\phi_{\nu+1\mu} \qquad (33)$$

and the η-field

$$\eta_{\nu\mu}=\xi^u_{\nu-1\mu-1}+\xi^u_{\nu\mu}+\xi^u_{\nu\mu-1}+\xi^l_{\nu-1\mu-1}+\xi^l_{\nu\mu}+\xi^l_{\nu-1\mu} \qquad (34)$$

The operator L corresponds to the difference scheme in Figure 7(a), the operator Δ to Figure 7(b), and the η-field is obtained from the ξ-field by applying a moving average summing over the shaded region of observations in Figure 7(c).

Rewriting (31) with appropriate changes in the meaning of the symbols λ and κ, we get

$$(L+\lambda\Delta)=3m^2\eta+\kappa \qquad (35)$$

where it should be recalled that the fields are defined periodically modulo m in the directions of both axes. Hence $M=L+\lambda\Delta$ can be expressed as a circulant four-dimensional array and has the $m\times m$ "eigenmatrices" with $\alpha,\beta=0,1,\ldots,m-1$

$$m^{\alpha\beta}=\{z^\alpha_\nu z^\beta_\mu;\,\nu,\mu=0,1,\ldots,m-1\}; \qquad z^\alpha_\nu=\frac{1}{\sqrt{m}}e^{2\pi i\nu\alpha/m} \qquad (36)$$

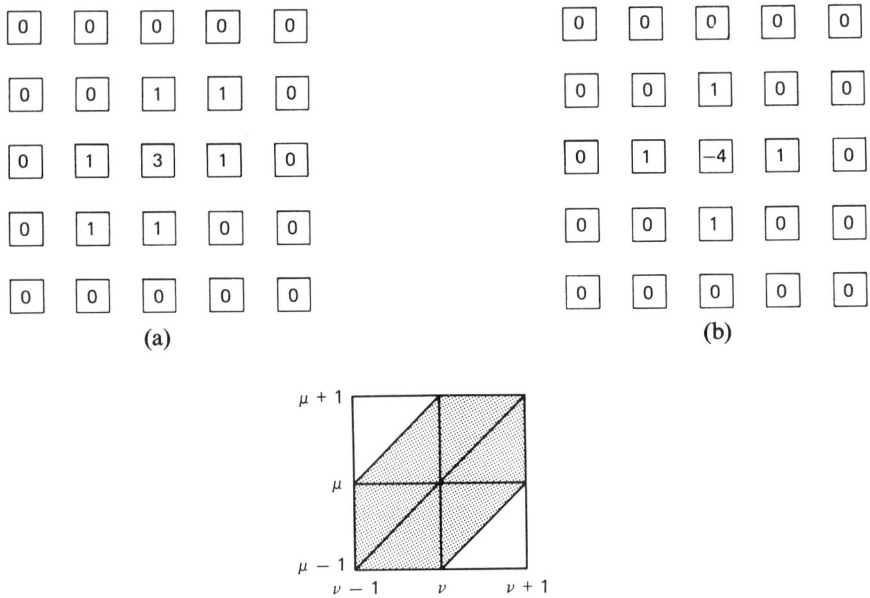

Figure 7

with the eigenvalues

$$\lambda^{\alpha\beta} = (m^{\alpha\beta})^T M m^{\alpha\beta} \tag{37}$$

Using (32) and (33), the $\lambda^{\alpha\beta}$ can be calculated in closed form to be

$$\lambda^{\alpha\beta} = \frac{1}{m}\left\{3 - 4\lambda + 2\cos\frac{2\pi}{m}(\alpha+\beta) + 2(1+\lambda)\cos\frac{2\pi\alpha}{m} + 2(1+\lambda)\cos\frac{2\pi\beta}{m}\right\} \tag{38}$$

The first Lagrange multiplier λ determines the mesh size of the sieve. To determine the value of the second Lagrange multiplier κ let us sum both sides of (35) over $\nu, \mu = 0, 1, 2, \ldots, m-1$ with the result, using (32)–(34),

$$9\Phi = 3m^2 \sum_{\nu,\mu=0}^{m-1} \eta_{\nu\mu} + m^2\kappa$$

where Φ denotes the sum of all the $\phi_{\nu\mu}$. Summing both equations in (27) over all (ν, μ)-values and adding the results, we get

$$1=\frac{1}{6m^2}6\Phi; \qquad \Phi=m^2$$

Summing (34) we get, in the same way,

$$\sum_{\nu,\mu=0}^{m-1}\eta_{\nu\mu}=3F^u+3F^l; \qquad F^u=\sum_{\nu,\mu=0}^{m-1}\xi^u_{\nu\mu}, \quad F^l=\sum_{\nu,\mu=0}^{m-1}\xi^l_{\nu\mu}$$

Recalling the definition (29) of the ξ's, and combining our results, gives us the equations

$$3(F^u+F^l)=3$$

so that, from

$$9\Phi=9m^2=9m^2+m^2\kappa,$$

we find, fortunately, that κ is simply zero. It may be remarked that with other boundary conditions κ could take (small) nonzero values.

Applying this together with (36) and (37) and (38) we can solve our equation (35) explicitly:

$$\begin{aligned}\phi^*_{\nu\mu}&=3m^2\sum_{\alpha,\beta=0}^{m-1}\lambda^{\alpha\beta}\sum_{r,s=0}^{m-1}\bar{z}^{\alpha}_{\nu}\bar{z}^{\beta}_{\mu}z^{\alpha}_{r}z^{\beta}_{s}\eta_{rs}\\&=3\sum_{\alpha,\beta,r,s=0}^{m-1}\lambda^{\alpha\beta}e^{(2\pi i/m)[\alpha(r-\nu)+\beta(s-\mu)]}\eta_{rs}\end{aligned}$$

Finally we get, using the linear form of α in each triangle in the $9m^2$ partition,

$$\alpha^*(x,y)=$$

$$\times\begin{cases}\phi^*_{\nu\mu}+m(\phi^*_{\nu+1\mu+1}-\phi^*_{\nu\mu+1})\left(x-\dfrac{\nu}{m}\right)+m(\phi^*_{\nu\mu+1}-\phi_{\nu\mu})\left(y-\dfrac{\mu}{m}\right);\\[2ex] \phi^*_{\nu\mu}+m(\phi^*_{\nu+1\mu}-\phi^*_{\nu\mu})\left(x-\dfrac{\nu}{m}\right)+m(\phi^*_{\nu+1\mu+1}-\phi_{\nu+1\mu})\left(y-\dfrac{\mu}{m}\right);\end{cases} \tag{39}$$

where the first equation holds if $(x,y)\in T^u_{\nu\mu}$ and the second if $(x,y)\in T^l_{\nu\mu}$.

Algorithmically we are well off since it is enough to apply the two-dimensional FFT twice in order to implement (39). In order to learn about the sampling properties of the estimator (39), one should derive its asymptotic bias and variance which we have not done.

Returning to the one-dimensional problem, let us see what can be done alternatively by a Bayesian approach. Several authors have attempted this (see Note 8.2.F), and the following approach is close in spirit to their analyses. Let $\alpha(x)$ be given by the following stochastic process, where r is a positive integer and the mesh size $\mu = 1/r$:

$$\alpha(x) = r\frac{\alpha_k}{\alpha_1 + \alpha_2 + \cdots + \alpha_r}; \qquad \text{for } x \in I_k = \left(\frac{k-1}{r}, \frac{k}{r}\right), \quad k = 1, 2, \ldots, r$$

and the stochastic variables α_k are i.i.d. with the exponential frequency function e^{-x} on $x \geqslant 0$ and zero otherwise. Any such α is clearly a legitimate frequency function over $(0,1)$. This defines the *a priori* measure $\pi(d\alpha)$ in the function space of functions constant on the intervals $\{I_k\}$.

We then get, using the observations $x_1, x_2, \ldots, x_n$ and integrating with respect to $\pi(d\alpha)$,

$$E[L(\alpha)] = E[\alpha(x_1)\alpha(x_2)\cdots\alpha(x_n)]$$

$$= r^n \int_{\mathbb{R}^n} \frac{\alpha_1^{n_1}\alpha_2^{n_2}\cdots\alpha_r^{n_r}}{(\alpha_1 + \alpha_2 + \cdots + \alpha_r)^n} e^{-\alpha_1 - \alpha_2 - \cdots - \alpha_r} d\alpha_1 d\alpha_2 \cdots d\alpha_r \tag{40}$$

Here n_k is the number of observations in I_k. This sort of integral can be evaluated directly [see, e.g., Bochner (1932), p. 62], and we get

$$E[L(\alpha)] = r^n \frac{n_1!n_2!\ldots n_r!}{(n+r-1)!}(r-1)!$$

Similarly we get, for any $x \in (0,1)$, the expected value $E[\alpha(x)L(\alpha)]$ in the same form as in (40) with one more factor α_k with x as the subscript of I_k, $x \in I_k$, so that

$$E[\alpha L(\alpha)] = r^{n+1}\frac{n_1!n_2!\ldots(n_k+1)!\ldots n_r!}{(n+r)!}(r-1)!$$

Hence the Bayes estimator $\alpha^*(x)$ is given by

$$\alpha^*(x) = \frac{E[\alpha(x)L(\alpha)]}{E[L(\alpha)]} = r\frac{n_k+1}{n+r}; \qquad x \in I_k$$

Of course this estimator is also piecewise constant. We can now assert that it is consistent.

THEOREM 3

If r is made to tend to infinity as o(n) when n tends to infinity, then, for any $\alpha \in L_1$, *the Bayes estimator is consistent in the norm of* L_1.

Proof We have

$$\|\alpha^* - \alpha\|_1 = \sum_{k=1}^{r} \int_{I_k} \left| r\frac{n_k+1}{n+r} - \alpha(x) \right| dx$$

which can be dominated by

$$\sum_{k=1}^{r} (A_k + B_k) = \sum_{k=1}^{r} A_k + \sum_{k=1}^{r} B_k = A + B$$

with

$$A_k = \left| \frac{n_k+1}{n+r} - \frac{1}{r}\bar{\alpha}_k \right| = |C_k|$$

$$\bar{\alpha}_k = r\int_{I_k} \alpha(x)\, dx \tag{41}$$

$$B_k = \int_{I_k} |\bar{\alpha}_k - \alpha(x)| dx$$

The term B is not random and depends only upon r. Since $\alpha \in L_1$ for any $\delta > 0$ we can find a continuous function c so that $\alpha = c + \varepsilon$, $\|\varepsilon\|_1 \leqslant \delta$ [see, e.g., Titchmarsh (1939), p. 376]. Let T denote the linear operator that takes α into $\alpha - \bar{\alpha}$, where $\bar{\alpha}$ is equal to its average $\bar{\alpha}_k$ over I_k on I_k. We have, for any $f \in L_1$,

$$\|Tf\|_1 \leqslant \|f\|_1 + \|\bar{f}\|_1 = \|f\|_1 + \sum_k \int_{I_k} |\bar{f}_k| dx$$

$$\leqslant \|f\|_1 + \sum_k \int_{I_k} |f(x)|\, dx = 2\|f\|_1$$

Therefore the norm of T is at most 2. Hence

$$\|T\alpha\|_1 \leqslant \|Tc\|_1 + 2\|\varepsilon\|_1 \leqslant \|Tc\|_1 + 2\delta$$

Since c is continuous, the functions $\bar{c}$ for $r=1,2,3,\ldots$ are uniformly bounded by $\max_x|c(x)|$ and tend to $c(x)$ as $r\to\infty$. Hence $(Tc)(x)\to 0$ for all x boundedly so that $\|Tc\|\to 0$ as $r\to\infty$. This shows that $B\to 0$ as $r\to\infty$.

To deal with C_k in the first equation in (41), we make use of the fact that we have binomial stochastic variables, $n_k=B(n,(1/r)\bar{\alpha}_k)$, so that

$$\begin{cases} E(C_k)=\dfrac{1-\bar{\alpha}_k}{n+r} \\ \operatorname{Var}(C_k)=\dfrac{n}{r(n+r)^2}\bar{\alpha}_k\left(1-\dfrac{\bar{\alpha}_k}{n}\right) \end{cases}$$

From Schwarz's inequality we get, noting that $\bar{\alpha}_k$ can be both smaller and greater than 1,

$$E[|C_k|]\leqslant\left[E(C_k^2)\right]^{1/2}\leqslant\frac{\left[(1-\bar{\alpha}_k)^2+(n/r)\bar{\alpha}_k\right]^{1/2}}{n+r}$$

$$\leqslant\frac{1}{n+r}\left[1+\bar{\alpha}_k+\sqrt{\frac{n}{r}}\sqrt{\bar{\alpha}_k}\right]$$

Summing over k and using $\Sigma\bar{\alpha}_k=r$ together with Schwarz's inequality, again we get the crude bound since we assumed $r=o(n)$

$$E(A)\leqslant O\left(\frac{r+\sqrt{nr}}{n+r}\right)=o(1)$$

so that $E(A+B)\to 0$. Q.E.D.

It is clear that μ plays the same role here as the mesh size does for sieves. When $\mu\downarrow 0$, that is, $r\to\infty$, the support of the *a priori* measure π becomes everywhere dense in L_1.

3 PARAMETER IN l_2- AND L_2-SPACES

Since there is only one real separable Hilbert space up to isomorphisms, it appears that we can treat $A=l_2$ and $A=L_2$ together, but we shall see below that this is wrong. Say, anyway, that $x=\alpha+y$, where $\alpha\in A$ and y has a Gaussian distribution over $X=A$ with mean zero and covariance operator R (see Chapter 1, Section 2, for the full meaning of this).

In order that P_{α_1} and P_{α_0} be equivalent, we know that $\alpha_1-\alpha_0\in$ range($R^{1/2}$), as was shown in Chapter 5, Section 1. It is therefore natural to assume that α takes values in range($R^{1/2}$).

Since R is of the trace type, we introduce its eigenvalues λ_ν and eigenvectors ϕ_ν so that the abstract version of the Karhunen-Loéve expansion takes the form

$$y=\sum_{\nu=1}^{\infty}\sqrt{\lambda_\nu}\,y_\nu\phi_\nu;\qquad \{\phi_\nu\}\text{ orthonormal vectors in }X$$

Use $(x,\phi_\nu)=x_\nu$ as observables so that the Radon-Nikodym derivative can be written (see, Chapter 5, Section 1),

$$\frac{P_\alpha(dx)}{P_0(dx)}=\prod_{\nu=1}^{\infty}\exp\left[\frac{2\alpha_\nu x_\nu-\alpha_\nu^2}{2\lambda_\nu}\right]\tag{1}$$

where $\alpha_\nu=(\alpha,\phi_\nu)$. To maximize the likelihood (1) with respect to the parameter α is straightforward, and we get $\alpha_\nu=x_\nu$, or the ML solution $\alpha^*=x$, which is the obvious answer.

If we have access to an i.i.d. sample of observed vectors $x^{(1)},x^{(2)},\ldots x^{(n)}\in L_2$, we get similarly the obvious ML solution $\alpha^*=(x^{(1)}+x^{(2)}+\cdots+x^{(n)})/n=\bar{x}$. We have already seen in Chapter 4, Section 6, that $\bar{x}$ is consistent in the norm of L_2. Hence *no sieve is needed.*

This is correct but somewhat misleading, however, as is illustrated by the following case.

Example 1 Let $x=(x_1,x_2,\ldots)$ where the x are given as $x_\nu=(1+\alpha_\nu)\xi_\nu$, $\{\xi_\nu\}$ an i.i.d. sample from the exponential distribution on $\mathbb{R}$ with mean 1. If $\alpha=(\alpha_1,\alpha_2,\ldots)\in l_2$, then we can apply Kakutani's theorem via Theorem 5 in Chapter 3, Section 3. Indeed, for $f(x)=e^{-x}$ the condition in the theorem is satisfied since

$$\frac{1}{8}\left[\int_0^\infty x^2e^{-x}\,dx-1\right]=\frac{1}{8}<\infty$$

Hence the condition that the measures P_α and P_0 be equivalent reduces to just $\alpha\in l_2$.

Then the Radon-Nikodym derivative is

$$\frac{P_\alpha(dx)}{P_0(dx)}=\prod_{\nu=1}^{\infty}\frac{1}{1+\alpha_\nu}\exp\left[\frac{x_\nu\alpha_\nu}{1+\alpha_\nu}\right]\tag{2}$$

To maximize each factor in (2) we should choose

$$\alpha_\nu^* = x_\nu - 1 = (\xi_\nu - 1) + \alpha_\nu \xi_\nu \tag{3}$$

With probability 1 the sequence $\{\alpha_\nu \xi_\nu\} \in l_2$. The first term on the right-hand side of (3) is a.c. not in l_2: the maximum is therefore not attained in our parameter space. Hence in this example a sieve is needed, although we have the same sample space as before.

This shows that *the need for a sieve depends not just on the topology of the parameter space A, but also on the way the measure P_α depends upon α in the topology of A.*

Example 2 As a variation of the last case let the observed process be

$$x(t) = \int_{-1/2}^{t} \alpha(s)\,ds + W(t); \qquad t \in \left[-\tfrac{1}{2}, \tfrac{1}{2}\right]$$

where the real function $\alpha \in L_2[(-\frac{1}{2}, \frac{1}{2})]$ and W is the Wiener process with unit variance per time unit. To avoid unessential complications, we assume that α is even around $t=0$ so that its sine- Fourier coefficients will vanish. Symbolically we observe $\alpha(t) + \dot{w}(t)$ and want to estimate α, when we have an i.i.d. sample $x_1(t), x_2(t), \ldots, x_n(t)$; $\forall t \in [-\frac{1}{2}, \frac{1}{2}]$.

To do this it is convenient to employ the observables for $k = 1, 2, \ldots$:

$$x_k = \sqrt{2} \int_{-1/2}^{1/2} \cos 2\pi k t x(dt) + \alpha_k + w_k$$

where

$$\alpha_k = \sqrt{2} \int_{-1/2}^{1/2} \cos 2\pi k t \alpha(t)\,dt$$

$$w_k = \sqrt{2} \int_{-1/2}^{1/2} \cos 2\pi k t w(dt) \tag{4}$$

For $k=0$ we use $x_0 = x(\frac{1}{2}) - x(-\frac{1}{2})$ and the corresponding values for α_0 and w_0. We also have the sine-coefficients, but since they carry no information about α, we can leave them out. Then w_k has a Gaussian distribution with variance 1, and all the w_k are independent. The Radon-Nikodym derivative with respect to the measure with $\alpha = 0$ becomes

$$\frac{P_\alpha(dx)}{P_0(dx)} = \exp \sum_{k=0}^{\infty} \left(\alpha_k x_k - \tfrac{1}{2}\alpha_k^2\right) \tag{5}$$

With the same reasoning as above it follows that the measures P_α and P_0 are equivalent, since $\{\alpha_k\} \in l_2$.

To maximize (5) let us do it for every term in the sum, leading to the ML solution $\alpha_k^* = x_k$. But $x_k = \alpha_k + w_k$, where $\{\alpha_k\} \in l_2$ and $\{w_k\} \notin l_2$ with probability 1. Hence the ML solution cannot be attained in the present parameter space, so that a *sieve is needed.*

The Radon-Nikodym derivative in (5) can also be written as

$$\frac{P_\alpha(dx)}{P_0(dx)} = \exp\left[\int_{-1}^{1} \alpha(t)x(dt) - \frac{1}{2}\int_{-1}^{1} \alpha^2(t)\,dt\right]$$

The exponent is formally maximized by choosing $\alpha^*(t) = x'(t)$, but this derivative does not exist. This is just another way of saying that a sieve is needed; compare with the discussion in Chapter 6, Section 5.

Let us use one of our usual sieves

$$\mathcal{S}(\mu) = \left\{\alpha \,\middle|\, \int_0^1 [\alpha'(t)]^2 \leqslant \text{constant}\right\} \tag{6}$$

or with the mesh size μ, as given in

$$\sum_{k=1}^{\infty} k^2\alpha_k^2 \leqslant \frac{1}{\mu} \tag{7}$$

Maximizing (5) with equality in (7) as the constraint, we get the restricted ML-solution

$$\alpha_k = \frac{x_k}{1+\lambda k^2}; \qquad k=1,2,\ldots, \qquad \alpha_0 = x_0$$

where λ should be adjusted to satisfy

$$\sum_{k=1}^{\infty} \frac{k^2 x_k^2}{(1+\lambda k^2)^2} = \frac{1}{\mu} \tag{8}$$

Note that this sum converges a.c. for any $\lambda > 0$. It is a strictly decreasing function in λ, taking the value $+\infty$ at $\lambda = 0$ a.c., and tending to zero a.c. as $\lambda \to +\infty$. Hence there is a.c. a unique root λ of (8), and moreover $\lambda_0 = \lambda_0(\mu)$ is a monotonically increasing function of μ. Hence when the mesh size λ is made smaller, the value of λ will decrease. We can therefore use λ instead as our mesh size.

We then get, with α^* as the L_2-function associated with the Fourier coefficients α_k in (4), but now for a sample of size n,

$$\sum_{k=1}^{\infty} E_\alpha\{[\alpha_k^* - \alpha_k]^2\} = \lambda^2 \sum_{k=1}^{\infty} \frac{k^4\alpha_k^2}{(1+\lambda k^2)^2} + \frac{1}{n}\sum_{k=1}^{\infty} \frac{1}{(1+\lambda k^2)^2}$$

But the first sum on the right takes the summable sequence $\{\alpha_k^2\}$ and multiplies it by uniformly bounded coefficients

$$\frac{\lambda^2 k^4}{(1+\lambda k^2)^2} \leqslant 1$$

Each coefficient tends to zero as $\lambda \downarrow 0$, so that the first sum tends to zero with λ. The term for $k=0$ obviously tends to zero as n tends to infinity; it does not involve λ.

After choosing λ so small that the first sum is small enough, we let $n \to \infty$ with λ fixed. The second sum then tends to zero. Hence *the ML solution restricted by the sieve in* (6) *is consistent in the topology of* L_2. The reader may find it of interest to compare this sieve with the regularization discussed in Chapter 7, Section 6; they are not identical but related. We see in Chapter 9, Section 2, how this should be understood in a more general context.

We could also have chosen as a sieve $\mathcal{S}(\mu)$ the subspace of $L_2[(-\frac{1}{2},\frac{1}{2})]$ spanned by $\phi_k(t) = \sqrt{2}\cos 2\pi kt$, $k=0,1,2,\ldots,N$, $\mu = 1/N$, where as usual it is understood that we leave out the factor $\sqrt{2}$ for $k=0$. Writing $\alpha = \alpha_N + e$, where

$$\alpha_N(t) = \sum_{k=0}^{N} a_k\phi_k(t); \qquad a_k = (\alpha, \phi_k) = \int_{-1}^{1} \phi_k(t)\alpha(t)\,dt$$

we are led to the constrained ML-estimate

$$\alpha_N^*(t) = \sum_{k=0}^{N} a_k^*\phi_k(t); \qquad a_k^* = \int_{-1}^{1} \phi_k(t)x(dt)$$

with the expected norm of the error

$$E\left[\|\alpha_N^* - \alpha\|^2\right] = \sum_{k=0}^{N} E\left[(a_k^* - a_k)^2\right] + \sum_{k=N+1}^{\infty} a_k^2 \tag{9}$$

The first sum on the right-hand side in (9) is easily calculated, realizing that a_k^* is an unbiased estimate of $a_k, k=0,1,\ldots,N$; with variance 1. Hence (9) reduces to

$$N+1+\sum_{k=N+1}^{\infty} a_k^2$$

If we have some information of how a_k tends to zero, we can use this to choose the best, or at least a nearly best, value of N.

We can look at this problem in still another way, which seems to have considerable potential for generalization. With a single observation $x(t)$ introduce a partition of $[0,1]$ into intervals $i_k=((k-1)/N, k/N), k=1,2,\ldots,N$, and with the observables $x_k=\int_{i_k} x(dt)$ we can write down the finite-dimensional likelihood ratio (see Chapter 3, Section 1). Passing to the limit, we get after some reasoning of the type used in Chapter 3 the Radon-Nikodym derivative

$$\frac{P_\alpha(dx)}{P_0(dx)}=\exp\left[\int_0^1 \alpha(t)x(dt)-\frac{1}{2}\int_0^1 \alpha^2(t)\,dt\right] \tag{10}$$

Maximizing the exponent in (10) formally we get $\alpha^*=x'(t)$, which does not exist as an element in L_2, in agreement with what we have just seen.

Introduce the *Sturm-Liouville sieve*, more general than the one in (6),

$$\mathcal{S}(\mu)=\left\{\alpha \Big| \int_0^1 \{p(t)[\alpha'(t)]^2+q(t)\alpha^2(t)\}\,dt \leqslant \frac{1}{\mu}\right\} \tag{11}$$

where p and q are positive and C_1. We also assume that $\alpha(0)=\alpha(1)=0$, but other boundary conditions can be dealt with in a similar way.

Proceeding in the usual manner, we want to minimize

$$\frac{1}{2}\int_0^1 \alpha^2(t)\,dt-\int_0^1 \alpha(t)x(dt)+\frac{\lambda}{2}I(\alpha)$$

where I is the quadratic functional appearing in (11), and λ is a Lagrange multiplier.

We then get the Euler equations in the form of the stochastic differential equation

$$L\alpha=\lambda(p\alpha')'-(\lambda q+1)\alpha=-x'$$

with the self-adjoint differential operator L. Denoting the Green's function for this Sturm-Liouville problem by $G(t,s)$, we can write the resulting ML-solution restricted by the sieve (11) as

$$\alpha^*(t) = -\int_0^1 G(t,s)x(ds)$$

so that

$$E_\alpha[\alpha^*(t)] = -\int_0^1 G(t,s)\alpha(ds)$$

$$\operatorname{Cov}[\alpha^*(t_1,t_2)] = \int_0^1\int_0^1 G(t_1,s)G(t_2,s)\,ds \tag{12}$$

For simple choices of p and q the Green's function can be calculated in closed form, and the integrals in (12) can then be evaluated to get the bias and variance and their dependence upon the mesh size μ.

The same reasoning applies when we have several i.i.d. observations of the $x(t)$-process.

After studying the special parameter spaces of this and the two previous sections, we can begin to recognize an emerging theme. When sieves are needed, *their mesh size should be made successively smaller, as we get more information for the sample, but at a sufficiently slow rate*. What is "sufficiently slow" will depend upon the region of the parameter space we are in. *In the "outskirts" of the parameter space the rate must be very slow, while the more regular elements in the "central parts" of A can be handled by a rapidly decreasing mesh size.*

4 PARAMETER IN *C*-SPACES

Let the *regression function* α be piecewise continuously differentiable, $\alpha \in PC^1([0,1])$, and observe

$$x_\nu = \alpha\left(\frac{\nu}{n}\right) + z_\nu; \qquad \nu = 1,2,\ldots,n$$

where z_ν are i.i.d. and $N(0,1)$. The unrestricted ML-estimate of α is then given by solving

$$\sum_{\nu=1}^{n}\left[x_\nu - \alpha\left(\frac{\nu}{n}\right)\right]^2$$

so that the supremum is attained for any $\alpha^*_{ML} \in A$ with the values

$$\alpha^*_{ML}\left(\frac{\nu}{n}\right) = x_\nu \tag{1}$$

and arbitrary for the rest. It is certainly not possible to specify the α^*-function in the rest of $[0,1]$ in such a way that we can guarantee that it would converge in probability to α according to the uniform topology of A: a sieve is needed.

Introduce therefore the local sieve in A:

$$\mathbb{S}(\mu) = \left\{ \alpha \,|\, \alpha \quad \text{in } PC^1 \text{ and both } |\alpha| \text{ and } |\alpha'| \leqslant \frac{1}{\mu} \right\}$$

We show that this sieve makes ML estimation consistent. That $\sup_\alpha L(\alpha)$ is attained, is seen easily, since we have to solve for

$$\min_{\alpha \in \mathbb{S}(\mu)} \sum_{\nu=1}^{n} \left[x_\nu - \alpha\left(\frac{\nu}{n}\right) \right]^2$$

and this depends only upon the values of α at the points ν/n so that the likelihood function can be viewed as essentially a continuous function on a compact set in $\mathbb{R}^n$. At the points ν/n the derivative need not exist, between the points we can make α linear if μ is small enough for the given sample.

If $\beta \in C(0,1)$, let us consider the quadratic and linear forms S, and S_1, S_2, S_3:

$$\begin{aligned} S(\beta) &= \frac{1}{n} \sum_{\nu=1}^{n} \left[x_\nu - \beta\left(\frac{\nu}{n}\right) \right]^2 - 1 \\ &= \frac{1}{n} \sum_{\nu=1}^{n} \left[\alpha\left(\frac{\nu}{n}\right) - \beta\left(\frac{\nu}{n}\right) \right]^2 + \left(\frac{1}{n} \sum_{\nu=1}^{n} z_\nu^2 \quad 1 \right) + \frac{2}{n} \sum_{\nu=1}^{n} \left[\alpha\left(\frac{\nu}{n}\right) - \beta\left(\frac{\nu}{n}\right) \right] z_\nu \\ &= S_1 + S_2 + S_3 \end{aligned} \tag{2}$$

It is easily seen that, in probability when $n \to \infty$,

$$\begin{aligned} &S_1 \to \int_0^1 [\alpha(x) - \beta(x)]^2 dx = I(\beta) \\ &S_2 \to 0 \\ &S_3 \to 0 \end{aligned} \tag{3}$$

Note that S is $-(2/n)\ln L$ plus a constant. Equation (3) shows that with large probability S is close to $I(\beta)$.

As usual we make the mesh size μ tend to zero. Choose a value of μ big enough for $\alpha \in \mathbb{S}(\mu)$ to hold. This number is not known in advance. Then with $\beta = \alpha$ in (2) and (3) we see that

$$\frac{1}{n}\sum_{\nu=1}^{n}\left[x_\nu - \alpha\left(\frac{\nu}{n}\right)\right]^2 \to 1 \text{ in probability as } n \to \infty \tag{4}$$

However, the quantity based on the ML-solution α_μ^* constrained to $\mathbb{S}(\mu)$

$$\frac{1}{n}\sum_{\nu=1}^{n}\left[x_\nu - \alpha_\mu^*\left(\frac{\nu}{n}\right)\right]^2 = \frac{1}{n}\sum_{\nu=1}^{n}\left[x_\nu - \alpha\left(\frac{\nu}{n}\right)\right]^2 + \frac{2}{n}\sum_{\nu=1}^{n} z_\nu\left[\alpha\left(\frac{\nu}{n}\right) - \alpha_\mu^*\left(\frac{\nu}{n}\right)\right] + \frac{1}{n}\sum_{\nu=1}^{n}\left[\alpha\left(\frac{\nu}{n}\right) - \alpha_\mu^*\left(\frac{\nu}{n}\right)\right]^2 \tag{5}$$

is at most as large as (4) by the definition of α_μ^*. Also note that the middle term on the right-hand side of (5) can be written, after Abelian rearrangement, as

$$\frac{2}{n}\left[\sum_{\nu=1}^{n-1} Z_\nu\left[\alpha\left(\frac{\nu}{n}\right) - \alpha\left(\frac{\nu+1}{n}\right)\right] + Z_n\alpha(1)\right] + \text{ second term} \tag{6}$$

where $Z_\nu = z_1 + z_2 + \cdots + z_\nu$, and the second term looks the same but involves $\alpha_\mu^*(\nu/n)$ rather than $\alpha(\nu/n)$. We therefore can bound the expected value of the first term in (6), using $\alpha \in (\mu)$, by the expression

$$\frac{\text{constant}}{n^2}\sum_{\nu=1}^{n-1} E[|Z_\nu|] + \frac{\text{constant}}{n} E[|Z_n|] \tag{7}$$

where the constant depends upon the mesh size μ. But $Z_\nu = N(0, \nu)$ so that (7) is bounded by

$$\frac{\text{constant}}{n^2}\sum_{\nu=1}^{n-1} \sqrt{\nu} + \frac{\text{constant}}{\sqrt{n}} \to 0 \qquad \text{as } n \to \infty$$

Therefore the middle term on the right-hand side of (5) tends to zero in probability.

Hence, using (4) we have

$$\frac{1}{n}\sum_{\nu=1}^{n}\left[\alpha\left(\frac{\nu}{n}\right) - \alpha_\mu^*\left(\frac{\nu}{n}\right)\right]^2 \to 0 \text{ in probability}$$

Say that for a fixed δ there is a $x_0 \in [0,1]$ such that

$$\alpha(x_0)-\alpha_\mu^*(x_0)>\delta \tag{8}$$

Since

$$|\alpha_\mu^*(x_1)-\alpha_\mu^*(x_2)|\leqslant\frac{|x_1-x_2|}{\mu}$$

it follows that if we consider the ν for which

$$\left|\alpha\left(\frac{\nu}{n}\right)-\alpha(x_0)\right|<\frac{\delta}{2}$$

we have an inequality valid for these ν:

$$\begin{aligned}\alpha\left(\frac{\nu}{n}\right)-\alpha_\mu^*\left(\frac{\nu}{n}\right)&=\alpha\left(\frac{\nu}{n}\right)-\alpha(x_0)+\alpha(x_0)-\alpha_\mu^*(x_0)+\alpha^*{}_\mu(x_0)-\alpha_\mu^*\left(\frac{\nu}{n}\right)\\&\geqslant\frac{\delta}{2}-\frac{1}{\mu}\left|x_0-\frac{\nu}{n}\right|\end{aligned} \tag{9}$$

For the same ν consider those also satisfying

$$\left|\frac{\nu}{n}-x_0\right|<\frac{\mu\delta}{4} \tag{10}$$

and summing over these ν, denoting the sum by a prime,

$$\frac{1}{n}\sum_{\nu=1}^{n}\left[\alpha\left(\frac{\nu}{n}\right)-\alpha_\mu^*\left(\frac{\nu}{n}\right)\right]^2\geqslant\frac{1}{n}\Sigma'\left[\alpha\left(\frac{\nu}{n}\right)-\alpha_\mu^*\left(\frac{\nu}{n}\right)\right]^2\geqslant\frac{\delta^2}{16n}\#\{\nu\text{'s in }\Sigma'\} \tag{11}$$

But using (9) and (10), we get

$$\#\{\nu\text{'s in }\Sigma'\}\geqslant\delta n\min\left(\frac{\mu}{4},\frac{1}{\mu}\right)\delta n\,K$$

Hence (11) is at least $\delta^3 K/16$. Since the left-hand side of (11) can be made as small as desired, δ must also be small, so that (8), as well as the reversed inequality with $-\delta$ instead of δ, cannot hold for any x_0 unless δ is small. Note that the bounds do not depend upon x_0 and that α_μ^* converges uniformly toward α in probability and we have the following theorem.

THEOREM 1

The ML solution α_μ^* in $\mathbb{S}(\mu)$ will tend uniformly to α in probability if μ is made to tend to zero sufficiently slowly as n tends to infinity.

Actually it is enough that μ is made equal to the (unknown) constant $1/\|\alpha'\|$. It is conjectured that the same theorem holds for $A=C(0,1)$, but we have no proof of this.

We now study the implementation of the sieve technique to the present case of regression functions employing a global sieve, which is more convenient for computations (see Notes 8.4.A for references).

Consider the parameter space A consisting of periodic continuous functions on the one-dimensional torus $T^1=[0,2\pi)$ and embed it in $C([0,2\pi))$. We observe the values

$$x_\nu=\alpha\left(2\pi\frac{\nu}{n}\right)+\xi_\nu;\qquad \nu=0,1,2,\ldots,n-1$$

where ξ_ν are i.i.d. and $N(0,1)$. To estimate α consistently with maximum likelihood, we know that a sieve is necessary and we now use the following global one:

$$\mathbb{S}(\mu)=\left\{\alpha|\alpha\in PC^1(T^1)\text{with}\int_0^{2\pi}[\alpha'(x)]^2dx\leqslant\frac{1}{\mu}\right\}$$

We then have to solve the constrained ML-equation

$$\inf_{\alpha\in\mathbb{S}(\mu)}\sum_{\nu=0}^{n-1}\left[\alpha\left(2\pi\frac{\nu}{n}\right)-x_\nu\right]^2$$

We claim that the infimum is attained by a piecewise linear function. Indeed, if we assume to begin with that all the values $\alpha_\nu=\alpha(2\pi\nu/\mathrm{n})$ are fixed for an $\alpha\in\mathbb{S}(\mu)$, it is clear that the criterion used for this sieve

$$I(\alpha)=\int_0^{2\pi}[\alpha'(x)]^2dx$$

can be made smaller if α is not linear between, say, ν/n and $(\nu+1)/n$. This is an immediate consequence of Schwarz's inequality:

$$(\alpha_{\nu+1}-\alpha_\nu)^2=\left[\int_{2\pi\nu/n}^{2\pi(\nu+1)/n}\alpha'(x)dx\right]^2\leqslant\int_{2\pi\nu/n}^{2\pi(\nu+1)/n}dx\int_{2\pi\nu/n}^{2\pi(\nu+1)/n}[\alpha'(x)]^2dx$$
$$=\frac{2\pi}{n}\int_{2\pi\nu/n}^{2\pi(\nu+1)/n}[\alpha'(x)]^2dx$$

and equality is attained for the α' piecewise constant, which proves the assertion.

With such a piecewise function the problem reduces to solving the finite-dimensional minimum problem

$$\min \sum_{\nu=0}^{n-1} (\alpha_\nu - x_\nu)^2$$
$$I(\alpha) = \frac{n}{2\pi} \sum_{0}^{n-1} (\alpha_{\nu+1} - \alpha_\nu)^2 \phi \frac{1}{\mu} \tag{12}$$

where $\alpha_n = \alpha_0$ because of the periodicity of α. This consists of minimizing a continuous function over a compact set so that solutions exist.

If we fix the value of $I(\alpha)$ to, say, $I_0 \leqslant 1/\mu$, we can apply a Lagrange multiplier technique and get for the solution of the minimum, which can be attained as stated, the difference equation with a multiplier λ to be determined later:

$$\alpha_{\nu+1} - 2\alpha_\nu + \alpha_{\nu-1} = \lambda(\alpha_\nu - x_\nu); \qquad \nu = 0, 1, \ldots, n-1$$

or

$$\alpha_{\nu+1} - (\lambda+2)\alpha_\nu + \alpha_{\nu-1} = -\lambda x_\nu$$

With the familiar circulant matrix

$$M = \begin{Bmatrix} -\lambda-2 & 1 & 0 & 0 & \ldots & 1 \\ 1 & -\lambda-2 & 1 & 0 & \ldots & 0 \\ \vdots & & & & & \\ 1 & 0 & 0 & 0 & \ldots & -\lambda-2 \end{Bmatrix} = \{m_{\nu-\mu}\}$$

and the vectors $\alpha^n = \mathrm{col}(\alpha_0, \alpha_1, \ldots, \alpha_{n-1})$, $x = \mathrm{col}(x_0, x_1, \ldots, x_{n-1})$ we get

$$\alpha^n = -\lambda M^{-1} x \tag{13}$$

and the two expressions in (12) take the values with Euclidean norm

$$\|\alpha^n - x\|^2 = \|(I + \lambda M^{-1})x\|^2$$
$$I_0 = \frac{n}{2\pi} \|(I-T)\alpha^n\|^2 = \lambda^2 \frac{n}{2\pi} \|(I-T)M^{-1}x\|^2 \tag{14}$$

Here T is the cyclic shift matrix

$$T=\begin{Bmatrix} 0 & 1 & 0 & 0 & \dots & 0 \\ 0 & 0 & 1 & 0 & \dots & 0 \\ \dots & \dots & \dots & & & \\ 1 & 0 & 0 & 0 & \dots & 0 \end{Bmatrix}=\{t_{\nu-\mu}\}$$

which is also circulant, so that it can be diagonalized simultaneously with M by the matrix U:

$$\begin{aligned} U^T M U &= \operatorname{diag}[\mu_0, \mu_1, \dots, \mu_{n-1}] \\ U^T T U &= \operatorname{diag}[\kappa_0, \kappa_1, \dots, \kappa_{n-1}] \end{aligned} \tag{15}$$

where, with complex conjugate transpose and $z=e^{2\pi i/n}$,

$$U=\frac{1}{\sqrt{n}}\{z^{kl};\ k, l=0,1,2,\dots,n-1\}$$

But using the well-known spectral form for circulants, the μ and κ can be calculated explicitly:

$$\mu_\nu(\lambda)=\sum_{l=0}^{n-1} z^{\nu l} m_l = z^\nu-(\lambda+2)+z^{-\nu}$$

$$\kappa_\nu=\sum_{l=0}^{n-1} z^{\nu l} t_l = z^\nu$$

Inserting this into (14), we get

$$\|\alpha^n-x\|^2=\sum_{\nu=0}^{n-1}\left|1+\frac{\lambda}{\mu_\nu(\lambda)}\right|^2 |(U^T x)_\nu|^2$$

$$I_0=\frac{n\lambda^2}{2\pi}\sum_{\nu=1}^{n-1}\left|\frac{1-\kappa_\nu}{\mu_\nu(\lambda)}\right|^2 |(U^T x)_\nu|^2$$

since $\kappa_0=1$, so that the second sum need not include the term for $\nu=0$. Also note that for $\lambda>0$ these expressions make sense, since then $\mu_\nu(\lambda)=-4\sin^2(\nu\pi)/n-\lambda$ so that $\mu_\nu(\lambda)\neq 0$. Now consider I_0 as a function $I_0(\lambda)$ of λ. If $\lambda\to\infty$, it is easy to see that since $\mu_\nu(\lambda)\sim-\lambda$, we have

$$\lim_{\lambda\to\infty} I_0(\lambda)=\frac{n}{2\pi}\sum_{\nu=0}^{n-1}|1-\kappa_\nu|^2 |(U^T x)_\nu|^2$$

$$\lim_{\lambda\to\infty}\alpha_k^n=x_k$$

The second relation follows from $M^{-1} \sim -1/\lambda I$ and (13). This represents a small mesh size and no smoothing.

On the other hand, letting λ decrease, it is clear that $I_0(\lambda)$ increases and will become large as $\lambda \downarrow 0$. At the same time, noting that $\lambda\mu_\nu^{-1}(\lambda) \to 0$ except for $\nu = 0$, $\lambda\mu_0^{-1}(\lambda) = -1$, and using (13) together with (15),

$$\lim_{\lambda \to 0} \alpha_k^n = \frac{1}{n} \sum_0^{n-1} x_k = \bar{x}$$

so that the mesh size is large with extreme smoothing.

In other words, λ and μ vary in inverse directions to each other. Rather than operate with μ, we can use λ directly and calculate the optimal interpolation for the sieve as

$$\alpha^*\left(\frac{k}{n}\right) = -\frac{\lambda}{n} \sum_{\nu=0}^{n-1} \frac{1}{z^\nu - (\lambda+2) + z^{-\nu}} z^{\nu k} \sum_{l=0}^{n-1} z^{-\nu l} x_l \tag{16}$$

Note that this can be realized algorithmically by two applications of the fast Fourier transform so that the computational effort is moderate.

To gain a better insight into how this sieve works, let us extend it to a more general case in which the result is not at all intuitive. The steps in the analysis that are analogous to the last case will be left out from now on. Consider the sieve

$$\mathcal{S}(\mu) = \left\{ \alpha \,|\, \alpha \in PC^1(T^1),\, Q(\alpha) \leq \frac{1}{\mu} \right\} \tag{17}$$

with a regularity criterion based on a quadratic differential form

$$Q(\alpha) = \int_0^{2\pi} \{ p(x)[\alpha'(x)]^2 + q(x)\alpha(x) \} dx \tag{18}$$

Here p and q should be positive and, for example, C_1; both conditions could be relaxed. The Euler equations for the corresponding variational problem are then

$$L\alpha = \sum_{\nu=0}^{n-1} A\left[\alpha\left(\frac{2\pi\nu}{n}\right) - x_\nu \right] \delta\left(x - \frac{2\pi\nu}{n}\right) \tag{19}$$

where we have used the formally self-adjoint linear differential operator

$$L\alpha = (p\alpha')' - q\alpha$$

The natural boundary conditions are, of course, since we are on T^1, given by

$$\alpha(0)=\alpha(2\pi)$$

$$\alpha'(0)=\alpha'(2\pi)$$

We could now solve (19) using the Green's function G of the operator L, just as we have done before. It will be illuminating however, to see what happens when (19) is replaced by

$$L\hat{\alpha}=A\hat{\alpha}-\frac{A}{n}x_{\nu}\delta\left(x-\frac{2\pi\nu}{n}\right) \tag{20}$$

since this leads to a construction that seems to have potential for considerable generalization. In (20) the term $A\hat{\alpha}$ appears for all x, in contrast to (19), where it only occurs at the points $(2\pi\nu)/n$. This also accounts for the factor $1/n$ in the last term of (20), which is needed in order to compensate for this change. It seems intuitively that $\hat{\alpha}$, from (20), and α^*, the solution of (19), should differ little if n is large.

We apply the Mercer expansion converging absolutely and uniformly and get

$$G(x,y)=\sum_{\nu=1}^{\infty}\frac{1}{\lambda_{\nu}}\phi_{\nu}(x)\phi_{\nu}(y)$$

where λ_{ν} and ϕ_{ν} are the eigenelements of $L\alpha=\lambda\alpha$ and all λ are negative. Introducing the modified kernel

$$G_A(x,y)=\sum_{\nu=1}^{\infty}\frac{A}{A-\lambda_{\nu}}\phi_{\nu}(x)\phi_{\nu}(y) \tag{21}$$

we can solve (20) by the attractive formula

$$\hat{\alpha}(x)=\frac{1}{n}\sum_{k=0}^{n-1}G_A\left(x,\frac{2\pi k}{n}\right)x_k \tag{22}$$

Note that λ_{ν} is negative in the denominator of (21).

To find the asymptotic bias of $\hat{\alpha}$ as n tends to infinity, we calculate

$$a_n(x)=E[\hat{\alpha}(x)]=\frac{1}{n}\sum_{k=0}^{n-1}G_A\left(x,\frac{2\pi k}{n}\right)\alpha\left(\frac{k}{n}\right)$$

Since both G_A and α, are continuous, we get

$$a(x) = \lim_{n\to\infty} a_n(x) = \frac{1}{2\pi}\int_0^{2\pi} G_A(x,y)\alpha(y)\,dy$$

If we make A large, which corresponds to small mesh size, the coefficients in (21) tend to 1, so that G_A tends to the singular kernel $\delta(x-y)$. To make this statement precise, say that, for example, $\alpha \in C_2(T^1)$ so that α can be expanded in an absolutely and uniformly convergent series in terms of the eigenfunctions ϕ_ν [see Courant-Hilbert (1937), pp. 311–312]:

$$\alpha(x) = \sum_{\nu=1}^{\infty} \alpha_\nu \phi_\nu(x) \tag{23}$$

The asymptotic bias is then for $n\to\infty$ and fixed A, using (21),

$$b_A(x) = \alpha(x) - a(x) = \sum_{\nu=1}^{\infty} \frac{-\lambda_\nu}{A-\lambda_\nu}\alpha_\nu\phi_\nu(x)$$

Using the convergence properties of (23) together with the fact that the coefficients in (22) tend to zero for large values of A, it follows that $b_A(x)\to 0$ uniformly in x as $A\to\infty$. Hence if the mesh size is small enough, the asymptotic bias of $\hat{\alpha}$ will be small.

We can also obtain the asymptotic covariance of $\hat{\alpha}$. Since the α_ν are i.i.d. with common variance, say σ^2, we get from (22)

$$n\,\mathrm{Cov}[\hat{\alpha}(x_1), \hat{\alpha}(x_2)] = \frac{\sigma^2}{n}\sum_{k=0}^{n-1} G_A\left(x_1, \frac{2\pi k}{n}\right) G_A\left(x_2, \frac{2\pi k}{n}\right)$$

As n becomes large, this tends to

$$\frac{\sigma^2}{2\pi}\int_0^{2\pi} G_A(x_1,y)G_A(x_2,y)\,dy = \frac{\sigma^2}{2\pi} G_A^{(2)}(x_1,x_2)$$

expressed in the iterated kernel $G_A^{(2)}$. We have proved the following theorem.

THEOREM 2

If $\alpha \in C(T^1)$, the Sturm-Liouville sieve (17) *leads to a solution of the form* (22) *in terms of the Green's function of the differential operator appearing in the sieve definition. The asymptotic expectation and covariance are given by* *the expressions, see* (21) *defining* G_A,

$$\begin{cases} \lim_{n\to\infty} E[\alpha^*(x)] = \frac{1}{2\pi}\int_0^{2\pi} G_A(x,y)\alpha(y)\,dy \\ \lim_{n\to\infty} n\,\mathrm{Cov}[\alpha^*(x),\alpha(y)] = \frac{1}{2\pi}\int_0^{2\pi} G_A(x,z)G_A(z,y)\,dz \end{cases} \tag{24}$$

Remark 1 To make the bias small, we should make the mesh size small, which here means A large. At the same time, however, the last integral in (24) will become large. We notice *the balance between high resolution (small bandwidth) and small variance.*

We will encounter a surprising difficulty when we try to extend the result from the early part of this section to two dimensions, say for $X=D$, the unit disk. The quadratic differential form in (18) generalizes naturally to the *Dirichlet functional*

$$D(\alpha)=\int\int\left[\left(\frac{\partial\alpha}{\partial x}\right)^2+\left(\frac{\partial\alpha}{\partial y}\right)^2\right]dx\,dy$$

so that the Euler equations (19) will now be in terms of the Laplacian operator in two dimensions. Since our method employs the Green's function, we could expect to end up with an $\alpha^*(z)$, $z=(x,y)$, containing a linear combination of *logarithmic potentials* since in the present case they represent the singular part of the Green's function.

When we try to do this rigorously, we run into inconsistencies. The resolution of this paradox is that *even with this sieve the constrained ML solution is not consistent*. For the construction of the $\mathcal{S}(\mu)$ constrained *ML* solution we will need a function f such that $f(0)=1, f(z)=0$ for $|z|>\rho$, and such that the Dirichlet functional $D(f)\leqslant 1/\mu$. The role of this function will be made clear below. We make f depend on z only via $r=|z|$, radial symmetry, so that we should minimize

$$\frac{1}{2\pi}D(f)=\int_0^{\rho}[f'(r)]^2 r\,dr=I(f)$$

An elementary calculation for this minimum problem leads to the Euler equation $rf''=\text{constant}$.

Let us try therefore the solution of this differential equation

$$f(r)=1+ar+br\ln r$$

such that $f(\rho)=0$, which condition we write as an inner product in the form $(\xi, v)=-1$ with $\xi=\text{col}(a, b)$, $v=(\rho, \rho \ln \rho)$. We get

$$I(f)=\xi^T M \xi$$

with the matrix

$$M=\begin{Bmatrix} \int r\,dr, & \int r(1+\ln r)\,dr \\ \int r(1+\ln r)\,dr, & \int r(1+2\ln r+\ln^2 r)\,dr \end{Bmatrix}$$

Solving by a Lagrange multiplier for the minimum of $I(f)-2\lambda(\xi, \nu)$ we get, by the usual manipulations, the minimum

$$I(f)=\frac{1}{v^T M^{-1} v}$$

and after some calculations,

$$M^{-1}=\frac{8}{\rho^2}\begin{Bmatrix} \frac{1}{2}+\ln\rho+\ln^2\rho, & -\frac{1}{2}-\ln\rho \\ -\frac{1}{2}-\ln\rho, & 1 \end{Bmatrix}$$

so that

$$I(f)=\tfrac{1}{4}.$$

With this choice of the function f form the estimator

$$\alpha^*(z)=\sum_{i-1}^{n} x_i f(z-x_i) \tag{25}$$

With probability 1, if ρ is small enough, the supports of the terms in (25) do not overlap, when the sample has been fixed, so that

$$\alpha^*(z_i)=x_i; \qquad i=1,2,\ldots,n$$

Further $D(f)=\pi/2<1/\mu$ if $\mu<2/\pi$. Since α^* coincides with α at all the points x_i, it maximizes the likelihood. Also $\alpha^*(z)$ vanishes at $\partial(D)$ if μ is small enough, so that it satisfies the boundary conditions. Hence α^* is a constrained ML-solution, but since $\alpha^*(z_i)=x_i=\alpha(z_i)+n_i$ and has the variance equal to 1, it cannot possibly be consistent.

Therefore we must use a *finer sieve*, and we propose the one given by (see Note 8.4.B)

$$\int\int\left[\left(\frac{\partial^2\alpha}{\partial x^2}\right)^2+\left(\frac{\partial^2\alpha}{\partial y^2}\right)^2\right]dxdy \leqslant \frac{1}{\mu} \tag{26}$$

which leads to the *biharmonic equation*

$$\Delta^2\alpha = \frac{\partial^4\alpha}{\partial x^4} + 2\frac{\partial^4\alpha}{\partial x^2\partial y^2} + \frac{\partial^4\alpha}{\partial y^4} = 0.$$

To simplify the analytical expressions, we shall assume the domain to be T^2 and the boundary conditions $\alpha = \Delta\alpha = 0$ at $\partial(T^2)$. The more natural boundary conditions expressing periodicity can be handled similarly, but with the extended Green's function; $\lambda = 0$ is then an eigenvalue. Denote by G_0 the Green's function for the Laplacian and with $\Delta\alpha = 0$, $\alpha = 0$ at $\partial(T^2)$, which can be obtained as

$$G_0(\xi_1,\eta_1;\xi_2,\eta_2) = \frac{4}{\pi^2}\sum_{m,k=1}^{\infty}\frac{\sin(k\xi_1/2)\sin(m\eta_1/2)\sin(k\xi_2/2)\sin(m\eta_2/2)}{k^2+m^2}$$

for $0 \leqslant x_1, x_2, y_1, y_2 < 2\pi$, with uniform convergence in (ξ_1, ξ_2) outside of any neighborhood of (η_1, η_2) [see Courant-Hilbert (1937), I, p.335]. Analogously to (19) we get the Euler equations

$$\Delta^2\alpha = \sum_{\nu,\mu=0}^{n-1} A\left[\alpha\left(\frac{2\pi\nu}{n},\frac{2\pi\mu}{n}\right)-\xi_{\nu\mu}\right]\delta\left(x-2\pi\frac{\nu}{n}, y-2\pi\frac{\mu}{n}\right)$$

$$= \sum_{\nu,\mu=0}^{n-1} A_{\nu\mu}\delta\left(x-2\pi\frac{\nu}{n}, y-2\pi\frac{\mu}{n}\right)$$

so that

$$\Delta\alpha = \sum_{\nu,\mu=0}^{n-1} A_{\nu_\mu} G_0\left(x,y,2\pi\frac{\nu}{n},2\pi\frac{\mu}{n}\right)$$

Solving this we get

$$\alpha(x,y) = \frac{4}{\pi^2}\sum_{\nu,\mu=0}^{n-1}\sum_{m,k=1}^{\infty} A_{\nu\mu}\frac{\sin(kx/2)\sin(my/2)\sin(\pi k\nu/n)\sin(\pi m\mu/n)}{(k^2+m^2)^2}$$

But

$$A_{\nu\mu}=A\left[\alpha\left(\frac{2\pi\nu}{n},\frac{2\pi\mu}{n}\right)-\xi_{\nu\mu}\right]$$

so that we have to find, analogously to (16), the inverse to a four-dimensional array. But this array is a four-dimensional circulant and can be inverted by two finite Fourier transforms, each in two dimensions. Hence *we can obtain the ML solution with the sieve* (26) *as a biharmonic interpolation algorithmically implemented by applying the two-dimensional fast Fourier transform twice.*

It will be instructive to compare the sieve technique used in this chapter with *a Bayesian estimation method.* Returning to the one-dimensional case, say that $\alpha=\alpha(t)$ is a stationary and periodic Gaussian process on $T=[0,2\pi)$, with mean value 0 and continuous covariance function $r(t)$. The Egervary expansion gives, see Example 1 in Chapter 2, Section 4, with absolute and uniform convergence

$$r(t)=\sum_{-\infty}^{\infty} a_k e^{ikt} \tag{27}$$

where the spectral density $\{a_k\}\in l_1$ has $a_k\geqslant 0$.

Corresponding to (27) the Cramér representation of the process itself reduces to

$$\alpha(t)=\sum_{-\infty}^{\infty} Z_k e^{ikt} \tag{28}$$

where Z_k are the increments of the orthogonal process $Z(\lambda)$, associated with $\alpha(t)$, at the frequencies $\lambda=k$ of the spectrum. We have $E(Z_k)=0$, $E[|Z_k|^2]=\alpha_k$.

Here the function given in (28) should be seen as the parameter α, our unknown continuous regression function. We assume, similarly to what we did earlier, that we can observe

$$x_\nu=\alpha\left(2\pi\frac{\nu}{n}\right)+e_\nu; \qquad \nu=0,1,2,\ldots,n-1$$

where $\{e_\nu\}$ is an i.i.d. sample from $N(0,\sigma^2)$.

The Bayes estimate using least-squared error as the criterion is simply the expected value of the parameter conditioned by the observations. In our case this gives

$$\alpha^*(t)=E[\alpha(t)|\{x_\nu\}] \tag{29}$$

where $\{x_\nu\}$ stands for the whole set of observed values $x_0, x_1, \ldots, x_{n-1}$.

We assume that $\Sigma a_k^{1/2}<\infty$, and later specify the condition further. Then it can be shown by standard arguments that (28) and (29) lead to

$$\alpha^*(t)=\sum_{-\infty}^{\infty} e^{ikt}E[Z_k|\{x_\nu\}] \tag{30}$$

with a.c. uniform convergence, so that $\alpha^*\in C$ with probability 1.

To investigate the asymptotic properties of (30) for large sample size n, we follow Chow (1978) and consider the estimation error

$$\alpha(t)-\alpha^*(t)=\sum_{k\notin[-N,N]} e^{itk}Z_k$$
$$+\sum_{k\in[-N,N]} e^{itk}[Z_k-E(Z_k|\{x_\nu\}]-\sum_{k\notin[-N,N]} e^{itk}E(Z_k|\{x_\nu\})$$

Hence with the sup norm in C,

$$E[\|\alpha-\alpha^*\|]\leqslant\sum_{k\notin[N,N]} E|Z_k|+\sum_{k\in[-N,N]} E[|Z_k-E(Z_k|\{x_\nu\})|]$$
$$+\sum_{k\notin[-N,N]} E[E(|Z_k||\{x_\nu\})] \tag{31}$$

Let us dominate each of the three sums. We see, since $E[|Z_k|]\leqslant\{E[|Z_k|^2]\}^{1/2}$, that the first sum is at most

$$\sum_{k\notin[-N,N]} \sqrt{a_k} \tag{32}$$

and the third sum in (31) is of course bounded by the same expression.

To deal with the middle sum, we have using L_2-norm, using the fact that conditional expectation solves the minimum approximation error,

$$\|Z_k-E(Z_k|\{x_\nu\}]\|_2\leqslant\|Z_k-\frac{1}{n}\sum_{l=0}^{n-1} x\left(\frac{2\pi l}{n}\right)e^{-i(2\pi l/n)k}\|_2$$
$$=\|Z_k-\frac{1}{n}\sum_{l=0}^{n-1}\alpha\left(\frac{2\pi l}{n}\right)e^{-i(2\pi l/n)k}-\frac{1}{n}\sum_{l=0}^{n-1}e_le^{-i(2\pi l/n)k}\|_2 \tag{33}$$

The triangle inequality then tells us that (33) is at most equal to

$$\|Z_k - \frac{1}{n}\sum_{l=0}^{n-1} \alpha(2\pi l/n)e^{-i(2\pi l/n)k}\|_2 + \|\frac{1}{n}\sum_{l=0}^{n-1} e_l e^{-i(2\pi l/n)k}\|_2 \tag{34}$$

The last term is of course equal to $\sigma/\sqrt{n}$. To handle the first one, note that (28) gives us the important relation

$$\frac{1}{n}\sum_{l=0}^{n-1} \alpha\left(\frac{2\pi l}{n}\right)e^{-i(2\pi l/n)k} = \sum_{\nu=-\infty}^{\infty} Z_{k+\nu n}$$

Together with (34) this leads to the bound

$$\leqslant \|Z_k - \sum_{\nu=-\infty}^{\infty} Z_{k+\nu n}\|_2 + \frac{\sigma}{\sqrt{n}}$$

$$= \|\sum_{\nu\neq 0} Z_{k+\nu n}\|_2 + \frac{\sigma}{\sqrt{n}} = \left(\sum_{\nu\neq 0} a_{k+\nu n}\right)^{1/2} + \frac{\sigma}{\sqrt{n}} \tag{35}$$

This will enable us to show that the estimate will converge fast to the true value in expected sup norm if the spectral density has little energy at high frequencies. More precisely, say that

$$a_k = 0(|k|^{-p}); \qquad |k| \to \infty \tag{36}$$

where the exponent $p > 2$ expresses how fast a_k tends to zero for high frequencies.

Then the bounds in (31), (32), and (35) lead, after some elementary evaluations, to the bound

$$0(N^{-p/2+1}) + 0(Nn^{-1/2})$$

If we let N increase proportionally to $n^{1/p}$, the bound takes the form $n^{-1/2+1/p}$ and we have proved the following theorem (see Note 8.4.C).

THEOREM 3

For the stationary periodic Gaussian a priori measure with spectral density $\{a_k\}$ *the expected sup norm of the Bayesian estimate is bounded by*

$$E[\|\alpha - \alpha^*\|] = 0\left(\frac{1}{n^{1/2-1/p}}\right); \qquad n \to \infty \tag{37}$$

if (36) *holds with* $p > 2$.

This result shows that if p is large, we get almost the asymptotic decrease $n^{-1/2}$ that we are used to in classical estimation problems. It is conjectured that a sieve approach to the same estimation problem would lead to the same asymptotic decrease as in (37).

5 PARAMETER IN A BV-SPACE

Let α be a bounded measure on a sample space X associated with an inhomogeneous Poisson process. To be specific, let $X=[0,1]$, and say we observe the outcome of the Poisson process n times independently of each other. The outcomes, denoted by $x_1, x_2, \ldots, x_n$, will all be discrete measures associating nonnegative integers, the number of observed values in respective sets, to any disjoint Borel sets $B_1, B_2, \ldots, B_r$ of X. The values $x(B_1), x(B_2), \ldots, x(B_r)$ will be stochastically independent, each one Poisson distributed with mean $\alpha(B_i)$.

The unknown parameter α, a bounded measure, will naturally be considered as an element of $BV(X)$, and we want to estimate it, if possibly consistently in the topology of $BV(X)$.

Having observed the i.i.d. *random measures* $x_1, x_2, \ldots, x_n$ form the likelihood function

$$L=\sum_{i=1}^{n} L_i; \qquad L_i=\left(\prod_{j=1}^{n_i} L_{ij}\right)\exp[-\alpha(y_i)]$$

with

$$L_{ij}=\frac{\left[\alpha(y_{ij})\right]^{x_i(y_{ij})}}{x_i(y_{ij})!}e^{-\alpha(y_{ij})}; \qquad i=1,2,\ldots,n, \quad j=1,2,\ldots,n_i \tag{1}$$

Here x_i attributes mass $x(y_{ij})$ to the points y_{ij} and no others and $y_i=[0,1]-\{y_{i1}, y_{i2}, \ldots\}$. Note that the points y_{ij} can have multiplicities greater than 1 with positive probability if α has atoms.

Then

$$\ln L=\sum_{i=1}^{n}\left\{-\alpha([0,1])+\sum_{j}\left[\alpha(y_{ij})+x_i(y_{ij})\ln\alpha(y_{ij}) - \alpha(y_{ij})-\ln x_i(y_{ij})!\right]\right\}$$

Direct maximization of 1 with respect to the measure α gives the obvious

ML-estimate α^* attributing the mass $x_i(y_{ij})/n$ to the points y_{ij} and no others.

For an arbitrary continuous function f on $[0,1]$ we should consider, to study weak convergence of the measures α^*, the functional

$$f(\alpha^*)=\int_0^1 f(x)\alpha^*(dx)=\frac{1}{n}\sum_{i,j} f(y_{ij})x_i(y_{ij})$$

We have unbiasedness

$$E[f(\alpha^*)]=\int_0^1 f(x)\alpha(dx)=f(\alpha)$$

and

$$\operatorname{Var}[f(\alpha^*)]=\frac{1}{n}\int_0^1 f^2(x)\alpha(dx)\to 0 \text{ as } n\to\infty$$

which implies that $f(\alpha^*)\to f(\alpha)$ in probability. *Therefore α^* converges weakly [that is, in the weak* topology of $BV(X)$] to α in probability, and there is no need for a sieve in this case.*

If, however, we are interested in some other topology in $BV(X)$, say strong convergence, then it is conjectured that a sieve will be needed, but we shall not study this question here.

Instead let us examine the following variation of the previous problem, now with a Bayesian method. Define the *a priori* measure over α by assuming $\alpha(B_1),\dots,\alpha(B_r)$ to be independent, for disjoint Borel sets B_i, with the marginal measures given by $\alpha(B)$=stochastic variable with Γ-frequency function $f(x,p)=1/\Gamma(p)x^{p-1}e^{-x}, p=m(B)$. This will define the *a priori* measure completely (see Chapter 1, Section 1), (see Note 8.1.B for references).

To find the *a posteriori* distribution, introduce a partition of X into the intervals

$$b_k=\begin{cases}\left[\frac{k-1}{N},\frac{k}{N}\right); & k=1,2,\dots,N-1\\ \left[\frac{N-1}{N},1\right]; & k=N\end{cases}$$

and apply the limiting argument as in Example 7 of Chapter 3, Section 2 with N increasing. Since x_k=number of observations in b_k will have a Poisson distribution with mean $n\alpha(b_k)$, for fixed parameter α, it follows by an elementary calculation that $\alpha(b_k)=\alpha_k$ will have the *a posteriori* frequency function

$$\frac{\frac{1}{\Gamma(1/N)}\alpha_k^{1/N-1}e^{-\alpha_k}\left[(\alpha_k n)^{x_k}/x_k!\right]e^{-\alpha_k n}}{\int_0^\infty \text{same function } d\alpha_k}=\frac{(1+n)^{1/N+x_k}}{\Gamma(1/N+x_k)}\alpha_k^{1/N+x_k}e^{-\alpha_k(1+n)}$$

The *a posteriori* expectation of α_k, given the observations, then becomes

$$\frac{\Gamma(1/N+x_k+1)}{\Gamma(1/N+x_k)}\frac{1}{1+n}=\frac{x_k+1/N}{n+1} \tag{2}$$

As $N\to\infty$, the limiting α^* will be the mixture of a uniform distribution over $[0,1]$, expressed by the term $1/N$ on the right-hand side of (2), with weight $1/(n+1)$ and a discrete measure as before at the points of observation, with weight $x_k/(n+1)$. It can be shown that this α^* also converges weakly in distribution as n tends to infinity.

6 OPERATORS AS PARAMETER

Consider the Hilbert space l_2 as sample space, $X=l_2$, and let the probability measure over l_2 be Gaussian with mean value element equal to the zero vector and covariance operator α. Then α has to be of the trace type, $\mathrm{tr}(\alpha)<\infty$ (see Chapter 1, Section 2).

Having observed i.i.d. the elements $x^1, x^2, \ldots, x^n \in l_2$, we want to estimate the operator α and study *consistency in operator norm.* In order to facilitate the discussion we assume that all α in A are simultaneously diagonalizable, which is a severe restriction as far as the next theorem is concerned.

We can then assume that all α are diagonal matrices, represented as vectors formed from their diagonals, say with elements $\alpha_1, \alpha_2, \ldots,$ which are nonnegative numbers forming a summable sequence. Introduce the Radon-Nikodym derivative of P_α with respect to some P_β, β fixed, $\beta=(\beta_1, \beta_2, \ldots)$ with $\beta_\nu>0$ and $\Sigma\beta_\nu<+\infty$. Then

$$\frac{dP_\alpha(x)}{dP_\beta(x)}=\prod_{i=1}^{n}\prod_{\nu=1}^{\infty}\sqrt{\frac{\beta_\nu}{\alpha_\nu}}\exp\left[\frac{1}{2}\left(\frac{1}{\beta_\nu}-\frac{1}{\alpha_\nu}\right)(x_i^\nu)^2\right]$$

assuming, of course, that P_α is absolutely continuous with respect to P_β. Conditions for this have been given in Chapter 5.

Operating formally we are led to the estimate (see Notes 8.6.A)

$$V^* = \frac{1}{n}\left\{ \sum_{i=1}^{n} (x_i^\nu)^2;\ \nu = 1,2,\ldots \right\} \in A \text{ a.c.}$$

Denote the coordinates of the true parameter value by $V_1, V_2, \ldots$ and consider V^*, whose coordinates V_i^* are stochastically independent and distributed as

$$V_i^* \underset{\text{i.d.}}{=} \frac{V_i}{n}\chi_n^2(i)$$

But the *operator norm* of the error $V^* - V$ is

$$\|V^* - V\| = \max_i |V_i^* - V_i|$$

This is equal in distribution to

$$e_n = \max_i V_i \left|\frac{1}{n}\chi_n^2(i) - 1\right| \tag{1}$$

in terms of a χ^2-variable with n degrees of freedom. We can write (1) as $e_n = \max(e_n^+, -e_n^n)$ with

$$\begin{cases} e_n^+ = \max_i V_i\left[\frac{1}{n}\chi_n^2(i) - 1\right] \\ e_n^- = \min_i V_i\left[\frac{1}{n}\chi_n^2(i) - 1\right] \end{cases}$$

Consider the distribution of e_n^+. A similar treatment is also applied to e_n^-. We have, for a fixed $\varepsilon > 0$,

$$P\{e_n^+ < \varepsilon\} = \prod_{i=1}^{\infty} F_n\left[\frac{n}{V_i}(1+\varepsilon)\right]$$

where F_n is a χ_n^2 distribution function. Hence

$$-\ln P\{e_n^+ < \varepsilon\} = \sum_{i=1}^{\infty}\left\{-\ln F_n\left[n\frac{1+\varepsilon}{V_i}\right]\right\}$$

Now we can use a dominated convergence argument by appealing to Chebyshev's inequality with absolute moments of order 1. The mean deviation of χ_n^2 is

$$E[|\chi_n^2 - n|] = \frac{n^{n/2}e^{-n/2}}{2^{n/2-1}\Gamma(n/2)}$$

[see Johnson-Kotz (1970), p. 168]. Using Stirling's approximation, this leads to the asymptotic behavior for large n:

$$E[\chi_n^2 - n|] \sim \sqrt{\frac{n}{\pi}}$$

and we get the bound, asymptotically valid for large n,

$$1 - F_n\left[n\left(1 + \frac{\varepsilon}{V_i}\right)\right] = \frac{1}{\sqrt{n}} 0(V_i)$$

But since $V_i \to 0$ as $i \to \infty$, we have

$$\ln F_n\left[n\left(1 + \frac{\varepsilon}{V_i}\right)\right] \geqslant 2\left[F_n\left[\frac{n}{V_i}(1+\varepsilon)\right] - 1\right]$$

for i large enough so that

$$-\ln F_n\left[n\left(1 + \frac{\varepsilon}{V_i}\right)\right] < 2\left[1 - F_n\left[\frac{n}{V_i}(1+\varepsilon)\right]\right] = \frac{1}{\sqrt{n}} 0(V_i)$$

Hence

$$-\ln P\{e_n^+ < \varepsilon\} \leqslant \text{constant} \frac{1}{\sqrt{n}} \sum_{i=1}^{\infty} V_i = \frac{1}{\sqrt{n}} \text{constant tr}(\alpha) \to 0 \qquad \text{as } n \to \infty$$

or

$$P\{e_n^+ < \varepsilon\} \to 1 \qquad \text{as } n \to \infty$$

For the opposite inequality we have, with positive ε,

$$-\ln P\{e_n^- > -\varepsilon\} = \sum_{i=1}^{\infty} -\ln\left\{1 - F_n\left[n\left(1 - \frac{\varepsilon}{V_i}\right)\right]\right\}$$

However,

$$F_n\left[n\left(1-\frac{\varepsilon}{V_i}\right)\right]=\frac{1}{\sqrt{n}}0(V_i)$$

so that since the trace of the operator is finite,

$$-\ln P\{e_n>-\varepsilon\}\leqslant\frac{1}{\sqrt{n}}\text{constant}\sum_{i=1}^{n}V_i=\frac{1}{\sqrt{n}}\text{constant tr}(\alpha)\to 0\qquad\text{as } n\to\infty$$

which implies

$$P\{e_n^->-\varepsilon\}\to 1$$

Hence $e_n\to 0$ in probability so that the estimate converges in probability according to the operator topology to the true operator.

THEOREM 1

The direct ML-estimate of the covariance operator converges in probability according to operator norm to the true operator: for any $\varepsilon>0$,

$$P\{\|\alpha^*-\alpha\|>\varepsilon\}\to 0\qquad\text{as } n\to\infty$$

In this case therefore the sieve is not essential (see Note 8.6.A). One should investigate whether this remains true when the α in A are no longer assumed to be simultaneously diagonalizable; we do not have the answer to this question. Consider now instead the set A of bounded operators on $X=l_2$ with $\text{tr}(\alpha^T\alpha)<\infty$ and topologized by the Hilbert-Schmidt norm $\|\alpha_1-\alpha_2\|^2=\text{tr}[(\alpha_1-\alpha_2)^T(\alpha_1-\alpha_2)]$. Define the random element x to be observed as

$$x=\alpha z+n;\qquad n=(n_i)$$

where z is a *design vector* chosen by the observer, and where n is a Gaussian element in l_2 with mean zero, independent coordinates, and of the trace type so that the variances V_i of its coordinates satisfy

$$\sum_{i=1}^{\infty}V_i<\infty;\qquad \text{Var}(n_i)=V_i$$

To estimate α we should choose the design such that we observe x for many values of z, and so that the set of z has the linear span tending to the whole l_2 as n increases. For each z we replicate the observation r times.

In other words we observe

$$x^{\nu\mu}=\alpha z^{\mu}+n^{\nu\mu};\qquad \nu=1,2,\ldots,r,\quad \mu=1,2,\ldots,s \tag{2}$$

so that the sample size $n=rs$. The ML-estimate minimizes

$$\sum_{\nu,\mu}\|x^{\nu\mu}-\beta z^{\mu}\|_n^2;\qquad \text{with the new norm } \|z\|_n^2=\sum_{i=1}^{\infty}\frac{1}{\text{Var}(n_i)}z_i^2$$

$$z=(z_1,z_2,\ldots)$$

To simplify the analysis, we choose the z^{μ} as orthogonal unit vectors along the first s coordinate axes in l_2:

$$\begin{aligned} z^1&=(1,0,0,\ldots)\\ z^2&=(0,1,0,\ldots)\\ z^3&=(0,0,1,\ldots)\\ &\vdots \end{aligned} \tag{3}$$

The unrestricted ML-solution can then be calculated to be the matrix $\alpha^*=\{\alpha_{ij};\ i,j=1,2,\ldots\}$, where

$$\alpha_{ij}^*=\frac{1}{r}\sum_{\nu,\mu}x_i^{\nu\mu}z_j^{\mu};\ i,j=1,2,\ldots,s$$

For the other subscripts we get no information from the observations, and we arbitrarily make these entries equal to zero.

For $i, j=1,1,\ldots,s$ we have, using (3) and (3),

$$\begin{aligned} \alpha_{ij}^*&=\alpha_{ij}+\frac{1}{r}\sum_{\nu,\mu}n_i^{\nu\mu}z_j^{\mu}\\ &=\alpha_{ij}+\frac{1}{r}\sum_{\nu}n_i^{\nu j} \end{aligned}$$

Hence $\alpha_{ij}^*-\alpha_{ij}$ has expected value zero and

$$E\left[(\alpha_{ij}^*-\alpha_{ij})^2\right]=\frac{1}{r}V_i;\qquad i,j=1,2,\ldots,s$$

so that

$$E\left[\sum_{i,j=1}^{\infty}(\alpha_{ij}^{*}-\alpha_{ij})^{2}\right]=\frac{s}{r}\sum_{i=1}^{s}V_{i}+\sum_{i\text{ or }j>s}\alpha_{ij}^{2}$$

It follows that if we make s and r tend to infinity in such a way that $s=o(r)$, then $\alpha^{*}\to\alpha$ in probability.

THEOREM 2

For bounded operators in l_2 with the above Hilbert-Schmidt topology we get consistent ML-estimation if we use design vectors as in (3) *with the replication number r tending to infinity faster than s: $s=o(r)$.*

As a variation of this we consider now as the unknown parameter to be estimated an integral operator α of periodic convolution type. Consider, with α periodic with period 2π, Egervary's integral equation

$$x(t)=\int_{0}^{2\pi}\alpha(t-s)z(s)\,ds;\qquad t\in(0,2\pi] \tag{4}$$

where $z(\cdot)$ is a continuous even function on the one-dimensional torus $(0,2\pi]=T^{1}$ (see Example 1 in Chapter 2, Section 4). Obviously α maps $L_2(0,2\pi)$ into itself and, forming the Fourier transforms for $k=\ldots-1,0,1,\ldots,$

$$x_k=\int_{0}^{2\pi}e^{ikt}x(t)\,dt$$

$$z_k=\int_{0}^{2\pi}e^{ikt}z(s)\,dt$$

$$\alpha_k=\int_{0}^{2\pi}e^{ik\lambda}\alpha(t)\,dt$$

we have $x_k=\alpha_k z_k$, and the operator norm of α can be seen to be just (see Note 8.6.B) $\|\alpha\|=\max_k|\alpha_k|$. The vectors (x_k), (z_k), (α_k) can now be viewed as elements in l_2.

Let us observe the sample functions of

$$x(t)=\int_{0}^{2\pi}\alpha(t-s)z(s)\,ds+n(t);\qquad t\in T \tag{5}$$

where $n(t)$ is a stationary Gaussian process on T with mean value function identically zero and a continuous covariance function $r(t)$. The Mercer expansion of r can be written

$$r(t-s)=\sum_{k=-\infty}^{\infty} e^{ik(t-s)}V_k \tag{6}$$

using Egervary's representation as several times before. In (6) the eigenvalues V_k form a summable sequence and can be identified with the variances σ_k^2 of the orthonormal random variables z_ν in the Karhunen-Loève expansion

$$n(t)=\sum_k e^{ikt}z_k$$

In the present case $X=L_2(T)$ and A is embedded in the space of symmetric bounded operators $L_2(T)\to L_2(T)$ topologized with the operator norm.

Say that we choose the "design functions" z as follows, in analogy to the previous case. Select s orthonormal vectors in $L_2(T)$, $z^1, z^2,\dots, z^s$, and observe the corresponding value of $x(t)$ in (5), but replicated r times. Using the isometric relation between $L_2(T)$ and l_2, this amounts to the same thing as saying that we observe

$$x_k^{\nu\mu}=\alpha_k z_k^\nu + n_k^{\nu\mu}$$

with $\nu=1,2,\dots,s$; $\mu=1,2,\dots,r$, and $k=\dots 1,0,1,\dots$. Of course the unrestricted ML-estimate of α leads to

$$\alpha_k^*=\frac{\sum_{\nu,\mu} x_k^{\nu\mu}\bar{z}_k^\nu}{r\sum_\nu |z_k|^2} \tag{7}$$

for those k for which the denominator is positive, say the set K. For the other k the observations do not carry any information, and we put the corresponding $\alpha_k^*=0$.

We get directly

$$E(\alpha_k^*)=\begin{cases}\alpha; & \text{for } k\in K\\ 0: & \text{otherwise}\end{cases}$$

Also we get

$$\operatorname{Var}(\alpha_k^*)=\frac{V_k}{r^2\sum_\nu |z_k^\nu|^2}; \qquad k\in K$$

In the topology of A we then have

$$\|\alpha^* - \alpha\| = \max_k |\alpha_k^* - \alpha_k| = \max(u, v)$$

with

$$\begin{cases} u = \max\limits_{k \in K} |\alpha_k^* - \alpha_k| \\ v = \max\limits_{k \notin K} |\alpha_k| \end{cases}$$

To fix ideas say that we choose the z^ν vectors similarly to before, with zeros for all coordinates except

$$\begin{aligned} x_{-m}^{(1)} &= 1 \\ x_{-m+1}^{(2)} &= 1 \\ &\vdots \\ x_{m-1}^{s-1} &= 1 \\ x_m^s &= 1 \end{aligned}$$

where $s = 2m+1$ must be odd. This restriction can easily be removed and plays no role in the asymptotic analysis. In other words we input as test signals into our filtering black box the pure harmonics $e^{-imt}, \ldots, e^{imt}$.

Then

$$\sum_\nu |z_k^\nu|^2 = \begin{cases} 1; & \text{if } k \in K = \{-m; m+1, \ldots, m\} \\ 0; & \text{otherwise} \end{cases}$$

It is clear that $v \to 0$ as $s \to \infty$, since $\{\alpha_k\} \in l_2$, and we must show that we also have $u \to 0$ in probability. To prove that this is so, consider for an arbitrary positive ε the probability

$$P\{u \leq \varepsilon\} = P\left\{ \frac{\sigma_k}{r} |\eta_k| \leq \varepsilon; \qquad \forall k \in K \right\}; \qquad V_k = \sigma_k^2 \tag{8}$$

where $\{\eta_k\}$ is an i.i.d. sample from $N(0,1)$. Hence (8) can be expressed as the product

$$\prod_{k \in K} \left[2\Phi\left(\frac{r\varepsilon}{\sigma_k} \right) - 1 \right]$$

where Φ is the distribution function of $N(0,1)$. Using the asymptotic expression for Φ [see Cramér (1945), p. 374],

$$1-\Phi(x)=\frac{1}{\sqrt{2\pi}\,x}\left[e^{-x^2/2}+0(x^{-2})\right]; \qquad x\to+\infty$$

we get, since $\sigma_k\to 0$ for large k, and using an elementary inequality for the logarithm function,

$$\ln P\{u\leqslant\varepsilon\}=\sum_{k\in K}\ln[1-\delta_k]\geqslant-\sum_{k\in K}\frac{\delta_k}{1-\delta_k}$$

with

$$\delta_k=2\left(1-\Phi\left(\frac{r\varepsilon}{\sigma_k}\right)\right)=\frac{2\sigma_k}{\sqrt{2\pi}\,r\varepsilon}\left[e^{-r^2\varepsilon^2/2\sigma_k^2}+0\left(\frac{\sigma_k^2}{r^e\varepsilon^2}\right)\right]$$

Combining the last two inequalities, we get

$$\ln P\{u\leqslant\varepsilon\}\geqslant-\frac{\text{constant}}{r\varepsilon}\sum_{k\in K}\left[\sigma_k e^{-r^2\varepsilon^2/2\sigma_k^2}+0\left(\frac{\sigma_k^2}{r^2\varepsilon^2}\right)\right] \tag{9}$$

For small $x=\sigma_k/r\varepsilon$ the function xe^{-a/x^2} has infinite contact at $x=0$ so that it is certainly $0(x^2)$, say. Since $\Sigma\sigma_k^2<\infty$, it follows that (9) tends to zero if $r\to\infty$.

THEOREM 3

The unrestricted ML-solution (7) *converges according to the operator topology to* α *in probability.*

7 PARAMETER IN GROUPS

We begin by an almost trivial case. Let the sample space be the natural numbers, $x=\mathbb{N}$, with a probability measure P on X, with the discrete positive probabilities $p_1>p_2>p_3>\cdots$ Here P is fixed and known. Consider the parameter space A as made up of all permutations of X, $A=$*symmetric group* over $\mathbb{N}$. We observe $x_1, x_2, \ldots, x_n$, where $x_i=\alpha y_i$, and the y_i have been drawn i.i.d. according to P, and want to observe the unknown element α of the permutation group. The likelihood function is then

$$L=\prod_{\nu=1}^{\infty}p_\nu^{n_\nu}; \qquad \text{where } n_\nu=\#\{y_i=\nu\}$$

To get the ML-estimate we should choose the permutation β that solves $L=\max$, or equivalently,

$$\max_{\alpha\in A}\sum_{\nu=1}^{\infty} n_\nu \ln p_\nu; \qquad n_\nu = \#\{\alpha^{-1}x_i=\nu\} \tag{1}$$

But the numbers $\ln p_1, \ln p_2, \ldots$ form a decreasing sequence, and it is well known that the maximum of (1) is then attained by choosing α such that

$$\#\{\alpha^{-1}x_i=1\} \geqslant \#\{\alpha^{-1}x_i=2\} \geqslant \ldots \tag{2}$$

[see, e.g., Hardy-Littlewood-Polya (1964), chap. 10]. In the case of ties in (2) α is not uniquely determined; denote one of these determinations by α^*.

This case is easy to analyze for consistency. Indeed, we naturally use the discrete topology on A so that we should show that for any μ the value of $\alpha^*\mu$ tends to $\alpha\mu$ with the notation that $\alpha\mu$ denotes the natural number into which α, the permutation of $\mathbb{N}$, carries the natural number μ. But say that $\alpha\mu$ is the number k. We have

$$P\{\alpha^*\mu=k\} = P\{\alpha^*\alpha^{-1}k=k\} = P\{n_{(k)}=k\} \tag{3}$$

where $n_{(i)}$ are the ordered n_i in decreasing order. But $(1/n)n_\nu$ tends to $P(\alpha^{-1}y=\nu)=P(y=\alpha\nu)=p_{\alpha\nu}$ a.c. Hence $n_{(k)}$ will tend to k a.c., and the expression in (3) tends to 1.

It follows that *no sieve is needed* to ensure that the ML estimate is consistent in the present case. A more interesting group is the following.

Consider the sample space $X=[0,1]$ on which a strictly increasing frequency function $p(x)$ has been defined. Let A consist of *the group of all piecewise linear measure preserving transformations* (with respect to Lebesque measure) *of X onto itself*, and topologize the space by convergence in measure. Denote the topological group A by G_2. It is not a complete space. The slope of the line segments describing the graph of the transformation must be ± 1.

We observe $x_i=\alpha y_i$, $i=1,2,\ldots,n$, where the y_i are i.i.d. stochastic variables with frequency function p. What is the ML estimate of α?

The maximum of p is at $x=1$ so that we cannot make the likelihood L greater than $p^n(1)$. We can, however, get arbitrarily close to this upper bound. Indeed, we map the points $x_1, x_2, \ldots, x_n$ with short intervals around them into points close to 1 and complete the definition of α so that it is piecewise linear with slope 1 and one to one. The graph in Figure 1 (a) should suffice to illustrate the procedure. The x-values are indicated by arrows toward the x-axis. If the intervals surrounding each $x_{(i)}$ are made

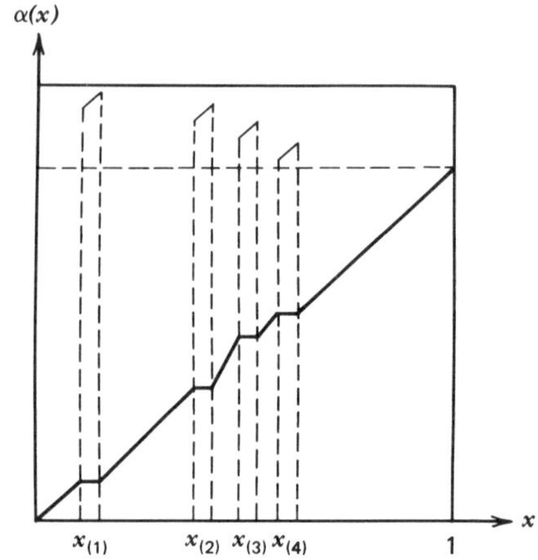

(a)

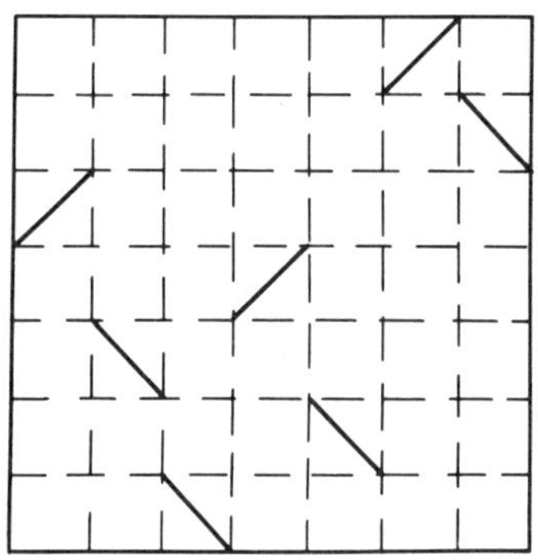

(b)

Figure 1

small, then $\alpha(x_{(i)})$ takes values close to 1. The supremum is not attained in A, and it is clear that the unrestricted ML solution does not converge to α in any reasonable sense.

To compensate for this, introduce the following sieve:

$$\mathbb{S}(\mu)=\{\alpha|\alpha\in A \text{ and } \alpha' \text{ existing and equal to } \pm 1 \text{ on}$$
$$\times \text{ each open interval } (\nu\mu,(\nu+1)\mu)\}$$

Without any serious loss of generality we have assumed here that $1/\mu$ is an integer so that ν ranges over the values $0,1,2,\dots,1/(\mu-1)$. A typical function from $\mathbb{S}(1/7)$ looks like Figure 1(b).

Given a mapping $\alpha \in A$, we can always approximate it arbitrarily well with convergence in Lebesque measure by $\alpha_\mu \in \mathbb{S}(\mu)$. The α_μ will only differ substantially from α at the break points of α in small neighborhoods. In other words we introduce regularizations of α (see Chapter 7, Section 2).

To get the restricted ML solution in $\mathbb{S}(\mu)$, introduce the numbers m_ν equal to the number of observed x-values in $I_\nu = (\nu\mu, (\nu+1)\mu)$. Since the number of elements in $\mathbb{S}(\mu)$ is finite for fixed μ, namely, $(1/\mu)!2^{1/\mu}$, the maximum is of course attained. The $\alpha_\mu^* \in \mathbb{S}(\mu)$ that maximizes the likelihood makes the sequence increasing. This is so for the same reason as before, since the probabilities associated with the intervals $y \in I_\nu$ are increasing when $p(x)$ is increasing.

Using what we know about the earlier discrete case, it follows that α_μ^* will converge, for fixed μ and increasing n, to the permutation that makes the probabilities

$$p_\nu^\mu = P(x \in I_\nu) = P(\alpha y \in I_\nu) = \int_{\alpha^{-1} I_\nu} p(x)\,dx \tag{4}$$

into an increasing sequence. But since $\alpha^{-1} I_\nu$ is close to $\alpha^{-1} I_\nu$ in measure, it follows that the probabilities on the right-hand side of (4) will be increasing if μ is small enough. Combining these statements we have this theorem (see Note 8.7.A).

THEOREM 1

For $A = G_2$ the ML-solution α_μ^* restricted to $\mathbb{S}(\mu)$ will converge in probability to α if μ is made to tend to zero slowly enough as n tends to infinity.

The parameter was here an (unknown) element of a (known) group. We now consider the case when *the group itself represents the unknown*: it is not completely specified.

Let G be a one-parameter Lie group, with parameter t, acting on sufficiently smooth functions defined on a two-dimensional manifold, which we choose for simplicity as T^2. The unknown parameter α to be estimated will now be chosen as the infinitesimal generator of G:

$$X(x,y)\frac{\partial}{\partial x} + Y(x,y)\frac{\partial}{\partial y} \tag{5}$$

where X and Y are C_1-functions of $(x,y) \in T^2$ which will be represented as $[0,1) \times [0,1)$. The group acts during a time interval $(0, \varepsilon)$ on T^2 and carries certain points ζ^ν into new positions z^ν, $\nu = 1, 2, \ldots, N$. The latter points are observed with random errors leading to the observations $Z^\nu = z^\nu + e^\nu$, where the errors e^ν are i.i.d. from, for example, a bivariate Gaussian distribution with means zero, correlation coefficient zero, and variances equal to σ^2. When N is made large and ε small, our task is to estimate consistently the infinitesimal generator given as a vector field $\alpha=(X,Y)$.

We are led to consider the differential equation $\dot{z}=\alpha(z)$, $Z\in T^2$, and the natural condition to impose on α in order that solutions exist and be unique is $\alpha\in Lip$. One would then choose the parameter space $A=Lip$ and show consistency in one of its natural norms. This has not been done, however, and the following is a less satisfactory approach to the problem.

The simplest design is to place the ζ^ν in a regular grid. With the coding $\nu=(k,l)$ we can write

$$\zeta^\nu=\left(\frac{k}{n},\frac{l}{n}\right);\qquad k,l=0,1,2,\ldots,n-1 \tag{6}$$

Recalling that we identify points modulo 1 in both x and y directions, (6) gives us $N=n^2$ points. The unconstrained ML solution asks for a generator α such that

$$\sum_\nu \|Z^\nu-z^\nu\|^2=\min \tag{7}$$

and z^ν is the value of the solution $z(t)$ of

$$\begin{cases}\dot{z}=\alpha(z)\\ z(0)=\zeta^\nu\end{cases} \tag{8}$$

evaluated at $t=\varepsilon$.

Consider Figure 2(a), in which a part of T^2 is shown and the points Z^ν are denoted by small crosses and the ζ^ν by dots. We now establish a disappointing fact that will force us to use a sieve: we can actually find a generator α which makes $z^\nu=Z^\nu$, when the Z^ν are given and distinct; the latter holds of course with probability 1.

To see that this can be done, connect each ζ^ν with Z^ν by nonintersecting arcs, say of smoothness C_2. Indeed, we can introduce one arc at a time and note that the set remains connected after leaving out these arcs. Then just go on adding a new arc, and so on. Some such arcs are indicated in the figure. Now define $X(x,y)$, $Y(x,y)$ in (5) as the components of tangent vectors along these arcs, and normalize the norm of the vector to be constant on each arc and equal to the length of the arc divided by ε. Complete the definition of X, Y arbitrarily for all the points that do not lie on any arc, but not violating the C_2 condition. With this α the value of the expression (7) is zero: we have an unconstrained ML solution.

When $N \to +\infty$, $\varepsilon \downarrow 0$, it is clear that *this estimate cannot be consistent*, the effect of σ via the random errors prevents convergence.

We mention in passing that in one dimension it is not always possible to reduce (7) to zero. The reason is that the topology of T^1 does not allow deleting an arc without destroying connectivity. Nevertheless, also the one-dimensional case requires a sieve, but we shall not study this problem here.

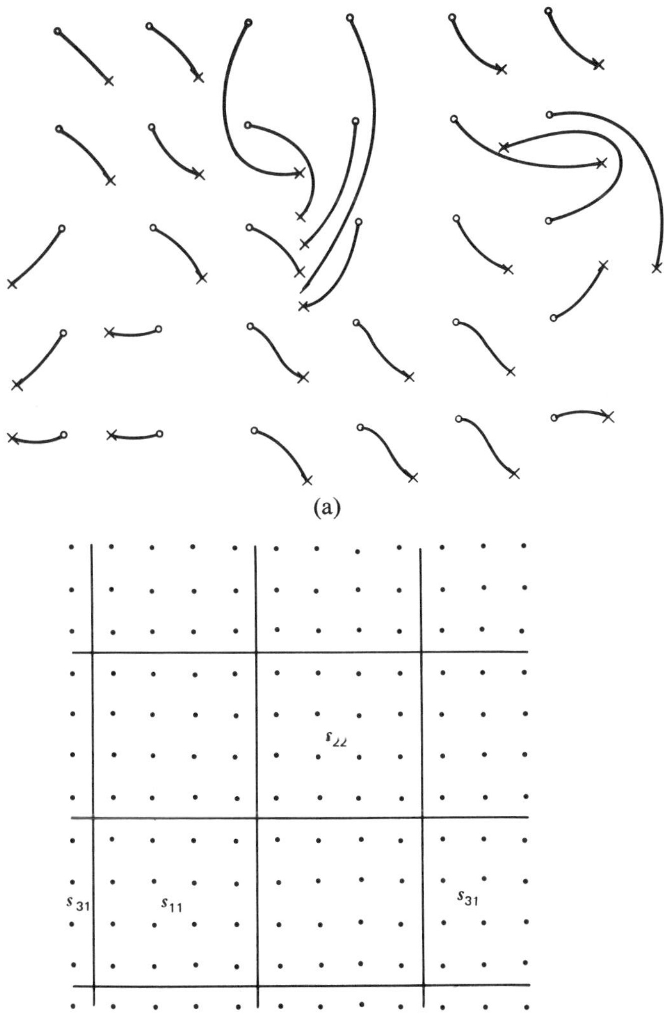

(a)

(b)

Figure 2

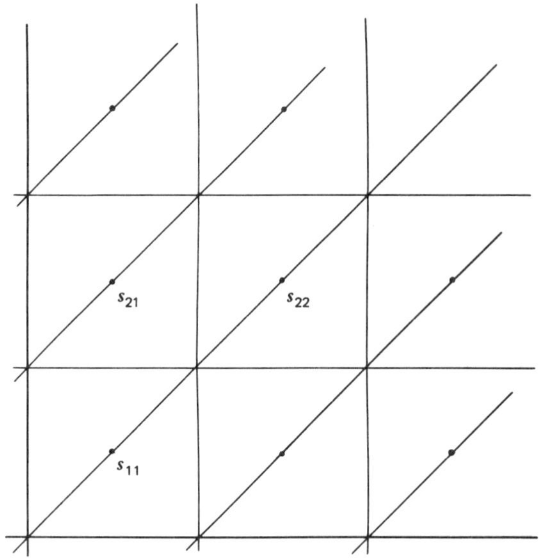

(c)

Figure 2 (Continued)

We therefore consider the following modified estimate α^*. Fix an even number m dividing n, which will play the role of the inverse of the mesh size, and consider the set of m^2 points $\zeta^\nu=(x,y)$ belonging to the square S_{ab}:

$$-\frac{m}{2}\leqslant nx-am<\frac{m}{2}$$
$$-\frac{m}{2}\leqslant ny-bm<\frac{m}{2}$$

Here a and b take the values $0,1,2,\ldots,(n/m)-1$. This is illustrated in Figure 2(b) for $n=12$, $m=4$. All observations Z^ν that belong to a certain ζ^ν in S_{ab} will be treated together in the construction of the estimate of the generator of the Lie group.

Consider all the points $s_{ab}=(a\frac{m}{n}, b\frac{m}{n})$ and the corresponding net of triangulation that we have used before in Section 2;[see Figure 2(c)]. We assume that in the sieve α is linear over each of these triangles and continuous in the whole T^2. We can then represent it as

$$\bar{\alpha}(x,y)=\sum_{a,b=0}^{(n/m)-1} L^{ab}(x,y)\alpha(s_{ab}) \tag{9}$$

where the L^{ab} are piecewise linear taking values between zero and one.

Let us first estimate the $\alpha(s_{ab})$ by solving for the 2-vector c that minimizes the sum

$$\sum_{\nu} \|Z^{\nu}-\zeta^{\nu}-\varepsilon c\|_2^2 \tag{10}$$

summed over ν with $\zeta^{\nu}\in S_{ab}$. This corresponds to treating α as approximately constant in the subset in question when ε is small. Note that it can very well happen that the observations Z^{ν} fall outside of S_{ab}; this will happen often if σ is big. If we write $d^{\nu}=(d_1^{\nu}, d_2^{\nu})=Z^{\nu}-\zeta^{\nu}$, the normal equations for minimizing (10) will be in coordinate form, $d^{\nu}=(d_1^{\nu}, d_2^{\nu})$, $c=(c_1, c_2)$

$$\sum_{\nu} (d_k^{\nu}-\varepsilon c_k)=0$$

so that since we sum over the m^2 points in S_{ab},

$$c^* = \frac{1}{\varepsilon m^2}\sum_{\nu} d^{\nu} = \frac{1}{\varepsilon m^2}\sum_{\nu}(Z^{\nu}-\zeta^{\nu}) \tag{11}$$

How does c^* behave asymptotically? This depends critically upon how fast we let m tend to infinity with n relative to ε. We have, using (8),

$$Z^{\nu}-\zeta^{\nu}=z^{\nu}-\zeta^{\nu}+e^{\nu}=\varepsilon\alpha(s_{ab})=o(\varepsilon)+e_{\nu}$$

Substituting this into (11), we get

$$c^*=\alpha(s_{ab})+\frac{1}{\varepsilon m^2}\sum_{\nu} e_{\nu}+o(1) \tag{12}$$

For a given $\alpha\in C_1$ we can approximate it by an $\bar{\alpha}$ of the form (9), and this uniform approximation error is $o(m/n)$ if $m/n\to 0$, which will be assumed from now on. We also know that the functions L^{ab} take values between 0 and 1, so that it only remains to make sure that $c^*-\alpha(s_{ab})$ tends uniformly to zero with large probability. With the aid of (12) we can guarantee this if we make

$$\frac{1}{\varepsilon m^2}\max_{a,b}|N_{ab}| \tag{13}$$

tend to zero in probability where we have $(n/m)^2$ i.i.d. Gaussian variables $N_{ab}=\Sigma_{\nu} e_{\nu}$ with mean zero and variance $\sigma^2 m^2$.

This leads to a simple exercise in elementary probability theory. If $(x_1, x_2, \ldots, x_n)$ is an i.i.d. sample from $N(0, \sigma_n^2)$, then

$$P(\max x_\nu \leqslant t) = \left(1 - \frac{G_n}{n}\right)^n \tag{14}$$

with

$$G_n = n\left[1 - \Phi\left(\frac{t}{\sigma_n}\right)\right]$$

where Φ is the density function of a standardized Gaussian distribution. Using again the classical asymptotic relation for the tail of a Gaussian distribution

$$1 - \Phi(x) \sim \frac{1}{\sqrt{2\pi}\, x} e^{-x^2/2}; \qquad x \to \infty$$

[see, e.g., Cramér (1945), p. 374], we should make

$$G_n \sim \frac{n\sigma_n}{\sqrt{2\pi}\, t} e^{-t^2/2\sigma_n^2}$$

tend to 0 for any $t > 0$ in order that (14) tend to 1. This happens if we let σ_n tend to zero slowly enough so that

$$\lim_{n\to\infty} \sigma_n^2 \ln n = 0 \tag{15}$$

Applied to the present case we have $(n/m)^2$ variables, each with variance $\sigma^2/(\varepsilon^2 m^2)$ when the first factor in (13) is incorporated into the variance corresponding to σ_n^2. We then need, to satisfy (15),

$$\frac{1}{\varepsilon^2 m^2} \ln\left(\frac{n}{m}\right) \to 0 \qquad \text{as } n \to \infty; \qquad \varepsilon \downarrow 0$$

and m increases, but so that $m = o(n)$ as was needed already. If, for example, we let m increase as $n^{1-\delta}$, $\delta > 0$, we could let ε tend to zero as An^{-p}, $A > 0$, $p < (1-\delta)$. Then convergence will take place.

This should only be considered as a tentative study of a problem with many ramifications: the statistical determination of the Lie algebra corresponding to a Lie group in the presence of measurement noise.

We again consider the estimation of an unknown group element α, but now for a compact group, which changes the problem drastically.

Let X be a second countable *topological group* with the group operation written in multiplicative form. We also assume that it is *compact* so that it has a well-defined normalized Haar measure that will be denoted by dx. With measures P_α absolutely continuous with respect to dx with density $p(\alpha^{-2}x)$ we can apply Theorem 1 of Chapter 7, Section 1, to study ML-estimation since our parameter space is compact.

This is all very well, but it is tempting to try alternatively to exploit the Fourier analysis that is now available to construct particular estimates. This will now be done in a concrete case.

Say that Q is a fixed Borel measure on X and that we observe $x=\alpha y$, where y has Q as its probability distribution. Here α is the unknown parameter so that the parameter space $A=X$.

Let $U_r(x)$, $r=0,1,2,\ldots$, be the set of unitary representations, irreducible and nonequivalent, as described in Chapter 1, Section 3. If d_r is the dimension of U_r and with the *group characters*

$$\chi_r(x)=\operatorname{tr} U_r(x)=\text{sum of eigenvalues of } U_r(x)$$

we define the *concentration function* (see Note 8.7.B) for $c_r>0$, $c_0=1$, as

$$\phi(x)=\frac{1}{2}\sum_{r=0}^{\infty} c_r\left[1+\frac{1}{d_r}\operatorname{Re}\chi_r(x)\right] \tag{16}$$

The role of this function will become clear later, but we start by proving some simple properties for it.

Lemma 1 The concentration function satisfies:

(a) $\phi(x)$ is nonnegative and equal to 1 if and only if $x=e$.
(b) It is invariant, $\phi(a^{-1}xa)=\phi(x)$ for all x and a.
(c) It is continuous.
(d) It is even, $\phi(x^{-1})=\phi(x)$.

Proof (a) Any $U_r(x)$ is unitary so that its eigenvalues are on the unit circle, and hence their real part is in $[-1,1]$, so that $\phi(x)\geqslant 0$. For $x=e$ we have $U_r(e)=1$, all eigenvalues equal to 1, so that $\phi(e)=1$. On the other hand if $\phi(x)=1$, we must have

$$\operatorname{Re}\chi_r(x)=d_r; \qquad \text{all } r$$

which implies that all the eigenvalues of $U_r(x)$ are equal to 1. But such a unitary matrix reduces to the identity I_r in $\mathbb{R}^{d_r}$. Since representations separate points, we have that $U_r(x)=I_r$ implies $x=e$.

(b) This follows from the fact that all group characters are invariant and the form of ϕ as a linear combination of group characters [see, e.g., Pontrjagin (1946), chap. IV].

(c) Since $|\chi_r(x)| \leqslant d_r$, the terms in (16) are bounded by c_r. The terms are continuous since the characters are, and the uniform convergence of (16) implies that ϕ is continuous.

(d) Since $U_r(x^{-1}) = U_r^{-1}(x) = U_r^*(x)$, we have $\chi_r(x^{-1})$ as the complex conjugate of $\chi_r(x)$. They have the same real part so that (16) implies that $\phi(x^{-1}) = \phi(x)$.

Q.E.D.

We use the concentration function to introduce the *centroid set* of a probability measure.

Definition 1 For a stochastic group element ξ with measure Q we define the centroid set as

$$\text{centroid}(\xi) = \left\{ x \,|\, E\left[\phi(x^{-1}\xi)\right] = \max, x \in X \right\}$$

It plays a role similar to that of mean value in topological vector spaces.

Lemma 2 The centroid set is a closed set with the following properties:

(a) Centroid $(a\xi) = a$ centroid (ξ).
(b) Centroid $(\xi^{-1}) = [\text{centroid } (\xi)]^{-1}$
(c) Centroid $(\xi) = \{x_0\}$ if $\xi = x_0$ a.c.
(d) Centroid $(\xi) = X$ if and only if $Q =$ Haar measure.
(e) If $\xi_n \to \xi$ in distribution then

$$\lim_{N\to\infty} \text{closure}\left[\bigcup_{n=N}^{\infty} \text{centroid}(\xi_n)\right] \subseteq \text{centroid}(\xi) \tag{17}$$

Proof Since ϕ is continuous, the family $\{\phi(x^{-1}a);\, a \in X\}$ is equicontinuous, and therefore $E[\phi(x^{-1}\xi)]$ is continuous. The centroid is closed and nonempty. To prove (a) note that $\phi(x^{-1}a\xi) = \phi[(a^{-1}x)^{-1}\xi]$ so that centroid $(a\xi)$ consists of all x such that

$$a^{-1}x \in \text{centroid}(\xi)$$

But this is just the relation in (a).

The centroid set of (ξ^{-1}) consists of all x such that

$$E\left[\phi(x^{-1}\xi^{-1})\right]=\max$$

Using (d) in Lemma 1, we know that $\phi(x^{-1}\xi^{-1})=\phi(\xi x)$. However, (b) in the same lemma shows that this is the same as $\phi(x\xi)$; choose $a=x^{-1}$. Hence centroid (ξ^{-1}) consists of all x^{-1}, $x\in$centroid (ξ), and this is statement (b).

Note that this implies that if ξ has symmetric distribution, then its centroid set is symmetric.

If ξ has all its probability mass in x_0, we have

$$E\left[\phi(x^{-1}\xi)\right]=\phi(x^{-1}x_0) \tag{18}$$

and (a) in Lemma 1 shows that the maximum value of (18) is 1 and is achieved uniquely for $x=x_0$: statement (c).

If $Q=dx$, we have (see Chapter 1, Section 3)

$$\hat{U}_r(Q)=E\left[Q_r(\xi)\right]=0$$

unless $r=0$ when (18) takes as value the identity. Hence, using (16),

$$E\left[\phi(x^{-1}\xi)\right]\equiv c_0+\frac{1}{2}\sum_{r=1} c_r \tag{19}$$

so that its maximum is achieved everywhere, centroid $(\xi)=X$.

On the other hand if centroid $(\xi)=X$, we have

$$E\left[\phi(x^{-1}\xi)\right]=\text{constant, say } c$$

or

$$\frac{1}{2}\sum_{r=0}^{\infty} c_r\left\{1+\frac{1}{d_r}\text{Re tr}\left[U_r(x^{-1})\hat{U}_r\right]\right\}\equiv c \tag{20}$$

Integrate this over X with respect to dx-measure,

$$\frac{1}{2}\sum_{r=0}^{\infty} c_r[1+\delta_r]=c_0+\frac{1}{2}\sum_{1}^{\infty} c_r=c$$

and compare with (19). Integrate (20) with respect to ξ-measure,

$$\frac{1}{2}\sum_{r=0}^{\infty} c_r\left\{1+\frac{1}{d_r}\text{Re tr}\left[(\hat{U}_r)^* U_r\right]\right\} \equiv c$$

and recall that $c_r > 0$. Then $\hat{U}_r = 0$ for $r \neq 0$ so that our measure coincides with Haar measure: statement (d).

If the measures of ξ_n and ξ are Q_n and Q, respectively with $Q_n \to Q$ weakly, it follows that

$$\lim_{n\to\infty} \psi_n(x) = \psi(x)$$

$$\psi_n(x) = E\left[\phi(x^{-1}\xi_n)\right]$$

$$\psi(x) = E\left[\phi(x^{-1}\xi)\right]$$

since ϕ is continuous. If z belongs to the set on the left-hand side of (17), there exists a sequence $z_1, z_2, \ldots$ tending to z and such that

$$z_k \in \text{centroid}(\xi_{n_k}); \qquad n_k \to \infty$$

We can show, however, that the sequence $\{\psi_n\}$ is equicontinuous, since ϕ is continuous, hence uniformly continuous, so that for any $\varepsilon > 0$ there exists a neighborhood N_ε of e such that

$$|\phi(u^{-1}y) - \phi(y)| < \varepsilon; \qquad y \in X, \quad u \in N_\varepsilon \tag{21}$$

Using (21) with $y = x^{-1}\xi_n$, we see that

$$|\psi_n(xu) - \psi_n(x)| = \left|\int_X \phi\left[(xu)^{-1}\xi_n\right] Q_n(d\xi_n) - \int_X \phi(x^{-1}\xi_n) Q_n(d\xi_n)\right| < \varepsilon$$

for $u \in N_\varepsilon$ for all n. Then $\{\psi_n\}$ is equicontinuous as stated. This implies that as $k \to \infty$,

$$\psi_{n_k}(z_k) = \max_y \psi_{n_k}(y) \to \psi(z)$$

so that $\psi(z)$ must equal the maximum of ψ: $z \in \text{centroid}(\xi)$. Q.E.D.

Consider now an i.i.d. sample $(x_1, x_2, \ldots, x_n)$ from P and let $\bar{x}$ denote a random element with the empirical distribution P_n^* of the sample. In other words $\bar{x}$, for fixed sample, is a discrete measure with n atoms, each with mas $1/n$. Form the random set

$$A_n = \text{centroid}(\bar{x})$$

THEOREM 2

As n tends to infinity we have, with probability 1,

$$\lim_{N\to\infty} \text{closure}\left(\bigcup_{n=N}^{\infty} A_n\right) \subseteq \alpha \text{ centroid}(y) \tag{22}$$

Proof The centroid of $x=\alpha y$ is given by the right-hand side of (22), recalling (a) in Lemma 2. We know that, with probability 1,

$$\int_X c_\nu(x) P_n^*(dx) \to \int_X c_\nu(x) P(dx); \qquad \text{for all } \nu = 1,2,\ldots \tag{23}$$

Here $\{c_\nu\}$ is some fixed sequence of continuous functions on X. Select this sequence so that it is dense in $C(X)$, which is possible since X is second countable. A standard argument then shows that with probability 1, $P_n^* \to P$ weakly. We can then appeal to Lemma 2, (e) which implies that (23) holds.

Q.E.D.

Remark 1 If centroid (y) consists of the single point m, then the set on the left of (22) also consists of a single point, say a. Then α is uniquely determined as am^{-1}: the random set estimates α consistently.

Remark 2 The centroid concept is modeled on that of a mean value, and it may be tempting to narrow it down so that it would always consist of a single point. If we want it to play a role similar to that of average in linear inference, however, this must be avoided. Say that y has a measure which is invariant with respect to a subgroup X_0 of X: $x_0 y \equiv y$ in distribution when $x_0 \in X_0$. Then it is clear that $(x_0\alpha)^{-1}y = \alpha^{-1}x_0^{-1}y \equiv \alpha^{-1}y$ in distribution so that α is not identifiable, except possibly modulo X_0. Therefore the centroid set should be allowed to have more than one element.

8 SETS AS PARAMETERS

A problem of considerable interest in *statistical geometry* is when *the unknown parameter is a set* [see Grenander (1978), chap. 5]. Let α take as values any closed bounded set in the plane whose boundary consists of a finite number of analytic arcs, and with positive area, $m(\alpha)>0$. It need not be connected or simply connected. Topologize the set of α by introducing neighborhoods of the form

$$N(\alpha_0)=\{\alpha|m(\alpha\Delta\alpha_0)<\varepsilon\}$$

where $m(\alpha\Delta\alpha_0)$ means the Lebesque area of the symmetric difference $(\alpha\cap\alpha_0^c)\cup(\alpha^c\cap\alpha_0)$. This parameter space is not complete.

Let $X=R^2$, and let the probability distribution for a given $\alpha\in A$ be the uniform distribution over α so that the frequency function is

$$f(x;\alpha)=\begin{cases}\dfrac{1}{m(\alpha)}; & x\in\alpha\\ 0; & \text{otherwise}\end{cases}$$

The likelihood function is then, for an i.i.d. sample $x_1, x_2, \ldots, x_n$,

$$L(\alpha)=\begin{cases}m^{-n}(\alpha); & \text{if all } x_i\in\alpha\\ 0; & \text{otherwise}\end{cases}$$

Obviously $L(\alpha)$ can be made arbitrarily large by concentrating the set α to small neighborhoods of the x_i so that the ML-estimate is undefined. *We therefore need a sieve*, and we shall construct it by employing sets with well-behaved boundaries.

The reader is recommended to compare the following with regularization of sets as in Chapter 7, Section 5.

Introduce the sieve $\mathbb{S}(\mu)$ whose elements both are subsets of A and can be written as

$$\beta_\mu(Z)=\bigcap_{D_\mu(y)\subset Z^c} D_\mu^c(y);\qquad Z\subset R^2 \tag{1}$$

where $D_\mu(z)$ denotes an open disk of radius μ centered at the point z, and Z is any point set. Intuitively this means that we construct sets by excluding disks of radius μ from the outside of Z. The cardinality of Z is arbitrary. For any $\alpha\in A$ we define its regularization as

$$\alpha_\mu=\beta_\mu(\alpha)$$

Since any $D_\mu^c(y)\supseteq\alpha$, if $D_\mu(y)\subset\alpha^c$ it follows that $\alpha\subseteq\alpha_\mu$. Due to the smoothness of $\delta\alpha$ it also follows that $m(\alpha_\mu-\alpha)\to 0$ as $\mu\to 0$. In Figure 1(a) the regularized set α_μ differs from α by a set whose boundary arcs are indicated by dashed lines.

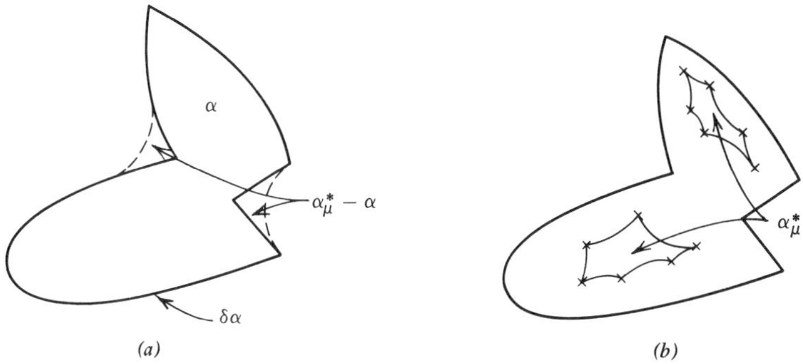

Figure 1

For a given set $\xi=(x_1, x_2,\ldots, x_n)$ of observed x-values the ML solution α_μ^* restricted to the sieve $\mathbb{S}(\mu)$ will then be attained, and since we are trying to remove as many disks as possible from the complement ξ^* in order to make the area as small as possible, it is actually given by $\beta_\mu(\xi)$. Since ξ is a finite point set, the boundary of α_μ^* consists of finite number of circular arcs. This is illustrated in Figure 1(b).

To prove consistency we first note that since $\beta_\mu(Z)$ is a monotonic set operation in terms of Z and since $\xi\subseteq\alpha$ a.c., we must have $\alpha_\mu^*\subseteq\alpha_\mu$. On the other hand we can write

$$m(\alpha_\mu-\alpha_\mu^*)=\int_{\alpha_\mu}[I(z)-I^*(z)]\,dz$$

where I and I^* are the indicator functions of α_μ and α_μ^*, respectively. Further we have

$$E[I^*(z)]=P\{z\in\alpha_\mu^*\}=1-P\{z\in\cup D_\mu(y)\}$$

with the union taken over those y for which no x_ν is in $D_\mu(y)$.

But if z is in the interior of α and recalling that the x are drawn from a uniform distribution over α, it follows that the probability that z is covered by any $D_\mu(y)$ not containing any x_ν tends to zero when μ is fixed and n tends to infinity. Hence

$$E[m(\alpha_\mu-\alpha_\mu^*)]=\int_\alpha+\int_{\alpha_\mu-\alpha}E[I(z)-I^*(z)]\,dz \tag{2}$$

where the first integral tends to zero by bounded convergence and the second can be made arbitrarily small since $m(\alpha_\mu - \alpha) \to 0$ as $\mu \to 0$. Combining the last statement with (2), we get the following theorem.

THEOREM 1

The sieve (1) *gives consistent ML estimation. For any positive* ε *we can find a* μ *such that*

$$P\{m(\alpha_\mu^* \Delta \alpha) < \varepsilon\} > 1 - \varepsilon$$

for n sufficiently large.

CHAPTER 9

General Parameter Space

1 GENERAL CONDITIONS FOR SIEVES

Armed with the knowledge about sieves in particular parameter spaces, which we gained in Chapter 8, we can now approach the question of how sieves should be selected in general. Our perspective will change: in the previous chapter the goal was not just to make constrained *ML* estimates consistent, but, more importantly, we studied the details of the construction and the analysis of the statistical behavior of particular estimates. Now, when we attempt greater generality, it will not be possible to be very specific as far as construction and analysis are concerned. Instead we will be limited to weaker results, to questions of consistency (in Section 4 also with efficiency), and how regularization in parameter space can be organized by means of sieves.

Given a sample space X, let dx be a σ-finite measure on some σ-algebra $\mathcal{Q}$ over X. All the probability measure P_α will be defined on the same $\mathcal{Q}$ and assumed to be absolutely continuous with respect to dx, so that

$$P_\alpha(E)=\int_E f(x;\alpha)dx; \qquad E\in\mathcal{Q}$$

Therefore we can deal with the set of frequency functions $\{f(\cdot;\alpha);\alpha\in A\}$ instead of the family $\mathcal{F}=\{P_\alpha;\alpha\in A]$, where of course we identify frequency functions that differ at most on a set dx-measure zero.

The parameter space A will be assumed to be a separable metric space with the metric d. This seems to be a natural level of generality. The topology will become important only when we come to the way the measures P_α depend upon α. We assume a number of conditions to hold.

Condition 1 The measures P_α shall be absolutely continuous relative to dx, and $\mathcal{F}=\{P_\alpha;\alpha\in A]$ will be given the topology of weak convergence.

The parameter space A shall be second countable Hausdorff, and for fixed but unknown α we observe an i.i.d. sample $x_1, x_2, \ldots, x_n$ from P_α.

So far the main restriction is that we require the observations to be i.i.d. This may seem to rule out many important cases, for example, when the x_t form a stationary stochastic process. Note, however, that we have left the nature of the elements in X unspecified. We can let them denote vectors in some $\mathbb{R}^d$ or let them be sample functions, and so on. This makes it possible to deal with stochastic processes, observed n times i.i.d., as well as other quite general probabilistic structures.

The product measure over X^n, induced by the common marginal measures P_α over X, will for convenience also be denoted by P_α when this will not give rise to any ambiguity.

Our aim is inference of the value of α, so that we shall of course assume the minimum condition that α be *identifiable*.

Condition 2

$$P_{\alpha_1} = P_{\alpha_2} \Rightarrow \alpha_1 = \alpha_2$$

which could also be expressed in terms of the frequency functions rather than the associated measures (compare with the discussion in Chapter 7, Section 1).

The way P_α varies as a function of α must be restricted, but we do this in Section 3.

Given a sample $x_1, x_2, \ldots, x_n$ we want to find estimates $\alpha_n^* = \alpha_n^*(x_1, x_2, \ldots, x_n)$, measurable with respect to the product σ-algebra $\mathcal{Q}^n$, such that $\{\alpha_n^*\}$ is consistent in probability. More precisely, for any neighborhood N_α of the true parameter value α, we want

$$\lim_{n\to\infty} P_\alpha[\alpha_n^* \in N_\alpha] = 1$$

or possibly with other modes of probabilistic convergence.

Also, if A has more structure, many other alternative modes of convergence (in A) have to be considered. In the important case when A forms a Banach space we have, for example, weak convergence in probability:

$$l(\alpha_n^*) \to l(\alpha) \quad \text{in } P_\alpha\text{-probability}; \qquad \forall l \in A^*$$

and strong convergence in probability:

$$\|\alpha_n^* - \alpha\| \to 0 \quad \text{in } P_\alpha\text{-probability}$$

We may have weak convergence in the mean, order $p \geqslant 1$,

$$\lim_{n\to\infty} E_\alpha\{[l(\alpha_n^*)-l(\alpha)]^p\}=0; \qquad \forall l \in A^*$$

or strong convergence in the mean:

$$\lim_{n\to\infty} E_\alpha[\|\alpha_n^*-\alpha\|^p]=0$$

Also we have various modes of convergence involving almost certain convergence.

And now the sieves. In all cases we have met, the subsets $\mathcal{S}(\mu)$ of A were indexed by a positive parameter, the mesh size μ, and were monotonic so that

$$\mathcal{S}(\mu_1)\subseteq\mathcal{S}(\mu_2); \qquad \text{if } \mu_1 \geqslant \mu_2$$

The limit as $\mu \downarrow 0$, in the set theoretic sense, then exists:

$$\mathcal{S}=\bigcup_{\mu>0}\mathcal{S}(\mu) \tag{1}$$

It should be big enough in A, although not necessarily equal to the entire parameter space.

Condition 3 The sieve should consist of a monotonic family $\{\mathcal{S}(\mu); \mu>0\}$ such that the closure of the set (1) is A:

$$\bar{\mathcal{S}}=A$$

In some instances we have met cases when we were motivated to extend A by adding new elements. It may then be convenient to let the sets in the sieve be subsets of an extended parameter space. This eventuality should be kept in mind in future work.

The likelihood function, for a fixed sample, is the function of α

$$L_n(\alpha)=\prod_{k=1}^{n} f(x_k;\alpha)$$

and we shall study restricted ML estimates α_μ^* realizing

$$L_n(\alpha_\mu^*)=\sup_{\alpha\in\mathcal{S}(\mu)} L_n(\alpha)$$

The sieve will be chosen so that such estimates exist, realizing the maximum, although this is not really needed for the analysis to follow. A case of great interest is when each $\mathbb{S}(\mu)$ is compact and $f(x;\alpha)$ is continuous in α for almost all x. Then such estimates obviously exist, although they need not be unique.

2 REGULARIZATION VIA SIEVES

The reader may have asked how the often occurring parameter approximations in Chapter 8 are related, if at all, to the regularization procedures in Chapter 7. The two concepts are clearly conceived in the same spirit, but we shall see in this section that in general they differ in one crucial respect. In one exceptional but important case they actually coincide.

To elucidate this, let us fix ideas by assuming that A is a separable Hilbert space with real scalars. The complex case also occurs in some of our special cases but can be dealt with in the same way after obvious modification. (See Note 9.2.A). We *measure the "degree of regularity" of a parameter value* α via a linear self-adjoint operator L with null space zero defined in a set D everywhere dense in A. Small values of

$$Q\alpha = \|L\alpha\|^2; \qquad \alpha \in D$$

shall mean highly regular α and vice versa. This should be compared with the discussion in Chapter 7, Section 6.

Writing L in its spectral representation,

$$L = \int_{-\infty}^{\infty} \lambda E(d\lambda)$$

expressed in terms of the spectral family $\{E(\lambda);\ -\infty < \lambda < \infty\}$, with $E(\lambda)$ continuous at $\lambda = 0$, we consider a regularization semigroup $\{T_\varepsilon;\ \varepsilon > 0\}$ of the form

$$T_\varepsilon = \int_{-\infty}^{\infty} K_\varepsilon(\lambda) E(d\lambda); \qquad \varepsilon > 0$$

where with pointwise multiplication

$$K_{\varepsilon_1}(\cdot) K_{\varepsilon_2}(\cdot) = K_\varepsilon(\cdot); \qquad \varepsilon = \varepsilon(\varepsilon_1, \varepsilon_2) \tag{1}$$

In (1) we have used the composition function $\varepsilon(\cdot,\cdot)$ discussed in Chapter 7, Section 2.

We are especially interested in two cases. The first is when we can write

$$K_\varepsilon(\lambda)=\exp[\varepsilon k(\lambda)]; \qquad \varepsilon>0$$

where

$$\int_{-\infty}^{\infty} k(\lambda)E(d\lambda)$$

represents the infinitesimal generator of our regularization semigroup [see Hille-Phillips (1957), pp. 583–605].

The second is when

$$k_\varepsilon(\lambda)=\begin{cases} 1; & \text{for } |\lambda|\leqslant \dfrac{1}{\varepsilon} \\ 0; & \text{otherwise} \end{cases} \tag{2}$$

so that $T_\varepsilon=E(\varepsilon+0)-E(-\varepsilon-0)$ is the projection down to the subspace $S(\varepsilon)$ associated with the increment of the spectral family over $[-1/\varepsilon, 1/\varepsilon]$.

Corresponding to this regularization measure we introduce two sieves.

Definition 1 The operator-induced sieve is given by the subset introduced above

$$\mathbb{S}(\mu)=\left\{\alpha|\alpha\in D, Q\alpha\leqslant\frac{1}{\mu}\right\} \qquad \mu>0 \tag{3}$$

Definition 2 The spectral family sieve is given by

$$\mathbb{S}(\mu)=S\left(\frac{1}{\mu}\right); \qquad \mu>0$$

Given an arbitrary parameter value α, not in D, we want to solve the optimal approximation problem

$$\|\hat{\alpha}-\alpha\|^2=\text{minimum for } \hat{\alpha}\in\mathbb{S}(\mu) \tag{4}$$

and denote the solution (when it exists and is unique) by $\hat{\alpha}_\mu$, following the notation used before. On the other hand if α is already in D, then it will belong to the operator-induced sieve $\mathbb{S}(\mu)$ when μ is small enough [see (3)] so that no regularization is needed for small mesh size.

THEOREM 1

For the operator-induced sieve (3) *the regularization* $\hat{\alpha}_\mu$ *of* α, $\alpha \notin D$, *can be written as*

$$\hat{\alpha}_\mu = T_\mu \alpha; \qquad T_\mu = \int_{-\infty}^{\infty} \frac{1}{1+a\lambda^2} E(d\lambda) \tag{5}$$

where a is chosen as the unique solution of

$$\frac{1}{\mu} = \int_{-\infty}^{\infty} \frac{\lambda^2}{(1+a\lambda^2)^2} N(d\lambda); \qquad N(d\lambda) = \| E(d\lambda)\alpha \|^2 \tag{6}$$

The operator T_μ *is quasihomogeneous in the sense*

$$T_\mu(k\alpha) = k(T_{\mu k^{-2}})(\alpha), \qquad \text{any real } k \tag{7}$$

Proof To show that (6) has a unique solution, consider the function of a for fixed α given by the integral on the right-hand side. Since N has bounded variation and the integrand is bounded, this integral converges for all positive a. The function is continuous and strictly decreasing. As $a \uparrow \infty$, the function tends to zero since $E(\lambda)$ is continuous at zero. On the other hand as $a \downarrow 0$, the function tends to $+\infty$ since α does not belong to D, and therefore

$$\int_{-\infty}^{\infty} \lambda^2 N(d\lambda) = +\infty$$

This shows that (6) has a unique solution c for any given positive mesh size, $\mu > 0$.

Consider the closed subspace spanned by all elements $E(\lambda)\alpha$, $-\infty < \lambda < \infty$. Project an arbitrary $\hat{\alpha}$ in $\mathbb{S}(\mu)$ down to this subspace and let β be the rest, orthogonal to the subspace. Then

$$\hat{\alpha} = \int_{-\infty}^{\infty} \phi(\lambda) E(d\lambda)\alpha + \beta \tag{8}$$

where ϕ is some function quadratically integrable with respect to N (see Note 9.2.B). Hence

$$L\hat{\alpha} = \int_{-\infty}^{\infty} \lambda\phi(\lambda) E(d\lambda)\alpha + \gamma; \qquad L\beta = \gamma \perp E(\lambda)\alpha \tag{9}$$

To see that the orthogonality holds, one can, for example, calculate the inner products directly. To calculate the solution (5) we use (8) and (9) and get

$$\|\hat{\alpha}-\alpha\|^2+a\|L\hat{\zeta}\|^2$$
$$=\int_{-\infty}^{\infty}\{[\phi(\lambda)-1]^2+a\lambda^2\phi^2(\lambda)\}N(d\lambda)+\|\beta\|^2+a\|\gamma\|^2 \tag{10}$$

Integrating the elementary inequality

$$[\phi(\lambda)-1]^2+a\lambda^2\phi^2(\lambda)\geqslant\frac{a\lambda^2}{1+a\lambda^2} \tag{11}$$

with respect to N-measure, we get from (10)

$$\|\hat{\alpha}-\alpha\|^2+a\|L\hat{\alpha}\|^2\geqslant a\int_{-\infty}^{\infty}\frac{\lambda^2}{1+a\lambda^2}N(d\lambda)+\|\beta\|^2+a\|\gamma\|^2 \tag{12}$$

Choosing $\hat{\alpha}_\mu$ as in (5) so that $\beta=\gamma=0$, we can satisfy, however, the equality

$$\|\hat{\alpha}_\mu-\alpha\|^2+a\|L\hat{\alpha}_\mu\|^2=a\int_{-\infty}^{\infty}\frac{\lambda^2}{1+a\lambda^2}N(d\lambda)$$

But (12) and the constraint $\|L\hat{\alpha}_\mu\|^2\leqslant 1/\mu$ imply that

$$\|\hat{\alpha}-\alpha\|^2\geqslant a\int_{-\infty}^{\infty}\frac{\lambda^2}{1+a\lambda^2}N(d\lambda)-\frac{a}{\mu}+\|\beta\|^2+a\|\gamma\|^2$$

and using (6) and (3) this is seen to be

$$=a^2\int_{-\infty}^{\infty}\frac{\lambda^4}{(1+a\lambda^2)^2}N(d\lambda)+\|\beta\|^2+a\|\gamma\|^2 \tag{13}$$

This shows that $\hat{\alpha}_\mu$, which has $\beta=\gamma=0$, attains the lower bound. Since equality in (11) and (12) is only attained for

$$\phi(\lambda)=\frac{1}{1+a\lambda^2}$$

a.e. with respect to N, it follows that $\hat{\alpha}_\mu$ is the unique solution.

If $\hat{\alpha}_\mu = T_\mu \alpha$ as described, we can solve the constrained minimum problem

$$\|\hat{\alpha} - k\alpha\|^2 = k^2 \|k^{-1}\hat{\alpha} - \alpha\|^2 = \min$$
$$\alpha \in \mathbb{S} \Leftrightarrow k^{-1}\hat{\alpha} \in \mathbb{S}(\mu k^{-2})$$

by

$$\hat{\alpha} = k T_{\mu k^{-2}} \alpha$$

so that (7) holds. Q.E.D.

Remark 1 One should not believe that T_μ is a linear operator. It is neither strictly homogeneous nor additive in general. Therefore this regularization is not of exactly the same type as the ones in Chapter 7.

Remark 2 For fixed α the measure N determines the speed of convergence $\hat{\alpha}_\mu \to \alpha$ as $\mu \downarrow 0$. Indeed, (13) gives us together with (6) the pair of equations

$$\|\hat{\alpha} - \alpha\|^2 = \int_{-\infty}^{\infty} \frac{a^2\lambda^4}{(1 + a\lambda^2)^2} N(d\lambda)$$

$$\frac{1}{\mu} = \int_{-\infty}^{\infty} \frac{\lambda^2}{(1 + a\lambda^2)^2} N(d\lambda)$$

If N is given by, for example, a power law or by an exponential distribution, the asymptotic rate of convergence of $\hat{\alpha}_\mu$ to α can be calculated in closed form.

Remark 3 We can find a good approximation α'_μ to the optimal regularization $\hat{\alpha}_\mu$ by putting

$$\alpha'_\mu = \int_{-\infty}^{\infty} e^{-a\lambda^2} E(d\lambda)\alpha$$

where a should be given the same value as determined for $\hat{\alpha}_\mu$ by (6). The approximation error is given by

$$\|\alpha'_\mu - \hat{\alpha}_\mu\|^2 = \int_{-\infty}^{\infty} d_1^2(\lambda) N(d\lambda)$$

where

$$d_1(\lambda)=e^{-a\lambda^2}-\frac{1}{1+a\lambda^2} \tag{14}$$

It should be compared to the rate with which $\hat{\alpha}_\mu$ approaches the true parameter value

$$\|\hat{\alpha}_\mu-\alpha\|^2=\int_{-\infty}^{\infty} d_2^2(\lambda)N(d\lambda)$$

where

$$d_2(\lambda)=1-\frac{1}{1+a\lambda^2}=\frac{a\lambda^2}{1+a\lambda^2} \tag{15}$$

[see (13)]. But

$$\left|\frac{d_1(\lambda)}{d_2(\lambda)}\right|=\frac{|(1+a\lambda^2)e^{-a\lambda^2}-1|}{a\lambda^2}\leqslant A<\infty, \qquad -\infty<\lambda<\infty \tag{16}$$

Indeed, the numerator is bounded and the denominator in the continuous function given by (16) is bounded away from zero, except at zero where the ratio takes the value zero. On the other hand at $\lambda=+\infty$ the ratio takes the value zero. Hence the ratio is bounded as stated in (16).

Applying (16) to the error relations (14) and (15) shows that

$$\|\alpha'_\mu-\hat{\alpha}_\mu\|\leqslant A\|\hat{\alpha}_\mu-\alpha\|$$

so that α'_μ is a reasonably good approximation. The interesting thing about α'_μ is that it is *given in terms of the regularization operator*

$$R_\varepsilon=\int_{-\infty}^{\infty} e^{-\varepsilon\lambda^2}E(d\lambda)$$

related to the Weierstrass kernel (Chapter 7, Section 2). It would then be more convenient to define the resulting regularization $R_\varepsilon\alpha$ directly in terms of the parameter ε rather than via μ.

Remark 4 When L is a differential operator, say on $\mathbb{R}$, it can be shown that the regularization can be conveniently expressed using the Green's function of the second-order differential operator appearing in the associated Euler equation of the extremum problem that the sieve gives rise to. This connects elegantly with several special cases encountered in Chapter 8.

Let us now examine the effect of using a spectral family sieve as in Definition 2. Then, for fixed mesh size, the sieve forms a subspace so that (4) is solved simply by projecting α down to the subspace

$$\hat{\alpha}_\mu = P_{\mathbb{S}(1/\mu)}\alpha$$

Hence *the optimal regularization*

$$T_\varepsilon = P_{\mathbb{S}(1/\varepsilon)}$$

is given in terms of the semigroup $\{T_\mu\}$ with composition function (see Chapter 7, Section 2)

$$\varepsilon(\varepsilon_1, \varepsilon_2) = \max(\varepsilon_1, \varepsilon_2)$$

Then $\hat{\alpha}_\mu$ can be obtained using a kernel of the type given in (2). We have already met a special case of this type, namely, Example 1 in Chapter 7, Section 6, and many more can be expected to occur in future work.

It is now clear why some of our regularization operators are linear while others are not. They will be linear, actually projections, when the sieve consists of subspaces. *With operator-induced sieves, however, the sieve is not linearly closed, and the resulting regularization operator will then not be linear in general.*

3 SIEVES IN GENERAL

The many special cases in Chapter 8 have shown that sieve constrained ML-estimates can have desirable statistical as well as analytical properties if the sieve is chosen well, and it is reasonable to believe that this statement can be made precise also for a more general situation. This is studied in this section.

The price we will have to pay in order to gain generality is, as usual, loss of strength in the proposition that can be asserted. In particular we only establish consistent estimation here, a weak but needed result. For more powerful propositions more stringent conditions would be required than the ones to be adopted below.

With the assumptions introduced in Section 1 and with the notation of Chapter 7, Section 1, we introduce the set of ML-estimators:

$$M_m^n = \{\alpha \in \mathbb{S}_m \mid L_n(x, \alpha) = L_n(x, \mathbb{S}_m)\}$$

and the maximum entropy set

$$\tilde{A}_m = \{\alpha | \alpha \in \mathbb{S}_m \text{ and } H(\alpha_0, \alpha) = H(\alpha_0, \mathbb{S}_m)\}$$

We write $c_m \to \alpha$ for $C_m \subset A$ if

$$\sup_{\beta \in C_m} d(\alpha, \beta) \to 0 \qquad \text{as } m \to \infty$$

and for $D \subset A$ we write $C_m \to D$ if

$$\sup_{\alpha \in C_m} \inf_{\beta \in D} d(\alpha, \beta) \to 0 \qquad \text{as } m \to \infty$$

A *sieve* is a sequence of compact subsets in A, but more will be required in order that it behave as wanted.

In addition to Conditions 1 and 2 and 3 in section 1 we now introduce the following.

Condition 4 For any m and $\beta \in \mathbb{S}_m$, $\varepsilon > 0$, the function $f[x, B_m(\beta, \varepsilon)]$ should be measurable in x, and for every m, $\beta \in \mathbb{S}_m$, there is an $\varepsilon > 0$ such that

$$E_{\alpha_0}\left[\ln f(x, B_m(\beta, \varepsilon)\right] < \infty \tag{1}$$

In (1) $B_m(\beta, \varepsilon)$ stands for the set of all α in $\mathbb{S}_m$ such that $d(\alpha, \beta) < \varepsilon$.

Condition 5 The function $f(x, \cdot)$ should be upper semicontinuous on $\mathbb{S}_m$, that is,

$$\lim_{\varepsilon \downarrow 0} f\left[x, B_m(\beta, \varepsilon)\right] - f(x, \beta) \text{ a.e. } dx$$

Condition 6 As $m \to \infty$, we should have $\tilde{A}_m \to \alpha_0$.

We then have the following important result.

THEOREM 1

EXISTENCE OF SIEVES, GEMAN-HWANG (1979). *Under Conditions 1–6 the set M_m^n is nonempty a.c., and for every sequence n_m increasing sufficiently fast we have consistency*: $M_{n_m}^m \to \alpha_0$ a.c.

Proof Conditions 4 and 5 imply that $H(\alpha_0,\alpha)$ and $L_n(x,\alpha)$ are upper semicontinuous in $\alpha, \alpha\in\mathbb{S}_m$. But $\mathbb{S}_m$ is compact so that $\tilde{A}_m$ and M_n^m are compact as well as not empty.

We first show that when n_m increases fast enough, $M_{N_m}^m$ eventually gets inside $B_m(\tilde{A}_m, 1/m)$. Condition 6 then finishes the proof.

Take $\alpha\in\mathbb{S}_m - B_m(\tilde{A}_m, 1/m)$ and $c>0$ such that $H(\alpha_0,\alpha)-H(\alpha_0,\mathbb{S}_m)< -c$. Fix $\beta_m\in\tilde{A}_m$. For any ε small enough

$$E_{\alpha_0}[\ln f(x, B_m(\alpha,\varepsilon))] - E_{\alpha_0}[\ln f(x,\beta_m)] < -c$$

Using this we get

$$\begin{aligned}
&\frac{1}{n}\{\ln L_n(x, B_m(\alpha,\varepsilon)) - \ln L_n(x,\beta_m)\}\\
&\qquad = \frac{1}{n}\left\{\ln \sup_{\gamma\in B_m(\alpha,\varepsilon)} \prod_{i=1}^{n} f(x_i,\gamma) - \ln \prod_{i=1}^{n} f(x_i,\beta_m)\right\}\\
&\qquad \leqslant \frac{1}{n}\left\{\ln \prod_{i=1}^{n} f(x_i, B_m(\alpha,\varepsilon)) - \ln \prod_{i=1}^{n} f(x_i,\beta_m)\right\}\\
&\qquad = \frac{1}{n}\sum_{1}^{n} (\ln f(x_i, B_m(\alpha,\varepsilon)) - \ln f(x_i,\beta_m))\\
&\qquad \underset{n\to\infty}{\overset{\text{a.s.}}{\to}} E_{\alpha_0}[\ln f(x, B_m(\alpha,\varepsilon))] - E_{\alpha_0}[\ln f(x,\beta_m)] < -c
\end{aligned}$$

Hence with probability 1,

$$\frac{L_n(x, B_m(\alpha,\varepsilon))}{L_n(x,\beta_m)} < e^{-nc} \tag{2}$$

whenever n is large enough (n may depend on the sample).

For every $\alpha\in\mathbb{S}_m - B_m(\tilde{A}_m, 1/m)$ we can pick a $c_\alpha>0$ and an $\varepsilon_\alpha>0$ such that with probability 1 (2) holds for large n. The balls $B_m(\alpha,\varepsilon_\alpha)$ cover the compact set $\mathbb{S}_m - B_m(\tilde{A}_m, 1/m)$, so we can find a finite set $\alpha_1^m,\ldots,\alpha_p^m$ such that $\{B_m(\alpha_k^m, \varepsilon_k^m)\}$ covers $\mathbb{S}_m - B_m(\tilde{A}_m, 1/m)$. Then with $c = \min_{k=1,2,\ldots,p_m}\{c_{\alpha_k}\}$, we have, with probability 1,

$$\sup_{k=1,\ldots,p_m} \frac{L_n(x, B_m(\alpha_k^m, \varepsilon_k^m))}{L_n(x,\beta_m)} < e^{-nc}$$

for all n large enough. Hence for each m there exists S_m such that whenever

$n \geqslant S_m$,

$$P\left\{\sup_{k=1,\dots,p_m} \frac{L_n(x, B_m(\alpha_k^m, \varepsilon_k^m))}{L_n(x, \beta_m)} \geqslant 1\right\} \leqslant \frac{1}{m^2}$$

Finally, for any sequence $\{n_m\}$ such that $n_m \geqslant S_m$ for each m,

$$\left\{M_{n_m}^m \not\subset B_m\left(\tilde{A}_m, \frac{1}{m}\right) \text{ i.o.}\right\}$$

$$\subseteq \left\{\sup_{\alpha \in S_m - B_m(\tilde{A}_m, 1/m)} \frac{L_{n_m}(x, \alpha)}{L_{n_m}(x, \beta_m)} \geqslant 1 \text{ i.o.}\right\}$$

$$\subseteq \left\{\sup_{k=1,\dots,p_m} \frac{L_{n_m}(x, B_m(\alpha_k^m, \varepsilon_k^m))}{L_{n_m}(x, \beta_m)} \geqslant 1 \text{ i.o.}\right\}$$

and the latter has probability zero by the Borel-Cantelli lemma. Q.E.D.

This theorem gives conditions for a sieve to be powerful enough to produce consistent estimates. Unfortunately it does not say anything about the speed with which n_m should tend to infinity. In order to obtain such information we need further conditions [see Geman-Hwang (1979)].

THEOREM 2

We shall assume that the sieve $\mathbb{S}_m$ satisfies conditions (a)–(e).

(a) *For every m, every $\alpha \in \mathbb{S}_m$, and every $\varepsilon > 0$, $f(x, B_m(\alpha, \varepsilon))$ is measurable in x; and for every m and every $\alpha \in \mathbb{S}_m$*

$$\lim_{\varepsilon \to 0} f(x, B_m(\alpha, \varepsilon)) = f(x, \alpha) \text{ a.c. } dx$$

(b) *$H(\alpha_0, \mathbb{S}_m) < \infty$ for all m, and there exists $\xi_m \downarrow 0$ such that*

$$A_m = \{\alpha \in \mathbb{S}_m \mid H(\alpha_0, \alpha) \geqslant H(\alpha_0, \mathbb{S}_m) - \xi_m\} \to \alpha_0$$

(c) *There exists $\{\varepsilon_m\}$ such that*

$$E_{\alpha_0} \ln f(x, B_m(\alpha, \varepsilon_m)) < H(\alpha_0, \alpha) + \frac{\xi_m}{2}$$

for all $\alpha \in \mathbb{S}_m$.

(d) *There exists* $\{K_m\}$ *such that for any* $\alpha, \beta \in \mathbb{S}_m$

$$E_{\alpha_0}\left[\left(\ln \frac{f(x, B_m(\alpha, \varepsilon_m))}{f(x, \beta)}\right)^2\right] \leqslant K_m$$

(e) *Let* l_m *be such that*

$$\mathbb{S}_m \subseteq \bigcup_{k=1}^{m} B_m\left(\beta_k, \frac{\varepsilon_m}{2}\right)$$

for some $\beta_1, \ldots, \beta_{l_m} \in \mathbb{S}_m$.

Then, for any sequence n_m *satisfying*

$$\sum_{m=1}^{\infty} \frac{l_m K_m}{n_m \xi_m^2} < \infty$$

$$M_{n_m}^m \to \alpha_0 \text{ a.s.}$$

Proof Since $H(\alpha_0, \alpha)$ is upper semicontinuous in α, A_m is compact and nonempty. We shall show that

$$P\{M_{n_m}^m \cap A_m^c \neq \phi \text{ i.o.}\} = 0$$

that is, eventually $M_{n_m}^m \subseteq A_m$. But

$$\{M_{n_m}^m \cap A_m^c \neq \phi\} \subseteq \left\{ \sup_{\alpha \in \mathbb{S}_m - A_m} \prod_1^{n_m} f(x_i, \alpha) \geqslant \sup_{\alpha \in A_m} \prod_1^{n_m} f(x_i, \alpha) \right\}$$

$$\subseteq \left\{ \sup_{\alpha \in \mathbb{S}_m - A_m} \prod_1^{n_m} f(x_i, \alpha) \geqslant \prod_1^{n_m} f(x_i, \beta) \right\}$$

where $\beta \in \tilde{A}_m$. By (e) there exists $\alpha_1, \ldots, \alpha_{l_m} \in \mathbb{S}_m - A_m$ such that

$$\mathbb{S}_m - A_m \subseteq \bigcup_1^{l_m} B(\alpha_k, \varepsilon_m).$$

Then

$$\{M_{n_m}^m \cap A_m^c \neq \phi\} \subseteq \bigcup_{k=1}^{l_m} \left\{ \sup_{\alpha \in B_m(\alpha_k, \varepsilon_m)} \prod_1^{n_m} f(x_i, \alpha) \geqslant \prod_1^{n_m} f(x_i, \beta) \right\}$$

We will bound the probability of this latter set:

$$P\left\{\bigcup_{k=1}^{l_m}\left[\sup_{\alpha\in B_m(\alpha_k,\varepsilon_m)}\prod_1^{n_m} f(x_i,\alpha)\geqslant\prod_1^{n_m} f(x_i,\beta)\right]\right\}$$

$$\leqslant\sum_{k=1}^{l_m} P\left\{\prod_1^{n_m} f(x_i,B_m)(\alpha_k,\varepsilon_m))\geqslant\prod_1^{n_m} f(x_i,\beta)\right\}$$

$$=\sum_{k=1}^{l_m} P\left\{\frac{1}{n_m}\sum_1^{n_m}\ln\frac{f(x_i,B_m(\alpha_k,\varepsilon_m))}{f(x_i,\beta)}\geqslant 0\right\}$$

Observe that

$$E_{\alpha_0}\left[\ln\frac{f(x_i,B_m(\alpha_k,\varepsilon_m))}{f(x_i,\beta)}\right]<H(\alpha_0,\alpha_k)+\frac{\xi_m}{2}-H(\alpha_0,\beta)$$

$$<H(\alpha_0,\mathfrak{S}_m)-\xi_m+\frac{\xi_m}{2}-H(\alpha_0,\beta)=-\frac{\xi_m}{2}$$

Hence

$$P\left\{\bigcup_{k=1}^{l_m}\left[\sup_{\alpha\in B_m(\alpha_k,\varepsilon_m)}\prod_1^{n_m} f(x_i,\alpha)\geqslant\prod_1^{n_m} f(x_i,\beta)\right]\right\}$$

$$\leqslant\sum_{k=1}^{l_m} P\left\{\frac{1}{n_m}\sum_1^{n_m}\left(\ln\frac{f(x_i,B_m(\alpha_k,\varepsilon_m))}{f(x_i,\beta)}-E_{\alpha_0}\left[\ln\frac{f(x_i,B_m(\alpha_k,\varepsilon_m))}{f(x_i,\beta)}\right]\right)\geqslant\frac{\xi_m}{2}\right\}$$

$$\leqslant\frac{4l_mK_m}{n_m\xi_m^2}\quad\text{(Chebyshev's inequality)}$$

But

$$\{M_{n_m}^m\cap A_m^c\neq\phi\text{ i.o.}\}$$

$$\subseteq\left(\bigcup_{k=1}^{l_m}\left[\sup_{\alpha\in B_m(\alpha_k,\varepsilon_m)}\prod_1^{n_m} f(x_i,\alpha)\geqslant\prod_1^{n_m} f(x_i,\beta)\right]\right)\text{ i.o.}\}$$

Therefore $P\{M_{n_m}^m\cap A_m^c\neq\phi\text{ i.o.}\}=0$ (Borel-Cantelli lemma). Now combine with (b), and the assertion is proved. Q.E.D.

It should be mentioned that this result can be strengthened considerably by a more refined analysis (see Notes 9.3.A).

4 CRAMÉR-RAO'S INEQUALITY IN BANACH SPACES

The theorems in Section 3 are quite general but they do not give us any detailed information on the speed of convergence of the estimates to the true parameter value.

It seems logical to try to supply some such information by extending the Cramér-Rao inequality to abstract parameter spaces. The natural setting for this is to let A be embedded in a (real, separable) Banach space and to ask some regularity conditions for $f(x, \alpha)$. Without striving for any generality to speak of, we shall assume that Condition (1) below holds. It closely parallels the finite-dimensional case.

For a given but arbitrary parameter value α consider a subset $B(\alpha)$ of the Banach space such that when $u \in B(\alpha)$ we have $\alpha + tu \in A$ for t in some neighborhood $N(\alpha, u)$ of 0. We demand that for α-values to be used in the derivation, the following conditions hold:

(a) $f(x, \alpha + tu), u \in B(\alpha)$, should be differentiable with respect to t in $N(\alpha)$.

(b) Its derivative $\lambda_n(x, \alpha)$ at $t=0$ should be dominated by a fixed L_1-function $g_1(x)$.

(c) The quantity $\|\alpha^*(x)\| \lambda_n(x, \alpha)$ should also be dominated by a fixed L_1-function $g_2(x)$.

(d) $0 < E[\lambda^2(u, x)/f^2(x, \alpha)] < \infty$.

We can then state a direct generalization of the classical Cramér-Rao inequality.

THEOREM 1

For an unbiased estimate α^* and a bounded linear functional $l \in A^*$ we have, if (1) *holds, the bound*

$$\mathrm{Var}_\alpha[l(a^*)] \geqslant \frac{l^2(u)}{E_\alpha\left\{\left[\dfrac{\lambda_u(x, \alpha)}{f(x, \alpha)}\right]^2\right\}} \tag{2}$$

Proof We follow the proof in the finite-dimensional case. Differentiating the identity

$$\int_X f(x, \alpha + tu)\,dx = 1, \qquad u \in B(\alpha), \quad t \in N(\alpha, u) \tag{3}$$

with respect to t, we get, for $t=0$,

$$\int_X \lambda_u(x,\alpha)dx=0 \tag{4}$$

where we have used condition (1b) to justify the change in the order of differentiation and integration [see Cramér (1945), pp. 67–68].

We can treat the identity, expressing that α^* is unbiased,

$$E_{\alpha+tu}[l(\alpha^*)]=\int_X l(\alpha^*)f(x,\alpha+tu)dx=l(\alpha)+tl(u) \tag{5}$$

in the same way and get, now appealing to condition (1c),

$$\int_X l(\alpha^*)\lambda_u(x,\alpha)dx=l(u) \tag{6}$$

Subtract $l(\alpha)$ times (4) from (6) on both sides of the respective equations, and use Schwarz's inequality. This yields

$$l^2(u)\leqslant\int_X l^2[\alpha^*(x)-\alpha]f(x,\alpha)dx\int_X\left[\frac{\lambda_u(x,\alpha)}{f(x,\alpha)}\right]^2 f(x,\alpha)dx \tag{7}$$

Noting that the first integral on the right-hand side of (7) represents the variance of the unbiased estimate α^*; the inequality (7) implies (2). Q.E.D.

Remark 1 The role of the element u in (2) may not be obvious since it docs not appcar cxplicitly in thc usual trcatmcnt of thc casc of a rcal parameter α. In any given concrete case we use the fact that we have considerable freedom of choosing u. We shall try to find an element u such that the right-hand side of 92) becomes as large as possible.

Remark 2 Conditions (1b) and (1c) are used only to justify the exchange of the integration and differentiation in (3) and (5). Sometimes it is easier to justify this by direct calculation, then by verifying the conditions.

Remark 3 In order to use (2) to define efficiency we should specify for which class of linear functionals we want equality to hold. The class should be wide enough, otherwise the resulting notion of efficiency will be inadequate.

A few examples will make it clearer how we should choose u, and we start with one we have met several times in Part II, but now viewed from a different perspective.

Example 1 Consider the sample space $X=l_2$ with a Gaussian distribution with mean value vector $(\alpha_i;\ i=1,2,\ldots)$ and diagonal covariance matrix with entries $(\sigma_i^2; i=1,2,\ldots)$ in the main diagonal. With dx corresponding to this Gaussian measure for $\alpha=0$ we get the Radon-Nikodym derivative

$$f(x,\alpha)=\exp\left\{-\frac{1}{2}\sum_{i=1}^{\infty}\frac{2x_i\alpha_i-\alpha_i^2}{\sigma_i^2}\right\}$$

under the condition for absolute continuity established in Chapter 5, namely,

$$\sum_{i=1}^{\infty}\frac{\alpha_i^2}{\sigma_i^2}<\infty$$

The obvious estimate $\bar{\alpha}=x$ is unbiased and has

$$\operatorname{Var}_\alpha[l(\bar{\alpha})]=\sum_{i=1}^{\infty}l_i^2\sigma_i^2 \tag{8}$$

if the functional l is represented as the inner product with the vector $(l_i;\ i=1,2,\ldots)$.

In this simple case, with $u=(u_i;\ i=1,2,\ldots)$ we should ask that

$$u\in B(\alpha)=\left\{u\,\Big|\,\sum_{i=1}^{\infty}\frac{u_i^2}{\sigma_i^2}<\infty\right\}$$

Then an easy calculation shows that

$$\frac{\lambda_u(x,\alpha)}{f(x,\alpha)}=\sum_{i=1}^{\infty}\frac{\alpha_i-x_i}{\sigma_i^2}u_i$$

The theorem then gives the bound

$$\mathrm{Var}_\alpha[l(\alpha^*)] \geqslant \frac{l^2(u)}{\sum_{i=1}^{\infty} \frac{u_i^2}{\sigma_i^2}} = \frac{\left(\sum_{i=1}^{\infty} l_i u_i\right)^2}{\sum_{i=1}^{\infty} \frac{u_i^2}{\sigma_i^2}} \tag{9}$$

To make the right-hand side of (9) as big as possible, we choose $u_i = l_i \sigma_i^2$, which is just the condition for Schwarz's inequality to become an equality. The right-hand side then gives us the same expression as in (8), which shows, not surprisingly, that the estimate is efficient.

Example 2 Let X consist of binary sequences with stochastically independent entries. With $x=(x_i;\ i=1,2,\ldots)$ let

$$P_\alpha(x_i=1)=1-P_\alpha(x_i=0)=\alpha_i$$

with the α_i-sequence summable. It follows from the Borel-Cantelli lemma that a.c. the x-sequence has only a finite number of entries equal to 1. Hence both X and A can be embedded in l_1.

The Radon-Nikodym derivative

$$f(x,\alpha)=\frac{P_\alpha(dx)}{P_{\alpha^0}(dx)}=\prod_{i=1}^{\infty}\left(\frac{\alpha_i}{\alpha_i^0}\right)^{x_i}\left(\frac{1-\alpha_i}{1-\alpha_i^0}\right)^{1-x_i}$$

is well defined, the infinite product converges a.c., if we assume

$$(\alpha_i=0)\Leftrightarrow(\alpha_i^0=0) \qquad \text{and } (\alpha_i=1)\Leftrightarrow(\alpha_i^0=1) \tag{10}$$

ensuring absolute continuity according to the likelihood ratio convergence theorem. The obvious estimate is again of course $\bar{\alpha}=x$ for which, with a functional of the form with arbitrary N,

$$l(\alpha)=\sum_{i=1}^{N} l_i\alpha_i$$

we get the variance

$$\mathrm{Var}_\alpha[l(\bar{\alpha})]=\sum_{i=1}^{N} l_i^2\alpha_i(1-\alpha_i) \tag{11}$$

Also

$$\frac{\lambda_u(x,\alpha)}{f(x,\alpha)}=\sum_{i=1}^{\infty}\frac{x_i-\alpha_i}{\alpha_i(1-\alpha_i)}u_i$$

and

$$E_\alpha\left[\frac{\lambda_u^2(x,\alpha)}{f^2(x,\alpha)}\right]=\sum_{i=1}^{\infty}\frac{u_i^2}{\alpha_i(1-\alpha_i)}$$

so that we arrive at the bound

$$\operatorname{Var}_\alpha[l(\alpha^*)]\geqslant\frac{\left(\sum_{i=1}^{N}l_iu_i\right)^2}{\sum_{i=1}^{\infty}\frac{u_i^2}{\alpha_i(1-\alpha_i)}}$$

To make the right-hand side as big as possible, we choose $u_i=l_i\alpha_i(1-\alpha_i)$; $i=1,2,\ldots,N$, and equal to zero otherwise, which will not violate (10) for $\alpha+tu$ if t is small enough. It leads to the bound

$$\sum_1^N l_i^2\alpha_i(1-\alpha_i)$$

which is the expression in (11), so that $\bar{\alpha}$ is efficient for these functionals.

Example 3 Consider now the inference situation described in the beginning of Chapter 8, Section 5, for simplicity with $n=1$. Then X can be considered as consisting of finite counting measures on $[0,1]$ and α is a finite measure on the same intervals. We therefore embed A as well as X in $BV([0,1])$.

Using (8.5.1) we can calculate $\lambda_u(x,\alpha)$ and get

$$\frac{\lambda_u(x,\alpha)}{f(x,\alpha)}=\sum_k x(y_k)\frac{u(y_k)}{\alpha(y_k)}-u([0,1]) \tag{12}$$

where of course u should be chosen so that $\alpha+tu$ is a finite measure, and where $y_1,y_2,\ldots$ are the points realized by the Poisson process and $x(y_k)$ are the associated multiplicities. To satisfy condition (1d) we have to restrict u further, which will be done below.

We can calculate the expected value of the square of (12) by considering it as the limit of finite sums with a nonrandom large number of terms, all stochastically independent and proportional to Poisson distributed random variables. Since the variance of a Poisson distributed variable equals its mean, we get, after this calculation, the result as the Hellinger integral

$$E_\alpha\left[\frac{\lambda_u^2(\alpha,\alpha)}{f(x,\alpha)}\right]=\int_0^1\frac{u^2(dy)}{\alpha(dy)}$$

We now choose u such that this Hellinger integral is finite and positive.

Consider a functional l of the form

$$l(\alpha)=\int_0^1 c(y)\alpha(dy) \tag{13}$$

where $c\in C([0,1])$. We should then try to choose u so that the ratio

$$\frac{\left[\int_0^1 c(y)u(dy)\right]^2}{\int_0^1\frac{u^2(dy)}{\alpha(dy)}} \tag{14}$$

is as large as possible.

It is not difficult to show that this maximum problem is solved by $u(dx)=c(x)\alpha(dx)$, in which case the ratio in (14) takes the value

$$\int_0^1 c^2(x)\alpha(dx) \tag{15}$$

For the obvious estimate $\bar{\alpha}=x$, the sample element itself, a functional $l(\bar{\alpha})$ of the form in (13) has a variance equal to (15). This $\bar{\alpha}$ is therefore efficient for these functionals.

These examples are quite easy. The following one is slightly more complicated.

Example 4 With the sample space $X=[0,1]^n\subset\mathbb{R}^n$ consider an i.i.d. sample from [0, 1] with continuous frequency function α. Embed A in the Banach space $C([0,1])$ and calculate $\lambda_u(x,\alpha)$. Note that since $\alpha+tu$ should have integral 1, it follows that we must require that

$$\int_0^1 u(x)dx=0 \tag{16}$$

We get by direct calculation, $x=(x_1, x_2,\ldots, x_n)$,

$$\frac{\lambda_u(x,\alpha)}{f(x,\alpha)}=\sum_{i=1}^{u}\frac{u(x_i)}{\alpha(x_i)} \tag{17}$$

and since the terms in (17) have expected values equal to zero,

$$E_\alpha\left[\frac{\lambda_u^2(x,\alpha)}{f^2(x,\alpha)}\right]=n\int_0^1\frac{u^2(x)}{\alpha(x)}Dx \tag{18}$$

Consider now a bounded linear functional $l\in C^*([0,1])$. According to the Riesz representation it can be written in terms of a bounded signed measure V:

$$l(\alpha)=\int_0^1\alpha(x)V(dx)$$

This form is too general for the present purpose, since it is well known that no unbiased estimates exist of $l(\alpha)$ if V has atoms on the support of α. Therefore we are forced to assume that V is absolutely continuous with respect to Lebesque measure, say with derivative $v\in L_1$, to be further restricted later on.

The bound in (2) then becomes

$$\operatorname{Var}_\alpha[l(\alpha^*)]\geqslant\frac{\left(\int_0^1 u(x)v(x)dx\right)^2}{n\int_0^1\frac{u^2(x)}{\alpha(x)}dx}$$

where we should not allow u to be essentially nonzero on the set of values (if any) where α vanishes. To make the right-hand side large, keeping in mind the constraint (16), the Euler equation gives

$$u(x)=\alpha(x)\left[v(x)-\int_0^1\alpha(x)v(x)dx\right]$$

To make this u belong to $C([0,1])$ we demand that v be continuous, in which case the right-hand side of (18) becomes

$$\frac{1}{n}\left\{\int_0^1\alpha(x)v^2(x)dx-\left[\int_0^1\alpha(x)v(x)dx\right]^2\right\}$$

Note that the standard estimate

$$\bar{\alpha}=\frac{1}{n}\sum_{i=1}^{n} v(x_i)$$

realizes this bound.

Returning to the general case of the inequality (2) we notice that the bound in (2) tells us how small the variance of a linear functional $l(\alpha^*)$ can be, say $m(l)$. We may prefer, however, *more direct measures of the dispersion* of α^*, such as the expected value of $\|\alpha^*-\alpha\|^2$. One case of considerable interest when this can be obtained directly from $m(l)$ is when the parameter takes values in a Hilbert space.

Indeed when this is the case, consider a sequence of $l_i \in A^*$ given as a CON system of vectors in A. We then have for an unbiased α^*, if $R(\alpha^*)$ denotes the covariance matrix of α^* with diagonal elements $R_{ii}(\alpha^*)$,

$$\begin{aligned} E_\alpha[\|\alpha^*-\alpha\|^2] &= \mathrm{tr}[R(\alpha^*)] = \sum_{i=1}^{\infty} R_{ii}(\alpha^*) \\ &= \sum_{i=1}^{\infty} \mathrm{Var}_\alpha[l_i(\alpha^*)] \geqslant \sum_{i=1}^{\infty} m(l_i) \end{aligned}$$

This gives a bound for the direct measure of the size of the error of estimation.

Notes

INTRODUCTION

1.A As any practicing statistician knows, the three types of inference mentioned in Section 2 do not exhaust all the possibilities; there are many others. For example, in exploratory data analysis the model $\{P_\alpha\}$ need not be specified in advance. Or, when designing the experiment, the sample space X is not given to begin with.

1.B In pattern analysis we have the parameter space $A=\mathfrak{I}=$the image algebra, and the sample space $\mathfrak{I}^{\mathfrak{D}}=$the set of deformed images.

1 PROBABILISTIC BACKGROUND

1.1.A It has been shown that some restriction is needed [see, e.g., Halmos (1950)] in order to guarantee the validity of the construction in Kolmogorov's consistency theorem. It is enough to assume that V is a Polish space: a separable and complete metric space [see Neveu (1964), pp. 74–79].

1.1.B The process $\tilde{x}$ can (in general) also take $\pm\infty$ as values. In most cases this will not be needed, however, and causes no difficulty.

1.2.A Let $\mathcal{Q}_0$ be the σ-algebra generated by all cylinder sets with finite-dimensional base. We show that $\mathcal{Q}_0$ contains all open sets. The closed unit ball $S=\{x\,|\,\|x\|\leqslant 1\}$ can be expressed as

$$S=\bigcap_{x^*\in F} A_{x^*}=A;\ A_{x^*}=\{x\,|\,|x^*(x)|\leqslant 1\}$$

where $F=\{x^*\in X^*|\,\|x^*\|=1\}$. Indeed, if x is in A and $x^*\in F$ with

$x^*(x) = \|x\|$, then $\|x\| \leqslant 1$ so that $x \in S$. On the other hand if $x \in S$, it is obvious that $|x^*(x)| \leqslant 1$ for all $x^* \in F$ so that $x \in A$.

Introduce a sequence $x^*_1, x^*_2, \ldots$ everywhere dense in F, which is possible due to the separability condition, and put

$$E_i = \{x \,|\, |x_i^*(x)| \leqslant 1\}$$

$$E = \bigcap_i E_i$$

Then of course $S \subset E$. Also, if $x \in E$ but not in S, we must have

$$|x_i^*(x)| \leqslant 1$$

$$|x^*(x)| > 1 \text{ for some } x^* \in F$$

Approximating this x^* by x_i^*, we get a contradiction, and it follows that any ball belongs to $\mathcal{Q}_0$. A direct consequence is that $\mathcal{Q} = \mathcal{Q}_0$ as stated.

1.2.B For proofs and more information on these questions the reader should consult Prohorov (1956, 1961), Sazonov (1958). An early and pioneering work dealing with the situation in a Banach space can be found in Mourier (1953).

1.3.A Theorem 4 has been obtained under varying regularity conditions by several authors. The statement in the text seems to be the most satisfactory one.

2 GEOMETRY OF STOCHASTIC PROCESS

2.1.A The fundamental idea of this chapter—to look at stochastic processes as curves in a Hilbert space—goes back to Kolmogorov (1940a, b). A thorough follow-up of this idea was carried out in Karhunen (1947).

2.1.B If F is allowed to have jumps some care is needed in defining the behavior of $x(t)$ at a jump, but this will seldom be the case in the following.

2.1.C The construction in (13) proves the existence of a process satisfying (12). The term "orthogonal process" applies to any process for which (12) holds.

2.2.A There exists an extensive literature on stochastic differential equations of the more general type. The reader can consult, for example, Gikhman-Skorokhod (1972).

2.3.A The seminal paper on general representation theory is Karhunen (1947).

2.4.A To derive (14) in Example 3, it is convenient to find the boundary conditions for ϕ and ϕ' and express then in c_1 and c_2.

2.4.B Let $x(t)=u(t+iv(t)$ be a Gaussian complex process and form [see Cramér-Leadbetter (1967), p. 122]:

$$r(s,t)=E\left[x(s)\overline{x(t)}\right]=E\left[u(s)u(t)+v(s)v(t)\right]$$

$$+iE\left[-u(s)v(t)+u(t)v(s)\right]$$

Introducing the function

$$q(s,t)=E\left[x(s)x(t)\right]$$

we can write it as

$$q(s,t)=E\left[u(s)u(t)-v(s)v(t)\right]+iE\left[u(s)u(t)+u(t)v(s)\right]$$

If $q(s-t)$ only depends upon $s-t$, it follows that all the functions $E[u(s)u(t)]$, $E[v(s)v(t)]$, $E[u(s)v(t)]$, $E[v(s)u(t)]$ depend only upon $s-t$: the process is strictly stationary. A moment's reflection shows that the condition on q is not only sufficient but also necessary for a weakly stationary Gaussian process to be strictly stationary.

The above should be related to the condition in (1.1.3).

2.4.C The prediction error ν_n of x_{n+1}^* using the observations $(x_1, x_2, \ldots, x_n)$ tends to a limit ν_∞ as $\nu\to\infty$. For an absolutely continuous spectrum the limit is calculated in Chapter 2, Section 5. If we denote by μ_n the prediction error of the *unbiased predictor* $\hat{x}_{n+1}$, to be used in Chapter 4, it is obvious that $\mu_n \geqslant \nu_n$. What is not obvious, however, is that the limit $\mu_\infty = \lim_{n\to\infty}\mu_n$ coincides with ν_∞. Here is the germ of a proof requiring no regularity assumption for f except $\ln f \in L_1$. Try to relate the (finite past) predictor y_{n+1}^* of the process $y_t = x_t - x_{t-1}$ to the unbiased (finite past) predictor $\hat{x}_{n+1}$ of the x-process. The y-process has spectral density $|1-e^{i\lambda}|^2$ times the spectral density of the x-process. Use Theorem 1 in Chapter 5, Section 4, together with the fact that the integral of $\ln|1-e^{i\lambda}|^2$ is zero, and the result follows. More about this in Grenander-Rosenblatt (1954).

2.5.A Factorizations similar to (10) also play a role in the study of point processes and have given rise to statistical estimation problems [see Brillinger (1972) and Rice (1975)].

3 GENERAL INFERENCE IN ABSTRACT SAMPLE SPACE

3.1.A The theorem on likelihood ratio convergence is crucial for the following. It has been proved in various formulations [see, e.g., Striebel (1959)].

3.2.A An important early contribution to inference in stochastic processes was given in Mann (1952, 1953). A recent survey of the topics treated in this section was given in Feigin (1976).

3.2.B The fact that S_N in (5) converges seems to have been discovered by several authors. The earliest one is probably Cameron-Martin (1947) who proved convergence a.c.

3.2.C The derivation leading to (9) is from Grenander (1950), pp. 227–229.

3.2.D The unusual representation of the covariance function in Example 6 was derived by the author first via direct and elementary calculations of horrendous length. The simpler technique used in the text can perhaps be extended to other types of stochastic spline processes. Example 6 is suggestive. It indicates that it should be possible to study general (more general than linear) random splines analytically, as we did in this example. The author is not aware of any such work in the literature.

3.2.E The results in Examples 7 and 8 were given in Grenander (1950), pp. 224–227.

3.2.F This perpendicularity was noted in Grenander (1950), p. 221.

3.2.G For Example 10 see Jones (1963) and the related work in Roy (1973).

3.3.A The proof given in the text of Theorem 1 is from Grenander (1950), pp. 257–260.

3.3.B For further comments related to Theorem 2 see Hopf (1937), pp. 56–65.

4 LINEAR INFERENCE

4.1.A To see that the support of y, in the proof of Theorem 1, is in $\mathfrak{R}$, it is enough to note that in (5) of Chapter 2, Section 3, the possible vector values belong to the subspace spanned by the eigenvectors associated with nonzero eigenvalues.

A convenient way to unify the treatment of finite-dimensional linear estimation, and avoid the special cases arising when R is singular, is to use the notion of generalized inverse, in particular the Moore-Penrose inverse. An attractive introduction to the Moore-Penrose inverse can be found in Davis (1979), Sec. 2.8. In Rao (1966) the statistical use of the generalized inverse is discussed.

4.1.B Lemma 1 is from Grenander-Rosenblatt (1956), pp. 87–88.

4.1.C For the generalization of Theorem 4 to $p>1$ see Anderson (1971).

4.1D The last remark in Section 1 of Chapter 4 can be proved easily. If $A\prec B$, find a nonsingular matrix M diagonalizing A and B simultaneously, $M^TAM=D_A$, $M^TBM=D_B$. After the basis change $x=My$ we get $D_A\prec D_B$, which means that the diagonal entries of A must be at most equal to the corresponding entries of B. But for $D_A^{-1}=M^{-1}A^{-1}(M^{-1})^T$, $D_B^{-1}=M^{-1}B^{-1}(M^{-1})^T$, we have the opposite inequalities, which implies $A^{-1}\succ B^{-1}$.

4.2.A Theorem 2 is essentially the same as in Grenander (1950), p. 219.

4.2.B The necessary and sufficient condition that $\bar{x}$ be a BLUE was given in Grenander (1951a). Here and elsewhere we make use of the fact that for a second-order process there is a bijective correspondence between its measure in V^T and the distribution of the Fourier coefficients. A complete discussion of this topic can be found in Rao (1963); see also Fleisher-Kooharian (1958) for related comments. A general treatment of linear estimation for stochastic processes has been given in Holevo (1973); see also Holevo (1971, 1972).

4.2.C The interpolation result was given in Kolmogorov (1941).

4.2.D If x_ν are i.i.d. in the Hilbert space and $\alpha=E(x)$ is arbitrary in X, it follows, as mentioned at the end of the paragraph, that $\bar{x}$ is the BLUE of α. The proof can be modeled on the finite-dimensional one. If $\alpha^*=\Sigma_1^n c_t x_t$ with $\Sigma_1^n c_t=1$, we have

$$E_\alpha\{[\|\alpha^*-\alpha\|^2]\}=\frac{1}{n^2}E_\alpha\left\{\left\|\sum_1^n c_t(x_t-\alpha)\right\|^2\right\}$$

But this equals, since the x_t are independent,

$$\frac{1}{n^2}\sum_1^n c_t^2 E_\alpha\{\|x_t-\alpha\|^2\} = \frac{\operatorname{tr}(R)}{n^2}\sum_1^n c_t^2$$

Using Schwarz's inequality and the unbiasedness condition, we see that

$$E_\alpha\{\|\alpha^*-\alpha\|^2 > \frac{1}{n}\operatorname{tr}(R)$$

But the value on the right-hand side of the inequality can be attained by choosing $\alpha^* = \bar{x}$, which proves the assertion.

4.2.E Probabilistic limit theorems in infinite-dimensional spaces were given in the early publications by Mourier (1953) and Fortet-Mourier (1953).

4.3.A Theorems 3 and 4 are from Grenander (1951a).

4.3.B The estimate used in Example 2 has been studied by Vilenkin (1959) for more general processes.

4.3.C The reader interested in complete regularity should consult the valuable monographe by Ibragimov-Rozanov (1978) where the concept "nearly best estimate" is also studied.

4.3.D Lemma 4 can be strengthened considerably. One way of doing this is by using an expression for the norm of a Hankel form given in the important paper by Nehari (1957). To find a good candidate for the function $F(\theta)$ in this reference one could use the Taylor expansion of $g(z)$ leaving out the first h terms.

4.3.E The Riemann function is usually introduced for partial differential equations [see, e.g., Courant-Hilbert (1937), pp. 311–317].

4.3.F For more information on reproducing kernel Hilbert spaces, see Parzen (1963).

4.3.G The last derivation of this section is from Grenander-Pollak-Slepian (1959) which also gives explicit solutions for several other covariance kernels. For the kernel $(\sin kt)/kt$ see Slepian (1954).

4.4.A The bound (6) could be obtained directly by noting that, using the Fourier inversion formula,

$$L\|\bar{y}\|^2 = \int_{-\Delta}^{L}\left(1-\frac{|t|}{L}\right)r(t)\,dt \to \int_{-\infty}^{\infty} r(t)\,dt = 2\pi f(0)$$

which also proves (5) again.

To see that the covariances decrease geometrically in the proof of Theorem 4, write P factorized in terms of its zeros and use calculus of residues when evaluating r_ν as an integral along $|z|=1$.

A detailed proof of (51) can be found in Grenander-Rosenblatt (1956), pp. 238–240.

4.4.B Theorems 3–5 appeared in Grenander (1954). Theorems 6—9 were given in Grenander-Rosenblatt (1956), and Theorems 9 and 10 in Vitale (1973); see also Siddiqui (1958).

4.4.C An elegant analysis of this problem can be found in Holevo (1976, 1978).

4.4.D The interested reader should also consult Ibragimov-Rozanov (1978), chap. VII.

4.4.E The vector valued case has been treated in Rosenblatt (1956a).

4.4.F A systematic approach to such problems due to Rasulov (1976) exploits Rozanov's idea of nearly best estimates and is recommended to the reader interested in this topic.

4.5.A In Grenander (1950), pp. 244–247, the proof of the asymptotic optimality of $\bar{x}_L$ used a quite different approach. The conditions under which it is valid differ somewhat from (a)–(c) in Theorem 2. Asymptotic optimality for regression parameters in the case of the Ornstein-Uhlenbeck process was established in Mann-Moranda (1954); see Striebel (1958).

In Ylvisaker (1962, 1964) a thorough treatment based on reproducing kernel Hilbert space methods applied to regression problems can be found. The interested reader should consult Rosenblatt (1956a, b) and Adenstedt 4, (1975a, b). See also Adenstedt-Eisenberg (1974) for a different approach to the scalar case.

The case with continuous time and several regression parameters has been dealt with in Chiang Tse-Pei (1959) to which the interested reader is referred.

4.5.B The discretization relations (12), (15), and (18) are from Grenander (1951a), pp. 528–530.

4.6.A To see that we can find a compact K as described, let $x_1, x_2, \ldots$ be a sequence everywhere dense in X and consider the balls $B_n^k = \{x \mid \|x-x_n\| \leqslant 1/k\}$. For given $\varepsilon>0$ choose m_1 big enough so that

$$P\left\{\bigcup_{n=1}^{m_1} B_n^1\right\} > 1-\varepsilon$$

Then choose m_2 big enough to make

$$P\left\{\left[\bigcup_{n=1}^{m_1} B_n^1\right]\cap\left[\bigcup_{n=1}^{m_2} B_n^2\right]\right\}>1-\varepsilon$$

and so on. The set

$$K=\bigcap_{i=1}^{\infty}\left[\bigcup_{n=1}^{m_i} B_n^i\right]$$

is compact and has $P(K)\geqslant 1-\varepsilon$.

4.7.A To see that the $\lambda_\nu^{(n)}$ in Theorem 2 are bounded by M it is enough to remark that the quadratic form

$$z^TRz=\frac{1}{2\pi}\int_{-\pi}^{\pi}\left|\sum_1^n z_te^{it\lambda}\right|^2 f(\lambda)\,d\lambda\leqslant M\|z\|^2$$

4.7.B The two corollaries give a surprisingly simple answer to the question of feature selection, but they are of course limited to the stationary case. To be able to extend the method, we would need knowledge of the asymptotic eigenvalue distribution for other covariance operators. Say, for example, that the process is generated as the (weak) solution of a stochastic differential equation, linear and over a certain region, and driven by white noise. The differential equation could be ordinary or partial. The covariance function will be of the type studied at the end of Section 3. This makes it appear promising to use knowledge about the asymptotic distribution of eigenvalues of differential operators when the region is made to increase [see, e.g. Courant-Hilbert (1937), I, pp. 373–387].

4.7.C The interested reader should compare this with an important result on "eigenvector distribution" [see Grenander-Szegö (1957), pp. 105–114].

5 TESTING GAUSSIAN PROCESSES

5.1.A This is the method originally used [see Grenander (1950), pp. 214–221]. A more recent article is Bethoux (1962).

5.1.B Theorem 1 is of course valid only for Gaussian measures. If we consider shifting a Hilbert space on which some other measure has been given, a good deal is also known about equivalence/perpendicularity and Radon-Nikodym derivatives [see Skorokhod (1970)].

5.1.C The differentiality condition in connection with (8) can perhaps be shown to follow from the semigroup property inherent in (10).

5.1.D More about equivalence for the same covariance operator can be found in Ibragimov-Rozanov (1978), chap. III and IV.

5.2.A In addition to the early work of Feldman (1958) and Hajék (1958) the interested reader should consult Shepp (1966a, b), Yaglom (1963), Rozanov (1962), Kallianpur-Oodira (1963), and especially the monograph by Rozanov (1971).

5.2.B To calculate Radon-Nikodym derivatives for a family $\{P_\alpha\}$ one could first calculate the derivatives of P_α with respect to a fixed measure Q and then use this for the original purpose. The choice of Q is crucial. If Q is chosen as the probability measure of the Wiener process, attractive results have been obtained in Kailath (1967, 1971) and Kailath-Zakai (1971). It is interesting to note that one of the earliest results on the actual computation of the Radon-Nikodym derivatives was done for the Wiener process [see Cameron-Martin (1944)] with the shift supposed to have a derivative of bounded variation.

5.2.C Theorems 1 and 2 have been given different formulations by a number of authors. In addition to Feldman and Hájek, the reader may consult Gikhman-Skorokhod (1966), Rozanov (1971), and Shepp (1964, 1966a, b, 1977).

5.2.D In the proof of Theorem 4 we need to establish the existence of the limits D^- and D^+. To do this let $g(x)$ be a function continuous for $x \geqslant 0$ and with uniformly bounded first and second derivatives for $x > 0$. Then, with $h(x) = g(x)/x$,

$$h'(x) = \frac{1}{x^2}[xg'(x) - g(x)] = \frac{1}{x^2}\int_0^x [g'(x) - g'(y)]\,dy$$

But this is bounded by

$$\frac{\text{constant}}{x^2}\int_0^x (x-y)\,dy$$

which is uniformly bounded. Hence

$$h(x)-h(x')=\frac{g(x)}{x}-\frac{g(x')}{x'}=0(|x-x'|)$$

so that the limit exists. Use $g(x)$ for $[r(t,t+x)-r(t,t)]/x$ in the theorem.

5.2.E For rational spectra the condition in Theorem 5 for perpendicularity is actually necessary. This result can be found in Feldman (1960), Pisarenko (1961), and Pinsker (1964). For stationary Gaussian processes with general spectral density see Capon (1964), Feldman (1960, 1968), Rozanov (1963, 1964, 1971), and Gikhman-Skorokhod (1966).

5.3.A The linear transformation preceeding Theorem 1 can be constructed from a matrix whose rows are the eigenvectors of R_2 with respect to R_1. The $d_k^{(n)}$ will be the eigenvalues of R_2 with respect to R_1.

5.3.B The author had originally intended to use Laplace's method in the Fourier domain to prove Theorem 1 before he observed that the frequency functions (and not only their Fourier transform) could be written as high powers of a continuous function. This led to the above variation of an old theme in the theory of limiting distributions.

5.3.C D. E. McClure has informed the author that the finite-dimensional part of Theorem 2 can be obtained by a direct argument of the type developed in Okamoto (1961).

6 INFERENCE IN MARKOV PROCESSES

6.2.A The proof given in the text of Theorems 2 and 3 is a streamlined version of the one in Albert (1962).

6.2.B In (12) the expectation is not the conditional one, but the integral of $x(t)$ extended over the set $x(t)=k$. Hence the outer integration in (14) is over all $x(t)$ values which had been fixed in the inner integration.

6.2.C That $\text{Re}(\lambda_\nu)<0$, $\nu\neq 1$, is classical and can be seen as follows. If λ is an eigenvalue of Q and $x^T(Q-\lambda I)=0$, $x=(x_1,x_2,\ldots,x_n)$, $x\neq 0$, it follows that for the x_i of largest absolute value,

$$|x_i||q_{ii}-\lambda| = \left|\sum_{j\neq i} q_{ij}x_j\right| \leqslant |x_i||q_{ii}|: \qquad x_i \neq 0$$

Hence λ is contained in a circle of radius $|q_{ii}|$ and centered at $q_{ii}<0$. Therefore either $\lambda=0$ or $\mathrm{Re}(\lambda)<0$.

6.3.A Further information on inference in birth and death processes can be found in Moran (1953), Darwin (1956), Sverdrup (1965), and Wolff (1965). Sverdrup's paper deals with problems in measuring mortality-morbidity. A valuable survey paper by Dion-Keiding (1978) on inference in branching processes also contains an extensive bibliography on related questions. See also Keiding (1975) for birth-and-death processes, including correction note. ibid. (1978).

6.4.A Conditions for equivalence/perpendicularity of processes with independent increments were given by Skorokhod (1957). See also Segall-Kailath (1975) for the discontinuous type.

6.4.B The random variable $Q_N^2-2Z_N$ is of the form studied in Example 3 of Chapter 3, Section 2 [see Doob (1953), p. 395].

6.4.C To implement Theorem 1 would require some caution. Say, for simplicity, that $x(t)$ has no Gaussian component or trend. We would then approximate G by a step function G_p with a finite number p of steps and write for the approximate characteristic function

$$\phi_a(z,t) = \exp\left\{t\sum_{k=1}^{p} a_k(e^{izb_k}-1)\right\}$$

For small $t=\varepsilon$,

$$\frac{\phi_a(z,\varepsilon)-1}{\varepsilon} \cong \sum_{k=1}^{p} a_k(e^{izb_k}-1)$$

On the other hand having observed

$$\Delta_\nu = x\left(\frac{\nu}{n}\right) - x\left(\frac{\nu-1}{n}\right)$$

where $\varepsilon=1/n$, we would expect their empirical characteristic function

$$\phi_a^*\left(z, \frac{1}{\varepsilon}\right) = \frac{1}{n} \sum_{\nu=1}^{n} e^{iz\Delta_\nu}$$

to be close to $\phi_a(z, \varepsilon)$. This could then be used as a basis for estimating G_p. This procedure has not been tested numerically so that we cannot claim that it is numerically feasible.

6.5.A For diffusion processes and stochastic differential equations as in (1) the reader may consult Gikhman-Skorokhod (1972).

6.5.B Related results on the equivalence of diffusion processes can be found in Prohorov (1956), Skorokhod (1960), Girsanov (1960), and Duncan (1968).

For conditions ensuring equivalence to the Wiener process see Kailath (1967, 1971) and Kailath-Zakari (1971). Much relevant information can be found in the monographs by Lipster-Shiryayev (1977, 1978) from a perspective somewhat different from ours, with much attention given to control theory applications.

For an estimation of finite-dimensional parameters in diffusion processes see Brown-Hewitt (1975) and Lee-Kozin (1977).

6.5.C Other approaches, not likelihood based, for the estimation of the drift (infinitesimal mean) function of a diffusion process have been suggested in Banon (1978), Banon-Nguyen (1978), and Nguyen (1977).

An appealingly simple approach in Geman (1979) estimates $a(l)$ by

$$a_n^*(l) = \frac{1}{n} \sum_{k=1}^{n} \frac{x(R_{k+\varepsilon_k}) - l}{\varepsilon k}$$

where

$$R_1 = \inf\{t \geq 0 | x_t = l\}$$
$$R_{k+1} = \inf\{t \geq R_k + \varepsilon_k | x_t = l\}$$

and $\{\varepsilon_k\}$ is a sequence of positive numbers tending to zero sufficiently slowly in order that

$$\sum_1^{\infty} \frac{1}{k^2 \varepsilon_k} < \infty$$

Under mild conditions on the a- and b-functions, and that level crossing occurs at times tending to infinity, this guarantees that $a_n^*(l)$ is a. c. consistent as n tends to infinity.

All these references deal only with pointwise convergence of $a_n^*(l)$. (See also Ibragimov-Hasminskii (1977).

6.6.A Theorem 1 was given in Grenander (1950), pp. 255–257. It is believed that the conditions used can be relaxed considerably.

7 REGULARIZATION IN ABSTRACT PARAMETER SPACE

7.1.A The first rigorous study of ML-estimates under general conditions seems to be due to Doob (1934). For another result, related to Wald's proof, see Wolfowitz (1949).

7.1.B The role of the conditions in (11) deserves a comment. The first one says just that the entropy is finite; such a condition will occur repeatedly in the following chapters. The second one ensures that the likelihood function vanishes at the point of infinity for any given x. Without this condition, or something of the same quality, we would have needed a sieve to guarantee consistency of the ML solutions.

Note that the proof of consistency does not really use the context-free structure of the grammar and can be extended to more general types. While this is true, it is also true that the performance of these "grammar abduction" schemes is probably very poor; the convergence can be expected to be unacceptably slow. If special grammatical structure is assumed, much better performance can be achieved [see Grenander (1978), chap. 7].

7.1.C The concept of a sieve was first formalized in our work on pattern theory. The basic idea of a sieve, although not formalized, underlies early work in spectral estimation as well as estimation of other densities. The reader should also consult Perlman (1972), in particular sec. 3, for a similar approach. There are several alternative methods of dealing with the problem in question. The most well known is to start with an *a priori* measure over A and then do a Bayesian analysis. Under appropriate regularity conditions this amounts to working with *a posteriori* densities of the form $cp(\alpha)f(x;\alpha)$ where c will depend upon x. Another method is to use so-called penalty functions [see de Montricher-Tapia-Thompson (1975) and Tapia-Thompson (1978)] so that $\log L(x,\alpha)+f(\alpha)$ is maximized; f penalizes certain parts of parameter space more than others. This is related to Bayesian analysis using the mode of the *a posteriori* density. A third, more restricted method is to impose order conditions on the parameter, for example, that it be increasing when it is a function on some space or to use other qualitative constraints.

7.1.D It has been shown by examples that even for finite-dimensional parameter spaces the ML method is inappropriate unless modified. Such examples seem to be the exception rather than the rule; sometimes they even appear contrived. In our case, when A is infinite dimensional, it is the other way around: a sieve is usually required in order that the ML method be applicable.

7.3.A Uniform convergence of the spectral estimate was shown in Grenander-Rosenblatt (1956), pp. 262–263. A thorough discussion of spectral estimation was presented in Hannan (1960) for the scalar case and in Hannan (1970) for the vector valued case.

The periodogram can also be studied from the point of view of sieves. Say that x_t is stationary and Gaussian with mean zero and covariance operator R. For a given sample $x_1, x_2, \ldots, x_n$ let us estimate R using the sieve of covariance operators

$$\phi_\mu = \{\text{covariance operator } R \,|\, x_t \text{ periodic with period } 1/\mu;\ \mu = 1/n\}$$

The unitary transformation

$$\hat{x}_\lambda = \frac{1}{\sqrt{n}} \sum_{t=1}^{n} e^{2\pi i t\lambda/n} x_t; \qquad \lambda = 1, 2, \ldots, n$$

leads to i.i.d. Gaussian variables with mean zero and variances proportional to $\Delta F_\lambda \geqslant 0$ (see Example 5 in Chapter 1, Section 1, for a discussion of complex Gaussian variables]. Here ΔF_λ is the discrete spectral mass associated with the frequency $2\pi\lambda/n$ under a fixed $R_\mu \in \phi_\mu$. We are then led to estimating ΔF_λ simply by constant $\times |x_\lambda|^2$, in other words using the periodogram.

7.3.B For an expository article on curve estimates the reader is referred to Rosenblatt (1971), which contains a careful presentation dealing mainly, but not entirely, with pointwise properties of the estimates. Related work using splines to carry out the regularization has been reported in Cogburn-Davis (1974). The relation between spline regularization and Bayesian estimation was studied in Kimeldorf-

Wahba (1970). Further information about consistency and also on optimality in spectral estimation has been given in Parzen (1957).

7.4.A The study of Watson-Leadbetter has been continued in Davis (1977). Another reference, using Fourier analysis as do Watson and Leadbetter, is Pickands (1969).

Schuster (1969, 1970) contain information about uniform convergence of density estimates; see also van Ryzin (1969).

Regularization of frequency functions was started in Rosenblatt (1956a); see also Parzen (1962). Regularization by splines can be found in Wahba (1975).

A survey of "curve estimates" is given in Rosenblatt (1971); see also Bickel-Rosenblatt (1973).

7.4.B In this section we assumed that the frequency functions are quadratically integrable in order to be able to use the criterion in (1). An undesirable consequence of the procedure used is that we cannot guarantee that w be nonnegative.

7.5.A The mass in a small ball will of course tend to zero faster than any power of ε, and the faster, the closer the convergence exponent ρ is to 1, or, in other words, the faster the eigenvalues tend to zero. One can give a probabilistic interpretation of ρ as follows. We know that $x \in l_2$ with probability 1. Introduce ρ as

$$p = \inf\{q \mid x \in l_q \text{ with probability } 1\}.$$

We then have the following result. *The convergence exponent* ρ, $0<\rho<1$, *is equal to* $p/2$. To see that this is so, consider the l_q norm

$$\|x_q\|^q = \sum_{n=1}^{\infty} |x_n|^q = \sum_{n=1}^{\infty} \lambda_n^{q/2} |y_n|^q. \qquad (*)$$

Note that $\{|y_n|^q\}$ is a sequence of i.i.d. variables with some finite mean. If $q>2\rho$, it follows that $\{E|x_n|^q\}$ is summable so that the sum in $(*)$ converges with probability 1, which implies that $x \in l_q$ with probability 1. Hence $p \leqslant 2\rho$.

On the other hand if the sum in $(*)$ converges a.c., then

$$\sum_{n=1}^{\infty} \lambda_n^{q/2} \frac{|y_n|^q}{1+|y_n|^q} < \infty$$

The terms in this sum are a.c. unformly bounded. It then follows [see, e.g., Doob (1953), p. 111] that the sequence of their means is summable, so that

$$\sum_{n=1}^{\infty} \lambda_n^{q/2} < \infty$$

This means that $\rho \leqslant p/2$, which leads to the desired conclusion $\rho = p/2$.

7.5.B Because of the importance of the asymptotic relations derived in this section, a good deal of effort has been given to other proof techniques. The function theoretic argument is believed to have considerable potential for generalization; it is applicable, for example, when the zeros of $D(z)$ are in a sector of the complex plane.

One could also use Tauberian methods, but the usual Tauberian theorems of the power law type do not suffice. The author has applied a result of de Bruijn (1959) to this problem in Grenander (1979), from which we quote the following result. If the counting function $N(u) = \{n \mid \lambda_n \geqslant 1/u\}$ satisfies $N(u) \sim \lambda u^\rho$, $0 < \rho < 1$, as $u \to \infty$, then

$$\ln P\{\|x\|^2 \leqslant t\} \sim -Bt^{-1/(1-\rho)}; \qquad t \downarrow 0$$

with

$$B = (1-\rho)(K\rho^\rho)^{1/(1-\rho)}$$

$$K = \frac{2^{\rho-1}\pi\lambda}{\sin \pi\rho}$$

Note that this method also supplies the value of the constant B.

A related result, but for the radius tending to infinity, can be found in Hwang (1979). Then the crucial parameter is no longer the convergence exponent, but the size and multiplicity of the largest eigenvalue.

It is also possible to base the derivation on representation theorems for the distribution function of the radius of the ball. Such a technique has been used in Bates (1980).

7.6.A Strictly speaking, we should operate on periodic functions, so that instead of F we could consider $F(x) - x/(2\pi)$, where F is defined to increase by 1 over each period, but this will have no effect on the end result. Also, we can operate with D directly, multiplying by i.

Theorem 2 points to the possibility of dealing with still more general situations. One should try to extend it to Banach spaces and consider not only strong convergence but the weak and weak* topologies.

8 SIEVES IN PARAMETER SPACE

8.1.A Theorem 1 holds under a condition that is related to the convergence exponent of the sequency $\{\alpha_k\}$; the value $\frac{1}{2}$ is critical. Theorem 1 as well as many others of the special cases analyzed in Chapter 8 were presented in our Research Seminar in Pattern Theory during 1976–1977.

8.1.B A reader interested in the Bayesian approach to estimation of probability measures should read the pioneering paper by Ferguson (1973) [see also Ferguson (1974) and Antoniak (1974)]. The *a priori* measures used by these authors do not have absolutely continuous realizations, however, so that they do not correspond to $A = L_1$ strictly speaking.

8.2.A To deal with $X = (-\infty, \infty)$ the sieve in (1) should be modified by an additional condition requiring the frequency functions to be uniformly small in a neighborhood of $\pm\infty$.

8.2.B Boswell (1966) analyzes the estimation of Poisson intensities with monotonicity constraints.

8.2.C Theorem 2 has a companion result for the case when the corresponding intensity, namely, $\alpha/[1-F(x)]$, is known to be nondecreasing. The ML solution will then consist of arcs on each of which the intensity is constant. The precise form of the solution can be found in Grenander (1956), pp. 43–46.

8.2.D Related results, both for frequency functions and for intensities, are given in Barlow-Marshall-Proschan (1963), Brunk (1965), Marshall-Proschan (1965), Prakasa Rao (1969, 1970), Reiss (1973), and Wegman (1969, 1970).

8.2.E A related question is when $\alpha^* \in A$ but $P_{\alpha^*} \perp P_\alpha$. Again this is not serious, but it points to the possibility of improving α^*.

8.4.A Spline estimation for regression functions has been used recently by several authors, although with other motivations than the one advanced in this chapter. The reader is referred to papers mentioned under Chapter 7; in particular de Montricher-Tapia-Thompson (1975) is of interest in the present context, since it employs penalty functions related to the sieve criteria used in the text. For a related question see Watson (1964). A different approach was used in Nadaraya (1964, 1970).

8.4.B The biharmonic sieve seems a likely candidate for estimation in the two-dimensional case. However, we have not shown that it actually leads to consistent estimation. We believe that a proof of this can be obtained using a Green's function argument similarly to the proof of Theorem 2.

8.4.C Theorem 3 is a modified version of a result in Chow (1978). It seems plausible that the bounds can be replaced by asymptotic statements, but this would require a refined analysis.

8.6.A In the case of these estimators we have the situation described in Note 8.2.E. One could use, for example, the sieve of all α with $\alpha_1, \alpha_2, \ldots, \alpha_m$ free and $\alpha_n = \beta_n$ for $n = m+1, m+2, \ldots$. Letting $m = 1/\mu \to \infty$ at arbitrary rate, we get the essentially unrestricted estimate V^*. Since the rate is not essential, it is clear that the sieve is not really needed in the way it has been seen to be required many times earlier in the text.

8.6.B The convolution operator α appearing in (4) is normal with eigenvalues α_k. Therefore its norm is $\max|\alpha_k|$.

8.7.A The author conjectures that the result remains true if A is the full group of Borel-measurable measure preserving transformations of X onto itself.

8.7.B A related notion of dispersion function on groups was introduced and studied in Grenander (1963), pp. 76–78.

9 GENERAL PARAMETER SPACE

9.2.A The assumption that A be a Hilbert space is seen to be too restrictive when compared to the many special cases in Chapters 7 and 8. One should try to extend the discussion to the case when A is a Banach space, perhaps starting with the situation when L is bounded. As an example let α represent a piecewise constant regression function $\alpha(x)$, $x \in [0,1)$, with sup norm in A. Consider the sieve

$$\mathcal{L}(\mu) = \{\alpha \mid \alpha(x) \text{ constant on each interval } [(\nu-1)\mu, \nu\mu)\}$$

where $\mu = N^{-1}$, $\nu = 1,2,\ldots,N$. Then we get the regularization

$$\hat{\alpha}_\mu(x) = \frac{1}{2}\left[\inf_{x\in i_\nu} \alpha(x) = \sup_{x\in i_\nu} \alpha(x)\right]; \qquad i_\nu = [(\nu-1)\mu, \nu\mu)$$

As far as we know, this and similar cases have not been studied.

9.2.B The reader may prefer to bypass (8) and derive the inequality (12) by expressing $L\hat{\alpha}$, which is well defined, as

$$L\hat{\alpha} = \int_{-\infty}^{\infty} \psi(\lambda)E(d\lambda)\alpha + \gamma$$

where ψ is quadratically integrable with respect to N, and $\gamma \perp E(\lambda)\alpha$, all λ.

9.3.A The important paper by Wald (1949) employs the concept of *approximate ML-estimators*, later extended by Bahadur (1967), Huber (1967), Perlman (1972), and others. Although the term approximate ML-solution is not used explicitly, the discussion in Part III repeatedly uses this concept implicitly.

In Geman-Hwang (1979) the reader can find a sharpening of Theorem 2 using exponential bounds. This will lead to slower growth of n_m. We do not have to ask that the sample size increase as fast as stated for consistency to hold.

One can actually use finite sets as sieves, but in special cases this does not seem to lead to as attractive estimation procedures as for the sieves used in Chapter 8. Therefore finite sieves will be only of theoretical interest.

Bibliography

The literature on inference in stochastic processes is very large, and what follows represents only a small subset of it. We have listed only references of theoretical significance, leaving out most of the applied publications. Applications to communication engineering and control problems have therefore in general not been included.

On the other hand there are not as many papers dealing with inference in abstract parameter spaces; most of them are found in the following list.

Adenstedt, R. K. (1974): On large-sample estimation for the mean of a stationary random sequence, *Ann. Stat.*, **2**, pp. 1095–1107.

______ (1975a): Linear estimation of regression coefficients for vector-valued processes, *Quart. Appl. Math.*, **XXXIII**, pp. 39–46.

______ (1975b): Asymptotically efficient estimators for a constant regression with vector-valued stationary residuals, *Ann. Stat.*, **3**, pp. 1109–1121.

Adenstedt, R. K. and B. Eisenberg (1974): Linear estimation of regression coefficients, *Quart. Appl. Math.*, **XXXII**, pp. 317–327.

Albert, A. (1962): Estimating the infinitesimal generator of a continuous time, finite state Markov chain, *Ann. Math. Stat.*, **33**, pp. 727–753.

Alekseev, V. G. (1964): Some methods for the exact determination of a parameter of a Gaussian stochastic process, *Theory Prob. Appl.*, **IX**, pp. 466–468.

Anderson, T. W. (1971): *The Statistical Analysis of Time Series*, Wiley, New York.

Anderson, T. W. and L. A. Goodman (1957): Statistical inference about Markov chains, *Ann. Math. Stat.*, **28**, pp. 89–110.

Antoniak, C. E. (1974): Mixtures of Dirichlet processes with applications to Bayesian nonparametric problems, *Ann. Stat.*, **2**, pp. 1152–1174.

Aronszajn, N. (1950): Theory of reproducing kernels, *Trans. Am. Math. Soc.*, **68**, pp. 337–404.

Bahadur, R. R. (1967): Rates of convergence of estimates and test statistics, *Ann. Math. Stat.*, **38**, pp. 303–324.

______ (1971): *Some Limit Theorems in Statistics*, SIAM, Philadelphia.

Balakrishnan, A. V. (1976): *Applied Functional Analysis*, Applications of Math., Vol. 3, Springer, New York.

Banon, G. (1978): Nonparametric identification for diffusion processes, *SIAM J. Contr. Optim.*, **16**, pp. 380–395.

Banon, G. and H. T. Nguyen (1978): Sur l'estimation récurrente de la densité et de sa dérivée pour un processus de Markov, *C. R. Acad. Sci. Paris*, **286**, pp. 691–694.

Barlow, R. E., A. W. Marshall, and F. Proschan (1963): Properties of probability distributions with monotone hazard rate, *Ann. Math. Stat.*, **34**, pp. 375–389.

Bates, R. (1980): Gaussian measure of balls in Hilbert space, Ph.D. thesis, Brown University, Providence, R. I.

Baxter, G. (1956): A strong limit theorem for Gaussian processes, *Proc. Am. Math. Soc.*, **7**, pp. 522–527.

Bethoux, P. (1962): Tests et estimations concernent certaines fonctions aléatoires, en particulier laplaciennes, *Ann. Inst. Henri Poincaré*, **XVII**, pp. 255–321.

Bickel, P. J. and M. Rosenblatt (1973): On some global measures of the deviations of density function estimates, *Ann. Stat.*, **I**, pp. 1071–1095.

Bickel, P. J. and J. A. Yahav (1969): Some contributions to the asymptotic theory of Bayes solutions, *Z. Wahrsch. verw. Gebiete*, **11**, pp. 257–276.

Billingsley, P. (1961): *Statistical Inference for Markov Processes*, Univ. of Chicago Press, Chicago.

Blum, J. R. and J. Rosenblatt (1964): On random sampling from a stochastic process, *Ann. Math. Stat.*, **35**, pp. 1713–1717.

Bochner, S. (1932): *Vorlesungen über Fouriersche Integrale*, Akademische Verlagsgesellschaft, Leipzig.

Boswell, M. T. (1966): Estimating and testing trend in a stochastic process of Poisson type, *Ann. Math. Stat.*, **37**, pp. 1564–1576.

Brillinger, D. R. (1972): The spectral analysis of stationary interval functions, *Proc. 6th Berkeley Symp. Math. Stat. Prob.*, Vol. I, pp. 483–513.

Brown, B. M. and J. J. Hewitt (1975): Asymptotic likelihood theory for diffusion process, *J. Appl. Prob.*, **12**, pp. 228–238.

de Bruijn, N. G. (1959): Pairs of slowly oscillating functions occurring in asymptotic problems concerning the Laplace transform, *Nieuw Arch. voor Wiskunde*, **VII**, pp. 20–26.

Brunk, H. D. (1965): Conditional expectation given a σ-lattice and applications, *Ann. Math. Stat.*, **36**, pp. 1339–1350.

Cameron, R. H. and W. T. Martin (1944): Transformations of Wiener integrals under translations, *Ann. Math.*, **45**, pp. 386–396.

______ (1947): The behavior of measure and measurability under change of scale, *Bull. Am. Math. Soc.*, **45**, pp. 130–137.

Capon, J. (1964): Radon-Nikodym derivatives of stationary Gaussian measures, *Ann. Math. Stat.*, **35**, pp. 517–531.

Chiang, T. P. (1958): On the estimation of regression coefficients of a continuous parameter time series with a stationary residual, *Theory Prob. Appl.*, **IV**, pp. 373–390.

Chow, T. R. (1969): Equivalence of Gaussian stationary processes, *Ann. Math. Stat.*, **40**, pp. 197–202.

Chow, Y. S. (1978): Bayesian estimation for periodic and stationary Gaussian processes, Reports on Pattern Analysis, no. 73, Div. Appl. Math., Brown University, Providence, R.I.

Cogburn, R and H. T. Davis (1974): Periodic splines and spectral estimation, *Ann. Stat.*, **2**, pp. 1108–1126.

Courant, R. and D. Hilbert (1937): *Methoden der Mathematischen Physik*, Vols. I, II, Springer, Berlin.

Cramér, H. (1927): *Sannolikhetskalkylen och några av dess användningar*, Stockholm.

______ (1942): On harmonic analysis in certain functional spaces, *Ark. Mat. Astron. Fys.*, **28B**, 7 pp.

______ (1945): *Mathematical Methods of Statistics*, Alongvist & Wiksell, Stockholm.

Cramér, H. and M. R. Leadbetter (1967): *Stationary and Related Stochastic Processes*, Wiley, New York.

Darwin, J. H. (1956): The behaviour of a simple birth and death process, *Biometrika*, **43**, pp. 23–31.

Davis, K. B. (1977): Mean integrated square error properties of density estimates, *Ann. Stat.*, **5**, pp. 530–535.

Davis, P. (1979): *Circulant Matrices*, Wiley, New York.

Dion, J. P. and N. Keiding (1978): Statistical inference in branching processes, in *Branching Processes*, A. Joffe and P. Ney, Eds., Adv. in Prob. Related Topics, Vol. 5, Marcel Dekker, New York.

Dolph, C. L. and M. A. Woodbury (1952): On the relation between Green's functions and covariances of certain stochastic processes and its applications to unbiased linear prediction, *Trans. Am. Math. Soc.*, **72**, pp. 519–550.

Doob, J. L. (1934): Probability and statistics, *Trans. Am. Math. Soc.*, **36**.

______ (1953): *Stochastic Processes*, Wiley, New York.

Duncan, T. E. (1968): Evaluation of likelihood functions, *Inf. and Cont.*, **13**, pp. 62–74.

Dvoretzky, A., J. Kiefer, and J. Wolfowitz (1953a): Sequential decision problems for processes with continuous time parameter. Problems of estimation, *Ann. Math. Stat.*, 24, pp. 403–415.

______ (1953b): Sequential decision problems for processes with continuous time parameter. Testing hypotheses. *Ann. Math. Stat.*, 24, pp. 254–264.

Eisenberg, B. (1969): The relation of the equivalence conditions for the Brownian motion to the equivalence conditions for certain stationary processes, *Ann. Math. Stat.*, **40**, pp. 2143–2148.

Feigin, P. D. (1976): Maximum likelihood estimation for continuous-time stochastic processes, *Adv. Appl. Prob.*, **8**, pp. 712–736.

Feldman, J. (1958): Equivalence and perpendicularity of Gaussian processes, *Pac. J. Math.*, **8**, pp. 699–708.

______ (1959): Correction to equivalence and perpendicularity of Gaussian processes, *Pac. J. Math.*, **9**, pp. 1295–1296.

______ (1960): Some classes of equivalent Gaussian processes on an interval, *Pac. J. Math.*, **10**, pp. 1211–1220.

______ (1968): A clarification concerning certain equivalence classes of Gaussian processes on an interval, *Ann. Math. Stat.*, **39**, pp. 1078–1079.

Feller, W. (1957): *An Introduction to Probability Theory and Its Applications*, Vol. I, 2nd ed., Wiley, New York.

______ (1966): *An Introduction to Probability Theory and Its Applications*, Vol. II, Wiley, New York.

Ferguson, T. S. (1973): A Bayesian analysis of some non-parametric problems, *Ann. Stat.*, **1**, pp. 209–230.

_____ (1974): Prior distributions on spaces of probability measures, *Ann. Stat.*, **2**, pp. 615–629.

Fleisher, I. and A. Kooharian (1958): On the statistical treatment of stochastic processes, *Ann. Math. Stat.*, **29**, pp. 544–549.

Fortet, R. (1961): Hypothesis testing and estimation for random Laplacian functions, *Proc. 4th Berkeley Symp. Math. Stat. and Probl*, Vol. I, pp. 289–305.

Fortet, R. and E. Mourier (1953): Convergence de la répartition empirique vers la répartition théorique, *Ann. E. Norm. Sup.*, **60**.

Gantmacher, F. R. (1959): *The Theory of Matrices*, Vol. 1, Chelsea, New York.

Geman, S. (1979): On a common sense estimator for the drift of a diffusion, Reports in Pattern Analysis, no. 79, Brown University, Providence, R.I.

Geman, S. and C. R. Hwang (1979): Nonparametric maximum likelihood estimation by the method of sieves, Reports in Pattern Analysis, no. 80, Div. Appl. Math., Brown University, Providence, R. I.

Gikhman, I. I. and A. V. Skorokhod (1966): On the densities of probability measures in function spaces, *Russian Math. Surveys*, **21**, no. 6, pp. 83–156.

_____ (1972): *Stochastic Differential Equations*, Springer, Berlin.

Girsanov, I. V. (1960): On transforming a certain class of stochastic processes by absolutely continuous substitution of measures, *Theory Prob. Appl.*, **V**, pp. 285–301.

Gnedenko, B. V. and A. N. Kolmogorov (1954): *Limit Distributions for Sums of Independent Random Variables* (trans. from the Russian by K. L. Chung), Addison-Wesley, Reading, Mass.

Good, I. J. and R. A. Gaskins (1971): Nonparametric roughness penalties for probability densities, *Biometrika*, **58**, pp. 255–277.

Goursat, E. (1942): *Cours d'analyse mathématique*, Vol. II, Gauthier-Villars, Paris.

Grenander, U. (1949): Stochastic processes and integral equations, *Ark. Mat.*, **1**, pp. 67–70.

_____ (1950): Stochastic processes and statistical inference, *Ark. Mat.*, **1**, pp. 195–277.

_____ (1951a): On empirical spectral analysis of stochastic processes, *Ark. Mat.*, **1**, pp. 503–531.

_____ (1951b): On Toeplitz forms and stationary processes, *Ark. Mat.*, **1**, pp. 555–571.

_____ (1954): On the estimation of regression coefficients in the case of an autocorrelated disturbance, *Ann. Math. Stat.*, **25**, pp. 252–272.

_____ (1956): On the theory of mortality measurement, *Skand. Aktuarie Tidskrift*, pp. 1–55.

_____ (1963): *Probabilities on Algebraic Structures*, Wiley, New York.

_____ (1967): Syntax controlled probabilities, Reports in Pattern Analysis No. 2, Brown University, Providence, R.I.

_____ (1968): Eight lectures on statistical inference in stochastic processes, 66 pp., Tech. Report no. 2, Div. of Appl. Math., Brown University, Providence, R.I.

_____ (1974): Large sample discrimination between two Gaussian processes with different spectra, *Ann. Stat.*, **2**, pp. 347–352.

_____ (1976): *Pattern Synthesis. Lectures in Pattern Theory*, Vol. I, Springer, New York.

_____ (1978): *Pattern Analysis. Lectures in Pattern Theory*, Vol. II, Springer, New York.

_____ (1979): Gaussian measure of small balls in a Hilbert space, Reports in Pattern Analysis, no. 76, Div. Appl. Math., Brown University, Providence, R.I.

Grenander, U., H. O. Pollak, and D. Slepian (1959): The distribution of quadratic forms in normal variates: a small sample theory with applications to spectral analysis, *SIAM J*, **7**, pp. 374–401.

Grenander, U. and M. Rosenblatt (1954): An extension of a theorem of G. Szegö and its applications to the study of stochastic processes, *Trans. Am. Math. Soc.*, **76**, pp. 112–126.

_____ (1956): *Statistical Analysis of Stationary Time Series*, Almqvist and Wiksell, Stockholm.

Grenander, U. and G. Szegö (1957): *Toeplitz forms and their Applications*, Univ. California Press, Berkeley.

Hajék, J. (1958): On a property of normal distributions of any stochastic process, *Cz. Math. J.*, **8**, pp. 610–617.

_____ (1960): On a simple regression model in Gaussian processes, *Trans. 2nd Prague Conf. Inform. Theory Stat. Decision Funct. Random Processes*, pp. 185–198.

_____ (1962): On linear statistical problems in stochastic processes, *Cz. Math. J.*, **12**, pp. 404–444.

Halmos, P. R. (1950): *Measure Theory*, Van Nostrand, Princeton, N.J.

Hannan, E. J. (1960): *Time Series Analysis*, Methuen, London.

_____ (1970): *Multiple Time Series*, Wiley, New York.

Hanner, O. (1950): Deterministic and non-deterministic stationary stochastic processes, *Ark. Mat.*, **1**, pp. 161–177.

Hardy, G. H., J. E. Littlewood, and G. Polya (1964): *Inequalities*, Cambridge Univ. Press, Cambridge.

Harris, T. E. (1963): *The Theory of Branching Processes*, Springer, Berlin.

Herglotz, G. (1911): Über Potenzreihen mit positivem reellem Teil im Einheitskreis, *Ber. Verh. Kgl. Sächs. Ges. Wiss.* (Leipzig), **63**, p. 501.

Hewitt, E. and K. Stromberg (1975): *Real and Abstract Analysis*, Springer, New York.

Hille, E. and R. S. Phillips (1957): *Functional Analysis and Semi-groups*, *Am. Math. Soc. College Publ.*, **XXXI**, Providence, R.I.

Hobson, E. W. (1918): On Hellinger's integrals, *Proc. London Math. Soc.*, **2**, pp. 249–265.

Holevo, A. S. (1971): Almost periodic signal estimation, *Teor. Veroyatn.*, **XVI**, pp. 249–263.

_____ (1972): On estimating functions of the mean, *Teor. Veroyatn.*, **XVII**, pp. 573–577.

_____ (1973): On the general problem of mean estimation *J. Multivar. Anal.*, **3**, pp. 262–275.

_____ (1976): On the asymptotically efficient regression estimates in the case of degenerate spectrum, *Teor. Veroyatn.*, **XXI**, pp. 324–333.

_____ (1978): A regression problem for continuous time series, *Teor. Veroyatn.*, **XXIII**, pp. 762–771.

Hopf, E. (1937): *Ergodentheorie*, Ergebnisse der Math. Vol. 5, No. 2, Springer, Berlin.

Huber, P. J. (1967): The behavior of maximum likelihood estimates under nonstandard conditions, *Proc. 5th Berkeley Symp.*, Vol. I, pp. 221–233.

Hwang, C. R. (1979): Gaussian measure of large balls in a Hilbert space, Reports in Pattern Analysis, no. 75, Div. Appl. Math., Brown University, Providence, R.I.

Ibragimov, I. A. (1963): On estimation of the spectral function of stationary Gaussian process, *Theory Prob. Appl.*, **VIII**, pp. 366–401.

Ibragimov, I. A. and R. Z. Hasminskii (1977): A problem of statistical estimation in Gaussian white noise, *Sov. Math. Dokl.*, **18**, pp. 1351–1353.

Ibragimov, I. A. and Y. A. Rozanov (1978): *Gaussian Random Processes*, transl. by A. B. Aries, Applications of Mathematics, Vol. 9, Springer, New York.

Jagers, P. (1974): Almost sure convergence of general branching processes and functionals thereof, *J. Appl. Prob.*, **11**, pp. 471–478.

Jones, R. H. (1963): Stochastic processes on the sphere, *Ann. Math. Stat.*, **34**, pp. 213–222.

Johnson, N. L. and S. Kotz (1970): *Continuous Univariate Distributions*, Vol. I, Houghton-Mifflin, Boston.

Kadota, T. T. (1965): Optimum reception of binary sure and Gaussian signals, *Bell Sys. Tech. J.*, **XLIV**, pp. 1621–1651.

_____ (1967): Generalized optimum receivers of Gaussian signals, *Bell Sys. Tech. J.*, **VLXI**, pp. 577–591.

Kadota, T. T. and L. A. Shepp (1967): On the best finite set of linear observables for discriminating two Gaussian signals, *IEEE Trans. Inform. Theory*, **IT-13**, pp. 278–284.

Kailath, T. (1967): On measures equivalent to Wiener measure, *Ann. Math. Stat.*, **38**, pp. 261–263.

_____ (1971): The structure of Radon-Nikodym derivatives with respect to Wiener and related measures, *Ann. Math. Stat.*, **42**, pp. 1054–1067.

Kailath, T. and M. Zakai (1971): Absolute continuity and Radon-Nikodym derivatives for certain measures relative to Wiener measure, *Ann. Math. Stat.*, **42**, pp. 130–140.

Kakutani, S. (1948): On the equivalence of infinite product measures, *Ann. Math. Stat.*, **49**, pp. 214–224.

Kallenberg, O. (1975): *Random Measures*, Academic Press, New York.

Kallianpur, G. and H. Oodira (1963): The equivalence and singularity of Gaussian measures, M. Rosenblatt, Ed., *Proc. Symp. Time Series Analysis*, Brown University, pp. 279–291.

Karhunen, K. (1946): Zur Spektraltheorie stochastischer Prozesse, *Ann. Ac. Sci. Fennicae*, **AI**, 34.

_____ (1947): Über lineare Methoden in der Wahrscheinlichkeitsrechnung, *Ann. Ac. Sci. Fennicae*, **37**, pp. 1–79.

_____ (1950): Über die Struktur stationärer zufälliger Funktionen, *Ark. Mat.*, **1**, pp. 141–160.

Keiding, N. (1975): Maximum-likelihood estimation in the birth-and-death process, *Ann. Stat.*, **3**, pp. 363–372. With correction note (1978) in *Ann. Stat.* **6**, p. 472.

Kholevo, A. S. (1969): On estimates of regression coefficients, *Theory Prob. Appl.*, **14**, pp. 79–104.

Kiefer, J. and J. Wolfowitz (1956): Consistency of the maximum likelihood estimator in the presence of infinitely many incidental parameters, *Ann. Math. Stat.*, **27**, pp. 887–906.

Kimeldorf, G. S. and G. Wahba (1970): A correspondence between Bayesian estimation on stochastic processes and smoothing by splines, *Ann. Math. Stat.*, **41**, pp. 495–502.

Kolmogorov, A. N. (1933): Grundbegriffe der Wahrscheinlichkeitsrechnung, *Ergebnisse der Math.*, **2**, no. 3.

_____ (1933): Sulla determinazione empirica di una legge di distribuzione, *Giorn. Ist. d. Ital. Attuari*, **4**, pp. 83–91.

_____ (1940a): Curves in Hilbert space which are invariant with respect to a one-parameter group of motions, *Dokl. Akad. Nauk SSSR*, **26**, pp. 6–9.

_____ (1940b): Wiener spirals and some other interesting curves in Hilbert space, *Dokl. Akad. Nauk SSSR*, **26**, pp. 115–118.

_____ (1941): Stationary sequences in a Hilbert space (Russian), *Bull. Univ. Moscow*, **2**.

Kraft, C. and L. LeCam (1956): A remark on the roots of the maximum likelihood equation, *Ann. Math. Stat.*, **27**, pp. 1174–1177.

LeCam, L. (1970): On the assumptions used to prove asymptotic normality of maximum likelihood estimates, *Ann. Math. Stat.*, **41**, pp. 802–828.

Lee, T. S. and F. Kozin (1977): Almost sure asymptotics likelihood theory, *J. Appl. Prob.*, **14**, pp. 527–537.

Lipster, R. S. and A. N. Shiryayev (1977): *Statistics of Random Processes, General Theory*, trans. from the Russian by A. B. Aries, Springer, New York.

_____ (1978): *Statistics of Random Processes II*, trans. from the Russian by A. B. Aries, Springer, New York.

Loève, M. (1948): Fonctions aléatoires du second ordre, Suppl. to P. Lévy, *Processus stochastiques et mouvement Brownien*, Gauthier-Villars, Paris, pp. 299–352.

Magness, T. A. and J. B. McGuire (1962): Comparison of least squares and minimum variance estimates of regression parameters, *Ann. Math. Stat.*, **33**, pp. 462–470.

Mann, H. B. (1952): On estimation of parameters determining the mean value function of a stochastic process, *Sankhyā*, **12**, pp. 117–120.

_____ (1953): Introduction to the theory of stochastic processes depending upon a continuous parameter, *Nat. Bur. Stand. Appl. Math. Ser.*, no. 24.

_____ (1954): A theory of estimation for the fundamental random process and the Ornstein-Uhlenbeck process, *Sankhyā*, pp. 325–350.

Mann, H. B. and P. B. Moranda (1954): On the efficiency of the least squares estimates of parameters in the Ornstein-Uhlenbeck process, *Sankhyā*, **13**, pp. 351–358.

Marshall, A. W. and F. Proschan (1965): Maximum likelihood estimation for distributions with monotone failure rate, *Am. Math. Stat.*, **36**, pp. 69–77.

Maruyama, G. (1949): The harmonic analysis of stationary stochastic processes, *Mem. Fac. Sci. Kyusyn Univ. A*, **4**, pp. 45–106.

McClure, D. E. (1970): Feature selection for the analysis of line patterns, Reports in Pattern Analysis, no. 16, Div. Appl. Math., Brown University, Providence, R.I.

_____ (1975): Best linear observables for large sample discrimination of Gaussian processes with different spectra, Reports in Pattern Analysis, no. 37, Div. of Appl. Math., Brown University, Providence, R.I.

Mercer, T. (1909): Functions of positive and negative type and their connection with the theory of integral equations, *Trans. London Phil. Soc.* (*A*), **209**, pp. 415–446.

Meschkowski, H. (1962): *Hilbertsche Räume mit Kernfunktion*, Springer, Berlin.

von Mises, R. (1931): *Wahrscheinlichkeitsrechnung und ihre Anwendung in der Statistik und theoretischen Physik*, Deuticke, Leipzig.

de Montricher, G. F., R. A. Tapia, and J. R. Thompson (1975): Nonparametric maximum likelihood estimation of probability densities by penalty function methods, *Ann. Stat.*, **3**, pp. 1329–1348.

Moran, P. A. P. (1953): The estimation of the parameters of a birth and death process, *J. R. Stat. Soc. B*, **15**, pp. 241–245.

Mourier, E. (1953): Elements aléatoires dans un éspace de Banach, *Ann. Inst. Henri Poincaré*, **13**, pp. 161–244.

Nadaraya, E. A. (1964): On estimating regression, *Theory Prob. Appl.*, **9**, pp. 141–142.

_____ (1970): Remarks on nonparametric estimate for density functions and regression curves, *Theory Prob. Appl.*, **15**, pp. 134–137.

Nehari, Z. (1957): On bounded bilinear forms, *Ann. Math.*, **65**, pp. 153–162.

Neveu, J. (1964): *Bases mathématiques du calcul des probabilités*, Masson, Paris.

Neyman, J. and E. S. Pearson (1933): On the problem of the most efficient tests of statistical hypotheses, *Phil. Trans. R. Soc. A*, **231**, p. 289.

Nguyen, H. T. (1977): Estimation de la densité d'un processus de Markov stationnaire, à temps continue, *C. R. Acad. Sci. Paris*, **284**, pp. 1397–1400.

Okamoto, M. (1961): Discrimination for variance matrices, *Osaka Math. J.*, **13**, pp. 1–39.

Paley, R. E. A. C. and N. Wiener (1934): *Fourier Transforms in the Complex Domain*, Am. Math. Soc., New York.

Parzen, E. (1957): On consistent estimation of the spectrum of a stationary time series, *Ann. Math. Stat.*, **28**, pp. 329–348.

______ (1958): On asymptotically efficient consistent estimates of the spectral density of a stationary time series, *J. R. Stat. Soc. B*, **20**, pp. 303–322.

______ (1961): An approach to time series analysis, *Ann. Math. Stat.*, **32**, pp. 951–989.

______ (1962): On the estimation of a probability density and mode, *Ann. Math. Stat.*, **33**, pp. 1065–1076.

______ (1963): Probability density functionals and reproducing kernel Hilbert spaces, M. Rosenblatt, Ed., *Proc. Symp. Time Series Analysis*, Brown University, Providence, R.I. pp. 155–169.

Perlman, M. D. (1972): On the strong consistency of approximate maximum likelihood estimators, *Proc. 6th Berkeley Symp. Math. Stat. Prob.*, pp. 263–281.

Pickands III, J. (1969): Efficient estimation of a probability density function, *Ann. Math. Stat.*, **40**, pp. 854–864.

Pinsker, M. S. (1964): *Information and Information Stability of Random Variables and Processes*, transl. by A. Feinstein, Holden-Day, San Francisco.

Pisarenko, V. F. (1961): To the problem of detection of a signal in background noise (in Russian), *Radio-tekhnika: Elektronika*, **6**, pp. 514–528.

Pitcher, T. (1959): Likelihood ratios of Gaussian processes, *Ark. Mat.*, **4**, pp. 35–44.

______ (1961): Likelihood ratios for diffusion processes with shifted mean values, *Trans. Am. Math. Soc.*, **101**, pp. 168–176.

______ (1963): The admissible mean values of a stochastic process, *Trans. Am. Math. Soc.*, **108**, pp. 538–546.

Pólya, G. and G. Szegö (1925, 1954): *Aufgaben und Lehrsätze aus der Analysis*, Springer, Berlin.

______ (1972): *Problems and Theorems in Analysis*, Vols. I, II, English ed., Springer, New York.

Pontrjagin, L. (1946): *Topological Groups*, Princeton University Press, Princeton.

Prohorov, Yu. V. (1956): Convergence of random processes and limit theorems in probability theory, *Theory Prob. and Appl.*, **1**, pp. 157–214.

______ (1961): The method of characteristic functionals, *Proc. 4th Berkeley Symp. Math. Stat. Prob.*, pp. 403–419.

Rao, C. R. (1966): Generalized inverse for matrices and its application in mathematical statistics, in *Research Papers in Statistics*, Festschrift for J. Neyman, F. N. David, Ed., Wiley, London.

Rao, M. M. (1963): Inference in stochastic processes, I, *Theory Prob. Appl.*, **VIII**, pp. 266–281.

_____ (1966): Inference in stochastic processes II, *Z. für Wahrsch. verw. Gebiete*, **4**, pp. 317–335.

_____ (1967): Inference in stochastic processes III, *Z. für Wahrsch. verw. Gebiete*, **8**, pp. 49–72.

_____ (1974): Inference in stochastic processes IV, Predictors and projections, *Sankhyā*, **36**, pp. 63–120.

_____ (1975): Inference in stochastic processes V, Admissible means, *Sankhyā*, **37**, pp. 538–549.

_____ (1977): Inference in stochastic processes, VI, Translates and densities, in *Multivariate Analysis*, P. R. Krishnaiah, Ed., North Holland, Amsterdam, pp. 311–324.

Prakasa Rao, B. L. S. (1969): Estimation of a unimodel density, *Sankhyā A*, **31**, pp. 23–36.

_____ (1970): Estimation for distributions with monotone failure rate, *Ann. Math. Stat.*, **41**, pp. 507–519.

Rasulov, N. P. (1976): On asymptotically efficient estimates of regression coefficients under spectral density of noise degeneration, *Theory Prob. Appl.*, **XXI**, pp. 316–324.

Reiss, R.-D. (1973): On the measurability and consistency of maximum likelihood estimates for unimodal densities, *Ann. Stat.*, **1**, pp. 888–901.

Rice, J. (1975): Estimated factorization of the spectral density of a stationary point process, *Adv. Appl. Prob.*, **7**, pp. 801–817.

Riesz, F. and B. Sz. Nagy (1952): *Leçons d'analyze fonctionelle*, Acad. Sci. de Hongrie, Budapest.

Root, W. L. (1963): Singular Gaussian measures in detection theory, M. Rosenblatt, Ed., *Proc. Symp. Time Series Analysis*, Brown University, pp. 292–315.

Rosenblatt, M. (1956a): On estimation of regression coefficients of a vector-valued time series with a stationary disturbance, *Ann. Math. Stat.*, **27**, pp. 99–121.

_____ (1956b): Remarks on some nonparametric estimates of a density function, *Ann. Math. Stat.*, **27**, pp. 832–837.

_____ (1971): Curve estimates, *Ann. Math. Stat.*, **42**, pp. 1815–1842.

Roussas, G. G. (1965): Extension to Markov processes of a result by A. Wald about the consistency of the maximum likelihood estimate, *Z. Wahrscheinlichk.*, **4**, pp. 69–73.

Roy, R. (1973): Estimation of the covariance function of a homogeneous process on the sphere, *Ann. Stat.*, **1**, pp. 780–785.

Rozanov, Yu. (1962): On the density of one Gaussian measure with respect to another, *Theory Prob. Appl.*, **VII**, pp. 82–87.

_____ (1963): On the problem of the equivalence of probability measures corresponding to stationary Gaussian processes, *Theory Prob. Appl.*, **VIII**, pp. 223–231.

_____ (1964): On probability measures in functional spaces corresponding to stationary Gaussian processes, *Theory Prob. Appl.*, **IX**, pp. 404–419.

_____ (1966): On the densities of Gaussian distributions and Wiener-Hopf integral equations, *Russian Teor. Verojetn. i. Primen.*, **11**, pp. 170–179.

_____ (1971): *Infinite Dimensional Gaussian Distributions*, transl. of Russian original in *Proc. Steklov Inst. of Math.*, 1968, Amer. Math. Soc., Providence, R.I.

_____ (1977): *Innovation Processes*, Wiley, New York.

Rubin, H. and H. G. Tucker (1959): Estimating the parameters of a differential process, *Ann. Math. Stat.*, **3**, pp. 641–658.

Rudin, W. (1962): *Fourier Analysis in Groups*, Interscience, New York.

van Ryzin, J. (1969): On strong consistency of density estimates, *Ann. Math. Stat.*, **40**, pp.

1765–1772.

Sazonov, V. (1958): On characteristic functionals, *Theory Prob. Appl.*, **III**, pp. 188–192.

Schuster, E. F. (1969): Estimation of a probability density function and its derivatives, *Ann. Math. Stat.*, **40**, pp. 1187–1195.

______ (1970): Note on the uniform convergence of density estimates, *Ann. Math. Stat.*, **41**, pp. 1347–1348.

Segall, A. and T. Kailath (1975): Radon-Nikodym derivatives with respect to measures induced by discontinuous independent increment processes, *Ann. Prob.*, **3**, pp. 449–464.

Shepp, L. A. (1964): The singularity of Gaussian measures in function space, *Proc. Nat. Acad. Sci.*, **52**, pp. 430–433.

______ (1965): Distinguishing a sequence of random variables from a translate of itself, *Ann. Math. Stat.*, **36**, pp. 1107–1112.

______ (1966a): Radon-Nikodym derivatives of Gaussian measures, *Ann. Math. Stat.*, **37**, pp. 321–353.

______ (1966b): Gaussian measures in function space, *Pac. J. Math.*, **17**, pp. 167–173.

______ (1977): Correction to "Radon-Nikodym derivatives of Gaussian measures", *Ann. Prob.*, **5**, pp. 315–317.

Siddiqui, M. M. (1958): Covariances of least-squares estimates when residuals are correlated, *Ann. Math. Stat.*, **29**, pp. 1251–1256.

Simar, L. (1976): Maximum likelihood estimation of a compound Poisson process, *Ann. Math. Stat.*, **4**, pp. 1200–1209.

Skorokhod, A. V. (1957): On the differentiability of measures corresponding to random processes. I. Processes with independent increments. *Theory Prob. Appl.*, **II**, pp. 407–432.

______ (1960): On the differentiability of measures which correspond to stochastic processes II, Markov processes, *Theory Prob. Appl.*, **V**, pp. 40–49.

______ (1970): On admissible translation of measures in Hilbert space, *Theory Prob. Appl.*, **XV**, pp. 557–580.

______ (1974): *Integration in Hilbert Space*, Ergebnisse der Math. und ihrer Grenzgebiete, 79, Springer, New York.

Slepian, D. (1954): Estimation of signal parameters in the presence of noise, *IRE Trans. Inform. Theory*, **PGIT-3**.

______ (1958): Some comments on the detection of Gaussian signals in Gaussian noise, *IRE Trans. Inform. Theory*, **IT-4**, pp. 65–68.

Smirnov, N. V. (1939): Estimation of the difference of two empirical distribution functions from two independent samples (in Russian), *Bull, Mosk. Univ.*, **2**, no. 3.

Striebel, C. T. (1958): On the efficiency of estimates of trend in the Ornstein-Uhlenbeck process, *Ann. Math. Stat.*, **29**, pp. 192–200.

______ (1959): Densities for stochastic processes, *Ann. Math. Stat.*, **2**, pp. 559–567.

______ (1961): Efficient estimation of a regression parameter, *Ann. Math. Stat.*, **32**, pp. 1299–1313.

Sverdrup, E. (1965): Estimates and test procedures in connection with stochastic models for deaths, recoveries and transfers between different states of health, *Skand. Akt. Tid*, **52**, pp. 184–211.

Szegö, G. (1920, 1921): Beiträge zur Theorie der Toeplitzschen Formen I, II, *Math. Zeitsch.*, **6**, pp. 167–202; **9**, pp. 167–190.

Tapia, R. A. and J. R. Thompson (1978): *Nonparametric Probability Density Estimation*, Johns Hopkins Univ. Press, Baltimore.

Titchmarsh, E. C. (1939): *The Theory of Functions*, Oxford Univ. Press, Oxford.

Tortrat, A. (1971): *Calcul des probabilités et introduction aux processesus aléatoires*, Masson, Paris.

Varberg, D. E. (1961): On equivalence of Gaussian measures, *Pac. J. Math.*, **11**, pp. 751–762.

_____ (1967): Equivalent Gaussian measures with a particularly simple Radon-Nikodym derivative, *Ann. Math. Stat.*, **38**, pp. 1027–1030.

Vilenkin, S. Ya. (1959): On the estimation of the mean in stationary processes, *Theory Prob. Appl.*, **IV**, pp. 415–416.

Vitale, R. A. (1974): Representation of convex, compact subsets of the plane, Reports in Pattern Analysis, no. 12, Brown University, Providence, R.I.

_____ (1973): An asymptotically efficient estimate in time series analysis, *Quart. Appl. Math.* **XXX**, pp. 421–440.

Wahba, G. (1975): Interpolating spline methods for density estimation I. Equi-spaced knots, *Ann. Stat.*, **3**, pp. 30–48.

Wald, A. (1949): Note on the consistency of the maximum likelihood estimate, *Ann. Math. Stat.*, **20**, pp. 595–601.

Watson, G. S. (1964): Smooth regression analysis, *Sankhyā A*, **26**, pp. 359–372.

Watson, G. S. and M. R. Leadbetter (1963): On the estimation of the probability density, I, *Ann. Math. Stat.*, **34**, pp. 480–491.

Wegman, E. J. (1969): A note on estimating a unimodal density, *Ann. Math. Stat.*, **40**, pp. 1661–1667.

_____ (1970): Maximum likelihood estimation of a unimodal density, II, *Ann. Math. Stat.*, **41**, pp. 2169–2174.

Wold, H. (1938): *A Study in the Analysis of Stationary Time Series*, Almqvist & Wiksell, Stockholm.

Wolff, R. W. (1965): Problems of statistical inference for birth and death queuing models, *Oper. Res.*, **13**, pp. 343–357.

Wolfowitz, J. (1949): On Wald's proof of the consistency of the maximum likelihood estimate, *Ann. Math. Stat.*, **20**, pp. 601–602.

Yaglom, A. M. (1963): On the equivalence or perpendicularity of two Gaussian probability measures in function space, M. Rosenblatt, Ed., *Proc. Symp. Time Series Analysis*, pp. 327–348.

Ylvisaker, N. D. (1962): On linear estimation for regression problems on time series, *Ann. Math. Stat.*, **3**, pp. 1077–1084.

_____ (1964): Lower bounds for minimum covariance matrices in time series regression problems, *Ann. Math. Stat.*, **35**, pp. 362–368.

Index

Applied Probability and Statistics (Continued)

FLEISS • Statistical Methods for Rates and Proportions, *Second Edition*
GALAMBOS • The Asymptotic Theory of Extreme Order Statistics
GIBBONS, OLKIN, and SOBEL • Selecting and Ordering Populations: A New Statistical Methodology
GNANADESIKAN • Methods for Statistical Data Analysis of Multivariate Observations
GOLDBERGER • Econometric Theory
GOLDSTEIN and DILLON • Discrete Discriminant Analysis
GROSS and CLARK • Survival Distributions: Reliability Applications in the Biomedical Sciences
GROSS and HARRIS • Fundamentals of Queueing Theory
GUPTA and PANCHAPAKESAN • Multiple Decision Procedures: Theory and Methodology of Selecting and Ranking Populations
GUTTMAN, WILKS, and HUNTER • Introductory Engineering Statistics, *Second Edition*
HAHN and SHAPIRO • Statistical Models in Engineering
HALD • Statistical Tables and Formulas
HALD • Statistical Theory with Engineering Applications
HARTIGAN • Clustering Algorithms
HILDEBRAND, LAING, and ROSENTHAL • Prediction Analysis of Cross Classifications
HOEL • Elementary Statistics, *Fourth Edition*
HOLLANDER and WOLFE • Nonparametric Statistical Methods
JAGERS • Branching Processes with Biological Applications
JESSEN • Statistical Survey Techniques
JOHNSON and KOTZ • Distributions in Statistics
Discrete Distributions
Continuous Univariate Distributions—1
Continuous Univariate Distributions—2
Continuous Multivariate Distributions
JOHNSON and KOTZ • Urn Models and Their Application: An Approach to Modern Discrete Probability Theory
JOHNSON and LEONE • Statistics and Experimental Design in Engineering and the Physical Sciences, Volumes I and II, *Second Edition*
JUDGE, GRIFFITHS, HILL and LEE • The Theory and Practice of Econometrics
KALBFLEISCH and PRENTICE • The Statistical Analysis of Failure Time Data
KEENEY and RAIFFA • Decisions with Multiple Objectives
LANCASTER • An Introduction to Medical Statistics
LEAMER • Specification Searches: Ad Hoc Inference with Nonexperimental Data
McNEIL • Interactive Data Analysis
MANN, SCHAFER and SINGPURWALLA • Methods for Statistical Analysis of Reliability and Life Data
MEYER • Data Analysis for Scientists and Engineers
MILLER, EFRON, BROWN, and MOSES • Biostatistics Casebook
OTNES and ENOCHSON • Applied Time Series Analysis: Volume I, Basic Techniques
OTNES and ENOCHSON • Digital Time Series Analysis
POLLOCK • The Algebra of Econometrics
PRENTER • Splines and Variational Methods
RAO and MITRA • Generalized Inverse of Matrices and Its Applications
RIPLEY • Spatial Statistics
SCHUSS • Theory and Applications of Stochastic Differential Equations
SEAL • Survival Probabilities: The Goal of Risk Theory
SEARLE • Linear Models
SPRINGER • The Algebra of Random Variables
UPTON • The Analysis of Cross-Tabulated Data
WEISBERG • Applied Linear Regression
WHITTLE • Optimization Under Constraints